EXTREMAL GRAPH THEORY

Béla Bollobás

Chair of Excellence in Mathematics,
University of Memphis, Tennessee
and Fellow of Trinity College,
University of Cambridge, United Kingdom

DOVER PUBLICATIONS, INC.
Mineola, New York

Bibliographical Note

This Dover edition, first published in 2004, is an unabridged republication of
the work originally published by Academic Press, London, in 1978.

Library of Congress Cataloging-in-Publication Data

Bollobás, Béla.
    Extremal graph theory / Béla Bollobás.
       p. cm.
    Originally published: London ; New York : Academic Press, 1978, in series:
L.M.S. monographs ; 11.
    Includes bibliographical references and index.
    ISBN 0-486-43596-2 (pbk.)
    1. Graph theory. 2. Extremal problems (Mathematics) I. Title.

QA166.B66 2004
511'.5—dc22

                              2004041355

Manufactured in the United States of America
Dover Publications, Inc., 31 East 2nd Street, Mineola, N.Y. 11501

To My Parents

Preface

Extremal graph theory, in its strictest sense, is a branch of graph theory developed and loved by Hungarians. Its study, as a subject in its own right, was initiated by Turán in 1940, although a special case of his theorem and several other extremal results had been proved many years earlier. The main exponent has been Paul Erdös who, through his many papers and lectures, as well as uncountably many problems, has virtually created the subject. (In retrospect, it seems inevitable that it was Erdös who, when I was fourteen, introduced me to graph theory.)

In extremal graph theory one is interested in the relations between the various graph invariants, such as order, size, connectivity, minimum degree, maximum degree, chromatic number and diameter, and also in the values of these invariants which ensure that the graph has certain properties. Often, given a property $\mathscr{P}$ and an invariant μ for a class $\mathscr{H}$ of graphs, we wish to determine the least value m for which every graph G in $\mathscr{H}$ with $\mu(G) > m$ has property $\mathscr{P}$. Those graphs G in $\mathscr{H}$ without the property $\mathscr{P}$ and with $\mu(G) = m$ are called the *extremal graphs* for the problem. For instance, every graph of order n and size at least n contains a cycle, and the extremal graphs are the trees of order n. At a slightly less frivolous level, a graph of order $2u$ contains a triangle if the minimum degree is at least $u + 1$, and the only extremal graph is $K^{u,u}$, the complete bipartite graph. The prime example of an extremal problem is the following: given a graph F, determine $ex(n; F)$, the maximum number of edges in a graph of order n not containing F as a subgraph.

Having said this, I hasten to emphasize that in this book extremal graph theory is interpreted in a much broader sense, including in its scope various structural results and any relations among the invariants of a graph, especially

vii

those concerned with best possible inequalities. The chapter titles give a broad outline of the content of the text and, although most of the material appears here for the first time in a book, the topics covered in most standard treatises on graph theory are also dealt with in depth. The most notable omissions are algebraic graph theory, matroids and the problems of enumeration and reconstruction. The relative importance of the topics covered in the different chapters is not reflected in their lengths; otherwise the results concerning Hamiltonian cycles, colouring graphs on surfaces, and graphs without certain subgraphs would take up most of the space. Inevitably, the selection of material and its presentation have been greatly influenced by my personal preferences.

The readers are expected to have some familiarity with graph theory, though the book is self-contained. It has grown out of two Part III courses given at the University of Cambridge (1970/71 and 1975/76) and is intended for research students and professional mathematicians. It seemed desirable to expand the lecture notes into a book, since even expert graph theorists seem to be unaware of quite a few of the results which were proved years ago. I hope the book will help a little to stem the present tide of duplications. Although it is exciting to introduce new concepts and to find new problems, there is also merit in the continuity and development of a theory. The main aim of this book is to bring readers up to date with the results in a number of areas and to entice at least a few of them to continue the work. There is a false myth that extremal results are rapidly superseded. I hope that this book will help to make the myth a reality.

I would like to emphasize that the proofs of the results are important; though it is easy to flip through the book and pick out some results, in many cases it is more advantageous to be familiar with the methods than to know the results. The exercises at the end of each section vary a great deal in importance and difficulty. They contain many results and a few, marked with the sign '+', are really too difficult to be called exercises. In many cases hints are given to bring the problem within reach. Very easy problems are marked with the sign '−'. Unresolved questions are called Problems.

The end of a proof or the absence of a proof is indicated by the symbol ∎; the greatest integer less than or equal to x is denoted by $\lfloor x \rfloor$ and $-\lfloor -x \rfloor$ is denoted by $\lceil x \rceil$.

It is a great pleasure to acknowledge the generous help of Professors P. Erdős, G. A. Dirac, R. K. Guy, R. Halin, N. Sauer and C. Thomassen. My research students, Stephen Eldridge and Andrew Thomason, made many helpful suggestions. In addition Keith Carne, Andrew Cornford, Michael

Davies, Donald Duval, David Goto, Ian Holyer, David Jackson, Terry Lyons, Richard Mason, Richard Nowakowski, Jonathan Partington, Richard Pinch, John Rickard, Geoff Thorpe and Antony Wassermann helped me find some of the mistakes; for the many which undoubtedly remain I apologise. For six months while writing this book I enjoyed the hospitality of mathematicians at the University of Calgary to whom I am very grateful. I acknowledge with thanks the excellent typing of Mrs Joan Scutt and Miss Karen McDermid.

I am especially grateful to my wife, without whose patience and understanding this book would never have been written. My research students in analysis also had to put up with a lot during the later stages of producing the book.

Finally, I would like to thank the editor of the series, Professor P. M. Cohn, for his speed and efficiency in handling the manuscript and for his help with the proofs, often beyond the call of duty.

Cambridge   B.B.
January 1978

Contents

Basic Definitions

The brass band stirred themselves, took a deep breath and played
through the "International" three times without a break.

I. A. Ilf and E. P. Petrov; "The Twelve Chairs".

Some of the concepts occurring in this book have a set theoretical or topo-logical flavour. However, most of the structures we investigate are *finite* and every problem we discuss is free of set theoretical and topological difficulties. In view of this we try to avoid pretentious notations and keep the defini-tions as pedestrian as possible. Sometimes we carry this to the extent of abusing the notation slightly. It is unlikely that many of the readers are un-familiar with the basic concepts of graph theory but to make sure that we speak the same language we run through the definitions needed in the sequel. In order to help the reader familiarize himself with the definitions we shall mention a few results as well. These results are hardly more than simple observations. For the convenience of the reader some of the definitions will be repeated in the chapter they are most used.

Unless otherwise stated every set is finite. The number of elements of a set X is denoted by $|X|$. If $|Y| = r$ then we say that Y is an r-set. If furthermore $Y \subset X$ then Y is an r-subset of X. The set of r-subsets of a set X is denoted by $X^{(r)}$, i.e. $X^{(r)} = \{Y : Y \subset X, |X| = r\}$. A *graph* G is an ordered pair of disjoint sets (V, E) such that $E \subset V^{(2)}$ and $V \neq \emptyset$. The set V is the set of *vertices* of G and E is the set of *edges*. An edge $\{x, y\}$ is said to *join* the vertex x to the vertex y and is denoted by xy. We also say that x and y are *adjacent* vertices and the vertex x is *incident* with the edge xy. Two distinct edges with a common endvertex are *adjacent*. Two graphs are *isomorphic* if there exists a 1–1 corre-spondence between their vertex sets that preserves adjacency. Usually we do not distinguish between isomorphic graphs, unless we want to specify the vertices and edges. This is reflected in the convention that if G and H are iso-morphic graphs then we write $G \cong H$ or simply $G = H$.

The vertex set of a graph G is denoted by $V(G)$ and the edge set by $E(G)$; if there is no danger of ambiguity, these are abbreviated to V and E. Even more, if the letter G occurs without any explanation then it stands for an *arbitrary graph*. Instead of $x \in V(G)$ we usually write $x \in G$ to denote that x is a vertex of G. In the same spirit the number of vertices of G, called the *order* of G, is denoted by $|G|$. A graph of order 1 is said to be *trivial*. The number of edges of G, called the *size* of G, is denoted by $e(G)$. We use the notation G^n to denote an *arbitrary* graph of order n. Similarly $G(n, m)$ denotes an *arbitrary* graph of order n and size m. The class of graphs of order n is $\mathcal{G}^n$.

A graph $G' = (V', E')$ is a *subgraph* of a graph $G = (V, E)$ if $V' \subset V$ and $E' \subset E$. In this case we write $G' \subset G$. If $V' = V$ then G' is a *factor* of G. If $W \subset V$ then the graph $(W, E \cap W^{(2)})$ is said to be the subgraph *induced* or *spanned* by W, and is denoted by $G[W]$. We say that H is an *induced subgraph* of G if $H \subset G$ and $H = G[V(H)]$.

The set of vertices adjacent to a vertex $x \in G$ is denoted by $\Gamma(x)$ and $d(x) = |\Gamma(x)|$ said to be the *degree* of x. If it is not clear which is the underlying graph, we put the symbol of the graph into the *suffix* of the appropriate symbol. Thus, if H is an induced subgraph of G and $x \in H$ then

$$\Gamma_H(x) = \Gamma_G(x) \cap V(H) = \Gamma(x) \cap V(H) \quad \text{and} \quad d_H(x) = |\Gamma_H(x)|.$$

For $W \subset V(G)$ we put $\Gamma(W) = \cup \{\Gamma(x): x \in W\}$. The *minimum degree* of the vertices of G is denoted by $\delta(G)$ and the *maximum degree* is denoted by $\Delta(G)$. If $\delta(G) = \Delta(G) = k$, i.e. every vertex of G has degree k, then G is said to be *regular of degree k* or *k-regular*. A 3-regular graph is said to be *cubic*.

If $E' \subset E(G)$ then $G - E'$ denotes the graph resulting from G if we omit the edges belonging to E', i.e. $G - E' = (V(G), E(G) - E')$. Similarly, if $W \subset V(G)$ then $G - W$ is the graph obtained from G by the removal of the vertices belonging to W. Of course, if a vertex $x \in W$ was to be omitted then every edge incident with x was to be omitted as well, i.e. if $G = (V, E)$ then $G - W = (V - W, E \cap (V - W)^{(2)})$. If $W = \{x\}$ we usually write $G - x$ instead of $G - \{x\}$, analogously we may write $G - xy$ instead of $G - \{xy\}$, $xy \in E$. If $H \subset G$ then we may write $G - H$ instead of $G - V(H)$. If $xy \in V^{(2)} - E$ then the graph obtained from G by addition of the edge xy is $G + xy = (V, E \cup \{xy\})$. We might use similar notation for the addition of vertices.

If $V(G) = \{x_1, x_2, \ldots, x_n\}$ then $(d(x_i))_1^n$ is said to be a *degree sequence* of G. Usually we order the vertices in such a way that the degree sequence is monotone increasing or monotone decreasing. Clearly

$$\sum_1^n d(x_i) = 2e(G)$$

so if $(d_i)_1^n$ is a degree sequence of a graph then

$$\sum_1^n d_i \equiv 0 \quad (\text{mod } 2). \tag{0.1}$$

Let x and y be two not necessarily different vertices of G. By an x–y *walk* W we mean an alternating sequence of vertices and edges, say $x_1, \alpha_1, x_2, \alpha_2, \ldots, x_l, \alpha_l, x_{l+1}$, such that $x_1 = x$, $x_{l+1} = y$ and $\alpha_i = x_i x_{i+1} \in E(G)$, $1 \leqslant i \leqslant l$. We usually put $W = x_1 x_2 \ldots x_{l+1}$ since from this form it is clear which are the edges in the sequence. The *length* of this walk W is l. The vertex set of W is $V(W) = \{x_i : 1 \leqslant i \leqslant l + 1\}$ and the edge set of W is $E(W) = \{\alpha_i : 1 \leqslant i \leqslant l\}$. The walk above is a *trail* if all its edges are distinct and it is a *path*-or $x_1 - x_{l+1}$ *path* if all its vertices are distinct. A trail whose endvertices coincide is a *circuit*. If $l \geqslant 3$, $x_1 = x_{l+1}$ but the other vertices are distinct from each other and x_1 then we call the walk a *cycle*.

This cycle is usually denoted by $x_1 x_2 \ldots x_l$ (instead of $x_1 x_2 \ldots x_l x_1$). A path P and a cycle C are identified with the graphs $(V(P), E(P))$ and $(V(C), E(C))$. In particular, $x_1 x_2 \ldots x_{l+1}$ and $x_{l+1} x_l \ldots x_1$ denote the same path, so an x–y path is also a y–x path. Similarly $x_1 x_2 \ldots x_l$ and $x_2 x_3 \ldots x_l x_1$ denote the same cycle. An edge of the form $x_1 x_j (3 \leqslant j \leqslant l - 1)$ is a *diagonal* of this cycle. We denote by P^l a *path of length* l and by C^l a *cycle of length* l. We call C^3 a *triangle*, C^4 a *quadrilateral*, C^5 a *pentagon*, etc. A cycle is *odd* (*even*) if its length is odd (even).

If $P = x_1 x_2 \ldots x_{l+1}$ is a path, $u = x_i, v = x_j$ and $1 \leqslant i < j \leqslant l + 1$ then the $u - v$ *segment of* P is the $u - v$ path $x_i x_{i+1} \ldots x_{j-1} x_j$. We denote it by uPv. If P is an x–y path and Q is a y–z path then $xPyQz$ is the x–z walk obtained by stringing the two paths together. Similarly we may string together *segments* of paths to obtain a walk or a path with the self-explanatory notation $x_1 P_1 x_2 P_2 x_3 \ldots P_l x_{l+1}$, where $x_i x_{i+1} \in V(P_i)$, $i = 1, 2, \ldots, l$.

If $x \in X$ and $y \in Y$ then an x–y path is also said to X–Y *path*. Similarly $\alpha \in E(G)$ is a X–Y edge if $\alpha = xy$ and $x \in X$, $y \in Y$. The number of X–Y edges is denoted by $e(X, Y)$. If $X = \{x\}$ we usually write $e(x, Y)$.

A graph is *connected* if every pair of vertices are joined by a path. A *maximal connected subgraph* is said to be a *component* of the graph. A connected graph not containing cycles is a *tree*, and a graph without cycles (an *acyclic* graph) is a *forest*. Clearly a forest is a graph whose every component is a tree. A tree of order n has n–1 edges and a forest or order n with c components has n–c edges.

The *distance* between two vertices x and y, denoted by $d(x, y)$ is the *minimum length* of an x–y path. If there is no x–y path, i.e. x and y belong to different components, then we put $d(x, y) = \infty$. The *diameter* of a graph G is defined as

$$\text{diam } G = \max\{d(x, y): \quad x, y \in G\}.$$

A related concept is the *radius* of G, $\text{rad } G = \min_x \max_y d(x, y)$. The *girth* of G, $g(G)$, is the minimum length of a cycle in G and the *circumference* of G, $c(G)$, is the maximum length of a cycle. If G does not contain a cycle then the girth and circumference are usually not defined though one might put $g(G) = c(G) = \infty$.

It might be appropriate to remark here that, following recent custom, we use the words "maximum" and "maximal" with different meanings. "Maximal" refers to a maximal element of an ordered set, in which, unless otherwise stated, the ordering is given by *inclusion*. "Maximum" refers to an element of maximal size. Thus P is a *maximal* path of a graph G if it is not properly contained in any other path and Q is a *maximum* path of G if G does not contain a path R *longer* than Q (i.e. $e(R) \leqslant e(Q)$ for every path R in G).

A graph G is an *r-partite graph with vertex classes* $V_1, V_2, \ldots, V_r$ if $V = V(G)$ is the disjoint union $V_1, V_2, \ldots, V_r$ and every edge joins vertices belonging to different vertex classes. Instead of 2-partite we say *bipartite*. We denote by $G_r(n_1, n_2, \ldots, n_r)$ an arbitrary r-partite graph whose ith class contains exactly n_i vertices.

A *k-colouring* or simply *colouring* c of a set X with colours $c_1, c_2, \ldots, c_k$ is a function $c: X \rightarrow \{c_1, \ldots, c_k\}$. We usually consider what one might call a *proper* colouring of the vertices or edges of a graph. This is a colouring in which adjacent elements (i.e. vertices in the vertex colouring and edges in the edge colouring) are assigned *different* colours. If G has a (proper) k-colouring of the vertices then G is said to be *k-colourable*. The *chromatic number* of G is $\chi(G) = \min\{k: G \text{ is } k\text{-colourable}\}$. If $\chi(G) = r$ we say that G is *r-chromatic*. It is easily seen that if G is a minimal r-chromatic graph then

$$\delta(G) \geqslant r - 1. \tag{0.2}$$

For if $x \in G$, $d(x) \leqslant r - 2$, then a (proper) $(r - 1)$-colouring of $G - x$ can be extended to a (proper) $(r - 1)$-colouring of G. In particular,

$$\text{if } \chi(G) \geqslant r \text{ then } \delta(H) \geqslant r - 1 \text{ for some } H \subset G. \tag{0.3}$$

Note that a (proper) r-colouring of the vertices of G is exactly a way of considering G as an r-partite graph: the ith vertex class is the set of vertices coloured with the ith colour. This is the reason why a vertex class is often referred to as a *colour class*. In spite of the equivalence of the terms r-partite and r-colourable we use both since when speaking about an r-partite graph we usually fix the vertex classes but the colour classes of an r-colourable graph are almost never supposed to be given *a priori*.

It is easily seen (e.g. [E16]) that every graph G contains a bipartite graph $B = G_2(n_1, n_2)$ whose size is at least half the size of G:

$$e(B) \geqslant \tfrac{1}{2}e(G). \tag{0.4}$$

For let B be a *maximum* bipartite subgraph of G. We may assume that B is the *bipartite subgraph of G spanned by the classes V_1 and V_2, where $V_1 \cup V_2 = V$*. If $x \in V_1$ then x is joined to at least as many vertices in V_2 as in V_1 since otherwise $V_1 - \{x\}$ and $V_2 \cup \{x\}$ could be chosen for the vertex classes, giving a bipartite subgraph of larger size. Consequently $d_B(x) \geqslant \tfrac{1}{2}d(x)$, implying (0.4).

The reader may find it amusing to prove that if $e(G) > 0$ then there is a subgraph $B = G_2(n_1, n_2) \subset G$ such that $n_1 + n_2 = n$, $|n_1 - n_2| \leqslant 1$ and $e(B) > \tfrac{1}{2}e(G)$. In particular, we may require strict inequality in (0.4). It is obvious that similar results can be proved for r-partite graphs. The weakest of these states that $e\big(G_r(n_1, \ldots, n_r)\big) \geqslant (1 - 1/r)e(G)$ for some $G_r(n_1, \ldots, n_r) \subset G$.

In a number of cases we shall find it convenient to consider a class of graphs defined as follows. Let $d \geqslant 1$ and put

$$\mathcal{D}_d = \left\{ G : |G| \geqslant d, e(G) \geqslant d|G| - \binom{d+1}{2} + 1 \right\}.$$

Note that if $G \in \mathcal{D}_d$ then $|G| > d$ since $|G| = d$ would imply

$$\binom{d}{2} \geqslant e(G) \geqslant d^2 - \binom{d+1}{2} + 1 = \binom{d}{2} + 1.$$

Furthermore, if $G \in \mathcal{D}_d$ then

$$G \text{ contains a subgraph } H \text{ with } \delta(H) \geqslant d + 1. \tag{0.5}$$

To see this note that if $\delta(G) \leqslant d$, say $d(x) \leqslant d$, then $G - x \in \mathcal{D}_d$ since $|G| > d$. By repeated application of this reduction we must arrive at a subgraph with minimal degree at least $d + 1$ since otherwise we would arrive at a graph $G_0 \in \mathcal{D}_d$ with $|G_0| = d$.

There are a number of structures related to graphs. A *hypergraph* or *set system* H is a set V together with a family Σ of subsets of V. Naturally $x \in V$

is a vertex and $S \in \Sigma$ is an edge of the hypergraph H. If $\Sigma \subset V^{(r)}$ then H is said to be an *r-graph* or *r-uniform hypergraph*.

By definition a graph does not contain a *loop*, i.e. an edge joining a vertex to itself, and two distinct vertices are joined by at most one edge, i.e. the graph does not contain *multiple edges*. If we allow *multiple edges* then instead of a graph we obtain a *multigraph*. The number of edges joining a vertex x to y is the *multiplicity* of the edge xy. Sometimes a multigraph is allowed to have loops (of course multiple loops) but it is more customary to call such an object a *pseudograph*. If $G = (V, E^*)$ is a multigraph, the underlying graph H of G has vertex set V and two vertices are joined in H iff they are joined in G. One might describe G as a graph H in which certain specified edges are multiple edges.

A *directed graph* D is a set $V = V(D)$ together with a collection $\vec{E} = \vec{E}(D)$ of *ordered pairs of distinct elements of* V. Of course, V is the set of vertices and $\vec{E}$ is the set of directed edges. A directed edge $(x, y) \in \vec{E}$ is denoted by $\overrightarrow{xy}$. An *oriented graph* is a directed graph containing no symmetric pair of directed edges, i.e. in which at most one of $\overrightarrow{xy}$ and $\overrightarrow{yx}$ is an edge. In other words an oriented graph G is obtained from a graph G by ordering each edge of G. Then we say that $\vec{G}$ is obtained from G by *giving G an orientation* or simply that $\vec{G}$ is an *orientation* of G. Finally we remark that the definition of an infinite graph is the obvious one: $G = (V, E)$ is an *infinite graph* if V is an infinite set, $E \subset V^{(2)}$ and $V \cap E = \varnothing$. In order to emphasize that an object in question is a graph we might call it a *simple* graph. Most of the concepts mentioned above can be carried over immediately to directed graphs. Note however that an $x_0 - x_k$ path corresponds to a *directed* $x_0 - x_k$ path, i.e. to a path $x_0 x_1 \ldots x_k$ such that $x_i x_{i+1}$ is directed from x_i to x_{i+1}. Accordingly $d(x, y)$ is the minimum length of a directed $x - y$ path.

Let $G = (V, E)$, $G' = (V', E')$ be graphs. A map $\phi: V \to V'$ is said to be a *homomorphism* of G into G' if $xy \in E$ implies that $\phi(x) \phi(y) \in E'$. If ϕ is also 1–1 then it is an *embedding* of G into G': clearly ϕ gives an isomorphism between G and a subgraph (denoted by $\phi(G)$) of G'.

Let G be a pseudograph, i.e. multigraph in which loops are permitted. We say that a multigraph G' is an *elementary subdivision* of G if there is an edge of G joining $x \in G$ to $y \in G$ ($x = y$ may hold) such that G' is obtained from $G - xy$ by adding a new vertex and joining it to x and y. (Thus to obtain G' we replace an edge by a path of length 2.) We say that H is a *subdivision of* G or that H is a *topological G*, in notation $H = TG$, if H can be obtained from G by a sequence of elementary subdivisions. Note that the notation TG is analogous to G^n and $G(n, m)$, since it denotes an *arbitrary* subdivision of G.

In fact throughout the book we use the notation TG only in the case when G does not have a vertex of degree 2 joined to distinct vertices. Two multigraphs are said to be *homeomorphic* if they have isomorphic subdivisions. It is trivial to see that two multigraphs are homeomorphic iff the topological spaces naturally associated with them (cf. Ch V, §3) are homeomorphic.

Let H be a connected subgraph of a graph G. Add a new vertex x_H to the graph $G–H$ and join it to every vertex $y \in G–H$ for which G contains a $y–H$ edge. The resulting graph is denoted by G/H and it is said to be the graph obtained from G by *contracting H* (to a vertex). If $E(H) = xy$ then $G/xy = G/H$ is an *elementary contraction* of G. We say that L is a *contraction* of G, in notation $G > L$, if L can be obtained from G by a sequence of contractions (of connected subgraphs). L is a *subcontraction* of G, in notation $G \succ L$ if L is a contraction of a subgraph of G.

The *complement* of a graph $G = (V, E)$ is the graph $\overline{G} = (V, V^{(2)} - E)$. The *complete graph of order n, K^n*, has every pair of its n vertices adjacent. In other words K^n is the graph of order n and size $\binom{n}{2}$, that is $K^n = G(n, \binom{n}{2})$. Note that $K^3 = C^3$ is the triangle. We call K^4 a complete quadrilateral, etc. We have chosen the notation K^n instead of the more common K_n since we shall use capital letters with subscripts $(G_k, H_l, K_p,$ etc.) to denote *specific* graphs. Thus K_p might denote a *given* complete subgraph. The complement of K^n is the *empty* or *null* graph of order n: $E^n = \overline{K}^n = G(n, 0)$. The unique maximal graph $G_r(n_1, n_2, \ldots, n_r)$ is denoted by $K_r(n_1, n_2, \ldots, n_r)$. It has r vertex classes, the ith class has n_i vertices and every pair of vertices belong to distinct classes are joined by an edge. Clearly

$$e(K_r(n_1, \ldots, n_r)) = \sum_{1 \leqslant i < j \leqslant r} n_i n_j.$$

If $r = 2$ then the index r might be omitted and n_1, n_2 might become upper indices, e.g. $K_2(3, 4) = K(3, 4) = K^{3,4}$. The tree $K^{1,p}$ is the *star* of order $p + 1$. The maximum order of a complete subgraph of G is the *clique number* of G. We denote it by $cl(G)$.

The *union* of the graphs G_1 and G_2 is denoted by $G_1 \cup G_2$. If

$$(V(G_1) \cup E(G_1)) \cap (V(G_2) \cup E(G_2)) = \varnothing \tag{0.6}$$

then

$$V(G_1 \cup G_2) = V(G_1) \cup V(G_2)$$

and

$$E(G_1 \cup G_2) = E(G_1) \cup E(G_2).$$

If (6) does not hold then we add an index to the elements of $V(G_i) \cup E(G_i)$ to make (6) hold and define the union as before. Sometimes we emphasize this by saying that $G_1 \cup G_2$ is the *disjoint* union of G_1 and G_2; e.g. $K^3 \cup K^4$ is the complement of $K(3, 4)$. The only exception occurs when G_1 and G_2 are subgraphs of a given graph. Then, naturally, $G_1 \cup G_2$ is defined by

$$V(G_2 \cup G_2) = V(G_1) \cup V(G_2) \quad \text{and} \quad E(G_1 \cup G_2) = E(G_1) \cup E(G_2).$$

It will be clear from the text which of these cases is at hand.

The union of several graphs is defined analogously. The disjoint union of k copies of the same graph G is denoted by kG. Thus $kK^1 = kE^1 = E^k$.

The *join* $G_1 + G_2$ of G_1 and G_2 is obtained from $G_1 \cup G_2$ by joining *each vertex of G_1 to each vertex of G_2.* Thus $E^3 + E^4 = K(3, 4)$. The join of several graphs is defined analogously:

$$E^{n_1} + E^{n_2} + \ldots + E^{n_r} = K_r(n_1, n_2, \ldots, n_r).$$

In a number of graph constructions it is convenient to choose a prime for one of the parameters. In order to extend the construction to every possible value of the parameters one then uses a shallow or deep result about the distribution of primes. *Bertrand's postulate* claims that for every natural number $n > 3$ there is a prime between n and $2n - 2$. This was verified by Bertrand for $n < 3\,000\,000$ and proved by Tchebychev in 1850 (cf. [HW1; p. 373]). Furthermore, the quotient of consecutive primes tends to 1. In fact there are $0 < \eta < 1$ and $C_\eta > 0$ such that for every $n \geqslant 2$ there is a prime between n and $n + C_\eta n^\eta$. This was proved by Hoheisel [H27] ($\eta = 1 - 3300^{-1} + \varepsilon$), Ingham [I1] ($\eta = \frac{5}{8} + \varepsilon$), Montgomery [M32] ($\eta = \frac{3}{5} + \varepsilon$) and Huxley [H27] ($\eta = \frac{7}{12} + \varepsilon$) where $\varepsilon > 0$ is arbitrary. We shall make use of the fact that if n is sufficiently large then

$$\text{there is a prime between } \quad n - \tfrac{1}{10}n^{2/3} \text{ and } n. \tag{0.7}$$

If q is a prime power then there is a finite projective plane $PG(2, q)$ over the field of order q. We represent the points and lines of this plane by triples (a, b, c) and $[a, b, c]$ of elements of the ground field such that each triple has at least one non-zero element. If $\lambda \neq 0$ then (a, b, c) and $(\lambda a, \lambda b, \lambda c)$ represent the same point; similarly $[a, b, c]$ and $[\lambda a, \lambda b, \lambda c]$ represent the same line. A point (x, y, z) is on a line $[a, b, c]$ if $ax + by + cz = 0$.

Chapter I

Connectivity

Perhaps the most basic property a graph may possess is that of being connected. At a more refined level, there are various functions that may be said to measure the connectedness of a connected graph. In the first section we introduce some of these functions and establish their basic properties. The first significant theorem of this chapter, Menger's theorem, is the highlight of §2. This fundamental theorem of Menger, published in 1937, is the basis of almost every proof of connectivity properties.

The most sensitive functions measuring the connectivity of a graph G are the *local vertex-* and *edge-connectivity* functions $\kappa(x, y)$ and $\lambda(x, y)$, defined on $V(G)^{(2)}$. The first of these is the minimum of the number of *vertices* whose omission disconnects x from y, the second is the minimum of the number of *edges* needed to disconnect x from y. The bulk of the chapter is devoted to the study of the minima and maxima of these functions. We are especially interested in the minima of the local connectivities, the so called *(vertex-) connectivity* $\kappa(G) = \min \kappa(x, y)$ and *edge-connectivity* $\lambda(G) = \min \lambda(x, y)$. A graph is *k-connected* if its connectivity is at least k and *k-edge-connected* if its edge-connectivity is at least k.

The main aim of the theory of connectivity of graphs is to produce a body of theorems describing various properties of k-connected and k-edge-connected graphs. (In order not to be too monotonous, in most of the sequel we speak only about k-connected graphs.) If a graph has a k-connected factor then the graph itself is also k-connected. Thus it is natural to focus one's attention on k-connected graphs without a proper k-connected factor. These graphs are said to be *minimally k-connected*.

The structure of 2- and 3-connected graphs is satisfactorily understood. We present some of the results in §3. The main theorem of the section is

1

Tutte's characterization of 2- and 3-connected graphs. This beautiful theorem is very near to providing a practical catalogue of all 3-connected graphs.

Not surprisingly, for $k \geqslant 4$ much less is known about the structure of minimally k-connected graphs. However, there are a number of deep results, due mostly to Mader, that are valid for every $k \geqslant 3$; we present these in §4.

We would like to point out that there is another natural subclass of the k-connected graphs the description of whose members would be a major step towards a characterization of the k-connected graphs. A graph G is *critically k-connected* if G is k-connected but $G - x$ is *not* k-connected for each $x \in G$. Clearly every k-connected graph contains a critically k-connected subgraph. The structure of critically k-connected graphs is less known than the structure of minimally k-connected graphs and the former will not be discussed in the main body of the text. However we give some of the results among the exercises.

In §5 we present most of what is known about $\bar{\kappa}(G) = \max \kappa(x, y)$ and $\bar{\lambda}(G) = \max \lambda(x, y)$. It is somewhat surprising that the usual analogue between vertex- and edge-connectivity breaks down in this area; the problems concerning $\bar{\kappa}(G)$ are considerably harder. The theory rather badly misses a result concerning the maximal size of a graph G of order n with $\bar{\kappa}(G) = k$ (Problem 36). The present estimates are very poor for large values of k.

In spite of our wealth of information about minimally k-connected graphs we know very little about the l-connected $(l < k)$ factors of a k-connected graph. Let us mention here a special case of one of the problems posed in §6. At least how many edges is it possible to omit from a k-connected graph of order n if we want the resulting graph to be $(k - 1)$-connected?

1. ELEMENTARY PROPERTIES

A graph $G = (V, E)$ is said to be *connected* if for every pair of vertices there is a path joining them. Otherwise the graph is *disconnected*. The maximal connected subgraphs are called the *components* of the graph. The component containing a vertex $x \in G$ is denoted by $C(x)$. If it is desirable to emphasize that the component is a component of G, we write $C(x, G)$ instead of $C(x)$. Clearly a graph is the disjoint union of its components. The smallest components are the isolated vertices and isolated edges of a graph. An *isolated vertex* is a vertex of degree 0 and an *isolated edge* is an edge joining two vertices of degree 1.

Let V_1, V_2 be subsets of the vertex set V of a graph G. If $V = V_1 \cup V_2$,

$S = V_1 \cap V_2$ is a proper subset of both V_1 and V_2, $|S| = k$ and no vertex in $V_1 - S$ is adjacent to a vertex in $V_2 - S$ then we say that (V_1, V_2) is a *k-separator* with *separating set S*, or that the *vertices of S separate G*. If $X \subset V_1 - S$ and $Y \subset V_2 - S$ then we say that S separates X from Y. If $X = \{x\}$ and $Y = \{y\}$ then we say that S separates the vertices x and y. Clearly S is a separating set of a connected graph G if and only if $G - S$ is disconnected.

The *connectivity* (or *vertex-connectivity*) $\kappa(G)$ of a graph G is the minimum number of vertices whose removal results in a disconnected graph or in the trivial graph. In other words

$$\kappa(G) = \min\{|G| - 1, \quad k: G \text{ has a } k\text{-separator}\}$$

$$= \min\{|G| - 1, \quad |S|: S \subset V \text{ separates } G\}.$$

The *edge-connectivity* $\lambda(G)$ is defined analogously, only instead of vertices we remove edges. A graph is *k-connected* (resp. *k-edge-connected*) if $\kappa(G) \geqslant k$ (resp. $\lambda(G) \geqslant k$).

By definition connectedness, 1-connectedness and 1-edge-connectedness coincide. Clearly every k-connected (resp. k-edge-connected) graph has at least $k + 1$ vertices and K^{k+1} is the only k-connected (resp. k-edge-connected) graph of order $k + 1$. Furthermore it follows from the definition that if G is any graph, $x \in G$ and $e \in E(G)$ then

$$\kappa(G) - 1 \leqslant \kappa(G - x) \leqslant \kappa(G), \tag{1}$$

$$\lambda(G) - 1 \leqslant \lambda(G - e) \leqslant \lambda(G). \tag{2}$$

It is almost as simple to check that the minimal degree, the edge-connectivity and vertex-connectivity satisfy the following inequality:

$$\delta(G) \geqslant \lambda(G) \geqslant \kappa(G). \tag{3}$$

The following simple observation has some corollaries due to Bondy [B41] and Chartrand and Harary [CH3].

THEOREM 1.1. *Let G be a graph of order $n \geqslant k + 1 \geqslant 2$. If G is not k-connected then there are two disjoint sets of vertices V_1, V_2, $|V_1| = n_1 \geqslant 1$, $|V_2| = n_2 \geqslant 1$, $n_1 + n_2 + k - 1 = n$, such that the vertices of V_i have degree at most $n_i + k - 2, i = 1, 2$.* ∎

COROLLARY 1.2. *Suppose a graph G has vertices $x_1, \ldots, x_n$, $d(x_1) \leqslant d(x_2) \leqslant \ldots \leqslant d(x_n) \leqslant \Delta$, and*

$$d(x_j) \geqslant j \quad \text{for all} \quad j \leqslant n - \Delta - 1.$$

Then G is connected. ∎

COROLLARY 1.3. *Let G be a graph with vertices $x_1, \ldots, x_n$, $d(x_1) \leqslant d(x_2) \leqslant \ldots \leqslant d(x_n)$. Suppose for some k, $0 \leqslant k < n$,*

$$d(x_j) \geqslant j + k - 1, \quad j = 1, 2, \ldots, n - 1 - d(x_{n-k+1}).$$

Then G is k-connected. ∎

COROLLARY 1.4. *Let G be a graph of order n. Then*

$$\kappa(G) \geqslant 2\delta(G) + 2 - n.$$ ∎

As a slight extension of results of Chartrand and Harary [CH3] and Chartrand [C7] we show that Corollary 1.4 and inequality (3) contain almost all the restrictions on δ, κ and λ.

THEOREM 1.5. *Given integers n, δ, κ and λ, there is a graph G of order n such that $\delta(G) = \delta$, $\kappa(G) = \kappa$ and $\lambda(G) = \lambda$ if and only if one of the following conditions is satisfied.*

  (i) $0 \leqslant \kappa \leqslant \lambda \leqslant \delta < \lfloor n/2 \rfloor$.
  (ii) $1 \leqslant 2\delta + 2 - n \leqslant \kappa \leqslant \lambda = \delta < n - 1$.
  (iii) $\kappa = \lambda = \delta = n - 1$.

Proof. The relation $\kappa(K^n) = \lambda(K^n) = n - 1$, Corollary 1.4 and inequality (3) imply that if G is any graph then the numbers $|G| = n$, $\delta(G) = \delta$, $\kappa(G) = \kappa$ and $\lambda(G) = \lambda$ satisfy one of the relations in Theorem 1.5.

Thus we have to show that if (i), (ii) or (iii) are satisfied then there is a graph G with appropriate constants $n, \delta, \kappa, \lambda$.

Suppose (i) holds. Let $G_1 = K^{\delta+1}, G_2 = K^{n-\delta-1}$, $u_1, \ldots, u_\kappa \in G_1$ and $v_1, \ldots, v_\kappa \in G_2$. Let G be the graph obtained from $G_1 \cup G_2$ by adding the edges $u_1 v_1, \ldots, u_\kappa v_\kappa$ and $\lambda - \kappa$ other $u_i v$ ($v \in G_2$) edges. Then $|G| = n$, $\delta(G) = \delta$ and it is easily checked that $\kappa(G) = \kappa$ and $\lambda(G) = \lambda$.

Now suppose that (ii) holds. Let $G_1 = K^\kappa, G_2 = K^a, G_3 = K^b$ and $G_0 = G_1 + (G_2 \cup G_3)$, where $a = \lfloor (n - \kappa)/2 \rfloor$ and $b = \lfloor (n - 1 - \kappa)/2 \rfloor$. To

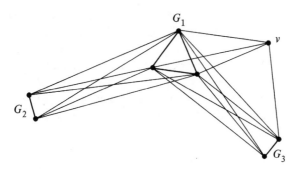

FIG. 1.1. A graph with $n = 8, \delta = \lambda = 4, \kappa = 3$.

construct a graph G with the required properties add a vertex v to G_0 and join it to the vertices of G_1 and $\delta - \kappa$ vertices of G_3 (see Fig. 1.1).

Finally, if (iii) holds, take $G = K^n$.   ∎

It was pointed out by Harary [H20] that given the number of vertices and edges of a graph, the largest connectivity possible can also be read out of inequality (3).

THEOREM 1.6. *For each n, m with $0 \leqslant n - 1 \leqslant m \leqslant \binom{n}{2}$*

$$\max \kappa(G) = \max \lambda(G) = \lfloor 2m/n \rfloor,$$

where the maxima are taken over all graphs $G = G(n, m)$.

Proof. For $n - 1 = m$ take a tree with n vertices. Thus we can suppose that $n \leqslant m$. As $\kappa \leqslant \lambda \leqslant \delta \leqslant \lfloor 2m/n \rfloor$, we have to show only that for every m and n there is a graph G with $|G| = n, e(G) = m, \delta(G) = \kappa(G) = \lfloor 2m/n \rfloor$. Furthermore, it suffices to show this for a given $\delta \geqslant 2$ and minimal m with $\delta = \lfloor 2m/n \rfloor$ since the addition of edges does not decrease the connectivity. Thus we can suppose that $2m - 1 \leqslant \delta n \leqslant 2m$. Let G be a graph obtained from C^n by joining vertices within distance $\lfloor \delta/2 \rfloor$ and adding some diagonals joining vertices at distance $\lfloor n/2 \rfloor$ on C^n in such a way that in G every vertex, except at most one, has degree δ. It is easily checked that $\kappa(G) = \delta$.   ∎

The decomposition of graphs into various parts according to their connectivity properties was first introduced and studied by Whitney [W18]

[W19]. A graph is said to be *separable* if it has a 1-separator. A vertex is said to be a *cutvertex* if its removal increases the number of components of a graph. Analogously a *bridge* is an edge whose removal increases the number of components. Thus x is a cutvertex of G if $\{x\}$ is a separating set of a component of G. Clearly a graph of order at least 3 is nonseparable if and only if it is 2-connected. A *block* of a graph is a maximal nonseparable nontrivial subgraph. Let I be the set of isolated vertices of a graph G and let $B_1, B_2, \ldots, B_l$ be the blocks of G. Then

$$G - I = \bigcup_1^l B_i,$$

any two blocks have at most one vertex in common and that vertex is a cutvertex of G. In particular

$$E(G) = \bigcup_1^l E(B_i)$$

is a partitioning of the edges of G. The *block graph* of G, denoted by $B(G)$, has vertex set $\{B_1, \ldots, B_l\}$ and $B_i B_j$ is an edge if and only if the blocks B_i and B_j have a cutvertex in common. The *block-cutvertex* graph of G, denoted by $bc(G)$, has vertex set $\{B_1, \ldots, B_l\} \cup \{c_1, \ldots, c_m\}$, where $c_1, \ldots, c_m$ are the cutvertices of G, and edge set $\{B_i c_j : c_j \in V(B_i)\}$. As a cycle of G containing a cutvertex is contained in a block, $bc(G)$ does not contain a cycle so it is a forest. The endvertices of this forest correspond to blocks of G so the distance between any two endvertices of the same tree of $bc(G)$ is even (cf. Ex. 6). The blocks corresponding to endvertices of $bc(G)$ are called the *end blocks* of G. Clearly a connected but not 2-connected graph has at least two end blocks.

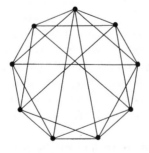

Fig. 1.2. A graph G with $|G| = 9$, $e(G) = 23$, $\lambda(G) = \kappa(G) = 5$.

If every edge of a graph is contained in a 2-connected subgraph (i.e. no edge is a block) then the blocks are exactly the maximal 2-connected subgraphs. A *k-component* of a graph is a *maximal k-connected subgraph*. Of course a graph might not have a k-component for $k \geqslant 2$, e.g. a tree does not have a 2-component. As observed by Harary and Kodama [HK1], two k-components can have at most $k - 1$ vertices in common (see Ex. 10).

2. MENGER'S THEOREM AND ITS CONSEQUENCES

The *local (vertex-) connectivity* $\kappa(x, y)$ of two non-adjacent vertices is the minimum number of *vertices* separating x from y. If x and y are adjacent vertices, their local connectivity is defined as $\kappa_H(x, y) + 1$ where $H = G - xy$. Analogously the local edge-connectivity $\lambda(x, y)$ of two vertices is the minimum number of *edges* separating x from y. Clearly for any graph G of order at least two we have

$$\kappa(G) = \min\{\kappa(x, y): x, y \in G, x \neq y\}$$

and

$$\lambda(G) = \min\{\lambda(x, y): x, y \in G, x \neq y\}.$$

The main aim of this section is to discuss the fundamental connections between $\kappa(x, y)$ (resp. $\lambda(x, y)$) and sets of x–y paths. Two paths in a graph are said to be *independent* if every common vertex is an endvertex of both paths. Two paths with no *vertex* (resp. no *edge*) in common are *strongly* (resp. *weakly*) *independent*. A set of *independent* (resp. *strongly independent*, resp. *weakly independent*) paths is a set of paths any two of which are independent (resp. strongly independent, resp. weakly independent).

It is obvious that if there are k independent x–y paths then $\kappa(x, y) \geqslant k$. Menger's theorem [M26] is the converse of this observation. We shall state this result in a form especially suitable for the beautiful simple proof due to Pym [P11]. (Another elegant proof is due to Dirac [D22], see Ex. 11.)

Let $X, Y \subset V$, where $G = (V, E)$ is a fixed graph. A *strict X–Y path* is an X–Y path whose only vertex in X (resp. Y) is its initial (resp. terminal) vertex. We say that X is *linked* to Y if $|X| = |Y|$ and there are $|X|$ strongly independent X–Y paths. A set $Z \subset V$ *detaches* X from Y in G if in $G - Z$ there is no X–Y path. Then we call Z a *detaching* set. A set Z detaching X from Y clearly separates X from Y iff $X \cap Z = Y \cap Z = \varnothing$.

THEOREM 2.1. *Denote by k the minimal number of elements in a detaching set. Then there exists a k-set $X_0 \subset X$ which is linked to a set $Y_0 \subset Y$.*

Proof. We apply induction on $e(G)$. The theorem being trivial for graphs of size 1, we suppose that $e(G) > 1$ and it holds for graphs of size less than $e(G)$. We can suppose without loss of generality that G is connected. We distinguish two cases.

Case (i). *Every detaching k-set coincides with either X or Y.* We may assume that $|X| = k$ and X is not contained in Y, so there exists a vertex $x_0 \in X - Y$. The set $X - \{x_0\}$ does not detach X from Y so there is an edge $x_0 y_0$ with $y_0 \notin X$. Put $G_0 = G - x_0 y_0$.

Now by the induction hypothesis we can suppose that in G_0 there is a minimal detaching $(k-1)$-set Z_0. As $Z_0 \cup \{x_0\}$ and $Z_0 \cup \{y_0\}$ are both detaching, they must be k-sets, i.e. $x_0 \notin Z_0$, $y_0 \notin Z_0$. Furthermore, $x_0 \in X-Y$ and $y_0 \notin X$ so $Z_0 \cup \{x_0\} = X$ and $Z_0 \cup \{y_0\} = Y$. Then the trivial paths with vertices in Z_0 and the path consisting of the edge $x_0 y_0$ link X to Y.

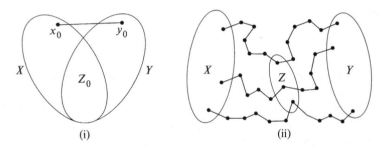

FIG. 2.1. Illustrations for Theorem 2.1.

Case (ii). *There is a detaching k-set Z different from X and Y.* Let G_1 (resp. G_2) be the subgraph of G spanned by those vertices that are on *strict X–Z* (resp. Z–Y) paths. In G_1 one needs at least k vertices to detach X from Z. As the vertices of Y–Z do not belong to G_1, by the induction hypothesis a k-set $X_0 \subset X$ can be linked to Z in G_1. Similarly Z can be linked to a k-set $Y_0 \subset Y$ in G_2. Then clearly one can link X_0 to Y_0. ∎

We can state now the usual form of Menger's theorem [M26] which is a special case of Theorem 2.1.

THEOREM 2.2. *Let* $x, y \in G$, $x \neq y$. *The maximum number of independent* x–y *paths is* $\kappa(x, y)$.

Proof. Apply Theorem 2.1 to the sets $X = \Gamma(x)$ and $Y = \Gamma(y)$. ∎

Though in the next chapter we discuss the existence of independent edges in bipartite graphs in detail, let us note here that P. Hall's theorem [H18] is a consequence of Theorem 2.1.

THEOREM 2.3. *Let* G *be a bipartite graph with vertex classes* X, Y. *Suppose*

$$|\Gamma(V)| \geqslant |V| \qquad for\ every \qquad V \subset X.$$

Then G *contains* $|X|$ *independent edges.*

Proof. Check that every set detaching X from Y has at least $|X|$ elements and apply Theorem 2.1. ∎

By considering the line graph of the graph, Theorem 2.1 implies the so called edge form of Menger's theorem.

THEOREM 2.4. *Let* $x, y \in G$, $x \neq y$. *The maximum number of weakly independent* x–y *paths is* $\lambda(x, y)$.

Proof. Let $H = L(G)$ be the *line graph* of G, i.e. $V(H) = E(G)$ and $\gamma\delta \in E(H)$ if and only if γ and δ are adjacent in G (see Harary [H19], p. 71). Let X (resp. Y) be the set of edges in G incident with x (resp. y). Then Theorem 2.1 applied to the sets X and Y in H gives the result. ∎

One might hope that these theorems of Menger could be extended to paths of length *at most* p. Thus if for $x, y \in G$, $x \neq y$, we put

$$S_p = S_p(x, y) = \min\{|W| : W \subset V - \{x, y\}, d_{G-W}(x, y) > p\},$$

$$I_p = I_p(x, y) = \max\{k : \exists k \text{ independent } x\text{–}y \text{ paths of length} \leqslant p\},$$

then clearly $S_p \geqslant I_p$ and one might hope that $S_p = I_p$. Unfortunately this is not true. Various counterexamples to the equality $S_p = I_p$ were given by Lovász, Neumann-Lara and Plummer [LNP1] (cf. Ex. 15) who also proved a number of positive results. Among others they gave lower bounds on S_p in terms of I_p and $d(x, y)$.

Theorems 2.2 and 2.4 imply the following characterization of k-connected and k-edge-connected graphs.

THEOREM 2.5. *A non-trivial graph is k-connected (resp. k-edge-connected) if and only if for any two vertices there are k disjoint (resp. k-edge-disjoint) paths joining them.* ∎

The next theorem and its consequences are due to Dirac [D12], [D13]. (The theorem was rediscovered by Halin and Jung [HJ1].)

Let U be a set of vertices of a graph G and let x be a vertex not in U. An *x–U fan* is a set of $|U|$ paths from x to U, any two of which have only the vertex x in common.

THEOREM 2.6. *A graph G is k-connected if and only if $|G| \geqslant k + 1$ and for any k-set $U \subset V(G)$ and vertex $x \in V(G) - U$, there is an x–U fan.*

Proof. (a) Suppose G is k-connected, $U \subset V(G), |U| = k$ and $x \in V(G) - U$. Let H be the graph obtained from G by adding a vertex y and joining y to every vertex in U. By Theorem 2.2 there are k independent x–y paths in H. Omitting the edges incident with y we obtain an x–U fan in G.

(b) Suppose $|G| \geqslant k + 1$ and S is a $(k - 1)$-set separating the vertices x and y. Then G does not contain an x-$(S \cup \{y\})$ fan. ∎

The straightforward proofs of the following two corollaries of Theorem 2.6 are left to the reader (Ex. 18).

COROLLARY 2.7. *If G is k-connected and $k \geqslant 2$, then for any set of k vertices there is a cycle containing all of them.* ∎

COROLLARY 2.8. *Let $|G| \geqslant 2k$. Then G is k-connected if and only if whenever V_1 and V_2 are disjoint k-sets of vertices, then V_1 is linked to V_2.* ∎

If $k > 2$, the condition in Corollary 2.7 is only necessary but not sufficient for a graph to be k-connected. However, it is obvious that for $k = 2$ the condition is also sufficient. Let us formulate this together with some other equivalent conditions.

THEOREM 2.9. *Let G be a graph of order at least 3. Then the following conditions are equivalent.*

   (i) *G is 2-connected.*
  (ii) *G has no cutvertex.*
 (iii) *Given any two vertices there is a cycle containing them.*
 (iv) *Given any vertex and any edge there is a cycle containing them.*
  (v) *Given any two edges there is a cycle containing them.* ∎

2-edge-connected graphs can be characterized analogously.

THEOREM 2.10. *Let G be a graph of order at least 3. Then the following conditions are equivalent.*

   (i) *G is 2-edge-connected.*
  (ii) *G has no isolated vertices or bridges.*
 (iii) *G is connected, has no isolated vertices and every edge is contained in a cycle.*
 (iv) *Given any two edges there is a circuit containing them.*
  (v) *Given any vertex and any edge there is a circuit containing them.*
 (vi) *Given any two vertices there is a circuit containing them.* ∎

The definition of connectivity, inequalities (1) and (2) of §1 and the last two results imply the following characterizations of k-connected and k-edge-connected graphs, pointed out by Halin [H12] and Lick [L8].

THEOREM 2.11. (a) *A graph G with at least k ($\geqslant 2$) vertices is k-connected if and only if for each $(k - 2)$-set U of vertices and for each pair x, y of vertices of $G - U$ there is a cycle in $G - U$ containing the vertices x and y.*

   (b) *A connected graph G with at least k ($\geqslant 2$) edges is k-edge-connected if and only if for each $(k - 2)$-set F of edges and for each pair e, f of edges of $G - F$ there is a circuit in $G - F$ containing the edges e and f.* ∎

3. THE STRUCTURE OF 2- AND 3-CONNECTED GRAPHS

One of the most important goals of the theory of k-connected graphs is to compile a list of all k-connected graphs. A natural way of achieving this would be to give some operations producing k-connected graphs from k-connected graphs such that every k-connected graph can be obtained from certain simple k-connected graphs by repeated applications of the operations. This task has been accomplished by Tutte [T15] for 3-connected graphs, by Dirac [D23] and Plummer [P6] for 2-connected graphs and by Slater [S22] for 4-connected graphs. Though structural results like these do not belong to extremal graph theory, we include some of them in this book since they are eminently suitable for tackling a number of extremal problems and since they go deeper than sheer numerical inequalities.

The addition of an edge does not decrease the connectivity of a graph so in order to describe all k-connected graphs it suffices to describe all *minimally k-connected graphs*, i.e. graphs that are k-connected but lose this property if we remove any of its *edges*. Actually the structure of minimally k-connected graphs is not necessarily more accessible than the structure of k-connected graphs but for $k = 2$ this seems to be the case. Minimally k-connected graphs will be investigated in the next section.

Let us start with the results concerning minimally 2-connected graphs. Most of these results were obtained independently by Dirac [D23] and Plummer [P6]. In the arguments that follow one should keep in mind the equivalence of the various conditions in Theorem 2.9.

Call an edge α of a 2-connected graph G *essential* if $G - \alpha$ is not 2-connected. Thus a 2-connected graph is minimally 2-connected if and only if every edge of it is essential. The following result states the most important property of the essential edges.

THEOREM 3.1. *Let* $xy \in E(G)$ *and* $H = G - xy$. *Then the following two assertions are equivalent.*

    (i) G *is 2-connected and* xy *is an essential edge.*

    (ii) H *has no isolated vertex,* x *and* y *are not cutvertices of* H, *the block-cutvertex graph of* H, $bc(H)$, *has a non-trivial* x–y *path,* x *belongs to the initial block and* y *belongs to the terminal block of* H (Fig. 3.1).

Proof. It is clear that (ii) implies (i) so we suppose that (i) holds and intend to prove (ii).

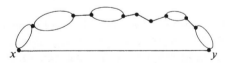

FIG. 3.1. The structure of G in Theorem 3.1.

It is immediate that H is connected (since G is 2-connected) and $bc(H)$ is a nontrivial tree (since H is connected and not 2-connected). If L is any graph, $u, v \in L$ and $M = L + uv$ then $L - u = M - u$ so if we join a vertex to a cutvertex u then u remains a cutvertex. Consequently x and y are not cutvertices of H.

To complete the proof it suffices to show that the tree $bc(H)$ does not contain an endvertex C (a block of H) such that neither x nor y is a vertex of C. Let c be the unique cutvertex of H in C. Then the addition of xy to H does not join a vertex of $C - c$ to a vertex of $H - C$, i.e. c is a cutvertex of H as well, contradicting the assumption that G is 2-connected. ∎

COROLLARY 3.2. *Let G be a 2-connected graph. An edge $\alpha = xy$ is an essential edge of G if and only if no cycle of $G - \alpha$ contains both x and y, i.e. α is not a diagonal of a cycle in G.*

In particular, a 2-connected graph is minimally 2-connected if and only if no cycle has a diagonal.

Proof. If no cycle of $G - \alpha$ contains both x and y then $G - \alpha$ is not 2-connected so α is an essential edge. The converse implication follows from Theorem 3.1. ∎

COROLLARY 3.3. *Every 2-connected subgraph of a minimally 2-connected graph is minimally 2-connected.* ∎

COROLLARY 3.4. *Let G be a 2-connected graph of order at least 4. Then G does not contain a triangle formed by essential edges.*

Proof. Suppose xyz is such a triangle and $u \in V(G) - \{x, y, z\}$. As G is 2-connected, it follows from Theorem 2.2 that there exist two independent paths from u to $\{x, y, z\}$, say a u–x path P_1 and a u–y path P_2, with $z \notin P_1 \cup P_2$. Then xy is a diagonal of the cycle uP_1xzyP_2u, contradicting Corollary 3.2. ∎

In order to formulate the structure theorem of minimally 2-connected graphs we need another concept. Call two vertices *compatible* if every edge joining two vertices of any path connecting them belongs to the path. Note that two non-adjacent vertices of C^l $(l \geq 4)$ are compatible.

THEOREM 3.5. *For each i, $0 \leq i \leq k (k \geq 1)$ let G_i be an edge $x_i x'_{i+1}$ or a minimally 2-connected graph containing compatible vertices x_i and x'_{i+1}. Let*

G be the graph obtained from $\overset{k}{\underset{0}{\cup}}\,G_i$ by identifying x_i with x'_i for $1 \leqslant i \leqslant k$ and joining x_0 to x'_{k+1}. Then G is minimally 2-connected.

Conversely, every minimally 2-connected graph can be obtained in the way described above.

Proof. The first part is an immediate consequence of Corollary 3.2.

To see the second part let G be a minimally 2-connected graph and let $xy \in E(G)$. As xy is an essential edge of G, by Theorem 3.1 (ii), $H - G - xy$ has blocks $B_0, B_1, \ldots, B_k\,(k \geqslant 1)$ such that B_{i-1} and B_i have a vertex x_i in common, $x \in B_0$, $y \in B_k$ and $x, x_1, x_2, \ldots, x_k, y$ are distinct vertices. Note that G contains an x_{i+1}–x_i path having only the vertices x_i and x_{i+1} in B_i. Therefore Corollary 3.2 implies that x_i and x_{i+1} are compatible vertices of B_i. Finally, by Corollary 3.3, each block B_i is either an edge (the edge $x_i x_{i+1}$) or it is minimally 2-connected. ∎

THEOREM 3.6. *Let G be a minimally 2-connected graph and let x, y be compatible vertices of G. Then every x–y path contains a vertex of degree 2.*

Proof. Suppose $P = x_0, x_1, \ldots, x_k\,(k \geqslant 2, x_0 = x, x_k = y)$ is a path not containing a vertex of degree 2. Omit the edges of P from G to obtain a graph H. Then H does not contain isolated vertices and every component of H contains at least two vertices of P. As x and y are compatible vertices, two vertices of P belonging to the same component of H are not adjacent on P. Let P_1 be an x_i–x_j path in a component of H such that $i < j$ (and so $i \leqslant j - 2$), $V(P) \cap V(P_1) = \{x_i, x_j\}$ and $j - i$ is minimal. Let P_2 be an x_{i+1}–x_l path in the component $C(x_{i+1})$ of H such that $V(P) \cap V(P_2) = \{x_{i+1}, x_l\}$. Then P_1 and P_2 are strongly independent paths since otherwise there would be an x_i–x_{i+1} path P_3 in H such that $V(P) \cap V(P_3) = \{x_i, x_{i+1}\}$. By the choice of P_1 either $l < i$ or $j < l$. If $l < i$ then $x_i x_{i+2}$ is a diagonal of $x_i P x_i P_1 x_j P x_{i+1} P_2 x_l$, contradicting Corollary 3.2. If $j < l$ then $x P x_i P_1 x_j P x_{i+2} P_2 x_l P x_k$ is an x–y path not containing the edge $x_i x_{i+1}$, contradicting the fact that x and y are compatible. ∎

THEOREM 3.7. *Let G be a minimally 2-connected graph which is not a cycle. Then every cycle in G contains two vertices of degree at least three which are separated on the cycle by two vertices of degree two.*

Proof. Let xy be an edge of a cycle C in G. By the proof of Theorem 3.5 the cycle C contains vertices $x_0 = x, x_1, \ldots, x_{k+1} = y$ $(k \geq 1)$ such that the arc of C between x_i and x_{i+1} is either an edge or a path connecting compatible vertices of a minimally 2-connected graph. Furthermore this latter case occurs at least once since G is not a cycle. Then $d(x_i) \geq 3$ and $d(x_{i+1}) \geq 3$ and by Theorem 3.6 these vertices are separated on the cycle by vertices of degree two.   ∎

As an immediate consequence of the last result we obtain another description of the structure of minimally 2-connected graphs.

THEOREM 3.8. *Let G be a minimally 2-connected graph that is not a cycle. Let $D \subset V(G)$ be the set of vertices of degree two. Then $F = G - D$ is a forest with at least two components. A component P of $G[D]$ is a path and the endvertices of P are not joined to the same tree of the forest F.*   ∎

As a consequence of Theorem 3.7 we see that every minimally 2-connected graph contains at least two vertices of degree two. In view of Theorem 3.8 we can say considerably more.

THEOREM 3.9. *A minimally 2-connected graph of order n contains at least $(n + 4)/3$ vertices of degree two. For $n \equiv -1, 0 \pmod 3$ the result is best possible.*

Proof. Let G be a minimally 2-connected graph of order n and let T be a tree of order t of $F = G - D$. Then at least $3t - 2(t - 1) = t + 2$ edges join T to D. As F has at least two components, if the order of F is f then at least $f + 4$ edges join F to D. Consequently

$$f + 4 = n - |D| + 4 \leq 2|D|,$$

as claimed.

Suppose now that $n \equiv -1 \pmod 3$, say $n = 3p - 1$ (the case $n \equiv 0 \pmod 3$ can be dealt with analogously). Take the disjoint union of two paths $x_1, x_2, \ldots, x_{p-1}$ and $y_1, y_2, \ldots, y_{p-1}$. Add vertices $z_0, z_1, \ldots, z_p$ and join z_i to x_i, $1 \leq i \leq p - 1$. Finally, join z_0 to x_1, y_1 and z_p to x_{p-1}, y_{p-1} (Fig. 3.2). It is easily checked that the obtained graph is minimally 2-connected.   ∎

For $n \equiv 1 \pmod 3$ the bound given by Theorem 3.9 can be improved slightly (see Ex. 20). Let us end the discussion of minimally 2-connected

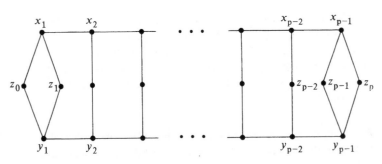

FIG. 3.2. A minimally 2-connected graph of order $3p - 1$ with $p + 1$ vertices of degree 2.

graphs with the following consequence of Theorem 3.8. We leave the simple computational proof to the reader.

THEOREM 3.10. *A minimally 2-connected graph of order $n \geq 4$ has size at most $2n - 4$. Given $n \geq 4$ the complete bipartite graph $K(2, n - 2)$ is the only minimally 2-connected graph of order n and size $2n - 4$.* ■

Let us turn now to 3-connected graphs. For $n \geq 3$ the *wheel of n spokes* is the graph $W^n = K^1 + C^n$. Naturally the vertex of K^1 is said to be the *centre* of the wheel and the edges incident with the centre are the *spokes*. Clearly $W^3 = K^4$ and W^n has a unique centre if $n \geq 4$. It is also obvious that every wheel is 3-connected. The following theorem, due to Tutte [T15], describes how every 3-connected graph can be obtained from a wheel by the application of certain operations. It is very surprising that such a beautiful characterization exists.

THEOREM 3.11. *A graph is 3-connected if and only if it is a wheel or can be obtained from a wheel by repeated applications of the following two operations.*
  1. *The addition of an edge.*
  2. *The replacement of a vertex x of degree ≥ 4 by two adjacent vertices x', x'' and joining every neighbour of x to exactly one of x', x'' in such a way that both x' and x'' will have degree ≥ 3. (This operation is called a splitting of the vertex x.)* ■

One of the most attractive proofs of this theorem, given by Halin [H13], makes use of some properties of minimally k-connected graphs. Rather than prove these properties in the special case $k = 3$, we postpone the proof of

Theorem 3.11 until after the proof of the general propositions in the next section (p.19) so that we can rely on them in the proof.

4. MINIMALLY k-CONNECTED GRAPHS

In the preceding section we described operations on k-connected graphs for $k = 2$ and 3 that produce again k-connected graphs and which are such that every k-connected graph can be obtained from certain k-connected graphs by the application of a sequence of them. It would be unreasonable to expect to find such *simple* operations for every k. As we mentioned already, Slater [S22] has found rather complicated and not too informative operations for $k = 4$. However, an entirely different line of attack by Halin [H15], [H9], [H10], [H11], [H12], [H13], [H14] and Mader [M5], [M6], [M7], [M10] has been rather successful. In these papers Halin and Mader proved successively stronger results about minimally k-connected graphs. The first results were obtained by Halin who also conjectured [H11] a number of extensions of those results. Subsequently all the conjectures were proved by Mader who obtained the strongest results so far, which, in many cases, are best possible. The aim of this section is to present these results.

Recall that a graph is said to be *minimally k-connected* if it is k-connected but omitting any of the edges the resulting graph is no longer k-connected. Every k-connected graph can be obtained from a minimally k-connected one by the addition of certain edges, i.e. every k-connected graph has a minimally k-connected subgraph with the same vertex set. The following two lemmas express elementary properties of minimally k-connected graphs. As we shall often deal with graphs obtained from another one by omitting an edge, for $xy \in E(G)$ we write $G_{xy} = G - xy$.

LEMMA 4.1. *Let G be a minimally k-connected graph, let $xy \in E(G)$ and let S be a separating $(k - 1)$-set of G_{xy}. Then $G_{xy} - S$ has exactly two components, one of which contains x and the other y.*

Proof. The graph $G_{xy} - S$ is disconnected but becomes connected if the edge xy is added to it. ∎

LEMMA 4.2. *A k-connected graph is minimally k-connected if and only if $\kappa(x, y) = k$ for every pair of adjacent vertices.*

Proof. The condition is obviously sufficient. Suppose now that G is a minimally k-connected graph and $xy \in E(G)$. Then $G_{xy} = G - xy$ can be separated by a $(k-1)$-set S. As S separates x from y in G_{xy} there are at most $k-1$ independent x–y paths in G_{xy}. Consequently $\kappa(x, y) \leqslant k$ but as $\kappa(G) = k$ we have $\kappa(x, y) = k$. ∎

It is clear that if each edge of a k-connected graph is incident with at least one vertex of degree k then the graph is minimally k-connected. The following example of Halin [H15] shows that this is far from being necessary.

Let $l > k \geqslant 3$. Take a tree T in which each vertex has degree 1 or at least l. Draw T in the closed unit disc in such a way that the boundary of the disc (the unit circle) contains the endvertices of T and no other point of the drawing. Let H be the graph obtained from T by joining the neighbouring vertices of the circle. Take $k - 2$ copies of H and identify the corresponding endvertices of the various copies of T (see Fig. 4.1). The graph obtained in this way is easily seen to be minimally k-connected.

This example shows that in a minimally k-connected graph a vertex of degree at least $k + 1$ might be arbitrarily far from the set of vertices of degree k. In the other direction it is not obvious at all that every minimally k-con-

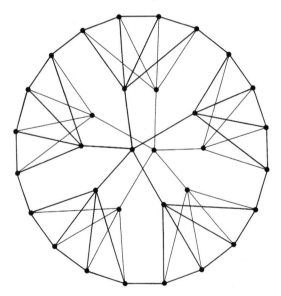

FIG. 4.1. A minimally 4-connected graph.

nected graph contains a vertex of degree k. This was first shown by Halin [H10], [H15].

THEOREM 4.3. *Let G be a minimally k-connected graph. Then $\delta(G) = k$.*

Proof. We present the simple proof given in [H15] though later we shall present much stronger results of Mader (Theorems 4.5 and 5.8), the proofs of which are independent of Theorem 4.3.

If $G = K^{k+1}$ there is nothing to prove so suppose that $|G| \geqslant k + 2$. Then G has separating k-sets. Let S be a separating k-set for which one of the components of $G - S$, say C, has minimum order. Put $D = G - S - C$, i.e. denote by D the union of the other components of $G - S$. Then D has at least as many vertices as C.

To prove the theorem it suffices to show that C consists of a single vertex c since then c can be joined only to vertices of S. Suppose C has at least two vertices. Then C contains an edge xy. As G is minimally k-connected, there is a $(k - 1)$-set T that separates x from y in $G_{xy} = G - xy$.

Now G_{xy} contains an x–S fan F_x for otherwise the proof of Theorem 2.6 implies that there is a set $T' \subset V(G) - \{x\}, |T'| = k - 1$, detaching x from D. However then $T^* = T' \cup \{y\}$ has k vertices and the component of $G - T^*$ containing x is strictly contained in C. This contradicts the minimality of C. Similarly G_{xy} contains a y–S fan F_y as well.

Let us show next that $V(D) \subset T$. If this is not so then there is a vertex $z \in V(D) - T$. Then there is a z–S fan F_z in G and so in G_{xy}. Consequently there are k independent x–z paths in G_{xy} and there are also k independent y–z paths. Thus T cannot separate x from y in G_{xy}, since it cannot separate x from z and z from y (Fig. 4.2).

Put $r = |S \cap T|$. For every vertex $u \in S - T$ either the x–u path in F_x or the y–u path in F_y contains a vertex of T and so T has at least $\frac{1}{2}(k + r)$ vertices in C. Thus D has at most $\frac{1}{2}(k + r)$ vertices and C has at least $\frac{1}{2}(k - r) + 2$ vertices, since $x, y \notin T$. In particular, $|C| > |D|$ and this contradicts the minimality of $|C|$. ∎

By now we are well prepared to present a proof of Theorem 3.11, postponed from the preceding section. Recall that G/xy is the graph obtained from G by contracting the edge xy. If G is 3-connected and no separating 3-set contains both x and y, then, clearly, G/xy is also 3-connected.

Proof of Theorem 3.11. Note first that the theorem has the following reformulation.

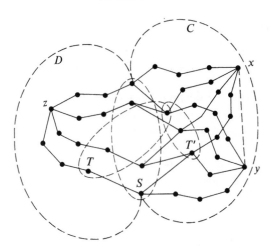

FIG. 4.2. The structure of G in the proof of Theorem 4.3.

Suppose G is minimally 3-connected and it is not a wheel. Then there is an edge xy, not contained in a triangle, such that G/xy is 3-connected.

We prove this by induction on G. For $|G| \leqslant 4$ there is nothing to prove. Assume now that $|G| > 4$ and let x_0 be a vertex of degree 3 guaranteed by Theorem 4.3. Write x_1, x_2 and x_3 for the vertices adjacent to x_0.

Suppose first that G contains the triangle $x_1 x_2 x_3$. Since no x_i is a cutvertex, we may assume that $G - x_0 - x_3$ contains an $x_1 - x_2$ path of length at least 2. Then $G - x_1 x_2$ contains three independent $x_1 - x_2$ paths, contradicting Lemma 4.2. Hence there are at most two edges of the form $x_i x_j$, $1 \leqslant i < j \leqslant 3$. We distinguish three cases, according to the number of such edges.

1. *No two of x_1, x_2 and x_3 are adjacent.* Then we may assume that in $G - x_0 - x_3$ there is a vertex y_{12} that separates x_1 from x_2, since otherwise $G/x_0 x_3$ is 3-connected. Now $G - x_0$ is 2-connected and in G there are three independent $y_{12} - x_3$ paths. Consequently $G - x_0$ has a cycle that contains two of the vertices x_1, x_2 and x_3, but not the third, say x_i. Then $G/x_0 x_i$ is 3-connected and $x_0 x_i$ is not contained in a triangle.

2. *G contains exactly one $x_i x_j$ edge, say $x_1 x_2$.* Note that there is no separating set of the form $\{x_0, x_3, y\}$ since otherwise $\{x_3, y\}$ is also a separating set, for the vertices in $\{x_1, x_2\} - \{y\}$ have to belong to the same component of $G - \{x_0, x_3, y\}$. Consequently $G/x_0 x_3$ is 3-connected and $x_0 x_3$ does not belong to a triangle.

3. *G contains exactly two $x_i x_j$ edges, say $x_1 x_2$ and $x_1 x_3$. Then $G - x_0$
$+ x_2 x_3$ is 3-connected* and either this graph or $G - x_0$ is minimally 3-con-
nected. Write H for the one that is minimally 3-connected. If H is a wheel
then a straightforward examination of the possible positions of x_1, x_2 and x_3
in this wheel shows that the result holds. If H is not a wheel then, by the induction hypothesis, it contains an edge yz not contained in a triangle (and so
different from $x_2 x_3$) such that H/yz is 3-connected. Then yz is not contained
in a triangle of G either, and G/yz is 3-connected. ∎

The structure results in the preceding section (including the result we have
just proved) imply that for $k = 2, 3$ each cycle of a minimally k-connected
graph contains at least two vertices of degree k. This is not true for $k \geqslant 4$ as
shown by the graph $P^k + E^{k-1}$ (see Fig. 4.3). This graph is easily seen to be

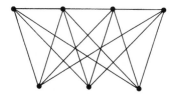

FIG. 4.3. The minimally 4-connected graph $P^4 + E^3$.

minimally k-connected and for $k \geqslant 4$ it contains cycles $C^3, C^4, \ldots, C^{k-1}$
having only one vertex of degree k. Mader [M7] proved that this example is
in a sense worst possible for a cycle of a minimally k-connected graph must
contain a vertex of degree k for every k. This powerful result (Theorem 4.5) goes
a long way towards characterizing the structure of minimally k-connected
graphs. As we shall see, it enables one to prove extremal results which are
analogous to the ones proved for $k = 2, 3$. Theorem 4.5 will be an immediate
consequence of a technical lemma.

LEMMA 4.4. *Let G be a minimally k-connected graph, and let ax, $ay \in E(G)$,
where $x \neq y$ and $d(a) \geqslant k + 1$. Let S be a $(k - 1)$-set separating G_{ax} and let
S' be a $(k - 1)$-set separating G_{ay}. For $z \in G$ put*

$$C_z = C(z, G_{ax} - S), \qquad C_z' = C(z, G_{ay} - S').$$

Then

$$|C_y'| < |C_a|.$$

Proof. For brevity we put $V_z = V(C_z)$, $V_z' = V(C_z')$. Thus $|C_z| = |V_z|$, $|C_z'| = |V_z'|$. Recall that by Lemma 4.1 we have

$$V_a \cup S \cup V_x = V, \tag{1}$$

$$V_a' \cup S' \cup V_y' = V. \tag{1'}$$

Put also (see Fig. 4.4)

$$T = (S \cap V_b') \cup (S \cap S') \cup (V_a \cap S').$$

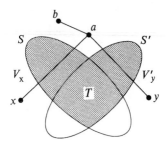

FIG. 4.4. Illustration to Lemma 4.4.

Our first aim is to show that

$$|V_x \cap S'| \leqslant |V_a' \cap S|. \tag{2}$$

Suppose (2) fails, i.e. $|S \cap V_a'| < |V_x \cap S'|$. Then

$$|T| < |S' \cap V_x| + |S' \cap S| + |S' \cap V_a| = k - 1.$$

Consequently

$$\Gamma(a) - T \cup \{x, y\} \neq \emptyset,$$

say

$$b \in \Gamma(a) - T \cup \{x, y\}.$$

Putting

$$B = V_a \cap V_a' - \{a\}$$

we have

$$\Gamma(B) \subset (V_a \cup S) \cap (V_a' \cup S') = (V_a \cap V_a') \cup T = B \cup \{a\} \cup T.$$

Naturally $b \in B - \{a\} \cup T$ and $x \notin B \cup \{a\} \cup T$. Consequently the set $T \cup \{a\}$ separates b from x. As $|T \cup \{a\}| \leq k - 1$ this is impossible and so (2) is proved.

Analogously to (2) we have

$$|V_y' \cap S| \leq |V_a \cap S'|. \tag{2'}$$

Put

$$D = V_x \cap V_y'.$$

Since $D \cap \{x, y\} = \varnothing$ by (1) we have

$$\Gamma(D) \subset (V_x \cup S) \cap (V_y' \cup S') = (V_x \cap V_y') \cup (S \cap V_y') \cup (S' \cap (V_x \cup S))$$
$$= D \cup (S \cap V_y') \cup (S' - V_a).$$

Therefore, by (2'), we get

$$|\Gamma(D) - D| \leq |S' \cap V_a| + |S' - V_a| = |S'| = k - 1.$$

This shows that $D = \varnothing$, i.e.

$$V_y' \subset V_a \cup S, \tag{3}$$

since otherwise $\Gamma(D) - D$, a set of at most $k - 1$ vertices, would separate the vertex $a \notin \Gamma(D) \cup D$ from the set D.

Finally, by (1), (2′) and (1′) we have

$$|V_y'| = |V_y' \cap V_a| + |V_y' \cap S| \leqslant |V_y' \cap V_a| + |V_a \cap S'|$$

$$= |V_a - V_a'| = |V_a - (V_b \cap V_a')| \leqslant |V_a - \{a\}| < |V_a|,$$

as claimed.   ∎

THEOREM 4.5. *Let G be a minimally k-connected graph and let T be the set of vertices of degree k. Then G − T is a (possibly empty) forest.*

Proof. Suppose $a_1 a_2 \ldots a_l$ is a cycle in $G - T$. Let S_i be a $(k - 1)$-set separating $G_{a_i a_{i+1}}$, $i = 1, \ldots, l$, where we put $a_{l+1} = a_1$. Put furthermore

$$n_i = |C(a_i, G_{a_i a_{i+1}} - S_i)|, \qquad i = 1, \ldots, l.$$

Then by Lemma 4.4 we obtain the contradiction

$$n_1 < n_2 < \ldots < n_l < n_1. \qquad\qquad ∎$$

COROLLARY 4.6. *Let H be a subgraph of a minimally k-connected graph G. Then $\delta(H) \leqslant k$.*

Proof. If H does not contain a vertex having degree k in G then H is a forest.   ∎

COROLLARY 4.7. *A minimally k-connected graph G is $k + 1$ colourable.*

Proof. Suppose $\chi(G) \geqslant k + 2$. Let H be a minimal subgraph of G such that $\chi(H) \geqslant k + 2$. The minimality of H implies $\delta(H) \geqslant k + 1$, contradicting Corollary 4.6.   ∎

According to Theorem 4.5 (or Corollary 4.6) a minimally k-connected graph G has a vertex of degree k. In fact, G has at least $k - 1$ vertices of degree k since otherwise we could separate the graph with the set T of vertices of degree k and one more vertex separating the forest $G - T$. Even more, Theorem 4.5 implies that the set of vertices of degree k is large in the sense that various parameters of the graph can be used to give a lower bound on the number of vertices of degree k. We present here one of the possible lower bounds and some others are left to the reader (Ex. 22–24).

THEOREM 4.8. *A minimally k-connected graph of order n has at least*

$$\frac{(k-1)n+2}{2k-1}$$

vertices of degree k.

Proof. Let G be a minimally k-connected graph of order n. Denote by T the set of vertices of degree k and put $U = V(G) - T$, $t = |T|$ and $u - |U|$. Then, by Theorem 4.5, $F = G[U] = G - T$ is a forest. Denote by c the number of components of this forest. Finally, denote by e' the size of $G[T]$, i.e. the number of edges joining vertices of degree k. As every vertex in T has degree k, there are $tk - 2e'$ edges joining T to U. On the other hand every vertex in U has degree at least $k + 1$ in G and there are $u - c$ edges in F so there are at least $u(k + 1) - 2(u - c)$ edges joining U to T. Consequently

$$tk - 2e' \geqslant u(k+1) - 2(u-c) = u(k-1) + 2c. \qquad (4)$$

As $u = n - t$, (4) gives

$$t(2k-1) \geqslant n(k-1) + 2c + 2e' \geqslant n(k-1) + 2. \qquad \blacksquare$$

The following example shows that Theorem 4.8 is essentially best possible. Take k copies of a path $x_1 x_2 \ldots x_l$ and $k - 1$ copies of a set of l independent vertices $y_1, y_2, \ldots, y_l$. Join each copy of y_i to each copy of x_i, $i = 1, 2, \ldots, l$. Finally, add two more vertices, y_0 and y_{l+1}, join y_0 to the k copies of x_1 and join y_{l+1} to the k copies of x_l. Denote by $G_{k,l}$ the graph obtained in this way (see Fig. 4.5). It is easily checked (Ex. 21) that this graph is minimally k-connected. If we denote its order by n then $n = (2k - 1)l + 2$ and it has $(k - 1)l + 2$ vertices of degree k. Note that this coincides with the bound given by Theorem 4.8. By modifying the graph $G_{k,l}$ slightly we see that if $v_k(n)$ denotes the minimal number of vertices of degree k in a minimally k-connected graph of order n then

$$\lim_{n \to \infty} \frac{v_k(n)}{n} = \frac{k-1}{2k-1}.$$

As the final problem of this section let us investigate the maximal size of a minimally k-connected graph of order n. The graph $K(k, n - k)$ is minimally k-connected if its order, n, is at least $2k$. Its size is $k(n - k)$. Mader

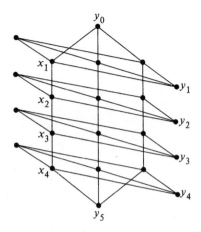

Fig. 4.5. The minimally 3-connected graph $G_{3,4}$.

[M10] determined the least order n_0 such that if $n \geqslant n_0$ then $K(k, n - k)$ is the unique minimally k-connected graph of order n and maximal size. Rather naturally, the proof is based on Theorem 4.5, but as the computations are non-trivial, we present the complete proof.

THEOREM 4.9. *Let G be a minimally k-connected graph of order n. If $n \geqslant 3k - 2$ then*

$$e(G) \leqslant k(n - k). \tag{5}$$

Furthermore, if $n \geqslant 3k - 1$ equality holds in (5) if and only if $G = K(k, n - k)$.

Remark. Both assertions are best possible. If $n = 3k - 3$ then the graph $C^{2k-1} + E^{k-2}$ shows that (5) need not hold. If $n = 3k - 2$ then in (5) equality holds for $C^{2k} + E^{k-2} \neq K(k, 2k - 2)$ as well.

Proof. Let us keep the notation of the proof of Theorem 4.8 and suppose G is a counterexample to one of the assertions of the theorem. In particular, $n \geqslant 3k - 2$ and $e(G) \geqslant k(n - k)$. Note that

$$e(G) = u - c + (n - u)k - e' = kn - u(k - 1) - c - e', \tag{6}$$

and the number of $T - U$ edges is

$$(n - u)k - 2e'.$$

Put also

$$m = (n - u)u - (n - u)k + 2e', \tag{7}$$

i.e. denote by m the number of $T - U$ edges missing from G. If xy is an edge of F, let S be a $(k - 1)$-set separating x from y in $G_{xy} = G - xy$. Put $S_U = S \cap U$, $S_T = S \cap T$, $T_x = T \cap V(C(x, G_{xy} - S))$, $U_x = U \cap V(C(x, G_{xy} - S))$, and define T_y, U_y analogously. Note that G does not contain $T_x - U_y$ and $T_y - U_x$ edges. If one of T_x and T_y is empty, say $T_x = \varnothing$ then $|U_x| \geqslant k + 1$. Consequently

$$m \geqslant 2(t - |S_T|) \geqslant 2(n - u - k + 1). \tag{8}$$

On the other hand, if $T_x \neq \varnothing$ and $T_y \neq \varnothing$ then

$$m \geqslant |U_x||T_y| + |U_y||T_x| \geqslant n - k - 1. \tag{9}$$

We discuss now several cases according to the value of u, the number of vertices of degree greater than k. In each of these cases our aim is to arrive at a contradiction. Not surprisingly, the case $u = k$ is the most difficult to deal with so we leave it to the end.

Case 1. $u \geqslant k + 1$. Then (6) gives

$$k(n - k) \leqslant e(G) = kn - u(k - 1) - c - e' \leqslant kn - u(k - 1) - 1.$$

Consequently equality holds everywhere in this sequence and so $c = 1$, $e' = 0$ and $u = k + 1$. In particular, $m = n - u (= n - k - 1)$. Let now xy be an end edge of F (i.e. $d_F(x) = 1$ and $xy \in E(F)$). Then $|\Gamma(x) \cap T| \geqslant k$ so $T_x \neq \varnothing$, say $z \in T_x$. Then z is joined to every vertex in $U - \{y\}$, implying $U_y = \{y\}$. Now $d_G(y) \geqslant k + 1$ gives $T_y \neq \varnothing$ and, as before, $U_x = \{x\}$. Consequently $S = S_U = U - \{x, y\}$. Thus at least $n - u$ edges of the form $x - T$, $y - T$ are missing from G, and y is joined to every vertex in T. Choosing another end edge we see that at least $n - u + 1$ $T - U$ edges are missing from G, contradicting $m = n - u$.

Case 2. $u = k - 1$. Then $G[S]$ is connected so $e' \geqslant n - u - 1 = n - k$. Consequently (6) gives

$$e(G) \leqslant kn - (k - 1)^2 - 1 - (n - k) \leqslant kn - k^2 + 3k - 2 - n,$$

contradicting the assumption that G is a counterexample.

Case 3. $u = k - d \leqslant k - 2$. Then (6) and (7) imply

$$k(n - k) \leqslant e(G) = k - d - c - \frac{m}{2} + (n - k + d)(k - d/2).$$

This gives $c < k - d$, so either (8) or (9) holds. Consequently

$$m \geqslant 2k - 3,$$

and so

$$k(n - k) \leqslant k - d - 1 - k + \frac{3}{2} + (n - k + d)(k - d/2)$$

$$\leqslant k(n - k) + \frac{d}{2}(3k - n - 2 - d) + \frac{1}{2} < k(n - k).$$

Case 4. $u = k$. As in the previous case, (6) and (7) imply

$$e(G) = k(n - k) + k - c - \frac{m}{2}.$$

Note now that if $c = k$ then $e(G) \geqslant k(n - k)$ implies that $G = K(k, n - k)$. Therefore $c < k$, i.e. there is an edge in F. If $n = 3k - 2$ then, as either (8) or (9) must hold,

$$m \geqslant 2k - 3.$$

Thus $e(G) \leqslant k(n - k) + \frac{1}{2}$, contradicting the assumption $e(G) \geqslant k(n - k) + 1$.
   If $n \geqslant 3k - 1$ then (8) cannot hold since then $m \geqslant 2k$ and $e(G) \leqslant k(n-k)-1$ follows. Consequently (9) holds for every edge xy of F:

$$m \geqslant |U_x||T_y| + |U_y||T_x| \geqslant n - k - 1 \geqslant 2k - 2, \tag{10}$$

and so

$$k(n - k) \leqslant e(G) \leqslant k(n - k) + k - \frac{2k - 2}{2} \leqslant k(n - k). \tag{11}$$

This implies that equality holds everywhere in (11) and (10). In particular $c = 1$, i.e. F is a tree. Furthermore, the only missing edges are given by (10), i.e. if $a \in T, b \in U$ then $ab \notin E(G)$ if and only if $a \in T_x$ and $b \in U_y$ or $a \in T_y$ and $b \in U_x$. Therefore the pairs $\{T_x, T_y\}, \{U_x, U_y\}$ are independent of the choice of the edge xy. In turn this gives that S is independent of xy and $S_U = \varnothing$.

Let xy and yz be edges of F. Then S separates x from y in G_{xy} but does not separate them in G_{yz}. This contradicts our previous conclusion that the partition $U_x \cup U_y = U$ is independent of the choice of $xy \in E(F)$.   ∎

The most powerful result so far concerning edge connectivity is a recent result of Mader [M15a]. Let $x_0 x_1$ and $x_0 x_2$ be edges of a multigraph G. Delete the edges $x_0 x_1, x_0 x_2$ and add an edge $x_1 x_2$ to G; we say that the new graph has been obtained from G by *lifting the pair* $\{x_0 x_1, x_0 x_2\}$. The operation of lifting several pairs is defined analogously. If in the new graph x_0 has degree 2 then we immediately omit x_0 and join the vertices adjacent to x_0. In this case we say that the vertex x_0 has been *dissolved* (by lifting pairs of edges). It is clear that if G_0 is obtained from G by lifting some pairs of edges incident with x_0 then

$$\lambda_{G_0}(x, y) \leqslant \lambda_G(x, y)$$

for every pair of vertices $x, y \in V(G) - \{x_0\}$. Proving an extension of a conjecture of Lovász [L25], very recently Mader [M15a] proved the deep and somewhat surprising result that, barring trivial exceptions, we can always find a lifting in which *equality holds* for every pair $\{x, y\}$. The proof is, of course, rather complex; here we only state the result.

THEOREM 4.10. *Suppose $x_0 \in G$ is not a cutvertex and $d(x_0) \geqslant 4$. Then one can lift a pair of edges at x_0 such that the new graph G_0 satisfies*

$$\lambda_{G_0}(x, y) = \lambda_G(x, y), \qquad x, y \in V(G) - \{x_0\}.$$

If x_0 has even degree then x_0 can be dissolved without affecting the local connectivity of any pair of vertices in $V(G) - \{x_0\}$.   ∎

5. GRAPHS WITH GIVEN MAXIMAL LOCAL CONNECTIVITY

The connectivity properties of a graph G are fairly well characterized by the functions $\kappa(x, y), \lambda(x, y)$ defined on the pairs of vertices. In the previous sections we investigated k-connected graphs, i.e. graphs in which $\kappa(x, y) \geqslant k$

holds for every pair of vertices. The aim of this section is to study graphs with a given *upper* bound on the function $\kappa(x, y)$ (resp. $\lambda(x, y)$). Put

$$\bar{\kappa}(G) = \max\{\kappa(x, y): x, y \in V(G), x \neq y\}.$$

Analogously, denote by $\bar{\lambda}(G)$ the *maximum* of the *local edge-connectivities*. For a given k we shall determine conditions on a graph G ensuring that $\bar{\kappa}(G) \geqslant k$(resp. $\bar{\lambda}(G) \geqslant k$). The first problem of this kind was considered in [B20], where the maximal size of a graph of given order was determined provided the maximum of its local connectivities is 3. We present this result together with an even simpler one, proposed by Erdős [S12] as a problem for undergraduates. Put

$$f(n; \bar{\kappa} \leqslant k) = \max\{e(G^n): \bar{\kappa}(G^n) \leqslant k\}$$

and

$$f(n; \bar{\lambda} \leqslant k) = \max\{e(G^n): \bar{\lambda}(G^n) \leqslant k\}.$$

If $n \leqslant k + 1$ then both of these functions are trivially equal to $\binom{n}{2}$. Therefore we shall always assume in the sequel that $n \geqslant k + 1$. Let $\mathscr{E}_2$ be the set of connected graphs whose blocks are triangles with the exception of at most one block which is an edge or a C^4. Note that if $G \in \mathscr{E}_2$ then every block of G is a triangle iff G has odd order. Furthermore, let $\mathscr{E}_3$ be the set of connected graphs whose blocks are wheels (see Fig. 5.1).

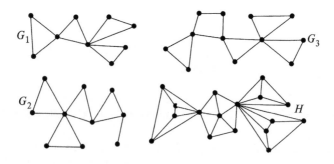

FIG. 5.1. $G_1, G_2, G_3 \in \mathscr{E}_2$ and $H \in \mathscr{E}_3$.

THEOREM 5.1. *Let $k = 2, 3$ and let G be a graph of order n such that $\bar{\kappa}(G) \leqslant k$. Then*

$$e(G) \leqslant \left\lfloor \frac{k+1}{2}(n-1) \right\rfloor$$

with equality iff $G \in \mathscr{E}_k$ or $n = 1$.

In particular, $f(n; \bar{\kappa} \leqslant 2) = \lfloor \frac{3}{2}(n-1) \rfloor$ and $f(n; \bar{\kappa} \leqslant 3) = 2n - 2$.

Proof. We apply induction on n. For $n \leqslant k + 1$ the assertion is trivial so suppose $n > k + 1$ and the assertion holds for smaller values of n.

If $G \in \mathscr{E}_k$ then $\bar{\kappa}(G) = k$ and if H is obtained from G by the addition of an edge then $\bar{\kappa}(H) = k + 1$. Assume now that

$$|G| = n, \quad \bar{\kappa}(G) \leqslant k \quad \text{and} \quad e(G) = \left\lfloor \frac{k+1}{2}(n-1) \right\rfloor.$$

By the preceding remark to prove the theorem it suffices to show that $G \in \mathscr{E}_k$.

(i) Let $k = 2$. Then $e(G) = \lfloor \frac{3}{2}(n-1) \rfloor$ so $\delta(G) \leqslant 2$. If $\delta(G) \leqslant 1$, say $d(x) \leqslant 1$ for some $x \in G$, then by the induction hypothesis $G - x \in \mathscr{E}_2$. This implies $G \in \mathscr{E}_2$, as required.

Suppose now that $d(a_0) = 2$ for some $a_0 \in G$, say $\Gamma(a_0) = \{a_1, a_2\}$. If $a_1 a_2 \in E(G)$ then $G - a_0 a_1 - a_1 a_2 - a_2 a_0 = H$ has at least 3 components, say H is the disjoint union of H_1, H_2 and H_3. Then $n = \Sigma_1^3 n_i$, where $n_i = |H_i| \geqslant 1, i = 1, 2, 3$. Trivially, $\bar{\kappa}(H_i) \leqslant \bar{\kappa}(H) \leqslant 2$ so

$$\lfloor \tfrac{3}{2}(n-1) \rfloor = e(G) = 3 + \sum_1^3 e(H_i) \leqslant 2 + \sum_1^3 \lfloor \tfrac{3}{2}(n_i - 1) \rfloor \leqslant \lfloor \tfrac{3}{2}(n-1) \rfloor.$$

Consequently $H_i \in \mathscr{E}_2$ and at most one of the H_i has a block which is not a triangle. This means exactly that $G \in \mathscr{E}_2$.

Finally, if $a_1 a_2 \notin E(G)$ then the graph $H = G - a_0 + a_1 a_2$ has order $n - 1, \bar{\kappa}(H) \leqslant 2$ and so

$$\lfloor \tfrac{3}{2}(n-2) \rfloor \geqslant e(H) = e(G) - 1 = \lfloor \tfrac{3}{2}(n-1) \rfloor - 1.$$

Consequently n is even and $H \in \mathscr{E}_2$. Thus G is obtained from a connected graph whose every block is a triangle by the subdivision of an edge into two edges. Therefore $G \in \mathscr{E}_2$; in fact G is connected and its blocks are triangles with the exception of one which is a C^4.

(ii) Let $k = 3$. Then $e(G) = 2n - 2$ and so $\delta(G) \leqslant 3$. Let us show that $\delta(G) = 3$. If this is not so then there is a vertex $x \in G$ such that $d(x) = \delta(G) \leqslant 2$.

Put $H = G - x$. Then $|H| = n - 1$, $\bar{\kappa}(H) \leqslant 3$ and so

$$2n - 4 \leqslant 2n - 2 - \delta(G) = e(H) \leqslant 2n - 4.$$

Therefore by the induction hypothesis $\delta(G) = d(x) = 2$ and $H \in \mathscr{E}_3$. It is easily checked that this contradicts the assumption $\bar{\kappa}(G) \leqslant 3$. Therefore $d(a_0) = 3$ for some $a_0 \in G$, say $\Gamma(a_0) = \{a_1, a_2, a_3\}$.

Suppose first that G contains the triangle $a_1 a_2 a_3$. Then the K^4 induced by the vertices a_0, a_1, a_2, a_3 is a block of G. Therefore the graph H obtained from G by omitting the edges $a_i a_j$, $0 \leqslant i < j \leqslant 3$, has at least four components, say H is the disjoint union of H_1, H_2, H_3 and H_4. Then $\sum_1^4 n_i = n$ where $n_i = |H_i|$ and so

$$2n - 2 = e(G) = 6 + \sum_1^4 e(H_i) \leqslant 6 + \sum_1^4 (2n_i - 2) = 2n - 2.$$

Consequently $H_i \in \mathscr{E}_3$ for each i which implies $G \in \mathscr{E}_3$.

Suppose next that G does not contain the triangle $a_1 a_2 a_3$, say $a_1 a_2 \notin E(G)$. Put $H = G - a_0 + a_1 a_2$. Then H has order $n - 1$, $\bar{\kappa}(H) \leqslant 3$ and $e(H) = 2n - 4$. Therefore by the induction hypothesis $H \in \mathscr{E}_3$. We leave it to the reader to check that this implies that a_1, a_2 and a_3 belong to the same block B of H and a_3 can be considered to be the centre of the wheel $B = W^l$ (Ex. 26). Then the blocks of G are the blocks of H different from B together with the block $B' = B + a_0 + a_0 a_1 + a_0 a_2 + a_0 a_3 - a_1 a_2$. Clearly $B' = W^{l+1}$ so $G \in \mathscr{E}_3$, as required.   ∎

Let G be a graph of order n containing a vertex of degree $n - 1$, $n - 2$ vertices of degree k and another vertex of degree $k - 1$ or k. Then clearly

$$\bar{\kappa}(G) \leqslant \bar{\lambda}(G) \leqslant k \quad \text{and} \quad e(G) = \left\lfloor \frac{k+1}{2}(n-1) \right\rfloor.$$

Therefore

$$f(n; \bar{\kappa} \leqslant k) \geqslant f(n; \bar{\lambda} \leqslant k) \geqslant \left\lfloor \frac{k+1}{2}(n-1) \right\rfloor. \tag{1}$$

In view of Theorem 5.1 one might be tempted to conjecture (and it was indeed conjectured by Bollobás and Erdős) that equality holds in (1). This conjecture was disproved by Leonard [L3] for $k = 4$ and then Mader [M11] constructed graphs disproving it for every $k \geqslant 4$.

Sørensen and Thomassen [ST1] determined $f(n; \bar{\kappa} \leqslant k)$ for $k = 4$. As the proof is rather long and contains the examination of many cases, we do not give it here.

THEOREM 5.2. *For* $n \geqslant 6, n \neq 7, n \neq 12$,

$$f(n; \bar{\kappa} \leqslant 4) = \lfloor \tfrac{8}{3}n \rfloor - 4. \qquad \blacksquare$$

Two of the extremal graphs are shown in Fig. 5.2.

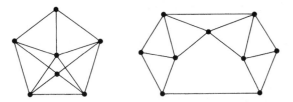

FIG. 5.2. Two extremal graphs for $f(n; \bar{\kappa} \leqslant 4)$.

At present we have only rather crude estimates for $f(n; \bar{\kappa} \leqslant k)$, $k \geqslant 5$. We start with a lower bound proved by Sørensen and Thomassen [ST1]. This lower bound is based on the construction given in the following lemma.

LEMMA 5.3, *Let* G_1, G_2, G_3 *be disjoint graphs with* $\bar{\kappa}(G_i) \leqslant k, i = 1, 2, 3.$ *Let* $x_i y_i \in E(G_i), i = 1, 2, 3,$ *such that* $\kappa(x_i, y_i) \leqslant k - 1.$ *Suppose furthermore that* G_1 *contains an edge* $x_1' y_1'$, *different from* $x_1 y_1$, *for which* $\kappa(x_1, y_1) \leqslant k - 1.$ *Let* G *be the graph obtained from* $\bigcup_1^3 G_i$ *by identifying the vertex* y_1 *with* x_2, *the vertex* y_2 *with* x_3, *and the vertex* y_3 *with* x_1 *(see Fig. 5.3). Then*

$$\bar{\kappa}(G) \leqslant k \qquad and \qquad \kappa(x_1', y_1'; G) \leqslant k - 1. \qquad \blacksquare$$

THEOREM 5.4. *If* $k \geqslant 4$ *then*

$$f(n; \bar{\kappa} \leqslant k) > \frac{k^2 + k - 2}{2k - 1}(n - k) = \left(\frac{k + 1}{2} + \frac{k - 3}{4k - 2} \right)(n - k).$$

Proof. The assertion is trivial for $n \leqslant k$ so we may suppose that $n \geqslant k + 1$. Let r be fixed, $0 \leqslant r \leqslant 2k - 2$, and let $m \geqslant 0$. We shall apply induction on m

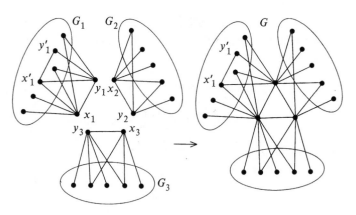

Fig. 5.3. Construction of graphs with given maximal local connectivity.

to construct a graph H_m of order $n = k + 1 + r + (2k - 1)m$ such that

$$e(H_m) > \frac{k^2 + k + 2}{3k - 1}(n - k), \quad \bar{\kappa}(H_m) \leqslant k,$$

and H_m contains an edge $z_1^m z_2^m$ for which

$$\kappa(z_1^m, z_2^m) \leqslant k - 1.$$

This will clearly imply the inequality of the theorem. Let H_0 be a graph of order $k + 1 + r$ such that one vertex of H_m has degree $n - 1$, one or two vertices have degree $k - 1$ and all other vertices have degree k. Let z_1^0 be a vertex of degree $k - 1$ and let $z_2^0 \in \Gamma(z_1^0)$. Suppose now that H_{m-1} $(m \geqslant 1)$ has already been constructed. Let G_i, $i = 1, 2$, be a K^{k+1} from which an edge has been omitted and let $x_i y_i \in E(G_i)$ where $d(x_i) = k - 1$. Furthermore, let $x_1' = x_1$ and $y_1' \in \Gamma(x_1') - \{y_1\}$. Construct H_m from G_1, G_2 and $G_3 = H_{m-1}$ as described in Lemma 5.3 with $x_3 = z_1^{m-1}$, $y_3 = z_2^{m-1}$. Then

$$e(H^m) = e(H^{m+1}) + k^2 + k - 2 > \frac{k^2 + k - 2}{2k - 1}((1 + r + (2k - 1)m)$$

and Lemma 5.3 implies that H_m has the other required properties with $z_1^m = x_1'$, $z_2^m = y_1'$. ∎

Starting with the graphs shown in Fig. 5.2 and repeatedly performing the operation given in Lemma 5.3, for every $m \geqslant 2$, $m \neq 4$, one can construct graphs $G(3m, 8m - 4)$ whose maximal local connectivity is 4. These graphs enable one to construct an extremal graph for $f(n; \bar{\kappa} \leqslant 4)$ for every $n \geqslant 5$, $n \neq 7$ and $n \neq 12$ (cf Ex. 27).

Suppose $\bar{\kappa}(G_1) \leqslant k$ and $\bar{\kappa}(G_2) \leqslant k$. Let G be obtained from the disjoint union of G_1 and G_2 by identifying a vertex of G_1 with a vertex of G_2. Then trivially $\bar{\kappa}(G) \leqslant k$. This implies that $f(n; \bar{\kappa} \leqslant k)/n$ tends to a limit as n tends to ∞ (cf. Ex. 37). Denote this limit by $\frac{1}{2}(k + 1) + \alpha_k$. It can be read out of the examples constructed by Mader [M11] that $\alpha_k \geqslant (1/3k)$ for $k \geqslant 4$ and Theorem 5.4 shows that $\alpha_k \geqslant (k - 3)/(4k - 2)$.

The following result from [B38] implies that, in fact, $\lim\limits_{k \to \infty} \alpha_k = \infty$.

THEOREM 5.5. *Suppose p is a prime power satisfying $p^2 + p + 1 \leqslant k$. Then*

$$f(n; \bar{\kappa} \leqslant k) > \frac{k(k + 1) - 2p^3 - 2p}{2(k + 1 - p)}(n - (p^2 + p)k).$$

Proof. The proof of this result is analogous to that of the previous theorem only instead of Lemma 5.3 we shall use the following lemma whose straightforward proof is also left to the reader (cf. Ex. 32).

LEMMA 5.6. *Let G_i, $1 \leqslant i \leqslant p^2 + p + 1$, be disjoint graphs with $\bar{\kappa}(G_i) \leqslant k$. Let $K_i = K^{p+1} \subset G_i$ and $K_0 = K^{p+1} \subset G_1$ be such that $V(K_0) \cap V(K_1) = \varnothing$, Suppose*

$$\kappa(x, y) \leqslant k - p^2 \quad \text{for } x, y \in K_i, \qquad 0 \leqslant i \leqslant p^2 + p + 1.$$

Let G be the graph obtained from $\bigcup\limits_{i = 1}^{p^2 + p + 1} G_i$ by identifying some of the vertices in $\bigcup\limits_{i = 1}^{p^2 + p + 1} V(K_i)$ in such a way that the sets

$$V(K_i), \qquad 1 \leqslant i \leqslant p^2 + p + 1,$$

will be the lines of a projective plane with $p^2 + p + 1$ points. Then

$$\bar{\kappa}(G) \leqslant k \quad \text{and} \quad \kappa(x, y; G) \leqslant k - p^2 \quad \text{for} \quad x, y \in K_0. \qquad \blacksquare$$

We may suppose that $n \geqslant k + 1$.

Let r be fixed, $0 \leqslant r < (p^2 + p)k - p^3$ and let $m \geqslant 0$. To prove the theorem we will show that there exists a graph H_m of order

$$n = p + 1 + r + \{(p^2 + p)k - p^3\}m$$

such that

$$e(H_m) > m\{(p^2 + p)\binom{k+1}{2} - p^3(p^2 + p + 1)\},$$

$$\bar{\kappa}(H_m) \leqslant k$$

and H_m contains a $\tilde{K}_m = K^{p+1}$ such that

$$\kappa(x, y) \leqslant k - p^3 \qquad \text{for } x, y \in \tilde{K}_m.$$

Let H_0 consist of a $\tilde{K}_0 = K^{p+1}$ and isolated vertices. Suppose now that $H_{m-1}(m \geqslant 1)$ has already been constructed. Let $G_i, 1 \leqslant i \leqslant p^2 + p$, be obtained from a K^{k+1} by selecting a $K_i = K^{p+1}$ in it and omitting p^3 of the edges incident with p of the vertices in K_i such that each of those p vertices will have degree $k - p^2$. Select a $K_0 = K^{p+1}$ in $G_1 - K_1$ and omit at most p^3 more edges from $E(G_1 - K_1) - E(K_0)$ such that p vertices of K_0 will have degree $k - p^2$. Construct H_m from $G_1, G_2, \ldots, G_{p^2+p}, G_{p^2+p+1} = H_{m-1}$ as described in Lemma 5.6 with $K_{p^2+p+1} = \tilde{K}_{m-1}$. Then

$$|H_m| = |H_{m-1}| + (p^2 + p)(k+1) - (p^2 + p + 1)p = |H_{m-1}| + (p^2 + p)k - p^3$$

and

$$e(H_m) = e(H_{m-1}) + (p^2 + p)\binom{k+1}{2} - (p^2 + p + 1)p^3.$$

Lemma 5.6 implies that H_m has the required properties with $\tilde{K}_m = K_0$.   ■

COROLLARY 5.7. *There is a constant $c > 0$ such that for $k \geqslant 4$ we have*

$$\alpha_k = \lim_{n \to \infty} \frac{1}{n} f(n; \bar{\kappa} \leqslant k) - \frac{k+1}{2} \geqslant ck^{\frac{1}{2}}.$$

Proof. For $4 \leqslant k \leqslant 200$ the existence of such a constant follows from Theorem 5.4. Suppose now that $k > 200$. Let p be a prime satisfying $\frac{1}{4}k^{\frac{1}{2}} < p \leqslant \frac{1}{2}k^{\frac{1}{2}}$ whose existence is guaranteed by Bertrand's postulate (§ 0).

Then by Theorem 5.6

$$\alpha_k \geqslant \frac{k(k + 1) - 2p^3}{2(k + 1 - p)} - \frac{k + 1}{2} = \frac{(p - 1)k - 2p^3 + p - 1}{2(k + 1 - p)} > \frac{1}{10} k^{\frac{1}{2}}. \qquad \blacksquare$$

The known best upper bound for $f(n; \bar{\kappa} \leqslant k)$, $k \geqslant 5$, is rather far from the lower bound given by Theorems 5.3 and 5.6. This (probably weak) upper bound is a consequence of the rather powerful Theorem 5.10 so we shall present it only after Theorem 5.10 (Corollary 5.12).

Let us turn now to the investigation of the functions $f(n; \bar{\lambda} \leqslant k)$, $k \geqslant 2$. It turns out that for this function inequality (1) is indeed best possible, so

$$f(n; \bar{\lambda} \leqslant k) = \left\lfloor \frac{k + 1}{2} (n - 1) \right\rfloor \quad \text{for} \quad n \geqslant k + 1.$$

For $k = 2$ and 3 this follows from Theorem 5.1, for $k = 4$ and 5 it was proved by Leonard [L4], [L5] and the general case was settled by Mader [M11]. In fact Mader proved the following stronger theorem.

THEOREM 5.8. *Let G be a graph of order $n \geqslant k + 1 \geqslant 3$ and suppose*

$$e(G) > \frac{k + 1}{2} (n - 1) - \frac{1}{2} \sigma_k(G), \tag{2}$$

where

$$\sigma_k(G) = \sum_{\substack{x \in G \\ d(x) \leqslant k}} (k - d(x)).$$

Then

$$\bar{\lambda}(G) \geqslant k + 1.$$

Proof. Suppose first that G has at most one vertex of degree at least $k + 1$. Then (cf. (1))

$$e(G) \leqslant \frac{k + 1}{2}(n - 1) - \frac{1}{2} \sigma_k(G),$$

contradicting (2). Therefore G contains two vertices of degree at least $k + 1$, say a_1 and a_2. This implies, in particular, that the theorem holds for $n = k + 1$. We shall apply induction on n. Suppose $n > k + 1$ and the result holds for smaller values of n. Suppose furthermore that, contrary to the assertion of the theorem, $\bar{\lambda}(G) \leqslant k$. Our aim is to arrive at a contradiction.

By Menger's theorem (Theorem 2.4) we can find a set E_0 of $\lambda(a_1, a_2)$ $\leqslant \bar{\lambda}(G) \leqslant k$ edges of G separating a_1 from a_2. Thus $G - E_0$ is the disjoint union of two graphs, say G_1 and G_2, such that $a_i \in G_i, i = 1, 2$.

Put

$$Z_i = V(G_i) - \bigcup_{xy \in E_0} \{x, y\},$$

$$U_i = V(G_i) - Z_i, \qquad i = 1, 2.$$

As E_0 is a minimal set of edges separating a_1 from a_2 each edge in E_0 has one endvertex in U_1 and the other in U_2. Consequently $d(a_i) \geqslant k + 1$ implies $|Z_i| \geqslant 1, i = 1, 2$.

The proof of the theorem hinges on the following inequality:

$$|U_1| + |U_2| \geqslant k + 2. \tag{3}$$

To prove (3) denote by $H_i(i = 1, 2)$ the graph obtained from G_i by adding to it the edges in E_0 (together with the endvertices of the edges, of course). Then $d(a_i) = d_{H_i}(a_i) \geqslant k + 1$ so $|H_i| \geqslant k + 2$. Therefore by induction

$$e(H_i) \leqslant \frac{k + 1}{2} (|H_i| - 1) - \frac{1}{2} \sigma_k(H_i), \qquad i = 1, 2. \tag{4}$$

Furthermore,

$$\sigma_k(H_1) + \sigma_k(H_2) = \sigma_k(G) + \sum_{x \in U_1} (k - d_{H_2}(x)) + \sum_{x \in U_2} (k - d_{H_1}(x))$$

$$= \sigma_k(G) + k(|U_1| + |U_2|) - 2|E_0|,$$

since

$$\sum_{x \in U_1} d_{H_2}(x) = \sum_{x \in U_2} d_{H_1}(x) = |E_0|.$$

Adding the two inequalities (4) we obtain

$$e(H_1) + e(H_2) = e(G) + |E_0| \leqslant \frac{k+1}{2}(n + |U_1| + |U_2| - 2) - \frac{1}{2}\sigma_k(G)$$

$$- \frac{k}{2}(|U_2| + |U_2|) + |E_0|$$

$$= \frac{k+1}{2}(n-1) - \frac{1}{2}\sigma_k(G) + \frac{1}{2}(|U_1| + |U_2|) - \frac{k+1}{2} + |E_0|.$$

Taking (2) into account inequality (3) follows immediately.

Let G_1' be the graph obtained from G by contracting G_2 to the vertex a_2. Define G_2' analogously. Then $|G_1'| + |G_2'| = n + 2$ and $|G_i'| < n$ since $|G_j'| = 1 + |Z_j| + |U_j| \geqslant 3$. As there are $\lambda(a_1, a_2) = \lambda(a_1, a_2; G)$ edge disjoint $a_1 - a_2$ paths in G and each edge in E_0 is in exactly one of these paths, if $x, y \in |G_i|$, $i = 1, 2$, then

$$\lambda(x, y; G_i') \leqslant \lambda(x, y; G) = \lambda(x, y). \qquad (4)$$

Either $k + 1 \leqslant |G_i'| < n$ and then by (4) and the induction hypothesis we have

$$e(G_i') \leqslant \frac{k+1}{2}(|G_i'| - 1) - \frac{1}{2}\sigma_k(G_i') \qquad (5)$$

or $|G_i'| \leqslant k$ and then

$$e(G_i') = \frac{k}{2}|G_i'| - \frac{1}{2}\sigma_k(G_i'). \qquad (6)$$

To complete the theorem we shall examine the three essentially different cases according to whether $|G_i'| \geqslant k + 1$ or $|G_i'| \leqslant k$.

(a) Suppose $|G_1'| \geqslant k + 1$ and $|G_2'| \geqslant k + 1$. Then

$$\sigma_k(G_1') + \sigma_k(G_2') \geqslant \sigma_k(G) + (k - |U_1|) + (k - |U_2|), \qquad (7)$$

since $d_G(a_i) \geqslant k + 1$ but $d_{G_i}(a_j) \leqslant |U_j|, j \neq i$. Therefore by (5) we have

$$e(G) - |E_0| + |U_1| + |U_2| = e(G_1') + e(G_2')$$

$$\leqslant \frac{k+1}{2}|G| - \frac{1}{2}(\sigma_k(G_1') + \sigma_k(G_2'))$$

$$\leqslant \frac{k+1}{2}(n-1) - \frac{1}{2}\sigma_k(G) - \frac{k-1}{2} + \frac{1}{2}(|U_1| + |U_2|). \qquad (8)$$

However, (7) cannot hold since then

$$e(G) \leqslant \frac{k+1}{2}(n-1) - \frac{1}{2}\sigma_k(G) - \frac{k-1}{2} - \frac{1}{2}(|U_1| + |U_2|) + |E_0|,$$

so by (2) and (3)

$$|E_0| > k,$$

contradicting the choice of E_0.

(b) Suppose now that $|G_1'| \geqslant k + 1$ and $|G_2'| \leqslant k$. Note that then

$$\sigma_k(G_1') + \sigma_k(G_1') \geqslant \sigma_k(G) + (k - |U_1|) + (k - |U_2|) + (k - |G_2'| + 1), \qquad (7')$$

since $d_G(a_2) \geqslant k + 1$ but $d_{G_2'}(a_2) \leqslant |G_2'| - 1$. Adding (5) with $i = 1$ and (6) with $i = 2$ and then taking (7') into account we obtain

$$e(G) - |E_0| + |U_1| + |U_2| = e(G_1') + e(G_2')$$

$$\leqslant \frac{k+1}{2}(|G_1'| - 1) + \frac{k}{2}|G_2'| - \frac{1}{2}\sigma_k(G_1') - \frac{1}{2}\sigma_k(G_2')$$

$$\leqslant \frac{k+1}{2}(n-1) - \frac{1}{2}\sigma_k(G) - \frac{k-1}{2} + \frac{1}{2}(|U_1| + |U_2|).$$

This is again inequality (8) which we know is false.

(c) Suppose finally that $|G_1'| \leqslant k$ and $|G_2'| \leqslant k$. Then, analogously to (7'), we have

$$\sigma_k(G_1') + \sigma_k(G_2') \geqslant \sigma_k(G) + (k - |U_1|) + (k - |U_2|) + (k + 1|G_1'|)$$

$$+ (k + 1 - |G_2'|).$$

Consequently by (6) we once again arrive at inequality (7) which does not hold:

$$e(G) - |E_0| + |U_1| + |U_2| = e(G_1') + e(G_2')$$

$$\leqslant \frac{k}{2}(n + 2) - \frac{1}{2}(\sigma_k(G_1') + \sigma_k(G_2'))$$

$$\leqslant \frac{k + 1}{2}(n - 1) - \frac{1}{2}\sigma_k(G) - \frac{k - 1}{2} + \frac{1}{2}(|U_1| + |U_2|). \quad \blacksquare$$

Let us return now to the problem of finding a pair of vertices with high local vertex connectivity. Though, as we have mentioned already, we know rather little about the maximal local connectivity of a graph of given order and size, Theorem 5.8 implies that the maximal local vertex connectivity in a graph is at least as large as the minimal degree. In fact Mader [M4], [M11], [M12], [M14], [M15] proved a number of stronger results than this, some of which we are going to mention in the sequel. In particular he proved that every non-empty graph contains two adjacent vertices whose local vertex connectivity is equal to the minimum of their degrees. (Note that this is another extension of Theorem 4.3.) This had been conjectured independently by Jung (see [M12]) and by Pelikán and Pósa (see [P3]). The proof of this result is based on a technical lemma we shall need in Chapter VII as well.

Let K be a complete subgraph of G, say $K = K^m$, and let $E_0 \subset E(G) - E(K)$. We shall say that (G, K, E_0) is an *admissible triple* if

(i) each edge in E_0 has one endvertex in $V(K)$,
(ii) there is a linear ordering of $V(K)$, say $V(K) = \{x_1, x_2, \ldots, x_m\}$, such that whenever $xx_i \in E_0$ we have $xx_j \in E(G) - E_0$ for every j, $1 \leqslant j < i$.

Recall that if G and H are graphs then $G - H = G - V(H)$.

LEMMA 5.9. *Let (G, K, E_0) be an admissible triple, and let $a_1, a_2 \in G - K$. Put*

$$k = \kappa(a_1, a_2; G).$$

Then either $\kappa(a_1, a_2; G - E_0) = k$ or $G - E_0$ contains $k - 1$ independent a_1-a_2 paths, say $P_1, P_2, \ldots, P_{k-1}$ and two more paths, say Q_1 and Q_2 such that Q_i joins a_i to an endvertex of an edge in E_0 and

$$V(Q_1) \cap V(Q_2) = \varnothing, \quad V(Q_i) \cap V(P_j) - \{a_i\}, \quad i = 1, 2, \quad j = 1, 2, \ldots, k - 1.$$

Proof. Let $R_1, R_2, \ldots, R_k$ be k independent a_1–a_2 paths. We may assume without loss of generality that $E(R_i) \cap E_0 \neq \emptyset, i = 1, 2, \ldots, k$. Furthermore, we may assume that R_i is a shortest $a_1 - a_2$ path in $G[V(R_i)]$. Then each R_i has exactly one vertex in K, x_{j_i}, that is incident with one or two edges in $E_0 \cap E(R_i)$. We may assume that $j_1 < j_2 < \ldots < j_k$.

Let P_j be the path obtained from R_{j+1} by replacing x_{j+1} with x_j. It is easily seen that the paths $P_1, P_2, \ldots, P_{k-1}$ together with two paths Q_1, Q_2 obtained from appropriate segments of R_1, have the required properties. ∎

THEOREM 5.10. *Let K be a complete subgraph of G with $E(G - K) \neq \emptyset$. Then there exists an edge $a_1 a_2 \in E(G - K)$ such that*

$$\kappa(a_1, a_2) = \min\{d(a_1), d(a_2)\}$$

Proof. We apply induction on the order of G. For $|G| = 1, 2$ there is nothing to prove so suppose $|G| > 2$ and the result holds for graphs of smaller order. Let $K^* = K^{m+1}$ be a maximal complete subgraph, a *clique*, of G containing K. We distinguish two cases according as $E(G - K^*)$ is empty or not.

Suppose first that $E(G - K^*) = \emptyset$. Let W be a minimal subset of $V(K^*)$ containing $V(K)$ such that

$$E(G - W) = \emptyset.$$

By the assumption of the theorem $V(K) \neq W$, say $x \in W - V(K)$. The choice of W implies that there exists an edge $xy, y \notin W$. Again by the choice of W we have

$$\Gamma(y) \subset W \subset V(K^*).$$

Consequently

$$\kappa(x, y) = d(y).$$

Suppose now that $E(G - K^*) \neq \emptyset$. Then we may assume without loss of generality that $K^* = K$. Put $V(K) = \{x_0, x_1, \ldots, x_m\}$. For $x \in \Gamma(x_0) - V(K)$ let

$$l(x) = \min\{i : xx_i \notin E(G)\}, \qquad 1 \leqslant i \leqslant m.$$

As K is a maximal complete subgraph of G, $l(x)$ is well defined. Put

$$E_0 = \{xx_{l(x)} : x \in \Gamma(x_0) - V(G)\},$$

$$G' = (V(G - x_0), E(G - x_0) \cup E_0),$$

$$K' = K - x_0.$$

Then (G', K', E_0) is an admissible triple and $E(G' - K') = E(G - K) \neq \varnothing$. Also by the construction

$$d(x) = d_G(x) = d_{G'}(x) \quad \text{for every} \quad x \in G - K.$$

As $|G'| = |G| - 1$, the induction hypothesis implies that there exists an edge $a_1 a_2 \in E(G' - K') = E(G - K)$ such that

$$\kappa(a_1, a_2; G') = K \quad \text{where} \quad k = \min\{d(a_1), d(a_2)\}.$$

By Lemma 5.9 either $\kappa(a_1, a_2; G) = k$ and then there is nothing to prove or else there exist paths $P_1, P_2, \ldots, P_{k-1}, Q_1, Q_2$ in G described in Lemma 5.9. Let P_k be the path in G obtained from $Q_1 \cup Q_2$ by adding to it the two edges joining x_0 to the endvertices of Q_1 and Q_2 different from a_1 and a_2. Then $\{P_1, P_2, \ldots, P_k\}$ is a set of k independent a_1–a_2 paths in G, showing $\kappa(a_1, a_2; G) = \min\{d(a_1), d(a_2)\}$.   ∎

COROLLARY 5.11. *If* $G \in \mathcal{D}_k, k \geq 1$, *i.e.* $|G| \geq k$ *and*

$$e(G) \geq k|G| - \binom{k+1}{2} + 1,$$

then G contains an edge xy such that

$$\kappa(x, y) \geq k + 1.$$

Proof. Apply Theorem 5.10 to a subgraph $H \subset G$ with $\delta(H) \geq k + 1$, whose existence is guaranteed by (0.5).   ∎

Putting together Corollaries 5.7 and 5.11 we have the following information on $f(n; \bar{\kappa} \leq k)$.

COROLLARY 5.12. *There is a constant $c > 0$ such that for $k \geqslant 4$*

$$\left(\frac{k+1}{2} + ck^{\frac{1}{4}} \right) n \leqslant f(n; \bar{\kappa} \leqslant k) \leqslant (k-1)n - \binom{k}{2}. \qquad \blacksquare$$

To conclude this section we present some results of Mader [M8] about k-connected subgraphs.

THEOREM 5.13. *Let $k \geqslant 1$, $p \geqslant -1$ and $G = G(n, m)$ where*

$$n \geqslant k + p + 2 \quad and \quad m > (k+p)(n-k+1).$$

Suppose each subgraph of order $k + p + 2$ of G has at most $(k+p)(p+3)$ edges. Then G contains a k-connected subgraph.

Proof. Let G_0 be a subgraph of G having minimal order under the conditions

$$|G_0| \geqslant k + p + 2 \quad and \quad e(G_0) > (k+p)(|G_0| - k + 1).$$

Then by the assumption of the theorem $|G_0| \geqslant k + p + 3$. Furthermore. the minimality of $|G_0|$ implies that $\delta(G_0) \geqslant k + p + 1$.

We claim that G_0 is k-connected. Suppose this is not so. Let (V_1, V_2) be a $(k-1)$-separator of G_0. Then G_0 is the edge disjoint union of two subgraphs, say G_1 and G_2, such that

$$V(G_1) = V_1, \qquad V(G_2) = V_2, \qquad |G_1| + |G_2| = |G_0| + k - 1.$$

As $\quad \delta(G_0) \geqslant k + p + 1$,

$$|G_i| \geqslant k + p + 2, \qquad i = 1, 2.$$

Consequently the minimality of G_0 implies that

$$e(G_i) \leqslant (k+p)(|G_i| - k + 1), \qquad i = 1, 2.$$

Therefore

$$e(G_0) = e(G_1) + e(G_2) \leqslant (k+p)(|G_1| + |G_2| - 2(k-1))$$

$$= (k+p)(|G_0| - k + 1),$$

contradicting the choice of G_0. $\qquad \blacksquare$

COROLLARY 5.14. *Let* $G = G(n, m)$, $n \geqslant 2k - 1 \geqslant 3$ *and*

$$m > (2k - 3)(n - k - 1).$$

Then G *contains a* k-*connected subgraph.*

Proof. Note that K^{2k-1} is the only graph of order $2k - 1$ and size at least $(2k - 3)k + 1$. Consequently either G contains a K^{2k-1} or it satisfies the conditions of Theorem 5.13 with $p = k - 3$. ∎

COROLLARY 5.15. *Let* G *have order* $n \geqslant k + 1 \geqslant 3$ *and size* $m > (k - 1)$ $(n - k + 1)$. *Suppose furthermore that the girth of* G *is at least* $g(k)$, *where* $g(k) = k + 1$ *for* $2 \leqslant k \leqslant 4$ *and* $g(k) = \max\{\frac{2}{5}k + 1, 6\}$ *for* $k \geqslant 5$. *Then* G *contains a* k-*connected subgraph.*

Proof. The assertion follows from Theorem 5.13 with $p = -1$ if we show that a graph H of order $k + 1$ and size $2k - 1$ is either k-connected or its girth is less that $g(k)$. This is trivial for $2 \leqslant k \leqslant 4$. To see it for $k \geqslant 5$ suppose the girth of H is $g \geqslant g(k)$ and let C be a cycle of length g in H. Then H does not contain a diagonal of C and a vertex not on C is joined to at most one vertex on C. Consequently $F = H - C$, a graph of order $k + 1 - g$, has at least $2k - 1 - (k + 1) = k - 2 > k + 1 - g$ edges. A graph of order s and size $s + 1$ either contains two cycles with at most one vertex in common or it contains two vertices joined by 3 independent paths so the girth of such a graph is at most $\lfloor \frac{2}{3}(s + 1) \rfloor$. Therefore the girth of F is at most $\lfloor \frac{2}{3}(k + 3 - g) \rfloor$ $< g$, and this is a contradiction. ∎

6. EXERCISES, PROBLEMS AND CONJECTURES

1^-. Every graph contains two vertices of equal degree.

2^-. If G is cubic then $\kappa(G) = \lambda(G)$.

3^-. Let $0 \leqslant k < n$. Determine the set of *maximal* graphs G^n with $\kappa(G^n) \leqslant k$.

4. Prove that if $\deg u + \deg v \geqslant p - 1$ for every pair of non-adjacent vertices then $\lambda(G) = \delta(G)$. Show that the result is best possible even among connected graphs (i.e. $p - 1$ cannot be replaced by $p - 2$). (Lesniak [L6])

5. Prove that if $\mathscr{A} = \{A_1, \ldots, A_n\}$ is a family of n distinct subsets of an

n-set S then there is an element $x \in S$ such that the sets $A_1 - \{x\}$, $A_2 - \{x\}$, ..., $A_n - \{x\}$ are all distinct.

[Hint. Define a graph G with vertex set $\mathscr{A}$ by joining A_i to A_j iff there exists an $x \in S$ such that either $A_i = A_j \cup \{x\}$ or $A_i \cup \{x\} = A_j$. Label the edge $A_i A_j$ with x and let $\mathrm{Lab}(G)$ be the set of labels of the edges. Show that there is a forest $F \subset G$ such that $\mathrm{Lab}(F) = \mathrm{Lab}(G)$.] (Bondy [B45].)

6. Show that a tree T is the block-cutvertex graph of some graph if and only if the distance between any two endvertices is even, i.e. T is a bipartite graph with all its endvertices belonging to the same class. (Gallai [G8], Harary and Prins [HP1].)

7. Show that if G is a graph without isolated vertices then $B(G)$, the block graph, and $bc(G)$, the block-cutvertex graph, determine each other.

8^-. Prove that a nontrivial connected graph has at least two vertices which are not cutvertices.

9. Prove that a graph G is the block graph of some graph H if and only if every block of G is a complete graph.

10^-. Prove that two k-components can have at most $k - 1$ vertices in common. (Harary and Kodama [HK1].)

11. Prove Menger's theorem (Theorem 2.2) as follows. Suppose it fails. Consider an appropriate graph containing vertices x and y such that k vertices are needed to separate x from y but one cannot find k independent x–y paths. Show first that $\Gamma(x) \cap \Gamma(y) = \varnothing$. Prove next that if W is a k-set separating x from y then either $W \subset \Gamma(x)$ or $W \subset \Gamma(y)$.

Deduce a contradiction from these. (Dirac [D22].)

12. Let $A = \{a_1, \ldots, a_p\}$ and $B = \{b_1, \ldots, b_q\}$ be disjoint sets of vertices of G such that

$$\kappa(a_i, b_j) \geqslant k \quad \text{for all} \quad i, j, \quad 1 \leqslant i \leqslant p, 1 \leqslant j \leqslant q.$$

Let $\lambda_1, \ldots, \lambda_p, \mu_1, \ldots, \mu_q$ be non-negative integers satisfying $\sum_1^p \lambda_i = \sum_1^q \mu_i = k$. Deduce from Menger's theorem (Theorem 2.2) that there exist k independent A–B paths in G such that λ_i of the paths start at a_i and μ_j of the paths end at b_j, $1 \leqslant i \leqslant p$, $1 \leqslant j \leqslant q$. (Dirac [D15].)

13. Show that there is a natural number s with the following property.

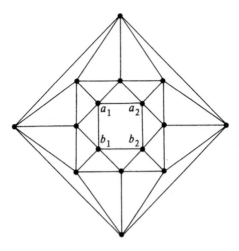

FIG. 6.1. A 5-regular 5-connected graph showing $s \geqslant 6$.

If $\kappa(G) \geqslant s$, $A = \{a_1, a_2\}$, $B = \{b_1, b_2\}$, $A \cup B \subset V(G)$ and $A \cap B = \varnothing$ then there exist disjoint paths joining a_i to b_i, $i = 1, 2$. (For an extension of this result, see Ex. VII. 5.)

By considering the graph in Fig. 6.1 show that $s \geqslant 6$. (Watkins [W12].)

14. PROBLEM. Suppose the distance between any two A–B paths is at most 1. Does there exist a vertex which is within distance 2 of every A–B path? (Gallai [G9].)

15. Let G be a graph and let $S_p(x, y)$, $I_p(x, y)$ be the functions defined after Theorem 2.4.

Show that $I_p(x, y) = S_p(x, y)$ whenever $p \leqslant 4$ but the graph in Fig. 6.2 has $I_5(x, y) = 1$ and $S_5(x, y) = 3$. (Lovász, Neumann–Lara and Plummer [LNP1].)

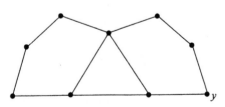

FIG. 6.2. A counterexample to $I_p = S_p$.

16^+. (Ex. 15 ctd.) Show that if $p \geqslant 2$ and $p \geqslant d(x, y)$ then

$$\frac{S_p(x, y)}{I_p(x, y)} \leqslant p - d(x, y) + 1.$$

(Lovász, Neumann-Lara and Plummer [LNP1].)

17. CONJECTURE (Ex. 15 ctd.) There is a constant $c > 0$ such that if G is any graph, $xy \in G$ and $d_G(x, y) = p \geqslant 2$ then

$$S_p(x, y) \leqslant cp^{\frac{1}{2}}I_p(x, y).$$

(Lovász, Neumann-Lara and Plummer [LNP1].)

18. Prove Corollaries 2.7 and 2.8.

19. Let G be a minimally 2-connected graph and let $x, y \in V(G), x \neq y$, $xy \notin E(G)$. Then G has a 3-colouring giving x and y the same colours and also a 3-colouring giving x and y different colours.
[Hint. Prove first that a tree of order at least 3 can be coloured with colours 1, 2 and 3 in such a way that any two fixed non-adjacent vertices get any pre-assigned colours.] (Dirac [D23].)

20. Prove that a minimally 2-connected graph of order $3m + 1$ has at least $m + 2$ vertices of degree 2. Note that this bound is best possible for $m \geqslant 1$ (cf. Theorem 3.9).

21. Prove that the graph $G_{k,l}$, constructed after Theorem 4.8 (see also Fig. 4.5), is minimally k-connected.

22. Let S be a separating k-set in a minimally k-connected graph G and let C be a component of $G - S$. Prove that C contains a vertex having degree k in G. (Mader [M7].)

23. Deduce from the previous exercise that a minimally k-connected graph has at least k vertices of degree k.

24^+. Let G be minimally k-connected. Prove that it has at least $\Delta(G)$ vertices of degree k.

25^-. Show that a k-connected subgraph of a minimally k-connected graph is minimally k-connected.

26^-. Let $H \in \mathscr{E}_3$ (cf. Theorem 5.1) and $a_1a_2 \in E(H)$. Suppose $a_0 \notin V(H)$, $G = H + a_0 + a_0a_1 + a_0a_2 + a_0a_3 - a_1a_2$ and $\bar{\kappa}(G) \leqslant 3$. Show that H has a block (wheel) B containing all of a_1, a_2, a_3 and a_3 can be considered to be the centre of the wheel B.

27. Let $m \geq 2$, $m \neq 4$. Show that there is a graph $G = G(3m, 8m - 4)$ such that $\bar{\kappa}(G) = 4$.
[Hint. Make use of the graphs shown in Fig. 5.2 and of the procedure given in Lemma 5.4.] (Sørensen and Thomassen [ST1].)

28. Suppose $n \geq g(G) \geq k + 2 \geq 6$ and $e(G) > (k/2)(n - k + 1)$. Prove that

$$\kappa(a, b) \geq k$$

for some $a, b \in V(G)$.(Mader [M11].)

29. A graph G is *critically k-connected* if $\kappa(G) = k$ and $\kappa(G - x) = k - 1$ for every $x \in G$. Prove that if G is critically 2-connected then $\delta(G) = 2$. (Kaugars [K4], see also Behzad and Chartrand [BC1, pp. 30–32].)

30^+. (Ex. 29 ctd.). Show that if G is critically k-connected $(k \geq 2)$ then $\delta(G) \leq (3k - 1)/2$. Give examples showing that the bound cannot be improved. (Chartrand, Kaugars and Lick [CKL2].)

31^-. Prove Lemma 5.6.

32. PROBLEM. Let $2 \leq l < k < n$. Determine the minimum of m for which every *k-connected* graph of order n has an *l-connected factor* of size *at most m*. Determine the analogous minimum for edge-connectivity.

33. PROBLEM. Let $1 \leq l < k < n$. Determine the maximal integer e with the following property. Every k-connected graph of order n contains e edges whose *omission* results in an *l*-connected graph. Determine the analogous maximum for edge-connectivity.

34. PROBLEM. Let $1 \leq k < l < n$. Determine the minimal integer e for which to every k-connected graph of order n it is possible to *add* at most e edges such that the resulting graph is *l*-connected. Determine the analogous minimum for edge-connectivity.

35. Show that for $k = 1, l = 2$ the minimum in Problem 34 is $n - 2$ for vertex-connectivity and $\lceil (n - 1)/2 \rceil$ for edge-connectivity.

36. PROBLEM. Let $k \geq 5$. Determine

$$\lim_{n \to \infty} f(n; \bar{\kappa} \leq k)/n.$$

37^-. Show that the limit in Problem 36 does exist.

Chapter II

Matching

A vertex and an edge are said to *cover* each other if they are incident. Given a graph G, the minimal number of vertices covering all the edges of G, i.e. the minimal number of vertices in a vertex cover, is called the *vertex covering number* and is denoted by $\alpha_0(G)$. The *edge covering number* of G, denoted by $\alpha_1(G)$, is defined analogously. The maximal number of independent vertices in G is denoted by $\beta_0(G)$ and is called the *vertex independence number* of G. The *edge independence number*, $\beta_1(G)$, is defined analogously.

It is easily seen that $\alpha_0(G) + \beta_0(G) = \alpha_1(G) + \beta_1(G) = n$ so only two of these four invariants are independent. The main aim of this chapter is to examine in detail the edge independence number and the possible sets of $\beta_1(G)$ independent edges in G. The vertex independence number $\beta_0(G)$ (or, what is essentially the same, the vertex covering number $\alpha_0(G)$) will be investigated in Chapter VI. In fact in our presentation it will be more natural to look at $cl(G) = \beta_0(\bar{G})$, the *clique number* of G. It is clear from the definition that the clique number of G is the maximum order of a complete subgraph of G.

It should be remarked that instead of covering one often speaks of *representation*, especially in connection with other subgraphs. Thus one might say that a set of vertices covering the edges represents the edges and one says that a set $W \subset V(G)$ *represents* the cycles of G if every cycle of G contains a vertex of W.

For obvious reasons a set of independent edges of G is also called a *matching* of G and a set of $\beta_1(G)$ independent edges is said to be a *maximum matching* of G. The sets of independent edges covering all the vertices of a graph are of particular interest. These are called the 1-*factors* of the graph. More generally, given a function $f : V(G) \to \mathbb{Z}^+$ (non-negative integers) an *f-factor* of G is a subgraph $H < G$ such that $d_H(x) = f(x)$ for every

$x \in V(G)$. If f is the constant function taking the value k then an f-factor is said to be a k-factor. Thus a k-factor of G is a k-regular spanning subgraph of G.

We say that G has a *factorization* into $F_1, F_2, \ldots, F_l$ if each F_i is a spanning subgraph of G, $E(F_i) \cap E(F_j) = \varnothing$ for $1 \le i < j \le l$ and $E(G) = \bigcup_{i=1}^{l} E(F_i)$. If each F_i is a k-factor then we say that G has a k-factorization.

In the first section we present the fundamental matching theorems, namely Philip Hall's theorem on bipartite graphs, Dilworth's theorem on partially ordered sets and Tutte's theorem on matchings in graphs. Of these Tutte's theorem, characterizing graphs with a 1-factor, is the most significant for us.

If a graph has a 1-factor then various rather weak conditions ensure that it has some other 1-factors. In §2 we examine the graphs with exactly one 1-factor. The main aim of the section is to determine the minimal number of 1-factors in a k-connected graph that possesses at least one.

The high point of §3 is another important theorem of Tutte, giving a necessary and sufficient condition for the existence of an f-factor.

Some natural extremal problems involving $\beta_1(G)$ are discussed in §4. We give lower bounds on $\beta_1(G)$, the maximal number of independent edges, in terms of the order, size, maximal and minimal degrees, vertex connectivity and edge connectivity.

In the last section we focus our attention on some natural factors of a graph (a path or a cycle and isolated vertices; a tree; a forest) and examine how many of these factors are needed to factorize a graph. Our aim is to represent a graph as the edge disjoint union of certain subgraphs, some of which might contain isolated vertices. This is why it is more appropriate to say that we wish to *cover* (the edges of) a graph, so the problems to be discussed are covering problems.

It seems that, due to the applicability of the two theorems of Tutte to be presented in the chapter, the natural problems concerning maximum matchings have been answered satisfactorily. In view of this it is even more frustrating that for $k \ge 3$ we know almost nothing about the existence of k-regular subgraphs with *arbitrary* vertex sets. A number of problems in this direction have been posed by Sauer, Erdös, Berge, Szemerédi and Chvátal but the existing methods seem powerless to attack them. Some of these problems will be presented in Chapter VII since they seem to be more related to the results and methods to be discussed there.

In §§1,2 we shall present several fairly satisfactory results about the number of 1-factors of graphs. Though multigraphs are mentioned only tangentially in this book, we would like to draw the reader's attention to the fact that we know very little about the number of 1-factors of multigraphs. In particular, it is not known whether or not Theorem 1.1 holds for multigraphs as well (Problem 13). We also draw attention to the fact that Van der Waerden's conjecture concerning permanents is equivalent to a conjecture about the number of 1-factors of regular multigraphs (Conjecture 14).

1. FUNDAMENTAL MATCHING THEOREMS

In Chapter I we applied Menger's theorem to deduce a theorem of Philip Hall (Theorem I.2.3), giving necessary and sufficient conditions for the existence of a matching in a bipartite graph covering all the vertices of a given class. The theorem is equivalent to a result of König [K9], discovered a few years earlier. There are a good number of other proofs and extensions of Hall's theorem; see, for example, Everett and Whaples [EW1], Halmos and Vaughan [HV1], Ryser [R15] and Rado [R3]. We start by presenting an extension proved by Marshall Hall [H16]. The proof we give is based on Halmos and Vaughan [HV1] and is due to Mann and Ryser [MR1]. Recall that $\Gamma(S) = \{y : x \in S, xy \in E(G)\}$.

THEOREM 1.1. *Let* $G = G_2(m, n)$ *be a bipartite graph with classes* U *and* $V, |U| = m, |V| = n$. *Suppose*

$$|\Gamma(S)| \geqslant |S| \text{ for every } S \subset U. \tag{1}$$

Then the following assertions hold.

(i) *G contains m independent edges,*

(ii) *If* $\delta \leqslant \delta_U(G) = \min\{d(x) : x \in U\}$ *then G contains at least*

$$r(\delta, m) = \prod_{1 \leqslant i \leqslant \min(\delta, m)} (\delta + 1 - i)$$

sets of m independent edges.

Proof. It clearly suffices to prove the second assertion. We apply induction on m. The case $m = 1$ is trivial so we proceed to the induction step.

Suppose that for each set $S \subset U$, $\emptyset \neq S \neq U$, we have

$$|\Gamma(S)| > |S|.$$

Fix a vertex $x \in U$ and then choose $y \in \Gamma(x)$. Note that, given x, there are at least δ choices for y. Put $G_y = G - \{x, y\} = G_2(m - 1, n - 1)$. This graph G_y satisfies the conditions of the theorem with $\delta - 1$ and $m - 1$ in place of δ and m. Consequently each G_y contains at least $r(\delta - 1, m - 1)$ sets of m independent edges. Adding xy to each of these sets we see that G contains at least $\delta r(\delta - 1, m - 1) = r(\delta, m)$ sets of m independent edges.

Suppose then that there is a set S, $\emptyset \neq S \neq U$ for which equality holds in (1). Let $s = |S|$, $G_1 = G[S \cup \Gamma(S)]$, $G_2 = G - S \cup \Gamma(S)$. Then G_1 satisfies the conditions of the theorem with s in place of m, and G_2 satisfies the conditions of the theorem with $\delta_2 = \max\{\delta - s, 1\}$ and $m - s$ in place of δ and m. As G_1 and G_2 are disjoint graphs, by the induction hypothesis G contains at least

$$r(\delta, s) \, r(\delta_2, m - s) = r(\delta, m)$$

sets of m independent edges. ∎

Let $A_1, A_2, \ldots, A_m$ denote subsets of a set A. A sequence $a_1, a_2, \ldots, a_m$ of distinct elements of S is said to be a *set of distinct representatives of* $\{A_i\}_1^m$ if $a_i \in A_i$, $i = 1, 2, \ldots, m$, A bipartite graph $G_2(m, n)$ with vertex classes U and V is naturally identified with the system $\{\Gamma(u) : u \in U\}$ of subsets of V. Then the sets of m independent edges of G are in one to one correspondence with the sets of distinct representatives of $\{\Gamma(u) : u \in U\}$. Conversely, a system of (not necessarily distinct) subsets of a set V indexed with a set U, $\{S_u : u \in U\}$, is naturally identified with the bipartite graph with vertex classes U, V (considered to be disjoint) and edge set $\{uv : v \in S_u, u \in U\}$. Using this natural identification of bipartite graphs and set systems described above, Theorem 1.1 goes into the following result.

THEOREM 1.1′. *Let* $A_1, A_2, \ldots, A_m$ *be subsets of a set* A, *each having at least* δ *elements. Suppose*

$$\left| \bigcup_{i \in S} A_i \right| \geqslant |S| \quad \text{for every } S \subset \{1, 2, \ldots, m\} \quad .$$

Then there are at least

$$r(\delta, m) = \prod_{1 \leqslant i \leqslant \min\{\delta, m\}} (\delta + z - i)$$

sets of distinct representatives of $\{A_i\}_1^m$. ∎

The theory of sets of distinct representatives (or theory of transversals) is outside the scope of this book; we refer the interested reader to the book of Mirsky [M30]. Some results concerning distinct representatives are posed as exercises at the end of this chapter. Here we would like to point out only that Philip Hall's theorem (Theorem I.2.3 or Theorem 1.1(i)) is contained in the following theorem of Dilworth [D5] on partially ordered sets (cf. Ex. 7). A subset C of a partially ordered set S is said to be a *chain* if for $x, y \in S$ either $x \leqslant y$ or $y < x$. A set $A \subset S$ is said to be an *antichain* if $x < y$ implies $\{x, y\} \subset A$.

THEOREM 1.2. *If every antichain in a finite partially ordered set S has at most m elements then S is the union of m chains.*

Proof. There are a good number of proofs besides the original one by Dilworth [D5], see for example, Dantzig and Hoffman [DH1], Fulkerson [F4], Gallai and Milgram [GM1], Ford and Fulkerson [FF1], Mirsky and Perfect [MP1] (where it is deduced from P. Hall's theorem), Perles [P4] and Tverberg [T19]. The following very simple proof is that of Tverberg.

We apply induction on $|S|$. If $S = \varnothing$ there is nothing to prove so suppose $|S| > 0$ and the theorem holds for sets with fewer elements. Let C be a maximal chain in S with maximal element c.

If no antichain in $S - C$ has m elements, the assertion follows from the induction hypothesis. Therefore we may assume that $S - C$ contains an antichain $A = \{a_1, \ldots, a_m\}$. Put

$$S^- = \{x \in S : x \leqslant a_i \text{ for some } i\},$$

$$S^+ = \{x \in S : x \geqslant a_j \text{ for some } j\}.$$

Note that $c \notin S^-$ since otherwise $c < a_i$ for some i and then $C \cup \{a_i\}$ is a chain strictly containing C. By the induction hypothesis $S^- = \bigcup_{1}^{m} C_i^-$ where C_i^- is a chain and $a_i \in C_i^-$.

Note next that a_i is the maximal element of C_i^-. For if this were not so then $a_i < x$ for some $x \in C_i^-$ and as $x \leqslant a_j$ for some j we would obtain the contradiction $a_i < a_j$. By symmetry, $S^+ = \bigcup_{1}^{m} C_i^+$, where C_i^+ is a chain with minimal element a_i. Then $S = S^- \cup S^+$ is the union of the chains

$$C_1^- \cup C_2^+, \quad C_2^- \cup C_2^+, \ldots, \quad C_m^- \cup C_m^+. \qquad \blacksquare$$

Before passing on to maximal matchings in graphs, we remark that regular bipartite graphs clearly satisfy the conditions of Theorem 1.1(i), so every regular bipartite graph has a 1-factorization.

Ore [O1] observed another consequence of Theorem 1.1(i).

COROLLARY 1.3. *Let $G = G_2(m, n)$ be a bipartite graph with vertex sets U and V, $|U| = m$, $|V| = n$. Suppose d is such that*

$$|\Gamma(S)| + d \geqslant |S| \text{ for every } S \subset U. \tag{2}$$

Then G contains a matching covering the vertices in U with the exception of at most d vertices. Furthermore,

$$\beta_2(G) = m - \max\{|S| - |\Gamma(S)| : S \subset U\}.$$

Proof. It clearly suffices to prove the first assertion. Let $H = G_2(m, n + d)$ be the graph obtained from G by adding d vertices to V and joining each of them to every vertex in U. Then (2) implies that H satisfies (1) so by Theorem 1.1(i) H contains m independent edges. By construction at most d of these edges do not belong to G. ∎

As we mentioned earlier, the key result of this chapter is Tutte's theorem [T9] characterizing graphs with a 1-factor.

Given a graph G, for the existence of a 1-factor it is clearly necessary that for every $S \subset V(G)$, $G - S$ has at most $|S|$ components of odd order. Tutte's (somewhat surprising) theorem is that this trivial condition is also sufficient. As one would expect, there are many proofs of Tutte's theorem in the literature, see for example Gallai [G1], [G6], [G7], Maunsell [M21], Kotzig [K11] and Edmonds [E1]. An elegant and simple proof of Tutte's theorem was discovered by Anderson [A3] and (independently though later) by Mader [M12]. This is the proof we give here.

Write $q(H)$ for the number of *odd* components of a graph H, that is for the number of components with an odd number of vertices. (Of course, the other components are called *even*.)

THEOREM 1.4. *A graph G has a 1-factor if and only if*

$$q(G - S) \leqslant |S| \quad \text{for all} \quad S \subset V(G). \tag{3}$$

Proof. As we have already pointed out, one of the implications is trivial so we have to prove only that (3) implies the existence of a 1-factor. We apply induction on $|G|$. For $|G| = 0$ there is nothing to prove. Suppose now that $|G| > 0$ and the result holds for graphs of order less than $|G|$. Let S_0 be a maximal subset of $V(G)$ for which

$$q(G - S_0) = |S_0|.$$

Suppose first that $S_0 = \emptyset$. Then G consists of even components. If we pick any vertex of G, the remaining graph has at least one (and so by (3) exactly one) odd component. However, this contradicts the maximality of $S_0 = \emptyset$.

Assume now that $S_0 \neq \emptyset$. Let $C_1, C_2, .., C_m$ $(m = |S_0| \geqslant 1)$ be the odd components and let $D_1, D_2, \ldots, D_k$ be the even components of $G - S_0$. As equality holds in (3) for $S = S_0$, each D_i satisfies (3) so by the induction hypothesis each D_i has a 1-factor.

Let $C = C_i$ for some i and let $c \in C$. We claim that $C' = C - c$ has got a 1-factor. For if this is not so then by the induction hypothesis

$$q(C' - S) > |S| \quad \text{for some} \quad S.$$

As

$$q(C' - S) + |S| \equiv |C'| \equiv 0 \qquad (\text{mod } 2),$$

this implies

$$q(C' - S) \geqslant |S| + 2.$$

Therefore

$$|S_0| + 1 + |S| \geqslant q(G - S_0 \cup \{c\} \cup S) = q(G - S_0)$$

$$-1 + q(C' - S) \geqslant |S_0| + 1 + |S|,$$

contradicting the maximality of S_0. Consequently for every i, $1 \leqslant i \leqslant m$, and every $c \in C_i$ the graph $C_i - c$ has got a 1-factor.

Let $H = G_2(m, m)$ be the bipartite graph with vertex classes $U = \{C_i : 1 \leqslant i \leqslant m\}$ and $V = S_0$ in which C_i is joined to $s \in S_0$ if and only if $s \in \Gamma(C_i)$. Let $A \subset U$ and let $B = \Gamma_H(A) \subset V$. Then (3) implies

$$|A| \leqslant q(G - B) \leqslant |B|,$$

so (1) is satisfied by H. Consequently by Hall's theorem H has a 1-factor, i.e. G contains m independent edges $s_i c_i$, $s_i \in S_0$, $c_i \in V(C_i)$. Adding to this set of edges a 1-factor of each $C_i - c_i$, $1 \leqslant i \leqslant m$, and a 1-factor in each D_i, $1 \leqslant i \leqslant k$, we obtain a 1-factor of G.   ∎

Berge [B6] observed that this result can be extended in a way similar to Ore's extension (Corollary 1.3) of Hall's theorem.

COROLLARY 1.5. *The maximal number of vertices covered by a matching in a graph G of order n is $n - d$, where*

$$d = \max\{q(G - S) - |S|: \ S \subset V(G)\}.$$

Proof. (i) It is clear that a matching can cover at most $n - d$ vertices.

(ii) By Theorem 1.4 the graph $H = G + K^d$ has got a 1-factor. Those edges of a 1-factor of H that belongs to G do not cover at most d vertices of G.   ∎

Another extension of the 1-factor theorem was deduced by Lovász [L12] from a more general and rather complicated result. However, the proof above can easily be adapted to give the extension. Alternatively, we can deduce this extension from the 1-factor theorem. Given a graph H and a set T, denote by $q(H \mid T)$ the number of odd components of H whose vertex set is contained in T.

COROLLARY 1.6. *Let G be a graph and let $T \subset V(G)$. Then G has a matching covering T iff*

$$q(G - S \mid T) \leqslant |S| \qquad \text{for all} \ \ S \subset V(G).$$

Proof. Add to G a K^n, where $n = |G|$, and join every vertex of this K^n to every vertex of $G - T$. Write H for the resulting graph. Then G has a matching covering T iff H has a 1-factor. To prove the result apply Theorem 1.4 to the graph G.   ∎

A similar trick gives the following consequence of Theorem 1.4.

COROLLARY 1.7. *Let G be a graph, $T \subset V(G)$. The maximal number of vertices in T covered by a matching in G is $|T| - d$, where*

$$d = \max\{q(G - S \mid T) - |S|: S \subset V(G)\}. \qquad ∎$$

Let us reformulate the essential part of Corollary 1.5 in terms of $\beta_1(G)$, the edge independence number.

COROLLARY 1.8. *Let $G = G^n$ and suppose $\beta_1(G) = \beta$. Then there is a set S of, say, s vertices such that there are $q = n + s - 2\beta$ odd components in $G - S$.* ∎

Note that in the last corollary we must have $n \geqslant s + q = n + 2s - 2q$ so $s \leqslant \beta$. Given sets $S, V, S \subset V, |S| = s, |V| = n$ and an integer β satisfying $s \leqslant \beta \leqslant n/s$, let G be a *maximal* graph with vertex set V such that $G - S$ contains $q = n + s - 2\beta$ odd components. If $C_1, C_2, \ldots, C_q$ are the odd components of $G - S$ then the maximality of G implies that $\bigcup_1^q V(C_i) = V(G) - S$, each C_i is complete, $G[S]$ is complete and G contains all the edges joining a vertex in S to a vertex in C_i, $1 \leqslant i \leqslant q$. The graph G cannot contain a $C_i - C_j$ edge, $1 \leqslant i < j \leqslant q$, so G contains only the edges enumerated above. Thus we have another consequence of Theorem 1.4.

COROLLARY 1.9. *Let G be a maximal graph of order n with $\beta_1(G) = \beta$ (i.e. if $G \subset H, E(G) \neq E(H)$ then $\beta_1(H) > \beta$). Then G is of the form*

$$K^s + \bigcup_1^q K^{2n_i + 1},$$

where $q = n + s - 2\beta, n_i \geqslant 0$ and $\sum_1^q (2n_i + 1) = n - s$. ∎

It is easy to see that $e(K^j + \bigcup_1^q K^{2n_i+1})$ attains its maximum for $K^{2\beta+1} \cup E^{n-2\beta-1}$ or $K^\beta + E^{n-\beta}$. Consequently we have the following result, first proved by Erdős and Gallai [EG3].

COROLLARY 1.10. *If $n > 2\beta$ then the maximal size of a graph G^n with $\beta_1(G^n) = \beta$ is*

$$\max\left\{\binom{3\beta + 1}{2}, \binom{\beta}{2} + \beta(n - \beta)\right\}.$$

If $2\beta + 1 \leqslant n < (5\beta + 3)/2$ then $K^{2\beta+1} \cup E^{n-2\beta-1}$ is the unique extremal graph; if $n > (5\beta + 3)/2$ then $K^\beta + E^{n-\beta}$ is the unique extremal graph. If $n = (5\beta + 3)/2$ then there are two extremal graphs, $K^{2\beta+1} \cup E^{n-2\beta-1}$ and $K^\beta + E^{n-\beta}$. ∎

2. THE NUMBER OF 1-FACTORS

Given a graph G denote by $F(G)$ the number of 1-factors of G. The aim of this section is to give good lower bounds on $F(G)$ provided $F(G) > 0$ and G has certain properties. We shall obtain conditions implying $F(G) > 1$. As the main result of the section we will give the best lower bound on $F(G)$ provided $F(G) > 0$ and G is k-connected.

Putting Theorem 1.1 together with some of the facts shown in the proof of Theorem 1.4 we obtain the following assertion.

THEOREM 2.1. *Let G be a graph with $F(G) > 0$ and let S_0 be a maximal subset of $V(G)$ satisfying*

$$q(G - S_0) = |S_0|.$$

Then $|S_0| = m \geqslant 1$, $G - S_0$ consists of m odd components, say $C_1, C_2, \ldots, C_m$, and for every i and for every $c_i \in V(C_i)$ we have $F(C_i - c_i) > 0$. Every set of m independent edges of the form

$$\{s_i c_i: \quad s_i \in S_0, c_i \in V(C_i), 1 \leqslant i \leqslant m\}$$

is contained in a 2-factor of G. If d is the minimum of $|\Gamma(C_i) \cap S_0|$ then $d \geqslant 1$ and

$$F(G) \geqslant r(d, s) = \prod_{i=1}^{\min(d, s)} (d + 1 - i).$$

If d' is the minimum number of edges joining C_i to S_0 then

$$F(G) \geqslant \tilde{r}(d', s),$$

where $\tilde{r}(d', s)$ denotes the minimum number of 1-factors in a bipartite multigraph $H = G_2(s, s)$ with a 1-factor in which d' is the minimum degree in the first vertex class. ∎

This theorem is the basis of the results we are going to present. Let us state some immediate corollaries. Note that $\tilde{r}(d', s) \geqslant d'$ can be read out of the proof of Theorem 1.1. (It is also a consequence of Ex. 6.1.)

COROLLARY 2.2. *Let G be a k-edge-connected graph with $F(G) > 0$. If there is a set $S_0 \subset V(G), |S_0| = s$, satisfying $q(G - S_0) = s$ then $F(G) \geqslant \tilde{r}(k, s)$. In particular $F(G) \geqslant k$ always holds.* ∎

The case $k = 2$ of this corollary was proved by Beineke and Plummer [BP1].

In particular if G is a connected graph and $F(G) = 1$ then G has got a bridge. In fact, observing that the components C_i in Lemma 2.1 are odd and that $F(G) = 1$ implies $d = 1$, we can say somewhat more.

COROLLARY 2.3. *Let G be a connected graph with $F(G) = 1$. Then G contains a bridge xy that is an edge of the 1-factor of G (and so $G - xy$ consists of two odd components).* ∎

Each component of $G - \{x, y\}$ in Corollary 2.3 must have exactly one 1-factor so Corollary 2.3 is equivalent to the following description of graphs with exactly one 1-factor.

COROLLARY 2.3′. *Let $\mathscr{F}_1$ be the minimal class of graphs containing the empty graph such that if G has a bridge xy and $G - \{x, y\} \in \mathscr{F}_1$ then $G \in \mathscr{F}_1$. Then $\mathscr{F}_1$ is the class of graphs having exactly one 1-factor.* ∎

Corollary 2.3 enables one to show that a graph with exactly one 1-factor cannot have large minimal degree (Lovász [L21]) or large size (Hetyei, see [L21]). We use $\log_2$ to denote the logarithm with base 2.

THEOREM 2.4. *The minimum degree of a graph of order $2n$ with exactly one 2-factor is at most $\lfloor \log_2(n + 1) \rfloor$. This bound is best possible for every n.*

Proof. Let us prove first that if $|G| = 2n$ and $F(G) = 2$ then $\delta(G) \leqslant \lceil \log_2(n+1) \rceil$. We apply induction on n. The case $n = 1$ being trivial we proceed to the induction step. Let xy be a bridge of G guaranteed by Corollary 2.3. Then either $\delta(G) = 1$ in which case we are home or $G - \{x, y\}$ has a component C of order at most $n - 1$ such that $F(C) = 1$. By the induction hypothesis there exists $c \in C$ with

$$d_C(c) \leqslant \left\lfloor \log_2\left(\frac{n - 1}{2} + 1\right) \right\rfloor \leqslant \lfloor \log_2(n + 1) \rfloor - 1.$$

To show the converse inequality we construct extremal graphs by induction. Take $G_0 = K^0$, $G_2 = K^2$. Suppose we have constructed $G_2, G_3, \ldots, G_{n-1}$, where $|G_k| = 2k$, $F(G_k) = 1$ and $\delta(G_k) = \lfloor \log_2(k+1) \rfloor$. If $n-1$ is even, say $n-1 = 2k$, to obtain G_n take $2(K^1 + G_k)$ and join the two vertices of the two K^1s (see Fig. 2.1). Then

$$\delta(G_n) = \delta(G_k) + 1 = \lfloor \log_2(k+1) \rfloor + 1 = \lfloor \log_2(2k+2) \rfloor = \lfloor \log_2(n+1) \rfloor.$$

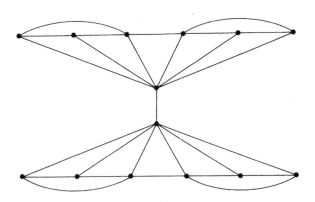

Fig. 2.1. The graph G_7 with $|G_7| = 14$, $\mathscr{F}(G_7) = 1$ and $\delta(G_7) = 3$.

If $n-1$ is odd, say $n-1 = 2k+1$, to obtain G_n take $(K^1 + G_k) \cup (K^2 + G_{k+1})$ and join the two vertices of the two K^1s. Then

$$\delta(G_n) = \delta(G_k) + 1 = \lfloor \log_2(k+1) \rfloor + 1 = \lfloor \log_2(2k+2) \rfloor = \lfloor \log_2 n \rfloor$$

$$= \lfloor \log_2(n+1) \rfloor,$$

since $n+1$ is odd. ∎

THEOREM 2.5. *Put* $f(n) = \max\{e(G): \ |G| = 2n, F(G) = 1\}$. *Then* $f(n) = n^2$.

Proof. By omitting a bridge of a graph G with $|G| = 2n$, $e(G) = f(n)$, $F(G) = 1$, we see that

$$f(n) \leqslant f(n-1) + 2n - 1.$$

As $f(1) = 1$ this gives $f(n) \leqslant n^2$.

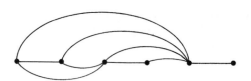

FIG. 2.2. The graph H_3 with $|H_3| = 6$, $F(H_3) = 1$ and $e(H_3) = 9$.

Conversely, put $H_1 = K^2$ and then define H_n inductively by $H_n = (H_{n-1} \cup K^1) + K^1$ (see Fig. 2.2.). Then $|H_n| = 2n$, $e(H_n) = n^2$ and $F(H_n) = 1$, showing $f(n) \geqslant n^2$. ∎

The main aim of this section is to determine at least how many 1-factors a k-connected graph ($k \geqslant 1$) must have provided it has at least one. Put

$$f(k) = \min\{F(G): \quad \kappa(G) \geqslant k \quad \text{and} \quad F(G) > 0\}.$$

We start with the following lemma of Lovász [L21].

LEMMA 2.6. *Let G be a k-connected ($k \geqslant 1$) graph with $F(G) > 0$. Then either $F(G) \geqslant k!$ or for every pair of vertices $\{x, y\}$, $x \neq y$, $F(G - \{x, y\}) > 0$.*

Proof. By Corollary 2.2 we may assume that $k \geqslant 3$. Suppose there exist $x, y \in G$, $x \neq y$, such that for $H = G - \{x, y\}$ we have $F(H) = 0$. Then by Tutte's theorem (Corollary 1.5) there is a set $S \subset V(H)$ such that

$$q(H - S) \geqslant |S| + 2.$$

Consequently

$$|S| + 2 \leqslant q(H - S) = q(G - S \bigcup \{x, y\}) \leqslant |S| + 2,$$

so there is a maximal subset S_0 *containing* $\{x, y\}$ for which

$$q(G - S_0) = |S_0|.$$

As $|S_0| \geqslant 2$ and $G - S_0$ has $|S_0|$ components, $\kappa(G) \geqslant k$ implies that $|S_0| \geqslant k$. Thus by Corollary 2.2 we have $F(G) \geqslant k!$, as claimed. ∎

The lemma above gives immediately a lower bound on $f(k)$ due to Zaks [Z1] which happens to be exact for odd k.

COROLLARY 2.7. *Let G be a k-connected $(k \geqslant 1)$ graph with $F(G) > 0$. Then either $F(G) \geqslant k!$ or*

$$F(G) \geqslant \Delta(G) f(k - 2). \tag{1}$$

Furthermore,

$$f(k) \geqslant k!! = \prod_{i=0}^{\lfloor (k-1)/2 \rfloor} (k - 2i). \tag{2}$$

Proof. To prove (1) note that if $F(G) < k!$ then $k \geqslant 3$ and by Lemma 2.6 every edge of G occurs in at least one 1-factor of G. If we pick a vertex x of degree $\Delta(G)$ there are $\Delta(G)$ different ways of choosing an edge xy and as $G - \{x, y\}$ is $(k - 2)$-connected and has a 1-factor, there are at least $f(k - 2)$ 1-factors of G containing xy.

Inequality (2) is immediate by induction on k if we make use of (1).   ∎

THEOREM 2.8. *If $k \geqslant 0$ then*
$$f(2k + 1) = (2k + 1)!! \tag{3}$$

If $k \geqslant 1$ then K^{2k+2} is the unique $(2k + 1)$-connected graph with $(2k + 1)!!$ 1-factors.

Proof. As $F(K^{2k+2}) = (2k + 1)!!$ for $k \geqslant 0$, (3) holds.

Let now G be a $(2k + 1)$-connected graph with $F(G) = (2k + 1)!!$, $k \geqslant 1$. Then by (1) and (3) we have $\Delta(G) = 2k + 1$ so G is $(2k + 1)$-regular. Furthermore, the proof of Corollary 2.7 implies that for every edge $xy \in E(G)$ we must have $1 \leqslant F(H) \leqslant (2k - 1)!!$, where $H = G - \{x, y\}$. Therefore either $k = 1$ and H is connected or $k \geqslant 2$ and H is $(2k - 1)$-regular. As there are exactly $4k$ edges joining $\{x, y\}$ to $V(H)$, in either case it follows that $|H| = 2k$. Consequently G is a $(2k + 1)$-regular graph of order $2k + 2$ and so $G = K^{2k+2}$, as claimed.   ∎

The considerably more difficult task of determining $f(2k)$ was accomplished by Mader [M12]. The rather complicated proof of Mader is based on results about certain special minimally $2k$-connected graphs. A simple and direct proof was given by Bollobás [B36] and this is the proof we present here.

Denote by $S_k(k \geqslant 0)$ the graph obtained from K^{2k+2} by omitting a 1-factor. Note that for $k \geqslant 1$ S_k is $2k$-connected, $2k$-regular and contains a 1-factor. Thus $f(2k) \leqslant F(S_k)$. We shall prove that for $k \geqslant 4$ the graph S_k is the unique $2k$-connected graph with $f(2k)$ 1-factors.

Let us calculate first $F(S_i)$. Note that $F(S_0) = 0$, $F(S_1) = 2$. Let $V(S_k) = \{x_1, x_2, \ldots, x_{k+1}, y_1, y_2, \ldots, y_{k+1}\}$ and suppose $x_i y_i \notin E(S_k)$, $i = 1, 2, \ldots, k + 1$. When choosing a 1-factor F of S_k, there are $2k$ ways of choosing the edge incident with x_1. By symmetry we may assume that this edge is $x_1 x_2$. If $y_1 y_2 \notin E(F)$ then the remaining k edges of F form a 1-factor of $S_k - \{x_1, x_2\} - y_1 y_2$ which is exactly an S_{k-1}. On the other hand if $y_1 y_2 \in E(F)$ then in order to complete $\{x_1 x_2, y_1 y_2\}$ to F we have to choose a 1-factor of $S_k - \{x_1, x_2, y_1, y_2\}$ which is an S_{k-2}. Consequently,

$$F(S_k) = 2k(F(S_{k-1}) + F(S_{k-2})).$$

Thus $F(S_k)$ is given recursively by the relations $F(S_0) = 0$, $F(S_1) = 2$ and

$$F(S_k) = 2k(F(S_{k-1}) + F(S_{k-2})). \tag{4}$$

In particular, $F(S_2) = 8$, $F(S_3) = 60$ and $F(S_4) = 544$.

In the proof of the next theorem we shall make essential use of an auxiliary function $g(2k, \Delta)$ where $k \geqslant 1$ and $\Delta \geqslant 2k$. Define $g(2k, \Delta)$ recursively by

$$g(2, \Delta) = 2,$$

$$g(2k, \Delta) = \min\{(2k)!, \quad \Delta g(2k - 2, \Delta_1): \quad 2k - 2 \leqslant \Delta_1 \leqslant \Delta$$

$$\text{and if} \quad \Delta \leqslant 2k + 2 \quad \text{then} \quad 2k - 1 \leqslant \Delta_1 \leqslant \Delta\}.$$

It is easily checked that

$$g(2, \Delta) = 2,$$

$$g(4, \Delta) = \min\{2\Delta, 4!\}$$

and for $k \geqslant 3$ we have

$$g(2k, \Delta) = \begin{cases} \frac{2}{3}\Delta(2k - 1)!! & \text{if} \quad 2k \leqslant \Delta \leqslant 2k + 2, \\ \min\{\frac{2}{3}\Delta(2k - 2)(2k - 3)!!, (2k)!\} & \text{if} \quad \Delta \neq 2k + 3. \end{cases}$$

Consequently, putting

$$g(2k) = \min_{\Delta \geq 2k} g(2k, \Delta), \quad k = 1, 2, \ldots,$$

we have $g(2) = 2, g(4) = 8, g(6) = 60, g(8) = 560$ and

$$g(2k) = \frac{4k}{3}(2k - 1)!! \quad \text{for} \quad k \geq 2. \tag{5}$$

THEOREM 2.9. *If* $k \geq 1$ *then*

$$f(2k) = F(S_k). \tag{6}$$

If $k \geq 4, |G| > 2k + 2, G$ *is* $2k$-*connected and* $F(G) > 0$ *then*

$$F(G) \geq \frac{4k}{3}(2k - 1)!! > F(S_k). \tag{7}$$

If $k \geq 2, |G| = 2k + 2, G \neq S_k$ *and* G *is* $2k$-*connected then*

$$F(G) \geq F(S_k) + F(S_{k-1}). \tag{7'}$$

In particular, for $k \geq 5$ *the graph* S_k *is the unique* $2k$-*connected graph with* $f(2k)$ *1-factors.*

Proof. Let S_k^+ be the graph obtained from S_k by the addition of an edge. If $|G| = 2k + 2$, $G \neq S_k$ and $\kappa(G) \geq 2k$ then $S_k^+ \subset G$. Consequently $F(G) \geq F(S_k^+) = F(S_k) + F(S_{k-1})$, proving (7').

To prove the main assertions of the theorem put

$$f(2k, \Delta) = \min\{F(G): |G| > 2k + 2, \kappa(G) \geq 2k, F(G) > 0 \text{ and } \Delta(G) = \Delta\},$$

$$\tilde{f}(2k) = \min_{\Delta \geq 2k} f(2k, \Delta).$$

Note that by (7') we have

$$f(2k) = \min\{F(S_k), \tilde{f}(2k)\}.$$

Our proof hinges on the fact that

$$f(2k, \Delta) \geq g(2k, \Delta) \quad \text{for every} \quad k \geq 1 \quad \text{and} \quad \Delta \geq 2k. \tag{8}$$

We prove (8) by induction on k. The case $k = 1$ follows from Corollary 2.2 so suppose $k > 1$ and (8) holds for smaller values of k. Suppose $|G| > 2k + 2$, $\kappa(G) \geqslant 2k$, $F(G) > 0$ and $\Delta(G) = \Delta$. Note that then $|G| \geqslant 2k + 4$. In order to prove (8) we have to show that

$$F(G) \geqslant g(2k, \Delta).$$

By Lemma 2.6 we may assume that every edge of G is contained in a 1-factor of G. Pick $x \in G$ with $d(x) = \Delta$. For $y \in \Gamma(x)$ put $H_y = G - \{x, y\}$. Then

$$F(G) = \sum_{y \in \Gamma(x)} F(G_y) \geqslant \Delta \min_{y \in \Gamma(x)} F(H_y). \tag{9}$$

Note that if $\Delta \leqslant 2k - 2$ then

$$|H_y - \Gamma(x)| \geqslant 2k + 2 - (\Delta - 1) \geqslant 1$$

so there is a vertex $z \in H_y$ not joined to x in G. As $\delta(G) \geqslant 2k$ this vertex z has degree at least $2k - 1$ in H_y so $\Delta_y \geqslant 2k - 1$. Also, by the choice of x we always have $2k - 2 \leqslant \Delta_y \leqslant \Delta$. Consequently, the induction hypothesis and inequality (9) imply $F(G) \geqslant g(2k, \Delta)$, proving (8). In turn, (8) gives

$$\tilde{f}(2k) \geqslant g(2k). \tag{8'}$$

The theorem follows easily from (8'). We know already that

$$F(S_k) = g(2k) \quad \text{for} \quad k \leqslant 3. \tag{10}$$

Let us show that

$$F(S_k) < g(2k) = \frac{4k}{3}(2k - 1)!! \quad \text{for} \quad k \geqslant 4. \tag{11}$$

To prove (11) we apply induction on k. As it holds for $k = 4$ we proceed to the induction step. By (4), (5), (10) and the induction hypothesis we have

$$F(S_k) = 2k\{F(S_{k-1}) + F(S_{k-2})\}$$

$$< 2k\left\{\frac{4(k-1)}{3}(2k-3)!! + \frac{4(k-2)}{3}(2k-5)!!\right\}$$

$$= \frac{4k}{3}(2k-5)!!\{2(k-1)(2k-3) + 2(k-2)\}$$

$$= \frac{4k}{3}(2k-5)!!\{(2k-1)(2k-3) + 1\} < \frac{4k}{3}(2k-1)!!.$$

To deduce equality (6) note that by (8'), (10) and (11)

$$F(S_k) = \min\{g(2k), F(S_k)\} \leqslant \min\{\tilde{f}(2k), F(S_k)\} = f(2k) \leqslant F(S_k).$$

Similarly (8') and (11) imply inequality (7):

$$F(G) \geqslant \tilde{f}(2k) \geqslant g(2k) > F(S_k). \qquad \blacksquare$$

With more care one can show that already for $k = 2$ and 3 the graph S_k is the unique $2k$-connected graph with $f(2k)$ 1-factors (see Mader [M22]).

3. f-FACTORS

Petersen [P5] was the first to investigate graphs with k-factors. Rado [R2] studied k-factors in bipartite graphs. Extending results of Tutte, Belck [B3] gave rather complicated conditions on the existence of a k-factor. The most important result in this direction was again obtained by Tutte [T10], who characterized the graphs G having an f-factor, where f is an arbitrary function $f: V(G) \to \mathbb{Z}^+$. Later Tutte [T11] deduced this result from Corollary 1.5. Another proof was given by Gallai [G1]. The main aim of this section is to present Tutte's characterization of graphs with f-factors, together with the proof from [T11].

In what follows we shall work with a fixed multigraph G and a fixed function $f: V(G) \to \mathbb{Z}^+$. We assume that G does not have loops and isolated vertices. If $S \subset V(G) = V$ and $H \subset G - S$ then $e(H, S)$ denotes the number of H–S edges, i.e. the number of edges having one endvertex in the graph H and the other in the set S. Given an ordered pair (D, S) of disjoint subsets of $V(G)$ and a component C of $G - D - S$, put

$$r(D, S; C) = e(C, S) + \sum_{x \in C} f(x).$$

We say that C is an *odd* or *even* component of $G - D - S$, *with respect to* (D, S), according as $r(D, S; C)$ is odd or even. The number of all odd components of $G - D - S$ is denoted by $q(D, S)$. If we wish to emphasize the dependence of r and q on f, we write $r_f(D, S; C)$ and $q_f(D, S)$. A trivial necessary condition for the existence of an f-factor of G is that $f(x) \leqslant d(x)$ for every $x \in G$. Let us assume that this holds. (In fact the condition is contained in the necessary and sufficient condition given in Theorem 3.2.) Denote the surplus degree of x by $s(x) = d(x) - f(x)$.

Given G and f let us construct a graph G^* that will have a 1-factor if and only if G has an f-factor. Let us associate two sets with each vertex of G; we choose all these sets *pairwise disjoint*. For $x \in G$ let $D(x)$ be a set consisting of $d(x)$ elements, one for each edge incident with x, say

$$D(x) = \{x_\alpha : \alpha \in E(G), \alpha \text{ is incident with } x\}.$$

Let $S(x)$ be a set of $s(x)$ elements, say

$$S(x) = \{x(i) : 1 \leqslant i \leqslant s(x)\}.$$

Then let $G^* = (V^*, E^*)$ be given by

$$V^* = V_1^* \cup V_2^*, \qquad E^* = E_1^* \cup E_2^*,$$

where

$$V_1^* = \bigcup_{x \in V} D(x), \qquad V_2^* = \bigcup_{x \in V} S(x),$$

$$E_1^* = \{x_\alpha y_\alpha : \alpha = xy \in E(G)\},$$

$$E_2^* = \{uv : u \in D(x), v \in S(x), x \in V\}.$$

Note that the map $\psi : E \to E_1^*$, defined by $\alpha = xy \xrightarrow{\psi} x_\alpha y_\alpha$ establishes a 1-1 correspondence between E and E_1^*. The set E_1^* consists of independent edges and the graph $G^*[V_1^*]$, whose edge set is exactly E_1^*, is such that if for each $x \in V$ we identify the vertices in $D(x) \subset V_1^*$ then we get back the original graph G (see Fig. 3.1).

Furthermore, $B(x) = G^*[D(x) \cup S(x)]$ is the complete bipartite graph with vertex classes $D(x)$ and $S(x)$. We call $B(x)$ the *bipartite graph* of $x \in V$.

The construction of G^* implies immediately the following lemma.

LEMMA 3.1. *G has an f-factor if and only if G^* has a 1-factor.*

Proof. Suppose G has an f-factor with edge set $F \subset E$. Then $\psi(F)$ consists of independent edges and in each set $D(x)$ exactly $s(x)$ vertices are not covered by $\psi(F)$. For each x one can add $s(x)$ independent $D(x)$–$S(x)$ edges covering the remaining vertices of $D(x)$. In this way we obtain a 2-factor of G^*.

Conversely, if G^* has a 2-factor with edge set $F^* \subset E^*$ then $\psi^{-1}(F^* \cap E_1^*)$ is the edge set of an f-factor of G. ∎

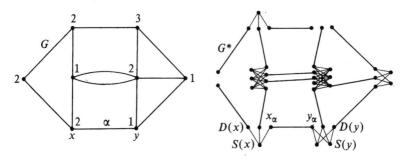

FIG. 3.1. A multigraph G, a function f and the graph G^* constructed from them.

In order to prepare the ground for the proof of the main theorem of this section, let us examine the graph G^* in more detail. We call a set $W \subset V^*$ *normal* if these two conditions are satisfied for every $x \in V$:

(i) $D(x) \cap W \neq \emptyset$ implies $D(x) \subset W$ and $S(x) \cap W \neq \emptyset$ implies $S(x) \subset W$,
(ii) $D(x) \cup S(x) \subset W$.

Note that if W is normal and $S(x) = \emptyset$, i.e. $s(x) = d(x) - f(x) = 0$, then $D(x) \cap W = \emptyset$.

Given a normal set $W \subset V^*$, define two subsets of V as follows:

$$D = \{x \in V: \ D(x) \subset W\},$$

$$S = \{x \in V: \ S(x) \subset W\}.$$

Then clearly

$$|W| = \sum_{x \in D} d(x) + \sum_{x \in S} s(x) = \sum_{x \in D} d(x) + \sum_{x \in S} \{d(x) - f(x)\}. \qquad (1)$$

Let H be a component of $G^* - W$. One of the following four cases must hold.

1. H consists of a single vertex in D, say $V(H) = \{x_\alpha\}$, $\alpha = xy \in E(G)$. Then

$$S(x) \cup \{y_\alpha\} \subset \Gamma(x_\alpha) \subset W$$

so $x \in S$ and $y \in D$. Conversely if $\alpha = xy \in E(G)$, $x \in S$ and $y \in D$ then the component of $G^* - W$ containing x_α consists only of the vertex x_α. Therefore the number of such components is equal to the number of D–S edges, i.e. to

$$\sum_{x \in S} \{d(x) - d_{G-D}(x)\} = \sum_{x \in D} \{d(x) - d_{G-S}(x)\}. \qquad (2)$$

2. $V(H) = \{x(i)\}$ for some $x \in V$ and $1 \leqslant i \leqslant s(x)$. Then $x \in D$. Conversely if $x \in D$ then for each $i, 1 \leqslant i \leqslant s(x)$, the component of $G^* - W$ containing $x(i)$ consists of this vertex alone. Therefore the number of these components is

$$\sum_{x \in D} s(x) = \sum_{x \in D} \{d(x) - f(x)\}. \qquad (3)$$

3. $\varnothing \neq E(H) \subset E_1^*$. As the edges in E_1^* are independent, $E(H)$ consists of a single edge so $|H| = 2$.

4. $E(H) \cap E_2^* \neq \varnothing$. Call these components H the *large* components of $G^* - W$. Note that if H contains an edge of a bipartite graph $B(x)$ then $D(x) \cup W, S(x) \subset W$ and as $B(x)$ is connected, $B(x) \subset H$. Put

$$M = \{x \in V: \quad B(x) \subset H\}.$$

The component H can be obtained from $\bigcup_{x \in M} B(x)$ by joining the bipartite graphs $B(x)$ with some edges to form a connected graph H_0 and then adding some vertices x_α to H_0 together with the edge $x_\alpha y_\alpha$, where $y_\alpha \in H_0$ and $\alpha = xy$. Note that in H the vertex x_α is joined only to y_α. Thus H is obtained from H_0 by adding to it all the edges of the form $x_\alpha y_\alpha$, where $y_\alpha \in H_0, \alpha = xy$ and $x \in S$, together with x_α. The set M is the vertex set of a component K_H of $G - D - S$. In fact, K_H is obtained from H_0 by collapsing each bipartite graph $B(x), x \in M$, to the single vertex x. Conversely, in this way we obtain *every* component of $G - D - S$. Therefore if H is a large component of $G^* - W$ then

$$|H| = \sum_{x \in K_H} \{d(x) + s(x)\} + e(K_H, S) \equiv \sum_{x \in K_H} f(x) + e(K_H, S) \pmod 2.$$

Consequently the number of large odd components of $G^* - W$ is equal to the number of components K_H of $G - D - S$ for which

$$\sum_{x \in K_H} f(x) + e(K_H, S) \equiv 1 \pmod 2.$$

By definition this number is

$$q(D, S).   (4)$$

Putting together (2), (3) and (4) we find that

$$q(G^* - W) = \sum_{x \in S} \{d(x) - d_{G-S}(x)\} + \sum_{x \in D} \{d(x) - f(x)\} + q(D, S).$$

Therefore by (1) we have

$$q(G^* - W) - |W| = q(D, S) - \sum_{x \in D} f(x) + \sum_{x \in S} \{f(x) - d_{G-D}(x)\}.   (5)$$

Armed with Lemma 3.1 and relation (5) we are ready to present the main result of the section. Define the *f-deficiency* def($D, S; f$) of an ordered pair (D, S) of disjoint subsets of $V(G)$ as

$$\text{def}(D, S; f) = q_f(D, S) + \sum_{x \in S} (f(x) - d_{G-D}(x)) - \sum_{x \in D} f(x).$$

The *f-deficiency* of G is the maximum *f*-deficiency of pairs; it is denoted by def($G; f$). Note that def($G; f$) is non-negative since def($\varnothing, \varnothing; f$) = $q_f(\varnothing, \varnothing) = 0$.

THEOREM 3.2. *A multigraph G has an f-factor if and only if* def($G; f$) = 0, *i.e.*

$$q(D, S) + \sum_{x \in S} \{f(x) - d_{G-D}(x)\} \leqslant \sum_{x \in D} f(x)   (6)$$

for all sets $D, S \subset V(G), D \cap S = \varnothing$.

Proof. Note that both the existence of an *f*-factor of G and inequality (6) with $D = \varnothing$, $S = \{x\}$, $x \in V$, imply that $f(x) \leqslant d(x)$. Therefore we may assume that $f(x) \leqslant d(x)$ for all $x \in V$. We assume also that G does not contain loops and $\delta(G) \geqslant 1$; we leave it to the reader to show that the result holds without these assumptions (Ex. 28). Furthermore, both an *f*-factor of G and inequality (6) with $D = S = \varnothing$ imply that $\sum_{x \in V} f(x) \equiv 0 \pmod{2}$. Let G^* be the graph constructed from G and f as described in the section preceding Lemma 3.1.

(a) Suppose first that (6) fails for some pair (D, S). Let D be a minimal set such that (6) fails for some $S \subset V - D$. If $f(x) = d(x)$ for some $x \in D$ then

(6) fails for the pair $(D - \{x\}, S \cup \{x\})$ as well, contradicting the minimality of D. Consequently for each $x \in D$ we have $s(x) > 0$, i.e. the bipartite graph $B(x)$ is connected. Define

$$W = \bigcup_{x \in D} D(x) \cup \bigcup_{y \in S} S(y).$$

Then $D(x) \subset W$ implies that $S(x) \neq \varnothing$ and $S(x) \cap W = \varnothing$, so the set W is a normal subset of $V^* = V(G^*)$. Consequently (5) and (6) give

$$|W| < q(G^* - W).$$

By Tutte's theorem on 1-factors (Theorem 1.4) the graph G^* does not have a 1-factor. Thus Lemma 3.1 implies that G does not have an f-factor.

(b) Suppose now that G does not contain an f-factor. Then, invoking Lemma 3.1, we see that for some $W \subset V^*$ the inequality

$$|W| < q(G^* - W) \tag{7}$$

must hold. Our aim is to show that there exists a *normal* set W satisfying (7) since then (5) implies that (6) fails for the pair (D, S) corresponding to W.

Let W be a *minimal* subset of V^* satisfying (7). We shall show that W is normal. Note that (7) implies

$$|W| \leqslant q(G^* - W) - 2 \tag{7'}$$

since

$$|G^*| = \sum_{x \in V} d(x) + \sum_{x \in V} (d(x) - f(x)) \equiv 0 \,(\text{mod } 2).$$

Inequality (7') shows that if $W' \subset W$, $W' \neq W$, then we must have

$$q(G^* - W') \leqslant q(G^* - W) - (|W| - |W'|) - 2 \leqslant q(G^* - W) - 3. \tag{8}$$

In order to show that W is normal, suppose first that $W \cap S(x) \neq \varnothing$ and $S(x)$ for some $x \in V$. Put $W' = W - S(x)$. Then at most one component of $G^* - W'$ is not equal to a component of $G^* - W$ so

$$q(G^* - W') \geqslant q(G^* - W) - 1,$$

contradicting (8).

Suppose now that $W \cap D(x) \neq \varnothing$, say $x_\alpha \in W \cap D(x)$, $\alpha = xy \in E(G)$. If $S(x) \subset W$ (in particular, if $W \cap S(x) \neq \varnothing$) then replacing W by $W' = W - \{x_\alpha\}$ at most one of the components of $G^* - W$ will be affected, again contradicting (8). Thus $S(x) \cap W = \varnothing$ and $S(x) \neq \varnothing$. Let us assume now that $D(x) \subset W$ holds as well. Then all the vertices in $S(x)$ belong to the same component C of $G - W$. When replacing W by $W' = W - \{x_\alpha\}$ only C and the component containing y_α are changed since $\Gamma(x_\alpha) = S(x) \cup \{y_\alpha\}$. Therefore

$$q(G^* - W') \geqslant q(G^* - W) - 2.$$

Since this contradicts (8) the set W is indeed normal. As we mentioned earlier, (7) and (5) imply that inequality (6) does not hold for the sets S, T belonging to this normal set W. ∎

An integer p is even or odd according as $p - 2\lfloor p/2 \rfloor$ is 0 or 1. Therefore the definition of $q(D, S)$ implies that

$$q(D, S) = \sum_{C \subset G - D - S} \{e(C, S) + \sum_{x \in C} f(x)\} - 2 \sum_{C \subset G - D - S} \lfloor \tfrac{1}{2}\{e(C, S) + \sum_{x \in C} f(x)\} \rfloor$$

$$= e(G - D - S, S) + \sum_{x \in G - D - S} f(x) - 2 \sum_{C \subset G - D - S} \lfloor \tfrac{1}{2}\{e(C, S) + \sum_{x \in C} f(x)\} \rfloor,$$

where the summations over C are over *all the components* of $G - D - S$. Note furthermore that

$$\sum_{x \in S} d_{G-D}(x) - e(G - D - S, S) = 2e(G[S]).$$

Consequently inequality (6) can be transformed into this form:

$$\sum_{x \in V - D} f(x) \leqslant \sum_{x \in D} f(x) + 2e(G|S|]) + 2 \sum_{C \subset G - D - S} \lfloor \tfrac{1}{2}\{e(C, S) + \sum_{x \in C} f(x)\} \rfloor.$$

Hence we arrive at the following form of Theorem 3.2.

THEOREM 3.2′. *A multigraph G has an f-factor if and only if for all sets $D, S \subset V(G)$, $D \cap S = \varnothing$, we have*

$$\sum_{x \in V - D} f(x) \leqslant \sum_{x \in D} f(x) + 2e(G[S]) + 2 \sum_{C \subset G - S - D} \lfloor \tfrac{1}{2}\{e(C, S) + \sum_{x \in C} f(x)\} \rfloor,$$

where the third summation is over all the components of $G - S - D$. ∎

Let G and f be as before, i.e. $f: V(G) \to \mathbb{Z}^+$. The graph G is said to be *f-soluble* is there exists a function $w: E(G) \to \mathbb{Z}^+$ such that

$$\sum_{y \in \Gamma(x)} w(xy) = f(x) \text{ for every } x \in V(G).$$

In order to formulate an elegant necessary and sufficient condition for *f*-solubility, due to Tutte [T11], let us introduce the following notations. For $D \subset V$ put

$$S(D) = \{x \in V - D : \Gamma(x) \subset S\}$$

and write $q_f(G - D)$ for the number of components C of $G - D$ of order at least 2 such that $\sum_{x \in C} f(x)$ is odd.

THEOREM 3.3. *A graph G is f-soluble if and only if*

$$q_f(G - D) + \sum_{x \in S(D)} f(x) \leqslant \sum_{x \in S} f(x)$$

for every $S \subset V$.

Proof. Add edges to G to obtain a multigraph G' such that $V(G) = V(G')$, $xy \in E(G')$ if and only if $xy \in E(G)$, and the multiplicity of an edge $xy \in E(G')$ is at least $d(x) + f(x)$. It is immediate that G is *f*-soluble if and only if G' has got an *f*-factor. We leave it to the reader to show (Ex. 28) that the result follows if we apply Theorem 3.2 to the multigraph G'.   ■

Given a graph G and a function $f: V(G) \to \mathbb{Z}^+$ with $0 \leqslant f(x) \leqslant d(x)$ for every $x \in G$, define $f': V(G) \to \mathbb{Z}^+$ by $f'(x) = d(x) - f(x)$. It is clear that G has an *f*-factor iff it has an *f'-factor*. Tutte [T19], [T21] showed that *f*-deficiencies and *f'*-deficiencies are connected by the following relation.

THEOREM 3.4. $\text{def}(D, S; f) = \text{def}(S, D; f')$.

Proof. If C is any component of $G - D - S$ then

$$r_f(D, S; C) + r_{f'}(S, D; C) = e(C, D \cup S) + \sum_{x \in C} d(x)$$

$$= 2 e(C, D \cup S) + 2e(C) \equiv 0 \qquad (\text{mod } 2).$$

Hence

$$q_f(D, S) = q_{f'}(S, D).$$

Furthermore,

$$\sum_{x \in S} \left(f(x) - d_{G-D}(x) + f'(x) \right) = \sum_{x \in S} e(x, D) = e(D, S)$$

$$= \sum_{x \in D} e(x, S) = \sum_{x \in D} \left(f(x) - d_{G-S} + f'(x) \right),$$

so $\operatorname{def}(D, S; f) = \operatorname{def}(S, D; f')$ as claimed. ∎

Neither form of the factor theorem (Theorems 3.2 and 3.2′) is easy to apply, in most cases it is rather tedious to check whether G and f satisfy the condition or not. In order to simplify the checking, it is advantageous to prove some properties of pairs with maximal f-deficiencies. Let (D, S) be an arbitrary pair of disjoint subsets of $V(G)$, let $x \in D$ and put $D_x = D - \{x\}$. We shall compare the f-deficiencies of (D, S) and (D_x, S).

Write $q(x)$ for the number of odd components of $G - D - S$ with respect to (D, S), that are joined to x by some edges. In $G - D_x - S$ all these components (and perhaps, some even components) are contained in a single component K; of course $x \in K$. The other odd components of $G - D - S$ remain odd components of $G - D_x - S$. If K is made up of x and the components $K_1, \ldots, K_l$ of $G - D - S$, then

$$r(D_x, S; K) = \sum_{i=1}^{l} e(K_i, S) + \sum_{i=1}^{l} \sum_{y \in K_i} f(y) + e(x, S) + f(x)$$

$$\equiv q(x) + e(x, S) + f(x) \pmod 2.$$

Define $\eta(x)$ as the number 0 or 1 according as the right-hand side above is even or odd. Then the description of the odd components of $G - D_x - S$ implies that

$$q(D, S) - q(D_x, S) = q(x) - \eta(x).$$

Hence

$$\operatorname{def}(D, S; f) - \operatorname{def}(D_x, S; f) = q(x) - \eta(x) - f(x) + e(x, S). \qquad (9)$$

Note that by the definition of $\eta(x)$ the right-hand side of (9) is even.

Now (9) implies that if (D, S) is a pair with maximal f-deficiency then the quantity on the right is non-negative; if (D_x, S) has maximal f-deficiency then the quantity is at least 0. If the expression on the right of (9) is 0 then x can be taken away from D or added to D_x without affecting the maximality. Thus if in the case $x \in G - D - S$ we define $q(x)$ and $\eta(x)$ for the pair $(D \cup \{x\}, S)$, then we obtain the following theorem, proved by Tutte [T21].

THEOREM 3.5. *Let (D, S) be a pair of maximal f-deficiency. If $x \in D$ then*

$$f(x) \leqslant q(x) - \eta(x) + e(x, S),$$

and if $x \in G - D - S$ then

$$f(x) \geqslant q(x) - \eta(x) + e(x, S).$$

If

$$f(x) = q(x) - \eta(x) + e(x, S)$$

then x can be taken away from D or added to D without affecting the f-deficiency of the pair. The number $f(x) - q(x) + \eta(x) - e(x, S)$ is even.   ∎

COROLLARY 3.6. *There is a pair (D, S) with maximal f-deficiency for which*

$$f(x) > 1 \quad \text{for every} \quad x \in S.$$

Proof. Let (D, S) be a pair of maximal f-deficiency in which S is minimal. Suppose $x \in S$ and $f(x) \leqslant 1$. Then (S, D) has maximal f'-deficiency and x can not be omitted from S without reducing the f'-deficiency. Consequently, by applying Theorem 3.4 to the pair (S, D) and function f', we see that

$$d(x) - 1 \leqslant f'(x) \leqslant q(x) - \eta(x) + e(x, S) - 2 \leqslant d(x) - 2.$$

This contradiction completes the proof.   ∎

We conclude the section with five applications of the foregoing results in which some of the details will be left to the reader. The first is a defect-form of the factor theorem from [BGW1].

Given a function $f: V(G) \to \mathbb{Z}^+$ and a subgraph $H \subset G$, we define the *f-deviation* of H as

$$\mathrm{dv}(f, H) = \sum_{x \in H} |f(x) - d_H(x)|.$$

If $f(x) \geqslant d_H(x)$ for every $x \in H$ then f *dominates* H and if $f(x) \leqslant d_H(x)$ for every $x \in H$ then H *dominates* f. It is easily seen that the minimum f-deviation of subgraphs of G can be attained on subgraphs dominated by f and if f is dominated by G then it can also be attained on subgraphs dominating f.

THEOREM 3.7. *The minimum f-deviation of a subgraph H of G is equal to the f-deficiency of G.*

Proof. We may assume without loss of generality that f dominates H. Apply induction on $\mathrm{def}(G; f)$. By Theorem 3.2 the result is true if either $\mathrm{def}(G; f) = 0$ or $\min_{H} \mathrm{dv}(H, f) = 0$, that is G has an f-factor; so we may assume that they are positive. Add a vertex x_0 to G and join it to every vertex of G. Write G^* for the resulting graph. Let f^* be the extension of f to $V(G^*)$ defined by $f^*(x_0) = 1$. Note there is a subgraph H of G majorized by f with $\mathrm{dv}(f, H) = d$ iff there is a subgraph H^* of G^* majorized by f^* with $\mathrm{dv}(f^*, H^*) = d - 1$. Therefore to complete the proof one has to show only that $\mathrm{def}(G^*: f^*) = d - 1$. We leave it to the reader to check this. ∎

COROLLARY 3.8. *If for every $D \subset V(G)$ we have*

$$\sum_{x \in V - D} \max\{f(x) - d_{G-D}(x), 0\} \leqslant \sum_{x \in D} f(x) + s,$$

then G contains a subgraph whose f-deviation is at most $s + n$.

Proof. Immediate from Theorem 3.7 and the definition of $\mathrm{def}(G; f)$. ∎

The next result was conjectured by Erdös and proved by Tutte [T21].

THEOREM 3.9. *Let G be a k-regular graph and let $0 \leqslant r < k$. Then G has a factor H in which every vertex has degree r or $r + 1$.*

Proof. It suffices to prove the theorem in the case when $|G|$ is even, since we may replace G by two copies of G. Suppose the theorem fails and let G be a counterexample of minimal even order $2m$. Take a new vertex x_0, join it to every vertex of G and attach $4m$ loops to x_0. Denote by G^* the resulting graph. Note that $d_{G^*}(x_0) = 10m$. Define $f: V(G^*) \rightarrow \mathbb{Z}^+$ by $f(x_0) = 4m$ and $f(y) = r + 1$ for every $y \in G$. It is clear that H exists iff G^* has an f-factor, so $\mathrm{def}(G^*; f) \leqslant 1$.

Let (D, S) have maximal f-deficiency. Note first that $x_0 \notin D$ since otherwise, by Theorem 3.5, we have

$$4m = f(x_0) \leqslant q(x_0) + e(x_0, S) - \eta(x_0) \leqslant 2m.$$

Similarly, by applying Theorem 3.5 to the pair (S, D) of maximal f'-deficiency, we see that $x_0 \notin S$. Consequently $G^* - D - S$ has only one component since x_0 is joined to every vertex of G^*. Hence $q(D, S) \leqslant 1$ and since $\mathrm{def}(D, S; f) \geqslant 1$, we have

$$0 \leqslant \mathrm{def}(D, S; f) - q(D, S) = \sum_{y \in S} \big(f(y) - d_{G^* - D}(y) \big) - \sum_{y \in D} f(y)$$

$$= (r + 1 - k)|S| + \sum_{y \in S} (k - d_{G^* - D}(y)) - (r + 1)|D|$$

$$= (r + 1 - k)|S| + e(D, S) - (r + 1)|D|,$$

that is

$$(r + 1)|D| + (k - r - 1)|S| \leqslant e(D, S). \tag{11}$$

Since $e(D, S) \leqslant k \min\{|D|, |S|\}$, inequality (11) implies that $|S| = |D|$ and $G[D \cup S]$ is a k-regular bipartite graph; it is also a component of G. As a k-regular bipartite graph has a 1-factorization, and so it has an r-regular factor for every r, $0 \leqslant r \leqslant k$, the choice of G implies that $D \cup S = \varnothing$. Now $\mathrm{def}(G^*; f) \leqslant 1$ implies that G^* itself is an odd component, so $\sum_{x \in G} f(x)$ is odd. However, this sum is $2m(r + 1) + 4m$. ∎

Of course, Theorem 3.2 can be applied directly to provide a criterion for the existence of a k-factor.

COROLLARY 3.11. *A multigraph G contains a k-factor if and only if*

$$q(D, S) + \sum_{x \in S} \{k - d_{G - D}(x)\} \leqslant k|D|$$

holds wherever $D, S \subset V(G), D \cap S = \varnothing$. ∎

Call a sequence $d_1, d_2, \ldots, d_n$ of non-negative integers *graphic* if it is a degree sequence of a graph G. Then the graph G is said to *realize* the sequence $(d_i)_1^n$. The following characterization of graphic sequences is due to Erdös and Gallai [EG2]. A slightly stronger variant of it was proved by Eggleton [E3].

COROLLARY 3.11. *A sequence $d_1 \geqslant d_2 \geqslant \ldots \geqslant d_n$ is graphic if and only if* $\sum_1^n d_i$ *is even and*

$$\sum_1^t d_i \leqslant t(t - 1) + \sum_{t+1}^n \min\{t, d_i\}$$

for every t, $1 \leqslant t \leqslant n$.

Proof. A sequence $(d_i)_1^n$ is graphic if and only if K^n contains an f-factor, where $V(K^n) = \{x_1, x_2, \ldots, x_n\}$ and $f(x_i) = d_i, 1 \leqslant i \leqslant n$. Apply Theorem 3.2 to $G = K^n$ and this function f. As $q(D, S) \leqslant 1$, inequality (6) is equivalent to the following:

$$\sum_{x \in S} f(x) \leqslant \sum_{x \in D} f(x) + (n - 1 - |D|)|S|.$$

This last inequality is most restrictive if $S = \{x_1, x_2, \ldots, x_s\}$ and

$$D = \{x_{n-d+1}, x_{n-d+2}, \ldots, x_n\}, \quad s + d \leqslant n,$$

in which case it becomes

$$\sum_1^s d_i \leqslant \sum_{n-d+1}^n d_i + (n - 1 - d)s.$$

The result follows if we note that for a fixed s the minimum of the right hand side under $0 \leqslant d \leqslant n - s$ is

$$\sum_{s+1}^n \min(d_i, s) + s(s - 1). \qquad \blacksquare$$

We must point out that the above proof of Corollary 3.11 uses unnecessarily heavy machinery. Corollary 3.11 can be proved directly or via the following assertion, which is implicit in [EG2] and can be found explicitly in Havel [H21] and Hakimi [H4]. The easy proof is left to the reader (Ex. 30).

THEOREM 3.12. *A sequence* $d_1 \geqslant d_2 \geqslant \ldots \geqslant d_n \geqslant 0$ *is graphic if and only if the sequence*

$$d_2 - 1, d_3 - 1, \ldots, d_{d_1+1} - 1, d_{d_1+2}, d_{d_1+3}, \ldots, d_n$$

is graphic. $\blacksquare$

4. MATCHING IN GRAPHS WITH RESTRICTIONS ON THE DEGREES

In this section we shall present bounds on $\beta_1(G)$, the maximal number of independent edges in a graph G, involving the minimal degree $\delta(G)$ and the maximal degree $\Delta(G)$. Some of the following parameters will also appear in our estimates: the order $|G|$, the size $e(G)$, the vertex connectivity $\kappa(G)$ and the edge connectivity $\lambda(G)$. Most of these inequalities are best possible and all of them are essentially best possible. We have already encountered a result of this nature: Corollary 1.10, which is an immediate consequence of Tutte's matching theorem, gives the maximal size of a graph of order n with at most β independent edges. Not surprisingly, the proofs of all the results of this section are also based on a version of Tutte's theorem, Berge's matching formula (Corollary 1.8). As we shall refer to it frequently, we restate it here in a form especially suitable for the applications that follow.

THEOREM 4.1. *Let G be a graph of order n with $\beta_1(G) = \beta$. Then there is a set S of, say s vertices such that there are $q = n + s - 2\beta$ odd components in $G - S$.* ∎

To start with we determine the greatest lower bound of the number of independent edges in a graph in terms of its order, minimal degree and maximal degree. Recall that G^n denotes a graph of order n; its two occurrences in the same context refer to the same graph G^n. We shall use G_i^n to denote a specific graph of order n. Let D be the set of all possible triples $(n, \delta(G^n), \Delta(G^n))$ with $\delta(G^n) \geqslant 1$. Since n is even if $\delta(G^n) = \Delta(G^n)$ is odd, it is easily seen that

$$D = \{(n, \delta, \Delta) \in \mathbb{N}^3 : \quad 1 \leqslant \delta \leqslant \Delta < n, n \text{ is even if } \delta = \Delta \text{ is odd}\}.$$

Define a function $m(n, \delta, \Delta)$ on D as follows;

$$m(n, \delta, \Delta) = \min\{\beta_1(G^n) : \quad \delta(G^n) = \delta, \Delta(G^n) = \Delta\}.$$

Some partial results about this function $m(n, \delta, \Delta)$ were obtained by Erdős and Pósa [EP1] ($\Delta = n - 1$) and by Weinstein [W14] ($\delta = 1$ and $\delta = 2$). Bollobás and Eldridge [BE1] determined the value of this function for every triple (n, δ, Δ).

It turns out to be natural to divide the domain D of $m(n, \delta, \Delta)$ into six parts. In order to make the proofs of the various lower bounds for $\beta_1(G)$ more systematic we shall define a function $m_0(n, \delta, \Delta)$ on D which we shall prove is equal to $m(n, \delta, \Delta)$. Put

$$
m_0(n, \delta, \Delta) =
\begin{cases}
\min\{\lfloor n/2 \rfloor, \delta\} & \text{if } \delta \leqslant \Delta - 2 \text{ and } n < \Delta + \delta, \\[2ex]
\left\lceil \dfrac{n\delta}{\delta + \Delta} \right\rceil & \text{if } \delta \leqslant \Delta - 2 \text{ and } n \geqslant \Delta + \delta, \\[2ex]
\left\lceil \dfrac{n\Delta}{2(\Delta + 1)} \right\rceil & \text{if } \delta = \Delta \text{ even or } \delta = \Delta - 1 \text{ odd}, \\[2ex]
\left\lceil \dfrac{n\delta + 1}{2(\delta + 1)} \right\rceil & \text{if } \delta = \Delta - 1 \text{ even}, \\[2ex]
\tfrac{1}{2}n & \text{if } \delta = \Delta \text{ odd and } n = \delta + 1.
\end{cases}
$$

Finally, when $\delta = \Delta$ odd and $n > \delta + 1$ there are integers u, k, r such that

$$ n = u(\delta + 1)^2 + (2k + 1)(\delta + 2) + r \quad \text{and} \quad 0 \leqslant 2k < \delta, 1 \leqslant r \leqslant 2\delta + 3 \quad (1) $$

and we define

$$ m_0(n, \delta, \Delta) = \tfrac{1}{2}\{n - u(\delta - 1)\} - k. \quad (2) $$

It is not hard to verify that (2) is well defined. For if $\delta = 1$ then $k = 0$ and if $\delta \geqslant 3$ then there are at most two choices for u in (1), say u_1 and $u_2 = u_1 + 1$, and the corresponding values of k are k_1 and $k_2 = k_1 - (\delta - 1)/2$.

THEOREM 4.2. $m(n, \delta, \Delta) = m_0(n, \delta, \Delta)$. *In words, every graph of order n with minimal degree δ and maximal degree Δ contains at least $m_0(n, \delta, \Delta)$ independent edges, and the result is best possible.*

Proof. In part (a) we prove that $m \geqslant m_0$ and in part (b) we construct various graphs G_i^m showing that $m \leqslant m_0$.

Part (a). Let G be a graph of order n and put $\delta = \delta(G)$, $\Delta = \Delta(G)$ and $\beta = \beta_1(G)$. Suppose $\delta \geqslant 1$. Denote by δ^* *the least even integer not less than δ.* Let S be a set of vertices whose existence is guaranteed by Theorem 4.1. Thus $G - S$ has $q = n + s - 2\beta$ odd components. Denote by a the number of these components with at most δ vertices. Clearly $q \leqslant n - s$ since $G - S$

has $n - s$ vertices, i.e.

$$s \leqslant \beta. \tag{3}$$

To ensure $\delta(G) = \delta$ each component must have at least $\delta + 1 - s$ vertices, so

$$q(\delta + 1 - s) \leqslant n - s. \tag{4}$$

As every component has at least one vertex and $q - a$ components have at least $\delta^* + 1$ vertices each, we have

$$a + (q - a)(\delta^* + 1) \leqslant n - s. \tag{5}$$

A component C with $|C| = 2k + 1 \leqslant \delta$ has at least $\delta - 2k$ edges joining each of its vertices to S. Thus at least $(2k + 1)(\delta - 2k) \geqslant \delta$ edges join C to S. On the other hand at most $s\Delta$ edges can leave S. Consequently

$$s\Delta \geqslant a\delta. \tag{6}$$

Inequalities (5) and (6) imply

$$s\Delta \geqslant a\delta \geqslant q(\delta^* - 1)\frac{\delta}{\delta^*} - n\frac{\delta}{\delta^*} + s\frac{\delta}{\delta^*},$$

i.e.

$$s\left(\Delta - \frac{\delta^* + 2}{\delta^*}\delta\right) \geqslant \delta\left(n - 2\beta\frac{\delta^* + 1}{\delta^*}\right). \tag{7}$$

Note that (4) is equivalent to

$$(n + s - 2\beta)(\delta + 1 - s) + s \leqslant n$$

and the minimum of the left-hand side is attained at one of the extremal values of s. Because of (3) these are $s = 0$ and $s = \beta$. Hence

$$\beta \geqslant \min\left\{\left\lfloor\frac{n}{2}\right\rfloor, \delta\right\}. \tag{8}$$

In what follows we make use of inequalities (3)–(8) to prove $m(n, \delta, \Delta) \geqslant m_0(n, \delta, \Delta)$. We examine separately the four cases $\delta \leqslant \Delta - 2$, $\delta \geqslant \Delta - 1$, $\delta = \Delta - 1$ even and $\delta = \Delta$ odd.

(i) Suppose $\delta \leqslant \Delta - 2$. Then the coefficient of s on the left-hand side of (7) is non-negative. Thus (3) implies

$$\beta\left(\Delta - \frac{\delta^* + 2}{\delta^*}\delta\right) \geqslant \delta\left(n - 2\beta\frac{\delta^* + 1}{\delta^*}\right)$$

and so

$$\beta \geqslant \left\lfloor\frac{n\delta}{\delta + \Delta}\right\rfloor. \tag{9}$$

(ii) Suppose $\delta \geqslant \Delta - 1$. Then the coefficient of s on the left-hand side of (7) is at most 0. Thus (7) implies

$$\beta \geqslant \left\lfloor\frac{n\delta^*}{2(\delta^* + 1)}\right\rfloor. \tag{10}$$

(iii) Suppose $\delta = \Delta - 1$ is even. We shall improve (10) slightly. Suppose

$$\beta = \frac{n\delta^*}{2(\delta^* + 1)} = \frac{n\delta}{2(\delta + 1)}.$$

Then n is a multiple of $\delta + 1$ and there is equality in all the inequalities from which (10) was derived. So $s = 0$ from (7) and $a = 0$ from (6). From (5) we deduce that $G = G - S$ consists of $q - a = q$ components of exactly $\delta + 1$ vertices each. But then $\Delta(G) \leqslant \delta < \Delta$. This contradiction shows that

$$\beta \geqslant \left\lfloor\frac{n\delta + 1}{2(\delta^* + 1)}\right\rfloor. \tag{11}$$

(iv) Suppose $\delta = \Delta$ is odd. If $n = \delta + 1$ then $G = K^n$ so $\beta = \frac{1}{2}n$. Let now $n > \delta + 1$ and choose u, k and r satisfying (1). Note that in this case $m(n, \delta, \Delta) = m(n, \delta, \delta)$ is given by (2). If $\delta = 1$ and $n > 2$ then G is a 2-factor so $\beta = \frac{1}{2}n = m_0$, as predicted by (2).

Thus we may also assume that $\delta \geqslant 3$ and $n \geqslant \delta + 3$. Since $\delta = \Delta$ is odd, each component of G must have even order. Therefore each odd component of $G - S$ is joined to S by an edge. Hence (6) can be improved to

$$s\delta \geqslant a\delta + q - a. \tag{12}$$

Let us assume now that $\beta \leqslant m_0(n, \delta, \Delta) - 1 = \frac{1}{2}\{n - u(\delta - 1)\} - k - 1$.

Then

$$q \geqslant u(\delta - 1) + 2(k + 1) + s. \tag{13}$$

With inequality (12) this yields

$$a \leqslant \frac{s\delta - q}{\delta - 1} \leqslant \{s\delta - u(\delta - 1) - 2(k + 1) - s\}/(\delta - 1) = s - u - \frac{2(k + 1)}{\delta - 1}$$

Since a is a non-negative integer,

$$0 \leqslant a \leqslant s - u - 1. \tag{14}$$

Taking into account the fact that $\delta^* = \delta + 1$, (5) and (13) give

$$(\delta + 1)a \geqslant s - n + q(\delta + 2) \geqslant (s - u)(\delta + 3) + \delta + 2 - r. \tag{15}$$

Inequalities (14) and (15) imply

$$r \geqslant 2(\delta + 2) + 1 + 2(s - u - 1) \geqslant 2(\delta + 2) + 1.$$

This contradicts the choice of r in (1) so

$$\beta \geqslant \tfrac{1}{2}\{n - u(\delta - 1)\} - k. \tag{16}$$

Inequalities (9), (10), (11) and (16) together with the trivial cases $\delta = \Delta = 1$ and $\delta = \Delta = n - 1$ is odd, imply exactly that

$$\beta \geqslant m_0(n, \delta, \Delta) \quad \text{so} \quad m(n, \delta, \Delta) \geqslant m_0(n, \delta, \Delta).$$

Part (b).

Case 1. $\delta \leqslant \Delta - 2$ and $n < \delta + \Delta$.

Since one always has $m(n, \delta, \Delta) \leqslant n/2$ there is nothing to prove unless $\delta \leqslant n/2$ so we shall assume that $\delta \leqslant n/2$. To construct a graph G_1^n divide the n vertices into three disjoint sets A, B and C containing $\Delta + \delta - n$, $n - \delta$ and $n - \Delta$ vertices respectively. Let c_0 be a fixed vertex in C. Join each vertex of B to each vertex of $A \cup C$ and join c_0 to each vertex of A. Then

$$\delta(G_1^n) = \delta, \qquad \Delta(G_1^n) = \Delta$$

and

$$m(n, \delta, \Delta) \leqslant \beta_1(G_1^n) \leqslant \delta = m_0(n, \delta, \Delta).$$

We note for later use that G_1^n is δ-connected.

Case 2. $\delta \leqslant \Delta - 2$ and $n \geqslant \delta + \Delta$.

Put $s = \lfloor n\delta/(\delta + \Delta) \rfloor$. We shall again require that G_2^n be δ-connected when $\delta \geqslant 2$. One can easily check that when $\delta \geqslant 2$ the graphs which we use in the construction of G_2^n are all δ-connected. Put $s = \lceil n\delta/(\delta + \Delta) \rceil$. Let H_s be the bipartite graph with vertex classes $X = \{x_i : 1 \leqslant i \leqslant s\}$ and $\{y_i : 1 \leqslant i \leqslant s\}$ in which y_i is joined to $x_i, \ldots, x_{i+\delta-1}$ where the subscripts are taken modulo s. Note that since $\delta \leqslant s$, this construction is indeed possible. Also H_s is δ-regular and if $\delta \geqslant 2$ then it is δ-connected. Add a new vertex y_{s+1} and join it to δ vertices in X. In the new graph, H_{s+1}, the degrees of two vertices of X differ by at most 1. Next add a vertex y_{s+2} to H_{s+1} and join it to δ vertices in X in such a way that in the new graph, H_{s+2}, the degrees of two vertices of X again differ by at most 1. Construct $H_{s+3}, \ldots, H_{n-s}$ similarly. By construction the vertices y_i in H_{n-s} have degree δ and the other vertices have degrees between δ and Δ. Since, clearly, $n - s \geqslant \Delta$, we may, if necessary, ensure that the maximal degree is Δ simply by joining x_1 to some more vertices y_i. Then $\delta \leqslant \deg y_i \leqslant \delta + 1$, for every $i, 2 \leqslant i \leqslant n - s$, and $\deg y_1 = \delta$. Thus the resulting bipartite graph, which we denote by G_2^n, satisfies $\delta(G_2^n) = \delta$ and $\Delta(G_2^n) = \Delta$ and so

$$m(n, \delta, \Delta) \leqslant \beta_1(G_2^n) = s = \left\lceil \frac{n\delta}{\delta + \Delta} \right\rceil.$$

Furthermore, if $\delta \geqslant 2$, G_2^n is δ-connected as promised.

Remark. It is necessary that $\delta \geqslant 2$ to ensure that we can make G_2^n δ-connected (cf. Theorem 4.4(i)).

Case 3. $\delta = \Delta$ even.

Let $n = b(\delta + 1) + c$ where $\delta + 1 \leqslant c < 2(\delta + 1)$. Let G_3^n be the disjoint union of a δ-regular graph with c vertices and b copies of $K^{\delta+1}$. Then $\delta(G_3^n) = \delta$, $\Delta(G_3^n) = \Delta$ and so

$$m(n, \delta, \Delta) \leqslant \beta_1(G_3^n) \leqslant \tfrac{1}{2}b\delta + \lceil \tfrac{1}{2}c \rceil = \left\lceil \frac{n\delta}{2(\delta + 1)} \right\rceil.$$

Case 4. $\delta = \Delta - 1$ odd.

Replace δ by $\delta + 1 = \Delta$ and take the corresponding graph G_3^n. Omit an edge to obtain G_4^n. Then

$$\delta(G_4^n) = \delta, \qquad \Delta(G_4^n) = \Delta$$

and so

$$m(n, \delta, \Delta) \leqslant \beta_1(G_4^n) \leqslant \left\lceil \frac{n\Delta}{2(\Delta + 1)} \right\rceil.$$

Case 5. $\delta = \Delta - 1$ even.

Let $n = b(\delta + 1) + c$ where $\delta + 1 < c \leqslant 2(\delta + 1)$. Add an edge to a δ-regular graph with c vertices to obtain a graph H_5. Let G_5^n be the disjoint union of H_5 and b copies of $K^{\delta+1}$. Then $\delta(G_5^n) = \delta, \Delta(G_5^n) = \Delta$ and so

$$m(n, \delta, \Delta) \leqslant \beta_1(G^n) \leqslant \tfrac{1}{2}b\delta + \left[\tfrac{1}{2}c\right] = \left\lceil \frac{n\delta + 1}{2(\delta + 1)} \right\rceil.$$

Case 6. $\delta = \Delta$ odd.

Let n be given by (1). We shall construct G_6^n as the disjoint union of certain other graphs.

Let p and d be odd natural numbers with $p > \delta$ and $p > d$. Denote by $\mathbb{Z}_p$ the integers modulo p. Construct a graph $L_{p,d}$ with vertex set $\mathbb{Z}_p$ as follows. Join $x, y \in \mathbb{Z}_p$ iff $x - y = \pm 1, \ldots, \pm\tfrac{1}{2}(\delta - 1)$. Then join x to $\tfrac{1}{2}(p + 2x - 1)$ for $x = 1, 2, \ldots, \tfrac{1}{2}(p - d)$. Then $L_{p,d}$ has $p - d$ vertices of degree δ and d vertices of degree $\delta - 1$.

Denote by M the graph obtained from the disjoint union of δ copies of $L_{\delta+2,1}$ by adding one vertex and joining it to the vertices of degree $\delta - 1$. Since each $L_{\delta+2,1}$ has exactly one such vertex, M is δ-regular. M has $(\delta + 1)^2$ vertices and $\beta_1(M) = \tfrac{1}{2}\{(\delta + 1)^2 - (\delta - 1)\}$.

Similarly we construct a graph N as follows. Take the disjoint union of $L_{\delta+1+r, \delta-2k}$ and $2k$ copies of $L_{\delta+2,1}$. Again add one vertex and join it to the vertices of degree $\delta - 1$. Clearly N is a δ-regular graph with $(2k + 1)(\delta + 2) + r$ vertices and $\beta_1(N) + \tfrac{1}{2}\{(2k + 1)(\delta + 2) + r\} - k$.

Finally denote by G_6^n the disjoint union of N and u copies of M. Clearly G_6^n is δ-regular so

$$m(n, \delta, \Delta) \leqslant \beta_1(G_6^n) = \tfrac{1}{2}\{n - u(\delta - 1)\} - k. \qquad \blacksquare$$

Inequality (8) was first proved by Erdös and Pósa [EP1]. We restate it here since in some applications it can be used instead of Theorem 4.2.

COROLLARY 4.3. *If* $|G| \geqslant 2\delta(G)$ *then* $\beta_1(G) \geqslant \delta(G)$.   ∎

Most of the graphs constructed in the proof of Theorem 5.2 in order to show that $m(n, \delta, \Delta) \leqslant m_0(n, \delta, \Delta)$, are *disconnected*. Thus it is natural to expect that a graph G^n with $\delta(G^n) = \delta, \Delta(G^n) = \Delta$ and vertex (or edge) connectivity at least $\kappa \geqslant 1$ (or $\lambda \geqslant 1$) must have more independent edges than $m(n, \delta, \Delta)$. Bollobás and Eldridge [BE1] gave reasonably good lower bounds on the number of independent edges showing how far this is true.

Let $(n, \delta, \Delta) \in D$ and $1 \leqslant \lambda \leqslant \delta$. Define

$$m_e(n, \delta, \Delta, \lambda) = \min\{\beta_1(G^n)\colon \delta(G^n) = \delta,\; \Delta(G^n) = \Delta,\; \lambda(G^n) \geqslant \lambda\}.$$

Define $m_v(n, \delta, \Delta, \lambda)$ similarly, replacing $\lambda(G^n)$ by $\kappa(G^n)$. (Thus m_e refers to edge connectivity and m_v to vertex connectivity.) Clearly

$$m(n, \delta, \Delta) \leqslant m_e(n, \delta, \Delta, \lambda) \leqslant m_v(n, \delta, \Delta, \lambda).$$

Furthermore, if $\delta(G) = \Delta(G) = 1$ then $G = K^2$ so to avoid this trivial case we assume that if $\delta = 1$ then $\Delta \geqslant 2$.

In the proof of Theorem 4.2 we emphasized that if $\delta \leqslant \Delta - 2$ and $\delta \geqslant 2$ then the graphs G_1^n and G_2^n are δ-connected. Thus Theorem 4.2 has the following consequence.

THEOREM 4.4. *If* $\delta \leqslant \Delta - 2$ *and* $\delta \geqslant 2$ *then*

$$m_v(n, \delta, \Delta, \lambda) = m_e(n, \delta, \Delta, \lambda) = m(n, \delta, \Delta).$$ ∎

The estimation of $m_v(n, \delta, \Delta, \lambda)$ and $m_e(n, \delta, \Delta, \lambda)$ requires considerably more work than the proof of Theorem 4.2. Therefore we only state the following theorem of Bollobás and Eldridge [BE1] and for the tedious proof we refer the reader to the original paper. Recall that δ^* is the least even integer not less than δ. Let λ' be the least integer not less than λ which has the same parity as δ.

THEOREM 4.5. (i) *If* $\Delta \geqslant 2$ *then*

$$m_v(n, 1, \Delta, 1) = m_e(n, 1, \Delta, 1) = \left\lceil \frac{n-1}{\Delta} \right\rceil.$$

(ii) *Let* $2 \leqslant \delta \leqslant \Delta \leqslant \delta + 1$ *and let* δ *be odd if* $\delta = \Delta$. *Put.*

$$m_e(\delta, \Delta) = \frac{1}{2} \frac{\delta^*(\Delta\delta - \Delta - \delta) + 2(\delta - 1)}{\delta^*(\Delta\delta - \Delta - \delta) + \Delta(\delta - 1)}.$$

Then

$$\lceil m_e(\delta, \Delta)n - \tfrac{1}{2}\rceil \leqslant m_v(n, \delta, \Delta, 1) = m_e(n, \delta, \Delta, 1) \leqslant m_c(\delta, \Delta)n + \tfrac{5}{2}.$$

(iii) *If* $\delta = \Delta - 1, x > 1$ *and* $\delta \leqslant 3\lambda$ *then*

$$m_e(n, \delta, \Delta, \lambda) = \min\left\{ \left\lfloor \frac{n}{2} \right\rfloor, \ \left\lfloor \frac{n\delta}{2\delta + 1} \right\rfloor \right\} = m_c(n, \delta, \Delta, \lambda).$$

(iv) *If* $\delta = \Delta - 1, \lambda > 1$ *and* $\delta > 2\lambda$ *then*

$$m_e(n, \delta, \Delta, \lambda) \geqslant \min\left\{ \left\lfloor \frac{n}{2} \right\rfloor, \ \frac{n}{2}\frac{\delta^*(\delta + 1) + 2\lambda}{(\delta^* + 1)(\delta + 1) + \lambda} \right\}$$

$$\geqslant m_v(n, \delta, \Delta, \lambda) - \frac{\delta + 2 - \lambda}{2}.$$

(v) *If* $\delta = \Delta$ *and if* $\lambda > 1$ *when* δ *is odd then*

$$m_e(n, \delta, \Delta, \lambda) \geqslant \min\left\{ \left\lfloor \frac{n}{2} \right\rfloor, \ \frac{n}{2}\frac{\delta\delta^* + 2\lambda'}{\delta(\delta^* + 1) + \lambda'} \right\}$$

$$\geqslant m_v(n, \delta, \Delta, \lambda) - \frac{\delta + 1 - \lambda'}{2}. \qquad \blacksquare$$

Note that (v) implies that a k-connected k-regular graph of order n has $\lfloor n/2 \rfloor$ independent edges and if $k \geqslant 3$ and k is odd then a $(k - 1)$-edge-connected k-regular graph has a 1-factor. In particular the case $k = 3$ is Petersen's theorem [P5].

Let us turn now to the other set of problems to be discussed in this section. Let $f(\beta, \Delta)$ be the maximal size of a graph G with $\Delta(G) \leqslant \Delta$ and $\beta_2(G) \leqslant \beta$. The determination of $f(\beta, \Delta)$ was proposed by Erdös and Rado [ER2]. They proved in particular that $f(k, k) \leqslant 2k^2 - k$. (The hypergraph version of this problem concerns the so called delta systems and has been studied extensively. However, this is outside the scope of this book.) Following improvements by Abbott and Hanson, Sauer [S9] determined $f(\beta, \Delta)$ for $\Delta \leqslant 2\beta$, provided Δ is odd or it divides 2β. The determination of $f(\beta, \Delta)$ was completed by Chvátal and Hanson [CH4]. In fact they determined all the values of the function

$$f(n, \beta, \Delta) = \max\{e(G): \ |G| = n, \Delta(G) \leqslant \Delta, \beta_1(G) \leqslant \beta\}.$$

We only state this result of Chvátal and Hanson. The nature of the proof is, of course, fairly similar to the proof of Theorem 4.2. Since for $n \leqslant 2\beta$ clearly $f(n, \Delta, \beta) = \lfloor n\Delta/2 \rfloor$ it suffices to determine f for $n \geqslant 2\beta + 1$.

THEOREM 4.6. *Let* n, β, Δ *be natural numbers,* $n \geqslant 2\beta + 1$. *Put* $\Delta_0 = \lfloor \Delta/2 \rfloor$

(i)   *If* $\Delta \leqslant 2\beta$ *and* $n \leqslant 2\beta + \lfloor b/\Delta_0 \rfloor$ *then*

$$f(n, \beta, \Delta) = \begin{cases} \min\left\{\lfloor n\Delta/2 \rfloor, \beta\Delta + \left\lfloor \dfrac{2(n-\beta)}{\Delta+3} \right\rfloor \dfrac{\Delta-1}{2}\right\} & \text{if } \Delta \text{ is odd} \\[4mm] n\Delta/2 \text{ if } \Delta \text{ is even.} \end{cases}$$

(ii)   *If* $\Delta \leqslant 2\beta$ *and* $n \geqslant 2\beta + \lfloor b/\Delta_0 \rfloor$ *then*

$$f(n, \beta, \Delta) = \beta\Delta + \lfloor \beta/\Delta_0 \rfloor \lfloor \Delta/2 \rfloor.$$

(iii)   *If* $\Delta \geqslant 2\beta + 1$ *then*

$$f(n, \beta, \Delta) = \begin{cases} \max\left\{\binom{2\beta+1}{2}, \left\lfloor \dfrac{\beta(n+\Delta-\beta)}{2} \right\rfloor\right\} & \text{if } n \leqslant \beta + \Delta, \\[4mm] \beta\Delta \text{ if } n \geqslant \beta + \Delta. \end{cases}$$ ∎

Since $f(\beta, \Delta) = \max_n f(n, \beta, \Delta)$, the results of Sauer [S9] can clearly be recovered from this theorem. To conclude this section we show how $f(\beta, \Delta)$ can be determined in the case $\Delta > 2\beta$, since the proof is virtually trivial (see [B30]). In fact this furnishes a proof of the second case of Theorem 4.6(iii).

If $\Delta > 2\beta$ and $n \geqslant \beta + \Delta$ then it is easy to construct a graph G of order n with $\Delta(G) = \Delta$ that contains β vertices of degree Δ such that every edge is incident with exactly one of these vertices. Thus $f(n, \beta, \Delta) \geqslant \beta\Delta$. To prove the converse inequality let G be a graph of order n such that $\beta_1(G) = \beta$ and $\Delta(G) \leqslant \Delta$.

Let $S, |S| = s$, be the set whose existence is guaranteed by Theorem 4.1. As $G - S$ has $q = n + s - 2\beta$ odd components, $s \leqslant \beta$. It is clear that $G - S$ cannot have more edges than the graph consisting of $q - 1$ isolated points and whose qth (and last) component is a complete subgraph. Thus $G - S$ has at most $\binom{2\beta - 2s + 1}{2} = (\beta - s)(2\beta - 2s + 1)$ edges. As $\Delta(G) \leqslant \Delta$, G has at most

$$s\Delta + (\beta - s)(2\beta - 2s + 1) = s(2s + \Delta - 4\beta - 1) + 2\beta^2 + \beta$$

edges. In the range $1 \leqslant s \leqslant \beta$ the maximum of this function, which is exactly $\Delta\beta$, is attained at the unique point $s = \beta$. ∎

5. COVERINGS

Let $f: V(G) \to \mathbb{Z}^+$ be a function satisfying $0 \leqslant f(x) \leqslant d(x)$ for every $x \in V(G)$. Let $F \subset G$ be a graph of maximal size satisfying $d_F(x) \leqslant f(x)$, $x \in V(G)$, and let $H \subset G$ be a graph of minimal size satisfying $f(x) \leqslant d_H(x)$, $x \in V(G)$. As a consequence of the defect form of the f-factor theorem, we have seen (Corollary 3.8) that

$$e(F) + e(H) = \sum_{x \in G} f(x).$$

In the particularly simple case when $f \equiv 1$ we have $\alpha_1(G) = e(H)$ and $\beta_1(G) = e(F)$ so $\alpha_1(G) + \beta_1(G) = n$. Gallai [G2] was the first to notice this relation, together with a similar relation connecting the vertex covering number $\alpha_0(G)$ and the vertex independence number $\beta_0(G)$. These relations have very simple proofs that do not rely on the powerful f-factor theorem.

THEOREM 5.1. *If G is a graph of order n without isolated vertices then*

$$\alpha_0(G) + \beta_0(G) = n \quad and \quad \alpha_1(G) + \beta_1(G) = n.$$

Proof. (a) V_0 is a set of independent vertices if and only if $V - V_0$ is a set of vertices covering the edges. Therefore $\alpha_0 = \min|V - V_0| = n - \max|V_0| = n - \beta_0$.

(b) Let E_1 be a set of $\beta_1(G)$ independent edges. These β_1 edges cover $2\beta_1$ vertices. As G does not have isolated vertices, the remaining $n - 2\beta_1$ vertices can be covered with at most $n - 2\beta_1$ edges. Therefore $\alpha_1 \leqslant \beta_1 + n - 2\beta_1 = n - \beta_1$.

Suppose now that E_1^* is a set of α_1 edges covering the vertices. Then E_1^* cannot contain an edge whose endvertices are incident with other edges in E_1^*. Consequently E_1^* is the edge set of, say k, independent stars in G. A star with l edges covers $l + 1$ vertices so $|E_1^*| + k = \alpha_1 + k = n$. Selecting one edge from each of these stars we obtain k independent edges so $\beta_1 \geqslant k = n - \alpha_1$. ∎

König [K9] showed that if G is bipartite then $\alpha_0(G) = \beta_1(G)$. This can be read out of Theorem 1.2.2; in fact it is trivially equivalent to Hall's theorem (I.2.3).

Hedetniemi [H33] pointed out that the first equality in Theorem 5.1 can be generalized to a wide variety of graph invariants defined in terms of so called *monotone* properties. Call a property P of finite graphs *monotone* if whenever a graph has P so does every subgraph of it. Given a graph G, a set $S \subset V(G)$ is said to be a P-set of G if $G[S]$ has property P. Denote by $\beta_0(G; P)$ the *maximal* number of vertices in a P-set of G. Call $\bar{S} \subset V(G)$ a $\bar{P}$-set of G if whenever $G' \subset G$ and G' does not have property P then $V(G') \cap \bar{S} \neq \varnothing$. Denote by $\alpha_0(G; P)$ the *minimal* number of vertices in a $\bar{P}$-set of G.

THEOREM 5.2. *If P is a monotone property of graphs then for every graph G of order n we have*

$$\alpha_0(G; P) + \beta_0(G; P) = n.$$

Proof. Note that X is a P-set iff $\bar{X}$ is a $\bar{P}$-set. ∎

A good number of properties of graphs are monotone, e.g. P_1: "G does not contain K^p" (p fixed); P_2: "G does not contain a subgraph G_0" (G_0 fixed); P_3: "G does not contain a subgraph homeomorphic to G_0" (G_0 fixed); P_4: "$\chi(G) \leqslant k$" (k fixed), etc. It is not surprising that most of these examples are of the form "G does not contain a subgraph isomorphic to one of the graphs $\{F_1, F_2, \ldots\}$" since by definition *every* monotone property can easily be formulated in this way. Simply take $\{G : G$ does not have $P\}$ as the forbidden family of graphs. In Chapter VIII we return to the study of monotone properties and then we will prove nontrivial results about them.

Our main aim in this section is to investigate rather different kind of covering problems. In the sequel we say that a graph G is *covered* by a set $\{G_1, G_2, \ldots, G_k\}$ of subgraphs if $E(G) = \bigcup_1^k E(G_i)$. We also require that these subgraphs be edge disjoint but we shall emphasize this on most occasions. We are mainly interested in two essentially different problems: covering by paths and cycles and covering by forests.

Gallai conjectured that every connected graph of order n can be covered with $\lceil n/2 \rceil$ paths. This conjecture is still open, but the following weaker form of it was proved by Lovász [L15].

THEOREM 5.3. *A graph of order n can be covered by $\lfloor n/2 \rfloor$ edge disjoint paths and cycles.*

Proof. For brevity we shall call a set Σ of edge disjoint paths and cycles covering the edges of a graph G a *covering set* of G or a *covering k-set* where $k = |\Sigma|$. Denote by $w(G)$ the number of *nonisolated* vertices of G. We will prove that for every graph G there is a covering set of at most $\lfloor w(G)/2 \rfloor$ elements. This is trivially equivalent to the theorem.

We prove this assertion by induction on $z(G) = 2e(G) - w(G)$. If $z(G) \leqslant 0$ there is nothing to prove, for then $z(G) = 0$ is the union of a 1-factor and isolated vertices. Suppose now that $z(G) > 0$ and the result holds for all graphs H with $z(H) < z(G)$.

If there is a vertex $y \in G$ of even positive degree, let $G_1 = G$ and $x \in \Gamma(y)$. If there is no such vertex, choose an edge $xz \in E(G)$ and let G_1 be the graph obtained from G by subdividing xz by a vertex y. It is easily checked that if G_1 has a covering $\lfloor w(G_1)/2 \rfloor$-set then G has a covering $\lfloor w(G)/2 \rfloor$-set. Furthermore, $z(G_1) \leqslant z(G)$. Therefore, replacing G by G_1, if necessary, we may assume that there is an edge $xy \in E(G)$ such that $d(y)$ is even.

Let $G^* = G - \{xz \in E(G): d(z) \text{ is even}\}$. Then $w(G^*) \geqslant w(G) - 1$ and $e(G^*) \leqslant e(G) - 1$ so $z(G^*) \leqslant z(G) - 1$. Thus by the induction hypothesis G^* has a covering $\lfloor w(G^*)/2 \rfloor$-set, say Σ^*. Put

$$U = \Gamma_G(x) = \{z \in V(G): xz \in E(G)\},$$

$$W = \{c \in U: d(z) \text{ is even}\} = \{w_1, w_2, \ldots, w_t\}.$$

We know that $\varnothing \neq W \subset U$, so $t \geqslant 1$, and the vertices in U have odd degrees in G^*. Consequently for each $u \in U$ there is at least one *path* $P(u) \in \Sigma$ *starting* with u. For each i, $1 \leqslant i \leqslant t$, define a sequence $u_{i,0}$, $u_{i,1}$, $u_{i,2}$, ... of vertices of U as follows. Put $u_{i,0} = w_i$. Having defined $u_{i,j}$, if $x \notin P(u_{i,j})$, end the sequence with $u_{i,j}$, otherwise let $u_{i,j+1}$ be the vertex preceding x on $P(u_{i,j})$ when starting from $u_{i,j}$.

Clearly $u_{i,j}x \in E(G^*)$ if and only if $j \geqslant 1$. Furthermore, if $u_{i,j} = u_{k,l}$ then we must have $i = k$, $j = l$. This implies that each sequence $u_{i,0}$, $u_{i,1}, \cdots$ terminates, say in u_{i,r_i}. Put $W' = \{u_{ij} : 1 \leqslant i \leqslant t, 0 \leqslant j \leqslant r_i\}$.

To prove the theorem, for each path or cycle $Q \in \Sigma^*$ we shall define a path or cycle $\phi(Q)$ in G in such a way that $\Sigma = \{\phi(Q) : Q \in \Sigma^*\}$ is a covering set of G. If $Q \in \Sigma^*$ and $Q \notin \{P(u) : u \in W\}$, put $\phi(Q) = Q$. Suppose now that $Q \in \{P(u) : u \in W'\}$, say $Q = P(u_{i,j})$. Denote the other endvertex of Q by $v_{i,j}$. In order to define $\phi(Q)$ we distinguish four cases (see Fig. 5.1).

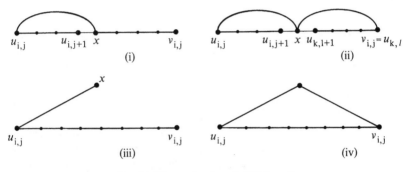

FIG. 5.1. The construction of $\phi(Q)$ from Q.

(i) If $x \in Q$ and $v_{i,j} \notin W'$ put $\phi(Q) = u_{i,j+1}Qu_{i,j}xQv_{i,j}$.
(ii) If $x \in Q$ and $v_{i,j} \in W'$, say $v_{i,j} = u_{k,l}$, put $\phi(Q) = u_{i,j+2}Qu_{i,j}xu_{k,l}Qu_{k,l+1}$.
(iii) If $x \notin Q$ and $v_{i,j} \notin W'$ put $\phi(Q) = xu_{i,j}Pv_{i,j}$.
(iv) Finally, if $x \notin Q$ and $v_{i,j} \in W'$, let $\phi(Q)$ be the cycle $u_{i,j}Qv_{i,j}xu_{i,j}$.

It is easily checked that Σ is a covering set of G, i.e. it consists of paths and cycles of G and each edge of G belongs to exactly one member of Σ. ∎

COROLLARY 5.4. *A graph of order n with at most one vertex of even degree can be covered by $\lfloor n/2 \rfloor$ edge disjoint paths.*

Proof. If Σ is a covering set guaranteed by Theorem 5.3, then each vertex of odd degree is the endvertex of a path in Σ, so Σ consists of paths. ∎

COROLLARY 5.5. *A graph G with $n_0 \geqslant 1$ vertices of even degree and n_1 vertices of odd degree can be covered by $n_0 + (n_1/2) - 1$ edge disjoint paths. These paths can be chosen in such a way that each vertex of even degree, with the exception of exactly one, is the endvertex of exactly two paths.*

Proof. By the previous corollary we may assume that $n_0 \geqslant 2$. Let G^* be obtained from G by adding $n_0 - 1$ new vertices to G and joining each new vertex to exactly one vertex of even degree, different new vertices to different vertices of G. By Corollary 5.4, G^* can be covered with $\lfloor \frac{1}{2}(n_0 + n_1 + n_0 - 1) \rfloor = n_0 + (n_1/2) - 1$ edge disjoint paths, since n_1 is even. The parts of these paths belonging to G form a required covering.   ∎

COROLLARY 5.6. *K^{2k} can be covered by k edge disjoint Hamiltonian paths and K^{2k+1} by k Hamiltonian cycles.*

Proof. The first assertion follows from Corollary 5.4. To see the second, take a covering of K^{2k+1} by k edge disjoint paths and cycles. A path contains at most $2k$ edges and a cycle at most $2k + 1$, with equality if and only if the cycle is Hamiltonian. As $e(K^{2k+1}) = k(2k + 1)$, each member of the covering must be a Hamiltonian cycle.   ∎

As we shall see, the problem of covering a graph with forests is closely connected to the factorization of a graph into a certain number of connected subgraphs. We start with the second problem. Clearly a graph G can be factored into k connected subgraphs, i.e. G is the union of k edge disjoint connected spanning subgraphs, if and only if G contains k edge disjoint spanning trees. It is more convenient and natural to discuss this problem for *multigraphs*.

Let $G = (V, E)$ be a multigraph and let $\mathcal{P}$ be a partition of V into disjoint non-empty sets $V_1, V_2, \ldots, V_p$. Put $|\mathcal{P}| = p$ and denote by $H = G|\mathcal{P}$ the multigraph of order p whose vertices are $V_1, V_2, \ldots, V_p$ and the number of edges joining vertex V_i to vertex $V_j \, (1 \leqslant i < j \leqslant p)$ is equal to the number of $V_i - V_j$ edges of G. If G contains k edge disjoint spanning trees then clearly

$$e(G|\mathcal{P}) \geqslant k(p - 1) = k(|\mathcal{P}| - 1).$$

Tutte [T14] and Nash-Williams [N1] proved that this obvious necessary condition is also sufficient. The original proofs were rather complicated but the result is an easy consequence of a theorem of Edmonds [E2] on matroids (see also [GM2]). Since we do not introduce matroids, the proof will not be given here. (For a proof in the language of graph theory see [B38].)

THEOREM 5.7. *Let G be a multigraph such that* $e(G|\mathscr{P}) \geqslant k(|\mathscr{P}| - 1)$ *for every partition* $\mathscr{P}$ *of V. Then G contains k edge disjoint spanning trees.* ∎

The *arboricity* $a(G)$ of a multigraph G is the minimal number of forests needed to cover G. The following formula for $a(G)$, due to Nash-Williams [N2], is a consequence of Theorem 5.7.

THEOREM 5.8. *Let G be a non-trivial multigraph and let* e_m *be the maximal size of a subgraph of order m in G. Put* $k = \max_{1 < m \leqslant n} \lceil e_m/(m - 1)\rceil$. *Then* $a(G) = k$.

Proof. Suppose $F_1, F_2, \ldots, F_l$ are edge disjoint forests whose union is G. Let $H \subset G, |H| \geqslant 2$, and put $F_i' = F \cap H, i = 1, 2, \ldots, l$. Then H is the union of the edge disjoint forests $F_1', F_2', \ldots, F_l'$, so

$$e(H) = \sum_1^l e(F_i') \leqslant \sum_1^l (|F_i'| - 1) \leqslant l(|H| - 1).$$

This implies that $a(G) \geqslant k$.

To prove the converse inequality, which is the non-trivial part of the theorem, let G be a maximal multigraph satisfying $k = \max_{1 < m \leqslant n} \lceil e_m/(m - 1)\rceil$. It suffices to show that G can be covered with m forests. The maximality of G implies immediately that $e(G) = k(n - 1)$. Let $\mathscr{P}$ be any partition of V into V_1, $V_2, \ldots, V_p$. Put $G_i = G[V_i]$, $n_i = |V_i|$. Note that by the definition of k, $e(G_i) \leqslant k(n_i - 1)$. Hence

$$e(G|\mathscr{P}) = e(G) - \sum_1^p e(G_i) \geqslant k(n - 1) - k \sum_1^p (n_i - 1) = k(p - 1).$$

Theorem 5.7 implies that G contains k edge disjoint spanning trees. These trees cover G since $e(G) = k(n - 1)$. ∎

6. EXERCISES, PROBLEMS AND CONJECTURES

1. Let $G = G_2(m, n)$ be a bipartite graph with vertex classes $U, V, |U| = m$, $|V| = n$. Suppose G contains a set of m independent edges. Prove that there

is a vertex $x \in U$ such that every edge incident with x is contained in a set of m independent edges. Using this, deduce the second part of Theorem 1.1 from the first part. (M. Hall [H16].)

2^-. Check that the proof of Theorem 1.1 implies the following extension of Theorem 1.1 as well.

Let $A_1, A_2, \ldots, A_m$ be subsets of A, $|A_1| \leqslant |A_2| \leqslant \ldots \leqslant |A_m|$. Put $d = \min\{|A_1|, m\}$. Suppose $\left| \bigcup_{i \in S} A_i \right| \geqslant |S|$ for every set $S \subset \{1, 2, \ldots, m\}$. Then $\{A_i\}_1^m$ has at least

$$\prod_{i \leqslant d} (|A_i| + 1 - i)$$

sets of distinct representatives. (Rado [R4].)

3. Apply Tychonov's theorem (the product of compact spaces is compact in the product topology) to deduce from Theorem 1.1 (or Theorem I.2.3) the following extension of it.

Let $\{A_u : u \in U\}$ be an infinite family of finite sets. Suppose

$$\left| \bigcup_{u \in S} A_u \right| \geqslant |S| \quad \text{for every finite subset } S \text{ of } U.$$

Prove that $\{A_u : u \in U\}$ has a set of distinct representatives. (Everett and Whaples [EW1], Halmos and Vaughan [HV1])

4. (Ex. 3 ctd.) Prove also that there are at least $d!$ sets of distinct representatives where $d = \min_{u \in U} |A_U|$.

5^-. Check that Corollary 2.2 holds for multigraphs as well.

6. Let $G = G_2(m, n)$ be a multigraph with $\beta_1(G) = m$. Let F be the set of edges of G not contained in any maximal matching. Prove that $|F| \leqslant m - 1$.

7^-. Show that Dilworth's theorem (Theorem 1.2) implies immediately the theorem of Philip Hall (Theorem I.2.3 or Theorem 1.1(i)).

8. Let $A = \bigcup_1^m A_i$ be a partition of A into disjoint sets. Let $B_1, B_2, \ldots, B_m$ be non-empty subsets of A. Suppose for every k, $1 \leqslant k \leqslant m$, the union of any k of the sets A_i contains at most k sets B_j. Then the sets A_i can be renumbered in such a way that

$$A_i \cap B_i \neq \emptyset, \qquad i = 1, 2, \ldots, m.$$

(P. Hall [H13])

9^-. Let $\bigcup_1^m A_i = \bigcup_1^m B_i$ be partititions into disjoint n-sets ($|A_i| = |B_i| = n$). Then the sets A_i can be renumbered in such a way that

$$A_i \cap B_i \neq \emptyset, \qquad i = 1, 2, \ldots, m.$$

(Van der Waerden [V1], König [K10].)

10^-. Let H and K be subgroups of order n of a (not necessarily finite) group G. Prove that there is a set $R \subset G$ such that $G = \bigcup_{r \in R} Hr$ and $G = \bigcup_{r \in R} Kr$ are the right coset decompositions for H and K, respectively. Prove that instead of the right coset decomposition for K we could have taken the left coset decomposition.

11. Let G be an abelian group of order n. Let $a_1, a_2, \ldots, a_n$ be a sequence of elements of G, possibly with repetitions. Prove that there exist two permutations $g_1, \ldots, g_n$ and $g_1', \ldots, g_n'$ of G such that $a_i = g_i + g_i'$, $1 \leqslant i \leqslant n$, iff $\sum_1^n a_i = 0$. (M. Hall [H17])

12. Let $A = (a_{ij})$ be an n by n stochastic matrix, i.e. $a_{ij} \geqslant 0$, $\sum_{j=1}^n a_{ij} = 1$ for every i and $\sum_{i=1}^n a_{ij} = 1$ for every j. Prove that $A = \sum_1^N \lambda_k P_k$ where $\lambda_k \geqslant 0$, $\sum_1^N \lambda_k = 1$ and $P_1, P_2, \ldots, P_N$ are permutation matrices. (König [K10].)

13. PROBLEM. Does Theorem 1.1 hold for multigraphs? In particular, if G is an n by n bipartite r-regular multigraph ($1 \leqslant r \leqslant n$) then is it true that $F(G) \geqslant r\,!$?

14. CONJECTURE. If G is a kn-regular n by n bipartite multigraph, where k and n are natural numbers, then

$$F(G) \geqslant k^n n!.$$

(Hajnal and Sós [HS1].)

15^-. Show that Conjecture 13 is equivalent to the well known conjecture of Van der Waerden: the permanent of a stochastic n by n matrix is at least $n! \, n^{-n}$.

16. Show that the graph G in Conjecture 14 has a 1-factor that occurs with

multiplicity at least k^n. Note that this is equivalent to the fact that the permanent in van der Waerden's conjecture has a member which is at least n^{-n}. (Hajnal and Sós [HS1].)

17. Suppose that every k-regular k by k bipartite multigraph has at least $k!$ 1-factors $(k = 1, 2, \ldots)$. Deduce that the permanent of a stochastic matrix is at least e^{-n}. [Hint. Take direct sums of matrices.]

18. Let $N = \{1, 2, \ldots, n\}$, let $A_k(k \in N)$ be finite sets and let $b_k(k \in N)$ be non-negative integers. Suppose

$$\left| \bigcup_{k \in L} A_k \right| \geqslant \sum_{k \in L} b_k \qquad (*)$$

for all subsets L of N. Then there exist disjoint sets $B_k \subset A_k$ $(k \in N)$ such that $|B_k| = b_k$.

19⁻. Use Tychonov's theorem on the product of compact spaces to show that the assertion of Ex. 18 holds if N is an arbitrary (infinite) set and $(*)$ is satisfied for every finite subset L of N. (Rado [A1])

20⁺. The assertion of Ex. 19⁻ can be used to prove the following result in measure theory.

Let X be a non-atomic measure space in which the measure of a set is denoted by $|Y|$. Let $(E_\alpha)_{\alpha \in A}$ be a family of measurable subsets of X with *finite* measure and let $(\lambda_\alpha)_{\alpha \in A}$ be a family of non-negative numbers indexed by the same infinite set A. Then the following two assertions are equivalent.

(i) There is a family $(X_\alpha)_{\alpha \in A}$ of measurable subsets such that for all $\alpha, \beta \in A$, $\alpha \neq \beta$ one has $X_\alpha \subset E_\alpha, |X_\alpha| = \lambda_\alpha, |X_\alpha \cap X_\beta| = 0$.

(ii)
$$\left| \bigcup_{\alpha \in B} W_\alpha \right| \geqslant \sum_{\alpha \in B} \lambda_\alpha$$

for all finite subsets B of A. [The proof is entirely analytical and is based on the Krein–Milman theorem.] (Bollobás and Varopoulos [BV1].)

21. Prove Corollary 1.9 by induction on β in the following way and give an alternative proof of Tutte's theorem.

Let $E_0 = \{x_1 y_1, x_2 y_2, \ldots, x_\beta u_\beta\}$ be a set of independent edges in $G, V_0 = \{x_1, y_1, \ldots, x_\beta, y_\beta\}$.

(i) Prove that if $x_1 a, x_2 b \in E(G)$, $a, b \in V - V_0$, $a \neq b$, then $d(x_1) = n - 1$. To do this, suppose x_1 is not adjacent to z. In $G + x_1 z$ there is a set F_0 of

$\beta + 1$ independent edges. Consider the paths and cycles in G with all the edges in $E_0 \cup F_0$. (These are called $E_0 - F_0$ alternating.)

By (i) we may suppose that every vertex in V_0 is adjacent to at most one vertex in $V(G) - V_0$.

(ii) Prove that if $x_1 z \in E(G)$, $z_1 \in V - V_0$, then $y_1 z \in E(G)$.

(iii) Prove that if $x_1 x_2 \in E(G)$ and $x_1 z \in E(G)$, $c \in V - V_0$, then $x_2 c \in E(G)$.

22^-. Enumerate all maximal graphs of even order not containing a 1-factor, (Read it out of Corollary 1.9. This answers a question raised by Kotzig [K12].)

23. Prove that if any $k \leqslant \frac{3}{2}n$ vertices of a graph of order $2n$ are adjacent to at least $\frac{4}{3}k$ vertices then the graph contains a 2-factor. (Anderson [A3].)

24^+. Let G be a 2-connected graph containing a 1-factor. Prove that there are at least $\delta(G)$ vertices such that every edge incident with at least one of these vertices occurs in a 1-factor of G. (This is a sharper form of a conjecture of Zaks [Z1], proved by Mader [M13].)

25. Let E_0 be a set of independent edges of G. Call a path P in G E_0-alternating if exactly one of any two consecutive edges of P is in E_0. An E_0-alternating path is augmenting if its end-edges do not belong to E_0. Call E_0 unaugmentable if there is no augmenting E_0-alternating path.

Prove that E_0 is a maximum matching (i.e. $|E_0| = \beta_1(G)$) iff E_0 is unaugmentable. (Berge [B5]; see also Norman and Rabin [NR3].)

26. Let G be a graph without isolated vertices. Show that a maximum matching in G may be extended to a minimum cover and every minimum edge cover of the vertices of G contains a maximum matching. (Norman and Rabin [NR3].)

27. (Ex. 26 ctd.) Prove the following assertions.

 (i) A minimal cover of G is minimum iff it contains a maximum matching of G.

 (ii) A maximal matching of G is maximum iff it is contained in a minimum cover of G. (Lewin [L7].)

28. Show that Theorem 3.2 holds even if G contains loops or $\delta(G) = 0$.

29. Complete the proof of Theorem 3.3.

30. (i) Prove Theorem 3.12.

 (ii) Deduce Corollary 3.11 from Theorem 3.12.

(See Erdös and Gallai [EG2], Hakimi [H4] and Havel [H21].)

31. Prove that a $2k$-regular graph is the union of edge disjoint 2-factors. (Petersen [P5].)

32. Show that the Petersen graph (Fig. III.1.2) is not the union of edge-disjoint 1-factors. (Petersen [P5].)

33[+]. (The exercise is about a graph one might consider a dual of a k-factor. The proof is not hard only not too short.)

Let $G = G^n$, $\Delta = \Delta(G)$, $\delta = \delta(G)$ and $\sum_{x \in G} (\Delta - d(x))$. Denote by $\text{reg}(G)$ the minimal integer N for which there is a Δ-regular graph H of order N containing G.

Prove that if $\text{reg}(G) = n + m$ then m is the minimal integer such that $m \geqslant S/\Delta$, $m^3 - (\Delta + 1)m + S \geqslant 0$, $m \geqslant \Delta - \delta$ and $(m + n)\Delta$ is even.

Prove furthermore that $\text{reg}(G) \leqslant 2n$ and for each $n > 3$ there exists a graph G such that $\text{reg}(G) = 2n$. (Erdös and Kelly [EK1].)

34. Prove that if $(d_i)_1^n$ and $(d_i - k)_1^n$ are both graphic sequences then there is a graph containing a k-factor whose degree sequence is $(d_i)_1^n$. (Kundu [K14]; for a shorter, algorithmic proof, see Kleitman and Wang [KW1]. The result was conjectured by Rao and Rao [RR1].)

35. Show by explicit constructions that

$$a(K^n) = \lceil n/2 \rceil \quad \text{and} \quad a(K^{m,n}) = \left\lceil \frac{mn}{m + n - 1} \right\rceil.$$

(Beineke [B2])

36. CONJECTURE. Every 4-regular graph contains a 3-regular subgraph. (N. Sauer and C. Berge, see [E28].)

37. Show that a sequence $d_1 \geqslant d_2 \geqslant \ldots \geqslant d_n \geqslant 0$ is graphic iff $\sum_1^n d_i$ is even and

$$\sum_1^t d_i \leqslant \sum_{t+1}^n d_i + \sum_1^t \min\{d_i, t - 1\}$$

for every t, $1 \leqslant t < n$ (cf. Corollary 3.8).

38. PROBLEM. A *measure graph* $G = (V, E)$ is a (not necessarily finite) graph such that there is a measure space (V, Σ, m) on V such that

$$\Gamma(X) \in \Sigma \quad \text{whenever} \quad X \in \Sigma$$

and

$$E^* = \{(x, y) \in V \times V : xy \in E\}$$

is a measurable subset of $V \times V$. The *degree function* or *marginal* of G, d: $V \to \mathbb{R}^M$, is defined by $d(x) = m(\Gamma(x))$.

Extend the Erdős–Gallai theorem (Corollary 3.11) to a characterization of the degree functions associated with an atomless measure space (V, Σ, m). (Nash-Williams [N7])

39. CONJECTURE. (See Problem 38.) Let $V = [0, 1]$ with the Lebesgue measure and let $d: [0, 1] \to [0, 1]$ be a non-increasing function. Then d is a degree function iff

$$\int_0^x (f(t) - x)\, dt \leqslant \int_x^1 \min(f(t), x)\, dt \quad \text{for every } x.$$

40. PROBLEM. (See Problem 38). Formulate and prove a measure graph version of Hall's theorem (Theorem I.2.3). (Nash-Williams [N7])

Chapter III

Cycles

In this chapter we tackle two closely related sets of questions. We investigate the set of integers occurring as cycle lengths in a graph and we discuss the relation of the cycles to each other.

In connection with the cycle lengths we are especially interested in the length of a shortest cycle, the girth, and in the length of a longest cycle, the circumference of the graph. It is easily seen that if δ and g are natural numbers and n is sufficiently large then most graphs $G(n, \delta n)$ contain a large subgraph of girth at least g. This implies that for given δ and g there is a graph G with $\delta(G) \geqslant \delta$ and $g(G) \geqslant g$. The main aim of §1 is to give bounds on the minimum order $n(g, \delta)$ of such a graph G.

In §2 we present results concerning graphs without a given number (say k) of vertex disjoint cycles and §3 is devoted to corresponding results about edge disjoint cycles. For $k = 2$ the structure of these graphs can be determined and the desired results follow from the structure theorem, but for $k \geqslant 3$ more direct methods have to be used which do not usually produce exact results. It should be emphasized that the vertex disjoint cycles present essentially different difficulties from those presented by edge disjoint cycles, though formally the questions we look at are very similar. Another problem to be discussed in §3 is the representability of all cycles by a given number of vertices.

In §4 we prove results about the circumference of a graph. The section centres around the most important unsolved problem concerning cycles, the Hamiltonian cycle problem. Given a graph G, a *Hamiltonian* cycle is a cycle of length $|G|$ and a *Hamiltonian* path is a path of length $|G| - 1$. If a graph contains a Hamiltonian cycle then it is said to be *Hamiltonian*. Graph theorists have been searching for a long time for conditions ensuring that a graph is Hamiltonian. Most of the known results are descendants of a theorem of Dirac [D7] published in 1952. In the rest of the section we give bounds on the circumference in terms of the order, size, connectivity and minimal degree of the graph.

We shall see in §5 that for most choices of n and m the set of numbers occurring as cycle lengths in *every* $G(n, m)$ contains a large block of integers. One of the aims of the section is to present results about this block. Among others we give conditions ensuring that this block is as large as possible, i.e. $\{3, 4, \ldots, n\}$. In this case the graph is said to be *pancyclic*. If $m \leqslant \lfloor n^2/4 \rfloor$ then only even integers occur in this block since $K(\lfloor n/2 \rfloor, \lceil n/2 \rceil)$ does not contain odd cycles. However, the block does exist for most values of $m \leqslant \lfloor n^2/4 \rfloor$ as well in the sense that every $G(n, m)$ contains a C^{2l} for l in a large interval depending on n and m.

The most important unsolved problem concerning cycles seems to be the determination or good estimation of the minimal order $n(g, \delta)$ of graphs of given girth g and minimal degree δ (Conjecture 13). At present the upper bound on $n(g, \delta)$ is almost the square of the lower bound. Though the gap is immense, narrowing it is likely to require a radically new idea. As it will be clear from the results to be presented, better bounds on $n(g, \delta)$ would find immediate applications in other areas.

The results concerning $c(G)$, the circumference, are only slightly more satisfactory. As we have already mentioned, almost all results in this area originate in Dirac's theorem. In particular, most of the conditions implying $c(G) \geqslant c$ are preserved under the addition of edges to G. A result of a different nature has been proved recently by Bollobás and Hobbs and extended by Jackson: a 2-connected r-regular graph of order n is Hamiltonian even if r is significantly less than $n/2$. It would be interesting to prove similar results for k-connected graphs (Conjecture 36). More generally, it would be desirable to prove results relating the degree sequence of a graph to its circumference, connectivity and girth (Problem 38).

Another rather neglected set of unsolved problems concern C^l-saturated graphs (Problem 39). A graph G is C^l-saturated if it does not contain a cycle of length l but the addition of any edge to G creates a cycle of length l.

1. GRAPHS WITH LARGE MINIMAL DEGREE AND LARGE GIRTH

Given integers δ and $g, \delta \geqslant 2, g \geqslant 4$, it is obvious that if n is fairly small there is no G^n satisfying $\delta(G^n) \geqslant \delta$ and $g(G^n) \geqslant g$. It is less obvious that if n is large there is a G^n such that $\delta(G^n) \geqslant \delta$ and $g(G^n) \geqslant g$. The *minimal number of vertices* of such a graph is denoted by $n(g, \delta)$. The aim of this section is to give bounds on $n(g, \delta)$. Note that $n(g, 2) = g$ and C^g is the only G^g satisfying $\delta(G^g) \geqslant 2$ and $g(G^g) \geqslant g$. Therefore in the sequel we shall always suppose $\delta \geqslant 3$. As usual, an extremal graph is a graph G^n with $n = n(g, \delta)$, $\delta(G^n) \geqslant \delta$, $g(G^p) \geqslant g$.

Let us start by giving a simple proof of the fact that $n(g, \delta) < \infty$.

THEOREM 1.1. *Given $\delta \geqslant 3$ and $g \geqslant 3$ there exists a G^n, $n \leqslant (2\delta)^g$, with minimal degree at least δ and girth at least g. Thus $n(g, \delta) \leqslant (2\delta)^g$.*

Proof. Let $N = (2\delta)^g$. Consider all graphs with vertex set $\{1, \ldots, N\}$, having δN edges. Note that there are

$$\left(\!\! \binom{N}{2} \atop \delta N \right)$$

such graphs. One can form

$$\tfrac{1}{2}(l - 1)! \binom{N}{l} < \frac{1}{2l} N^l$$

cycles of length l from these vertices and

$$\left(\!\! \binom{N}{2} - l \atop \delta N - l \right)$$

of the graphs contain a given cycle of length l. Consequently the average number of cycles of length at most $g - 1$ contained in these graphs is less than

$$\sum_{l=3}^{g-1} \frac{1}{2l} N^l \left(\!\! \binom{N}{2} - l \atop \delta N - l \right) \left(\!\! \binom{N}{2} \atop \delta N \right)^{\!\!-1} < \sum_{l=3}^{g-1} (2\delta)^l < N.$$

Thus there is a graph $G(N, \delta N)$ that contains at most $N - 1$ cycles of length at most $g - 1$. Omit an edge from each such cycle to obtain a graph $G_0 = G(N, M)$ with $M > (\delta - 1)N$ and girth at least g.

Since $G_0 \in \mathcal{D}_{\delta-1}$ by (0·5) it contains a subgraph G with $\delta(G) \geqslant \delta$. By construction we also have $g(G) \geqslant g$ and $|G| \leqslant N = (2g)^\delta$. ∎

In fact, given δ and g, it is not too hard to construct explicitly a δ-regular graph of very large order (very much larger than $(2\delta)^g$) that has girth g and also has a number of pleasant properties (see Exx. 2 and 3.).

One can easily obtain a lower bound on $n(g, \delta)$ which happens to be exact at a number of places (see Tutte [T8]).

THEOREM 1.2.

$$n(g, \delta) \geqslant \begin{cases} 1 + \delta \dfrac{(\delta - 1)^{(g-1)/2} - 1}{\delta - 2} & \text{if } g \text{ is odd,} \\[3mm] \dfrac{(\delta - 1)^{g/2} - 1}{\delta - 2} & \text{if } g \text{ is even.} \end{cases} \tag{1}$$

Equality holds for $\delta = 3, g \in \{3, 4, 5, 6, 8\}$ *,and* $g = 4, \delta \geqslant 3$.

Proof. Take a graph G^n satisfying $g(G^n) \geqslant g$ and $\delta(G^n) \geqslant \delta$. Let x and y be adjacent vertices in G^n.

Suppose first that g is odd, $g = 2d + 1$. Note that the subgraph spanned by the vertices within distance d from x is a tree since a cycle in it would have length at most $2d$. Consequently for $0 \leqslant l < d$ there are at least $\delta(\delta - 1)^l$ vertices at distance $l + 1$ from x (Fig. 1.1) so

$$n \geqslant 1 + \delta \sum_{0}^{d-1} (\delta - 1)^l.$$

Suppose now that g is even, $g = 2d$. Then the subgraph spanned by the vertices within distance d from $\{x, y\}$ is a tree and the bound follows as before (Fig. 1.1).

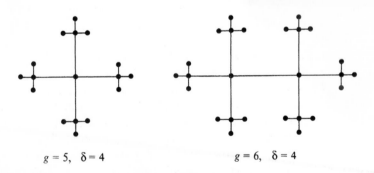

$$g = 5, \quad \delta = 4 \qquad\qquad\qquad g = 6, \quad \delta = 4$$

FIG. 1.1. Two trees in the proof of Theorem 1.2.

To see that equality holds for $\delta = 3$ and $g \in \{3, 4, 5, 6, 8\}$, note that $g(K^4) = 3, g(K^{3,3}) = 4$ and $g(G_l) = l, l = 5, 6, 8$, for the graphs G_l shown in Fig. 1.2. We have not drawn all the edges of G_6 and G_8, in addition to the edges shown a vertex with number i is adjacent to the two vertices whose pair of numbers contains i.

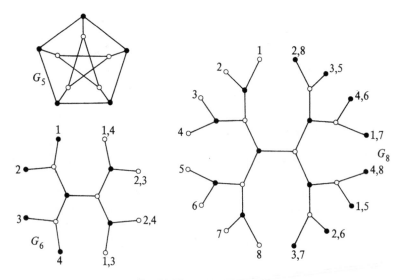

FIG. 1.2. The graphs G_5, G_6, G_8.

The graph G_5 is called the *Petersen graph* and G_6 the *Heawood graph*. To facilitate checking $g(G_l) = l$, note that the graphs G_6, G_8 are bipartite. Clearly $\delta(K^{\delta,\delta}) = \delta$, $g(K^{\delta,\delta}) = 4$ and $K^{\delta,\delta}$ has 2δ vertices, so $n(4, \delta) = 2\delta$, as claimed. ∎

Suppose that for some pair $\{g, \delta\}$ we have equality in (1) and let $G_0 = G^n$ be an extremal graph. The proof of Theorem 1.2 implies that G_0 is regular of degree δ. Furthermore, if $g = 2d + 1$ then G_0 has diameter d and n is also the maximal number for which there exists a graph G^n with $\Delta(G^n) \leqslant \delta$ and diam $G^n \leqslant d$. We call G_0 a *Moore graph of degree δ and girth g* or, if $g = 2d + 1$, a *Moore graph of degree δ and diameter d*.

Hoffman and Singleton [HS4] showed that if there exists a Moore graph of degree $\delta \geqslant 3$ and diameter 2 then $\delta = 3, 7$ or 57. Moore graphs of diameter 2 and degree 3 and 7 do exist (G_5 is a Moore graph of degree 3) but it is not known whether there is a Moore graph of degree 57 and diameter 2. Damerell [D1] and Bannai and Ito [BI1] proved independently that there is no Moore graph of degree $\delta \geqslant 3$ and diameter $\geqslant 3$. (Some partial results had been obtained by Friedman [F3].)

Let us turn now to the problem of the existence of Moore graphs of even girth. It is obvious that $K^{\delta,\delta}$ is the only Moore graph of degree δ and girth 4.

Feit and Higman [FH1] proved that if there is a Moore graph of degree $\delta \geqslant 3$ and girth $g = 2d \geqslant 5$ then $d = 3, 4$ or 6. Singleton [S21] gave an independent proof and, extending the results and methods of Kárteszi [K1, K2], showed also that for $d = 3$ and 4 there is a Moore graph for each finite projective geometry of dimension 2 and 3, respectively, with δ points on a line (see Exx. 5, 6 and 7).

The lower bound of $n(g, \delta)$ given by Theorem 1.2 is rather obvious but it is likely to be near to $n(g, \delta)$. In spite of this it is not too easy to improve considerably the poor upper bound of $n(g, \delta)$ given in Theorem 1.1. Let us start by showing that an upper bound of $n(g, \delta)$ for g odd implies an upper bound of $n(g + 1, \delta)$.

THEOREM 1.3. *If g is odd then*

$$n(g + 1, \delta) \leqslant 2n(g, \delta).$$

Proof. Take a graph H of order $n(g, \delta)$ such that $\delta(H) \geqslant \delta$ and $g(H) \geqslant g$. Let H' be a copy of H, disjoint from H. Define a bipartite graph G with vertex classes $V(H)$ and $V(H')$ as follows. If ab and $a'b'$ are corresponding edges of H and H' then let $a'b$ and ab' be edges of G. Then $|G| = 2n(g, \delta)$, $\delta(G) = \delta(H) \geqslant \delta$ and $g(G) \geqslant g(H) \geqslant g$. Furthermore, as H is bipartite, it does not have a cycle of length g. ∎

The first non-trivial upper bound on $n(g, \delta)$ was obtained by Erdős and Sachs [ES1], who also proved the existence of δ-regular graphs of girth at least g. (A weaker form of this result was rediscovered by Tutte [T17].) By exploiting the method of Erdős and Sachs, Sauer [S7] obtained the best upper bound on a (g, δ) so far. Independently of Sauer, Walther [W5], [W7] obtained only somewhat worse bounds. In order to avoid some tedious details, we prove here a slightly weaker form of the result due to Sauer [S7].

THEOREM 1.4.

$$n(g, \delta) \leqslant \begin{cases} 2\dfrac{(\delta - 1)^{g-1} - 1}{\delta - 2} & \text{if } g \text{ is odd}, \\[3mm] 4\dfrac{(\delta - 1)^{g-2} - 1}{\delta - 2} & \text{if } g \text{ is even}. \end{cases} \tag{2}$$

Proof. Note first that in view of Theorem 1.3 it suffices to prove Theorem 1.4 in the case when g is odd. In fact we shall prove the following slightly stronger version of this case.

THEOREM 1.4'. *Let*

$$m \geqslant \sum_{i=0}^{g-2} (\delta - 1)^i = \frac{(\delta - 1)^{g-1} - 1}{\delta - 2}$$

be an integer. Then there exists a δ-regular graph of order $2m$ and girth at least
g.

Proof. Apply induction on δ. The case $\delta = 2$ is trivial. Suppose $\delta > 2$ and the assertion holds for smaller values of δ. Let

$$\mathscr{G} = \{G^{2m} : g(G^{2m}) \geqslant g \text{ and } \delta - 1 \leqslant d(x) \leqslant \delta \text{ for all } x \in G^{2m}\}.$$

By the induction hypothesis $\mathscr{G} \neq \varnothing$. Let G be a *maximum* graph in $\mathscr{G}$ (i.e. one with *maximal number of edges*). To prove the theorem we shall show that G is δ-regular.

Suppose G is not δ-regular, i.e. it contains a vertex of degree $\delta - 1$. Since the sum of the degrees is always even, G contains at least two vertices of degree $\delta - 1$. Let x_1 and x_2 be two of those vertices. According to their distances from x_1 and x_2 we shall put the vertices of G into various subsets. When choosing the notation we consider $V = V(G)$ a metric space with the obvious distance function. Put $r = g - 2$ and let

$$B_i = \{x \in V : d(x, x_i) \leqslant r\}, \quad i = 1, 2,$$

$$S_i = \{x \in V : d(x, x_i) = r\}, \quad i = 1, 2,$$

$$B = B_1 \cup B_2,$$

$$C = V - B,$$

$$\partial B = \{x \in B : \min_{i=1,2} d(x, x_i) = r\},$$

$$T = S_1 \cap S_2.$$

If u, v are vertices of degree $\delta - 1$ then $d(u, v) \leqslant r = g - 2$ since otherwise $G + uv \in \mathscr{G}$, contradicting the maximality of G. Thus every vertex of degree $\delta - 1$ must be contained in $B_1 \cap B_2$, in particular $T \cup \{x_1, x_2\} \subset B_1 \cap B_2$

and so

$$|B_1 \cap B_2| \geqslant |T| + 2.$$

Noting that clearly

$$|B_i| \leqslant \sum_0^r (\delta - 1)^j \leqslant m, \quad i = 1, 2,$$

we obtain

$$|C| = 2m - |B| = 2m - |B_1| - |B_2| + |B_1 \cap B_2| \geqslant |T| + 2. \tag{3}$$

The vertices in C have degree δ, so by (3) it cannot happen that every vertex adjacent to a vertex in C belongs to T. Consequently there is an edge $y_1 y_2$ such that $y_1 \in C$ and $y_2 \notin T$. Then

$$y_2 \in C \cup (\partial B - T),$$

so we can suppose without loss of generality that $d(y_2, x_2) \geqslant r + 1 = g - 1$. We are now in a position to be able to alter G in such a way that the new graph H also belongs to G but has more edges than G. This graph H is obtained as follows:

$$H = G - y_1 y_2 + x_1 y_1 + x_2 y_2.$$

It is easy to check that $H \in \mathcal{G}$ and $e(H) = e(G) + 1$. This contradicts the maximality of G and so G is δ-regular, as claimed. ∎

The upper bound of $n(g, \delta)$ given in Theorem 1.4 is almost of the order of the square of the lower bound given in Theorem 1.2. Though Theorem 1.2 is almost trivial and Theorem 1.4 is certainly not, it is very likely that Theorem 1.2 gives the proper order of $n(g, \delta)$ (see Problem 11). The large number of results concerning small values of g and δ seem to support this belief (see Benson [B4], Brown [B50], [B53], Longyear [L10], Robertson [R10], Tutte [T17] and Wegner [W13]).

By choosing the sets in the proof of Theorem 1.4' more carefully and being less generous with the estimation of the degrees one can obtain the improvement on (2), mentioned before Theorem 1.4. The most significant improvement can be obtained for $\delta = 3$:

$$n(g, 3) \leqslant 2^{g-1}. \tag{4}$$

Simonovits [S18] constructed a 3-regular graph $G(l)$ of order 2^{l+1} which was claimed to have girth $2l + 1$. This would have shown that

$$\lim_{g \to \infty} g^{-1} \log_2 n(g, 3) = \tfrac{1}{2},$$

i.e. (1) is indeed very much nearer to the truth than (2) and its improvement, (4). Unfortunately $G(l)$ always contains a C^8 (in fact $g(G(l)) = 8$ for $l \geq 4$) and it is still not known whether $\lim_{g \to \infty} g^{-1} \log_2 n(g, 3) < 1$ (cf. Problem 12).

2. VERTEX DISJOINT CYCLES

Denote by Ω_k the family of graphs containing k *vertex disjoint* cycles and let Ω'_k be the family of graphs containing k *edge disjoint* cycles. Vertex disjoint cycles are said to be *independent* and edge disjoint cycles are called *weakly independent*. The family of graphs not belonging to Ω_k (resp. Ω'_k) is denoted by $\overline{\Omega}_k$ (resp. $\overline{\Omega}'_k$). Note that $\overline{\Omega}_1 = \overline{\Omega}'_1$ is just the family of *forests*.

The next two sections are devoted to some extremal problems concerning graphs in Ω_k and Ω'_k, e.g. for given values of n and k we are interested in the minimal value of m such that every $G(n, m)$ belongs to Ω_k. Of course, it is very convenient to rely on structural results, whenever they exist. Not surprisingly, the structural results proved so far concern the case $k = 2$. In the first part of this section we describe the structure of the graphs in $\overline{\Omega}_2$. The structure theorem will enable us to prove a number of extremal problems concerning Ω_2. In the second part of the section we investigate the families Ω_k, $k \geq 2$.

Let us start with a theorem of Dirac [D18]. The proof we present is due to Brown [B52] and is based on Tutte's theorem (Theorem I.3.11) on the structure of 3-connected graphs.

Let K^{n-} be the graph obtained from K^n by omitting an edge. Denote by $K(3, p)'$, $K(3, p)''$ and $K(3, p)'''$ the graph obtained from $K(3, p)$ by adding one, two and three edges, respectively, joining vertices in the class containing three vertices.

THEOREM 2.1. *Every 3-connected graph in $\overline{\Omega}_2$ is one of the following graphs:* W^k $(k \geq 3)$, K^5, K^{5-}, $K(3, p)$, $K(3, p)'$, $K(3, p)''$ and $K(3, p)'''$ $(p \geq 3)$.

Proof. If G is 3-connected then the graph obtained from G by adding an edge or splitting a vertex (cf. *p.* 16) is almost always in Ω_2. This intuitive statement is made precise by the following assertions. Each of these assertions can be proved by checking a few cases so we leave the proofs to the reader.

(i) If $G \in \Omega_2$ or $G = W^k$ $(k > 4)$ then by adding an edge or splitting a vertex we get a graph in Ω_2.

(ii) Suppose $G \in \overline{\Omega}_2$ is obtained from $K(3, p)$, $K(3, p)'$, $K(3, p)''$ or $K(3, p)'''$ by splitting a vertex (of degree at least 4). Then $G = K(3, p)$ or $G = K(3, p)'$ $(p \geqslant 3)$.

(iii) Suppose $G \in \overline{\Omega}_2$ is obtained from a 3-connected $K(3, p)$, $K(3, p)'$, $K(3, p)''$ or $K(3, p)'''$ by adding an edge. Then G is one of these graphs: K^5, $K^{5-} = K(3, 2)''$, $K(3, p)'$, $K(3, p)''$ and $K(3, p)'''$ $(p \geqslant 3)$.

(iv) If we split a vertex of K^5 we obtain a graph in Ω_2.

In view of these assertions our theorem is an immediate consequence of Tutte's theorem (Theorem I.3.11). For suppose $G_0 = W^k, G_1, \ldots, G_s$ is a sequence of graphs such that $G_s \in \overline{\Omega}_2$ and G_i $(1 \leqslant i \leqslant s)$ is obtained from G_{i-1} by adding an edge or splitting a vertex. By (i) each of the graphs G_{s-1}, $G_{s-2}, \ldots, G_0$ belongs to $\overline{\Omega}_2$. Then (ii), (iii) and (iv) imply that each of the graphs $G_1, G_2, \ldots, G_s$ occurs among the graphs listed in the theorem. ∎

Let us use this result to characterize the graphs in $\overline{\Omega}_2$. This characterization is due to Lovász [L12] who gave a proof independent of Theorems I.3.11 and III.2.1. Note that if G is a refinement of a multigraph H then G contains k independent (or weakly independent) cycles if and only if H does. Furthermore, if H has l vertices representing all the cycles then so does G. Thus it is rather natural to consider *multigraphs* without two independent cycles.

THEOREM 2.2. *Let G be a multigraph without two independent cycles. Suppose $\delta(G) \geqslant 3$ and there is no vertex representing all the cycles. Then one of the following six assertions holds (see Fig. 2.1).*

(a) *G has three vertices and multiple edges joining every pair of the vertices.*

(b) *G is a K^4 in which one of the triangles may have multiple edges.*

(c) *$G = K^5$.*

(d) *G is a K^{5-} such that some of the edges not adjacent to the missing edge may be multiple edges.*

(e) *G is a wheel whose spokes may be multiple edges.*

(f) *G is obtained from $K(3, p)$ by adding edges or multiple edges joining vertices in the first class.*

Proof. Clearly G must have at least 3 vertices and if it has exactly 3 vertices

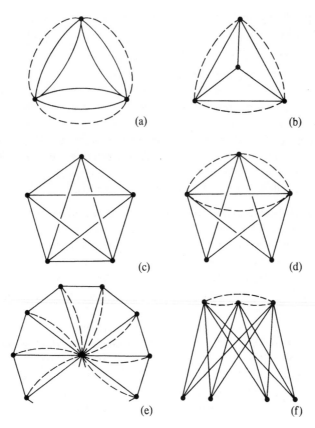

FIG. 2.1. Illustration to Theorem 2.2. A dotted line shows that there may be multiple edges joining the end vertices.

then it is one of the graphs given in (a). Suppose now that G has at least 4 vertices. If G is 3-connected then it is obtained from a graph listed in Theorem 2.1 by replacing some of the edges by multiple edges. We leave it to the reader to check that the graphs obtained in this way which still do not contain two independent cycles are exactly the graphs in $(b) - (f)$. Thus to complete the proof it suffices to show that G must be 3-connected.

Suppose this is not so, i.e. G has at least 4 vertices and it is not 3-connected. Let (V_1, V_2) be a 2-separator of G with separating set $V_1 \cap V_2 = S = \{s_1, s_2\}$. Let H_i be a component of $G[V_i]$. Then either H_i contains a cycle or there are at least four edges joining S to H_i and so there is an index $j_i \,(=1$ or 2) such that at least two edges join s_{j_i} to H_i. Consequently for each i there is an index j_i such that $G[V_i \cup \{s_{j_i}\}]$ contains a cycle Z_i, say. Clearly $j_1 = j_2$ since otherwise G contains two independent cycles. Let C be a cycle not containing s_{j_i}. Then C is independent of at least one of the cycles Z_1, Z_2.   ∎

THEOREM 2.3. *A multigraph G does not contain two independent cycles if and only if either it contains a vertex x_0 such that $G - x_0$ is a forest, or it can be obtained from a subdivision G_0 of a graph listed in Theorem 2.2 by adding a forest and at most one edge joining each tree of the forest to G_0. [In this context we consider the empty graph a forest.]*

Proof. The result is an immediate consequence of the following trivial observations.

1. If G does not contain two independent cycles then at most one of its components is not a tree.

2. If G has a vertex x of degree at most 1 then G contains two independent cycles if and only if $G - x$ does.

3. If G is 2-regular and does not contain two independent cycles then it is a cycle.

4. If $2 \leqslant \delta(G)$ and $3 \leqslant \Delta(G)$ then G is the subdivision of a multigraph H with $\delta(H) \geqslant 3$.   ∎

COROLLARY 2.4. *Let G be a graph of order $n \geqslant 6$ without two independent cycles. Then*

(i) *G has at most $3n - 6$ edges,*

(ii) *$\delta(G) \leqslant 3$, and*

(iii) *the cycles of G can be represented by three vertices.*   ∎

Later we shall generalize these assertions to graphs without k independent cycles. Note that (i) is an almost immediate corollary of Kuratowski's theorem (V.3.1) as well (Ex. 17).

It is trivial that the cycles of a planar graph $G \in \Omega_2$ can be represented by three vertices so it is not surprising that the original proof of (iii) was also based on Kuratowski's theorem (Ex. 18). For the definition of planarity and a short discussion of Kuratowski's theorem see Chapter V, § 3.

The final result [B18] concerning graphs without two independent cycles gives the maximal number of edges of $G^n \in \overline{\Omega}_2$ provided $g(G^n) \geqslant g$. We state this theorem only in the case $g \equiv 0 \pmod 4$ since this is the simplest and the

other cases are only slightly different. The result can be proved by applying
Theorem 2.3 and then discussing a number of cases; we leave this to the reader.
The original proof was, once again, based only on Kuratowski's theorem.

THEOREM 2.5. *Let k, l and n be natural numbers, $k \geqslant 3$ and $n = k(3l - 2) + 3 + r$.*
Then

$$\max\{m : \exists G = G(n, m) \in \overline{\Omega}_2, g(G) \geqslant 4l\} = \begin{cases} 3kl + r & \text{if} \quad 0 \leqslant r < 2l - 1, \\ 3kl + r + 1 & \text{if} \quad 2l - 1 \leqslant r < 3l - 2. \end{cases}$$

If $n = k(3l - 2)$ then there is a unique extremal graph. It is obtained from
$K^{3, k}$ *by subdividing each edge into l edges.* ∎

It is rather natural to ask at most how many edges a graph $G^n \in \overline{\Omega}_k$ can
have if its girth is at least g. Though for $k > 2$ rather little is known about this
problem, we shall return to it at the end of the section.

If a graph G^n has sufficiently many edges and n is sufficiently large compared
to k then clearly $G^n \in \Omega_k$. Erdős and Pósa [EP1] determined the maximum
size of a graph $G^n \in \overline{\Omega}_k$ provided n is sufficiently large compared to k. Even
more, they proved that there is a unique extremal graph.

THEOREM 2.6. *Let $k > 1$, $n \geqslant 24k$ and put $f(n, k) = (2k - 1)n - 2k^2 + k$*
$= (2k - 1)(n - k)$. Then every $G(n, f(n, k) + 1)$ contains k independent cycles.
If $G = G(n, f(n, k)) \in \overline{\Omega}_k$ then

$$G = K^{2k-1} + E^{n-2k+1}.$$

Let us restate Theorem 2.6 in a form more suitable for induction in spite
of being more complicated. Temporarily we shall use the notation $G(n, m)$
even when $m > \binom{n}{2}$, in which case it denotes K^n.

THEOREM 2.6'. *Every $G(n, n)$ contains a cycle $(k = 1)$. For $k > 1$ put*

$$g(n, k) = \begin{cases} f(n, k) + (24k - n)(k - 1) & \text{if} \quad n \leqslant 24k \\ f(n, k) & \text{if} \quad n \geqslant 24k. \end{cases}$$

Then if $3k \leqslant n \leqslant 24k - 1$ every $G(n, g(n, k))$ contains k independent cycles and
if $n \geqslant 24k$ and $l_0 \geqslant g(n, k)$ then every $G_0 = G(n, l_0)$ contains k independent
cycles except when $G_0 = K^{2k-1} + E^{n-2k+1}$.

Note that Theorem 2.6′ implies Theorem 2.6. Furthermore Theorem 2.6′ is trivial for $3k \leqslant n \leqslant 6k$ since then $g(n, k) \geqslant \binom{n}{2}$ so $G = K^n \supset K^{3k}$ which contains k independent triangles.

The proof of Theorem 2.6′ is based on the following lemma.

LEMMA 2.7. *Let $n \geqslant 6k$ and suppose that $G = G^n$ contains $2k$ vertices $x_1, \ldots, x_{2k}$ with $d(x_i) \geqslant n - k$ $(1 \leqslant i \leqslant 2k)$. Then G contains k independent quadrilaterals.*

Proof. Denote by $y_1, \ldots, y_{n-2k}$ the remaining vertices of G. Consider a maximal system of independent quadrilaterals of the form

$$x_{2i-1} y_{i(1)} x_{2i} y_{i(2)}, \qquad 1 \leqslant i \leqslant l, \ \ l \leqslant k.$$

To prove the lemma it suffices to show that $l = k$. Suppose $l < k$. Each of the vertices x_{2l+1}, x_{2l+2} is adjacent to at least $n - k$ vertices. Therefore at least $n - 2k \geqslant 4k$ vertices of G are adjacent to both x_{2l+1} and x_{2l+2}. As $2k + 2l \leqslant 4k - 2$, there are at least two vertices y_j not belonging to the quadrilaterals, which are joined to both x_{2l+1} and x_{2l+2}. This contradicts the maximality of the system of quadrilaterals.   ∎

Proof of Theorem 2.6′. We use induction on k. If $k = 1$, the result is trivial. Let $k = k' > 1, n = n'$ and suppose that Theorem 2.6′ holds for $k = k' - 1$ and also that it holds for $k = k'$ if $n < n'$. As we remarked after the statement of Theorem 2.6′, we may suppose that $6k < n$.

By Lemma 2.7 we may suppose that G_0 contains at most $2k - 1$ vertices of degree at least $n - k$, otherwise there is nothing to prove.

We may suppose without loss of generality that there is a vertex, say x_0, such that

$$2k \leqslant d(x_0) < n - k.$$

For if this is not so, then G_0 has at most $2k - 1$ vertices of degree at least $n - k$ and all other vertices have degree at most $2k - 1$. Consequently

$$e(G_0) \leqslant \tfrac{1}{2}\{(2k - 1)(n - 1) + (2k - 1)(n - 2k + 1)\} = f(n, k)$$

and equality can occur only if $G_0 = K^{2k-1} + E^{n-2k+1}$.

Let $x_1, \ldots, x_l$, $2k \leqslant l = d(x_0) < n - k$, be the vertices joined to x_0. Put $H = G_0[x_1, \ldots, x_l]$. To complete the proof of the theorem we consider two cases.

1. *Suppose first that* $\delta(H) < k$, *say* $x_1 \in H$, $d(x_1) < k$. We may suppose without loss of generality that x_1 is not joined to any of the vertices $x_{r+1}, \ldots, x_l$, where $r \leqslant k$. Denote by G_1 the graph obtained from $G_0 - x_0$ by adding the edges $x_1 x_{r+i}$, $1 \leqslant i \leqslant l - r$. Then G_1 has $n - 1$ vertices and

$$e(G_1) \geqslant e(G) - k,$$

i.e.

$$e(G_1) \geqslant g(n, k) - k \geqslant g(n - 1, k).$$

As G_1 cannot have $2k - 1$ vertices of degree $n - 2$, $G_1 \neq K^{2k-1} + E^{n-2k}$ and so by the induction hypothesis G_1 contains k independent cycles, say $C_1, \ldots, C_k$. At most one of these cycles, say C_1, contains one or two of the edges added to $G_0 - x_0$. If C_1 contains none of these edges, put $C_1^* = C_1$. If C_1 contains one of these edges, say $x_1 x_l$ then let C_1^* be the cycle obtained from C_1 by replacing the edge $x_1 x_l$ by the path $x_1 x_0 x_l$. Finally if C_1 contains two of these edges, say $x_1 x_{r+1}$ and $x_1 x_{r+2}$, then denote by C_1^* the cycle obtained from C_1 by replacing these edges by $x_0 x_{r+1}$ and $x_0 x_{r+2}$. In all these cases $C_1^*, C_2, \ldots, C_k$ are k independent cycles of G_0.

2. *Suppose now that* $\delta(H) \geqslant k$. Then, since $l \geqslant 2k$, by Lemma II.4.2 the graph H contains k independent edges, say $e_i = x_{2i-1} x_{2i}$, $1 \leqslant i \leqslant k$.

Note that the theorem follows easily if each edge e_i is contained in at least $k - 1$ triangles $x_{2i-1} x_{2i} y_t^{(i)}$, $1 \leqslant t \leqslant k - 1, 1 \leqslant i \leqslant k$, where the $y_t^{(i)}$ are different from $x_0, x_i, \ldots, x_{2k}$. For then it is easily seen that there are $k - 1$ triangles of the form $x_{2i-1} x_{2i} y_{t_i}^{(i)}$, $1 \leqslant i \leqslant k - 1$, and they are all independent of the triangle $x_0 x_{2k-1} x_{2k}$.

Thus we may assume that an edge e_i, say $e_1 = x_1 x_2$, is contained in at most $k - 2$ triangles in $G_0 - x_0 - x_3 - \ldots - x_{2k}$. Let us estimate the total number of edges incident with x_0, x_1 and x_2. First of all $d(x_0) < n - k$ by assumption. Secondly there are at most $4k - 3$ edges of the form $x_i x_j$, $1 \leqslant i \leqslant 2$, $2 \leqslant j \leqslant 2k$. Finally all but at most $k - 2$ vertices of $G_0 - x_0 - x_1 - \ldots - x_{2k}$ are joined to at most one of the vertices x_1 and x_2, so there are at most $n - k - 3$ edges of the form $x_i x_j$, $1 \leqslant i \leqslant 2, j > 2k$. Consequently there are at most

$$n - k - 1 + 4k - 3 + n - k - 3 = 2n + 2k - 7$$

edges in G_0, incident to x_0, x_1 and x_2. Thus for $k > 2$ we have

$$e(G_0 - x_0 - x_1 - x_2) \geqslant e(G_0) - 2n - 2k + 7 = (2k - 1)n - 2k^2$$

$$+ k - 2n - 2k + 7 = (2k - 3)(n - 3) - 2(k - 1)^2 + k + g(n, k)$$

$$- f(n, k) > g(n - 3, k - 1), \tag{1}$$

since it is easily checked that

$$g(n, k) - f(n, k) \geqslant g(n - 3, k - 1) - f(n - 3, k - 1).$$

For $k = 2$ we have

$$e(G_0 - x_0 - x_1 - x_2) \geqslant e(G_0) - 2n + 3 \geqslant n - 3. \tag{2}$$

By the induction hypothesis (1) and (2) imply that $G_0 - x_0 - x_1 - x_2$ contains $k - 1$ independent cycles, which are, obviously, all independent of the triangle $x_0 x_1 x_2$ in G_0. Thus G_0 contains k independent cycles. ∎

Dirac announced in [D26] that in a joint work with Justesen, to be published later, they proved the following extension of Theorem 2.6 to the range $3k \leqslant n < 24k$, conjectured by Erdös and Pósa.
If $6 \leqslant 3k \leqslant n = |G|$ and

$$e(G) \geqslant \max\left\{ \binom{3k-1}{2} + n - 3k + 2, (2k - 1)n - 2k^2 + k \right\}$$

$$= \max\left\{ \binom{3k-1}{2} + n - 3k + 2, f(n, k) \right\}$$

then $G \in \Omega_k$ or $G = K^{2k-1} + E^{n-2k+1}$.

Let us return to the problem of determining the maximal size of a graph $G^n \in \overline{\Omega}_k$ provided $g(G^n) \geqslant g$. Recall that Theorem 2.5 solves this problem for $g \equiv 0 \pmod 4$ and $k = 2$, but the solution is known [B18] for every g and $k = 2$. Furthermore, Theorem 2.6 provides the answer when $g = 3$, i.e. when the restriction on $g(G^n)$ is discarded. It is very likely that when $g \geqslant 4$ and n is large the structure of the extremal graphs is rather similar to the structure of the extremal graph given in Theorem 2.5. This conjecture, formulated exactly in §6 (Conjecture 22), is supported by the next theorem, which is a slight extension of a result of Erdös and Pósa [EP1]. The theorem, giving the answer in the case $g = 4$, will be proved by making use of the following simple lemma.

LEMMA 2.8. *Let C be a girdle of G^n, i.e. a cycle of length $g(G^n)$.*
If $g(G^n) \geqslant 5$ then

$$e(G^n - C) \geqslant e(G^n) - n.$$

If $g(G^n) = 4$ then

$$e(G^n - C) \geqslant e(G^n) - 2n + 4,$$

with equality iff each vertex of $G^n - C$ is joined to two nonadjacent vertices of the quadrilateral C.

Proof. Note that C is an induced subgraph since it is a girdle. Furthermore, if $x \in G - C$ is joined to $y, z \in C$ and $d_C(y, z) = d$ then G contains a cycle of length $d + 2$. Consequently $d + 2 \geqslant g(D)$ must hold. If $g(C) \geqslant 5$ then this is impossible since $d \leqslant \lfloor \tfrac{1}{2} g(C) \rfloor$. ∎

THEOREM 2.9. *Let $k \geqslant 2$ and $G = G(n, m)$ where $n \geqslant 4k$ and $m = (2k-1)(n - 2k + 1)$. If $g(G) \geqslant 4$ then G contains k independent cycles unless $G = K(2k - 1, n - 2k + 1)$.*

In particular, every graph of order n and size greater than $(2k - 1)(n - 2k + 1)$ contains a triangle or k independent cycles.

Proof. We apply induction on k. For $k = 2$ the assertion is contained in Theorem 2.5. Suppose now that $k > 2$ and the theorem holds for smaller values of k. Let C_1 be a girdle of G.

Assume first that $g = g(G) > 4$. Let C_2 be a girdle of $G - C_1$, let C_3 be a girdle of $G - C_1 - C_2$, etc. Assume that $F = G - C_1 - \ldots - C_s$ does not contain a cycle. If $s \geqslant k$ we are home so we may assume that $s \leqslant k - 1$. Then by Lemma 2.8 we have

$$e(F) \geqslant e(G) - sn = (2k - s - 1)n - (2k - 1)^2$$

and clearly

$$|F| = |G| - \sum_1^s |C_i| \leqslant n - 5s.$$

Since F is a forest,

$$e(F) \leqslant n - 5s - 1$$

and so

$$(2k - s - 1)n - (2k - 1)^2 \leqslant n - 5s - 1,$$

implying

$$(k - 1)n \leqslant (2k - 1)^2 - 5(k - 1) + 1.$$

As this inequality contradicts $n \geqslant 4k$, we see that $g = g(C^n) = 4$.

Now the second part of Lemma 2.8 can be applied to G and its girdle C_1. Thus

$$e(G - C_1) \geqslant m - 2n + 4 = (2k - 3)(n - 4 - (2k - 3)). \qquad (3)$$

By the induction hypothesis we may assume that

$$G - C_1 = K(2k - 3, n - 4 - (2k - 3)) = K(2k - 3, n - 2k - 1)$$

since otherwise $G - C_1 \in \Omega_{k-1}$ and so $G \in \Omega_k$. Note that equality holds in (3). By Lemma 2.8 every vertex of $G - C_1$ is joined to exactly two non-adjacent vertices of the quadrilateral $C_1 = x_1 x_2 x_3 x_4$. Let V_1 be the set of vertices of $G - C_1$ joined to x_1 and x_3 and let V_2 be the set of vertices of $G - C_1$ joined to x_2 and x_4. Then $V(C_1) \cup V_1 \cup V_2$ is a partition of $V(G)$. Since G does not contain a triangle, $G - C_1$ is a bipartite graph with vertex classes V_1 and V_2. As we know already that $G - C_1 = K(2k - 3, n - 2k - 1)$, we may assume that $|V_1| = 2k - 3$ and $G - C_1$ is the complete bipartite graph with vertex classes V_1 and V_2. Then $G = K(2k - 1, n - 2k + 1)$, as claimed, with vertex classes $V_1 \cup \{x_2, x_4\}$ and $V_2 \cup \{x_1, x_3\}$. ∎

There are a number of important results stating that graphs with certain degree sequences contain k independent cycles. The most beautiful of these is due to Corrádi and Hajnal [CH7], who proved that if $\delta(G) \geqslant 2k$ and $|G| \geqslant 3k$ then $G \in \Omega_k$. We shall mention some of these results among the Exercises of Chapter VI, after the proof of a powerful theorem of Hajnal and Szemerédi [HS3], generalizing the above-mentioned theorem of Corrádi and Hajnal.

3. EDGE DISJOINT CYCLES

We shall see that most of the problems concerning edge disjoint cycles are rather different from the corresponding problems concerning vertex disjoint cycles. In fact, there is surprisingly little interaction between the two sets of problems. However, Theorem 2.3, the structure theorem of multigraphs without two independent cycles, enables us to prove an analogous structure theorem. As in the previous section, we shall consider multigraphs with loops though the loops will never really come into play.

FIG. 3.1. The graphs $G \in \overline{\Omega}'_2$ with $\delta(G) \geqslant 3$.

THEOREM 3.1. *Let G be a multigraph without two edge disjoint cycles and with minimal degree $\delta(G) \geqslant 3$. Then G is one of the following three graphs: a triple edge, K^4 and $K^{3,3}$ (Fig. 3.1).*

Proof. Note first that G cannot contain a loop since $\delta(G) \geqslant 3$ implies that the graph obtained from G by deleting an edge (or loop) contains a cycle. Suppose now that G contains two vertices joined by at least two edges, say, both of the edges α and β join u to v. Then $G - \alpha - \beta$ does not contain a cycle so it has at most $n - 1$ edges, where $n = |G|$. On the other hand, $\delta(G) \geqslant 3$ gives $e(G) \geqslant 3n/2$. Consequently $n - 1 \leqslant (3n/2) - 2$ and so $n = 2$. This implies immediately that G is a triple edge.

Finally, if G does not contain a multiple edge then it is a (simple) graph without a vertex representing all the cycles. Therefore G is one of the graphs enumerated in Theorem 2.2. Of those only K^4 and $K^{3,3}$ are graphs without two edge-disjoint cycles. ∎

THEOREM 3.2. *A multigraph G does not contain two edge-disjoint cycles if and only if it can can be obtained from a subdivision G_0 of a loop or of a graph listed in Theorem 3.1 by adding a forest and at most one edge joining each tree of the forest to G_0.*

This structure theorem implies immediately a slight extension of a result of Erdös and Pósa [EP2].

COROLLARY 3.3 *Every $G(n, n + 4)$ contains two-edge-disjoint cycles. Furthermore, a graph $G = G(n, n + 3)$ does not contain two edge-disjoint cycles if and only if G can be obtained from a subdivision G_0 of $K^{3,3}$ by adding a forest and exactly one edge, joining each tree of the forest to G_0.* ∎

When investigating the size of a graph in $\overline{\Omega}'_k$ one might intend to search for an analogue of Theorem 2.6 for edge disjoint cycles. Thus, given n and k

one might wish to determine the maximal size of a graph $G^n \in \overline{\Omega}'_k$. It turns out that this problem is unsolved (cf. Problem 20) even in the weakest sense: the known bounds are just the bounds following from Theorems 1.2 and 1.4 in a trivial way. However, since a subdivision of a graph does not change the maximum number of edge disjoint cycles and increases the order and the size of the graph by the *same* amount, it is more natural to ask a slightly different question. Given k, what is the minimum of p if $G(n, n + p) \in \Omega'_k$ for *every multigraph* $G(n, n + p)$? It is not negligible either that non-trivial progress has been made in connection with this problem.

Given a natural number k put $p(k) = \min\{p: \text{every multigraph } G(n, n + p)$ contains k edge-disjoint cycles$\}$. Note that $p(2) = 4$ by Corollary 3.3. The main aim of this section is to present results concerning the function $p(k)$. We introduce some notation to be used *only* in this section. Denote by $\tilde{\mathscr{E}}_k$ the set of extremal multigraphs for $p(k)$, i.e.

$$\tilde{\mathscr{E}}_k = \{G = G(n, n + p(k) - 1): G \in \overline{\Omega}'_k\}.$$

Define further

$$\tilde{\mathscr{D}}_k = \{G \in \tilde{\mathscr{E}}_k: \delta(G) \geqslant 3\},$$

$$\tilde{\mathscr{C}}_k = \{G \in \tilde{\mathscr{E}}_k: G \text{ is 3-regular}\}.$$

Let us start with the following observation. If $H \in \mathscr{E}_k$ $(k \geqslant 1)$ and G is the union of H and $K^{3,3}$ such that they have one vertex in common then $G \in \overline{\Omega}'_{k+1}$ and G has $p(k) - 1 + 4$ more edges than vertices. Consequently $p(k + 1) \geqslant p(k) + 4$, $k = 1, 2, \ldots$. Note also that if $G = G(n, n + p(k + 1) - 1) \in \tilde{\mathscr{E}}_{k+1}(k \geqslant 1)$ and $g(G) = g$ then omitting the edges of a girdle of G we obtain a multigraph $H = G(n, n + p(k + 1) - 1 - g)$ with at most $k - 1$ edge disjoint cycles. Hence

$$p(k + 1) - 1 - g \leqslant p(k) - 1,$$

implying

$$p(k + 1) \leqslant p(k) + g(G) \quad \text{for} \quad G \in \tilde{\mathscr{E}}_{k+1}. \tag{1}$$

In particular, $g(G) \geqslant 4$ and so a multigraph in $\tilde{\mathscr{E}}_k (k \geqslant 2)$ cannot contain loops and multiple edges, i.e. $\tilde{\mathscr{E}}_k$ consists of (simple) *graphs*.

Given a graph G let G_0 be obtained from G by repeatedly deleting vertices of degree at most 1 and let H be the multigraph homeomorphic to G_0 whose every vertex of degree 2 is on a loop. We call H the *condensation* of G, H = cond G. Note that in this case G and H have the same maximum number of edge disjoint cycles and if $G = G(n, n + p)$, $H = H(n', n' + p')$ then $p \leqslant p'$ with equality if and only if H has exactly as many components as G (i.e. it does not have fewer). In this last case we say that G is a *dilution* of H, G = dil H. (N.B. A graph has infinitely many dilutions so dil H is certainly not unique.) By our remarks above, if $G \in \tilde{\mathscr{E}}_k(k \geqslant 2)$ then H = cond G is a *graph* in $\tilde{\mathscr{D}}_k$ and every graph $G \in \tilde{\mathscr{E}}_k$ is of the form G = dil H, $H \in \tilde{\mathscr{D}}_k$.

Note furthermore that if a graph $G' = G(n + 1, n + 1 + p')$ is obtained from a graph $G = G(n, n + p) \in \tilde{\mathscr{D}}_k$ by splitting a vertex of degree $d \geqslant 4$ into vertices of degree $\geqslant 3$ (cf. p. 16) then G' has at most as many edge disjoint cycles as G, $p = p'$ and G' has one less vertex of degree at least d than G. Thus $G' \in \tilde{\mathscr{D}}_k$. (Clearly G can be obtained from G' by contracting the new edge so G is an elementary contraction of G'.) By repeated applications of this process we obtain a graph in $\tilde{\mathscr{E}}_k$.

Denote by $g^{(3)}(n)$ the maximal girth of a 3-regular graph of order n. By definition

$$n(g^{(3)}(n), 3) \leqslant n$$

(see p. 103). Let

$$g_0^{(3)}(n) = \max\{2k, 2l + 1 : 2^{k+1} - 2 \leqslant n, \ 3 \cdot 2^l - 2 \leqslant n\}.$$

Then Theorem 1.2 can be reformulated as follows:

$$g^{(3)}(n) \leqslant g_0^{(3)}(n). \tag{2}$$

If we take into account that $\tilde{\mathscr{E}}_k \neq \varnothing$ and $G \in \tilde{\mathscr{E}}_k$ has order $2(p(k) - 1)$, since it is 3-regular and has $p(k) - 1$ more edges than vertices, then (1) and (2) imply

$$p(k) \leqslant p(k - 1) + g^{(3)}(2p(k) - 2) \leqslant p(k - 1) + g_0^{(3)}(2p(k) - 2). \tag{3}$$

We summarize these facts in the following theorem.

THEOREM 3.4. (i) $\tilde{\mathscr{E}}_k$ *consists of (simple) graphs and* $\tilde{\mathscr{E}}_k = \{\text{dil } G : H \in \tilde{\mathscr{D}}_k\}$.

(ii) $\tilde{\mathscr{D}}_k = \{G : \text{there exist } G_1, G_2, \ldots, G_n = G \text{ such that } G_1 \in \tilde{\mathscr{E}}_k, G_l \in \tilde{\mathscr{D}}_k \text{ and } G_{l+1} \text{ is an elementary contraction of } G_l\}$.

(iii) *If* $G \in \tilde{\mathscr{E}}_k(k \geqslant 2)$ *then* $g(G) \geqslant 4$.

(iv) *If* $p(k - 1) + g^{(3)}(2n) < n + 1$ *then* $p(k) \leqslant n$.

Proof. We have already proved (i), (ii) and (iii). To see (iv), note that $g^{(3)}(2n)$
$\leqslant g^{(3)}(2n - 2) + 1$, so if $n = p(k)$ then (3) implies

$$n - l \leqslant p(k - 1) + g^{(3)}(2n - 2l - 2)$$

for every $l \geqslant 0$.   ∎

Let us see first what we can say about $p(k)$ for small values of k. Part (i)
of the next result is simply a restatement of Corollary 3.3 but here we deduce
it from Theorem 3.4. Part (ii) is in Moon [M33], the construction in part (iii)
is due to Häggkvist [H2a].

THEOREM 3.5. (i) $p(2) = 4$ *and* $\tilde{\mathcal{D}}_2 = \tilde{\mathcal{C}}_2 = \{K^{3,3}\}$.
   (ii) $p(3) = 10$.
   (iii) $p(4) = 18$.

Proof. (i) As $K^{3,3} \in \overline{\Omega}'_2$, $p(2) \geqslant 4$. Let $G = G^n \in \tilde{\mathcal{C}}_2$. Then $n = 2p(2) - 2 \geqslant 6$.
Let C be a girdle of G, i.e. C has length $g = g(G) \geqslant 4$. Since $G - C$ does not
contain a cycle, it contains at most $n - g - 1$ edges. The 3-regularity implies
that there are at least $3(n - g) - 2(n - g - 1) = n - g + 2$ edges joining
$G - C$ to C. Therefore at least one vertex of $G - C$ is adjacent to at least two
vertices of C. This implies that $g \leqslant 4$ and so $g = 4$. Consequently, again by
3-regularity, exactly 4 edges join C to $G - C$. Hence

$$n - 4 + 2 \leqslant 2,$$

implying

$$n = 6.$$

It is easily checked that $K^{3,3}$ is the only $G(6, 9)$ without a triangle (this is also
a trivial case of Turán's theorem (Theorem VI.1.1)), so $G = K^{3,3}$.

As an elementary contraction of $K^{3,3}$ contains two edge-disjoint triangles,
the proof is complete by Theorem 3.4(ii).

(ii) Note that $g_0^{(3)}(20) = 6$ and so

$$p(2) + g^{(3)}(20) \leqslant 4 + 6 < 11.$$

Consequently, by Theorem 3.4(iv), $p(3) \leqslant 10$.

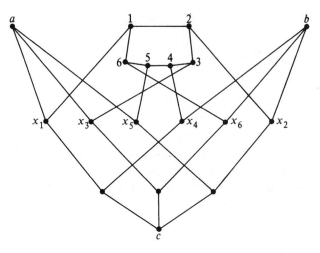

FIG. 3.2. The graph G_3 without 3 edge disjoint cycles.

To prove the converse inequality we show that the cubic graph G_3 = $G(18, 27)$ in Fig. 3.2 does not contain 3 edge disjoint cycles. Note that $G_3 - \{a, b, c\}$ is the hexagon $H = 12 \ldots 6$ together with three paths of length 4 joining diagonally opposite vertices. Therefore H is the only hexagon in $G_3 - \{a, b, c\}$. Furthermore, a hexagon in G_3 that contains c contains one of a and b as well. Thus a hexagon in G_3 is either H or it contains one of a and b, Consequently if G_3 contained three disjoint cycles then one of them would be H and the other two would be in $G_3 - H$. However, $G_3 - H$ is a dilution of $K^{3,3}$ so this is impossible.

(iii) As

$$p(3) + g_0^{(3)}(36) = 10 + 8 < 19,$$

Theorem 3.4 (iv) implies that $p(4) \leqslant 18$.

The converse inequality follows from the fact that the cubic graph of order 34, constructed by Häggkvist [H2a] and shown in Figure 3.3, does not contain three edge disjoint cycles. The not too short proof, based on a case examination, is left to the reader.   ∎

Let us estimate now $p(k)$ for large k. The following result, due to Erdös and Pósa [EP1], shows that the order of $p(k)$ is $k \log k$. (See also Simonovits [S18], Voss [V11] and Walther and Voss [WV1]). As before, $\log_2 x$ is the logarithm of x to base 2.

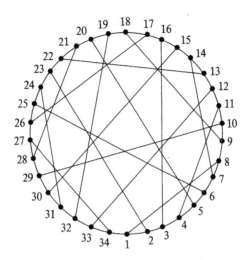

FIG. 3.3. A cubic graph of order 34 without three disjoint cycles.

THEOREM 3.6. $\frac{1}{2}k \log_2 k < p(k) \leq 2k(\log_2 k + \log_2 \log_2 k + 2) = L(k)$ *for all* $k \geq 2$. *Thus*

$$\frac{1}{2}k \log_2 k < p(k) < (2 + o(1))k \log_2 k.$$

Proof. Note that

$$g_0^{(3)}(2m) < 2 \log_2 m + 1,$$

and so

$$L(k - 1) + g_0^3(2L(k)) < L(k)$$

for $k \geq 4$. As

$$p(k) \leq L(k)$$

for $k \leq 4$, this inequality and Theorem 3.4(iv) imply

$$p(k) \leq L(k),$$

as asserted.

Given a natural number $k \geq 3$ let $g(k)$ be a natural number satisfying

$$\frac{k}{2} < \frac{2^{g(k)}}{g(k)} < k.$$

Then clearly

$$g(k) > \log_2 k.$$

Let n be the natural number for which

$$kg(k) - 2 \leqslant 2n < kg(k).$$

By Theorem 1.4′ there exists a 3-regular graph of order

$$2n \geqslant 2^{g(k)}$$

and girth at least $g(k)$. This graph has at most

$$\frac{2n}{g(k)} = k - 1$$

disjoint cycles and at least

$$n \geqslant \frac{kg(k)}{2} - 1 > \tfrac{1}{2}k \log_2 k - 1$$

more edges than vertices, so

$$p(k) > \tfrac{1}{2}k \log_2 k. \qquad\qquad \blacksquare$$

The results about $p(k)$ enable us to estimate the maximal girth of a graph $G(n, m)$.

THEOREM 3.7. (a) *If* $m \geqslant 2n$ *then*

$$g(G(n, m)) < 2 \frac{\log n}{\log\lfloor m/n \rfloor} + 2.$$

Conversely, there is a graph $G^* = G(n, m)$ *such that*

$$g(G^*) \geqslant \frac{\log n}{\log(2m) - \log(n - 1)}.$$

(b) *Let* $n < m < 2n$ *and* $p = m - n$. *Then*

$$g(G(n, m)) < \frac{2m}{p + o(p)} \log_2 p.$$

Conversely, there is a graph $G^*(n, m)$ such that

$$g(G^*) > \left\lfloor \frac{m}{3p} \right\rfloor \log_2 p.$$

Proof. (a) Put $k = \lfloor m/n \rfloor + 1$. Then $G(n, m)$ contains a subgraph H such that $\delta(H) \geqslant k$. If $g(G(n, m)) = g$ then $g(H) \geqslant g$ so by Theorem 1.2 we have

$$n > (k - 1)^{(g/2) - 1} = \lfloor m/n \rfloor^{(g/2) - 1}.$$

Therefore

$$g < 2 \frac{\log n + \log \lfloor m/n \rfloor}{\log \lfloor m/n \rfloor} = 2 \frac{\log n}{\log \lfloor m/n \rfloor} + 2.$$

Conversely, let $k = \lfloor 2m/(n - 1) \rfloor$ and

$$g = \left\lfloor \frac{\log n}{\log(k - 1)} \right\rfloor \geqslant \frac{\log n}{\log(2m) - \log(n - 1)}.$$

Then, by Theorem 1.4, there exists a k-regular graph of order $2\lfloor n/2 \rfloor$ and girth at least g. This graph has at least

$$\left\lfloor \frac{n}{2} \right\rfloor k > \frac{2nm}{2(n - 1)} > m$$

edges and $n - 1$ or n vertices. Consequently it contains a subgraph $G(n, m)$ with

$$g(G(n, m)) \geqslant \frac{\log n}{\log(2m) - \log(n - 1)}.$$

(b) Let k be the maximal natural number for which

$$L(k) = 2k \{ \log_2 k + \log_2 \log_2 k + 2 \} \geqslant p.$$

Then

$$\log_2 p = (1 + o(1)) \log_2 k$$

and

$$k \geqslant \tfrac{1}{2} \frac{p + o(p)}{\log_2 p}.$$

As in $G(n, m)$ there are $p = m - n$ more edges than vertices, it contains at least k edge disjoint cycles. Thus a shortest of these cycles has at most

$$\frac{2m}{p + o(p)} \log_2 p$$

edges.

To show the second inequality recall that by Theorem 1.4 there exists a cubic graph H of order $2p = 2(m - n)$ and girth at least $\lfloor \log_2 p \rfloor + 1 > \log_2 p$. Put $k = \lfloor m/3p \rfloor$ and replace each edge of H by a path of length k. The graph G obtained in this way has $(3k - 1)p \leqslant n$ vertices and $3kp$ edges. Furthermore $g(G) > k \log_2 p$. Let G^* be a graph obtained from G by adding $n - (3k - 1)p$ vertices and joining each vertex to a vertex of G. Then

$$G^* = G(n, m) \quad \text{and} \quad g(G^*) > k \log_2 p = \left\lfloor \frac{m}{3p} \right\rfloor \log_2 p. \qquad \blacksquare$$

Theorem 3.6 can also be used to obtain a result about the number of vertices needed to represent the cycles of a graph.

Recall that if $W \subset V(G)$ then we say that W *represents the cycles* of G if $G - W$ is acyclic. If a graph G contains k vertices representing the cycles then clearly $G \in \overline{\Omega}_{k+1}$. Conversely, for a given $k (\geqslant 1)$ denote by $r(k)$ the *minimal number such that every* $G \in \overline{\Omega}_{k+1}$ *contains at most* $r(k)$ *vertices representing the cycles*. We know from Corollary 2.4(iii) that $r(1) \leqslant 3$ and $G = K^5 \in \overline{\Omega}_2$ shows that $r(1) = 3$. It is not immediately obvious that $r(k)$ is finite for every k. We shall show that this is so and will give bounds on the function $r(k)$. The first bounds on $r(k)$, showing that $r(k)$ is of the order $k \log k$, were proved by Erdös and Pósa [EP2]. These bounds were sharpened by Simonovits [S18] and Voss [V10, V11] (see also Walther and Voss [VW1]).

THEOREM 3.8.

$$\tfrac{1}{4}k \log_2 k < r(k) \leqslant 2L(k + 1) + k = (4 + o(1))k \log_2 k.$$

Proof. (i) By the proof of Theorem 3.6 there exists a cubic graph $G \in \overline{\Omega}_{k+1}$ with $n \geqslant k \log_2 k$ vertices. Then if W is a set of r vertices such that $G - W$ is acyclic, $G - W$ has at most

$$n - r - 1$$

edges. As there are at most $3r$ edges of G incident with vertices in W, and G

has $\frac{3}{2}n$ edges,

$$n - r - 1 + 3r \geqslant \tfrac{3}{2}n.$$

Therefore

$$\tfrac{1}{4}k \log_2 k < \frac{n}{4} + \tfrac{1}{2} \leqslant r \leqslant r(k),$$

as claimed.

(ii) Let $G \in \overline{\Omega}_{k+1}$. We may suppose that G contains a cycle since otherwise the empty set represents the cycles of G. In a cycle every vertex has degree 2 so G contains a *maximal* subgraph H in which every vertex has degree 2 or 3. (If $x \in H$ then $d_H(x) = 2$ or 3.) A vertex of degree at least 3 is a *branchvertex*; the set of branchvertices of a graph F is $W(F)$. Put $B = W(F)$.

Suppose H contains s cycles without branchvertices. These cycles are disjoint, so

$$s \leqslant k,$$

and we can represent these cycles by a set S of s vertices.

The rest of H contains at most $k - s$ disjoint cycles. Consequently, by the proof of Theorem 3.6,

$$|S \cup B| \leqslant s + 2L(k - s + 1) \leqslant 2L(k + 1)$$

$$= 4(k + 1)\{\log_2(k + 1) + \log_2 \log_2(k + 1) + 2\} = (4 + o(1))k \log_2 k.$$

We say that a vertex z of H *belongs to* Z if $z \notin B$ and there is a cycle C_z in G such that $V(C_z) \cap V(H) = \{z\}$. The maximality of H implies that if $z, w \in Z$ then C_z and C_w are disjoint cycles of G. Consequently there are at most k vertices in Z.

It is easily checked that the set

$$S \cup B \cup Z$$

represents all cycles of G so

$$r(k) \leqslant 2L(k + 1) + k = (4 + o(1)) \, k \log_2 k.$$

as claimed.   ∎

Instead of considering *all* graphs of the form $G(n, n + p)$ and asking how many edge disjoint cycles they have to contain, we may restrict our attention to the graphs $G(n, n + p)$ that belong to a certain class, and ask the same question. Here we discuss this question for the class of *planar* graphs. (For the terminology and basic properties of planar graphs see Ch. V § 3.) Put

$p_0(k) = \min\{p$: every planar multigraph $G(n, n + p)$ contains k edge disjoint cycles$\}$.

Dirac and Erdös [DE1] proved $p_0(2) = 3$, Moon [M33] proved $p_0(3) = 7$ and $p_0(4) = 11$ and, using a construction due to Thomassen and relying on the validity of a weak form of the 4CC, Häggkvist [H2a] determined $p_0(k)$ for every k.

THEOREM 3.9. $p_0(k) \geqslant 4k - 5$ *for every* $k \geqslant 2$. *Furthermore, if every planar graph of order n contains at least n/4 independent vertices then* $p_0(k) = 4k - 5$ *for every* $k \geqslant 2$.

Proof. (i) Let G_2 be the plane graph K^4. Then $G_2 \notin \Omega'_2$ and $e(G_2) - |G_2| = 2 = 4.2 - 6$. Suppose now that G_k is a cubic plane graph with an exterior cycle C_k of length 3 or 4 and $e(G_k) - |G_k| = 4k - 6$. Place G_k in the interior of a cycle $C_{k+1} = y_1y_2y_3y_4$, subdivide each edge of C_k by x_1, x_2, x_3, x_4 and add the edges x_iy_i, $1 \leqslant i \leqslant 4$. The obtained plane graph G_{k+1} (see Fig. 3.4) is also cubic and $e(G_{k+1}) - |G_{k+1}| = 4(k + 1) - 6$. It is easily checked that $G_{k+1} \notin \Omega'_{k+1}$. This shows that $p_0(k) \geqslant 4k - 5$ for every $k \geqslant 2$.

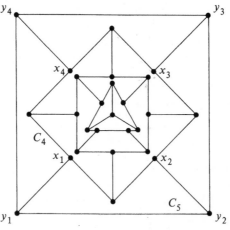

FIG. 3.4. The cubic plane graph G_5 without five disjoint cycles.

(ii) Let $G = G(n, n + p_0(k) - 1) \notin \Omega'_k$ be a plane graph. We may assume that G is connected and cubic, so $G = G(2(p_0(k) - 1), \ 3(p_0(k) - 1))$. By Euler's formula the map determined by G has $p_0(k) + 1$ faces, i.e. the geometric dual G^* of G has order $p_0(k) + 1$. Now if we assume that every plane graph G^n has at least $n/4$ independent vertices (in particular, if we assume the 4CC), then G^* has at least $(p_0(k) + 1)/4$ independent vertices. The cycles of G corresponding to these vertices are edge disjoint, so $(p_0(k) + 1)/4 \leqslant k - 1$, giving $p_0(k) \leqslant 4k - 5$.   ∎

4. THE CIRCUMFERENCE

Recall that the *circumference* of a graph G, denoted by $c(G)$, is the length of a longest cycle in G. We shall obtain estimates of $c(G)$ in terms of the order and the size of G, and also in terms of the degrees of the vertices of G. It turns out that in some cases we are able to take into account the degrees of all the vertices and not only the minimal and maximal degrees, as we usually do.

Before turning to the general problem of estimating the circumference of a graph let us give various conditions ensuring that a graph is Hamiltonian.

Following Nash-Williams [N3] we call a graphic sequence $(d_k)_1^n$ *forcibly Hamiltonian* if every graph with this degree sequence is Hamiltonian. Dirac [D7] (see also [N4]), Ore [O2], Pósa [P9] (see also [N4]) and Bondy [B41] gave successively weaker conditions on a sequence to be forcibly Hamiltonian. Each of these conditions is such that whenever a sequence $d = (d_k)_1^n$ satisfies it then so does every sequence $d' = (d'_k)_1^n$ majorizing d. Naturally we say that d' *majorizes d* or is *greater than d* if $d_k \leqslant d'_k$ for every k, $1 \leqslant k \leqslant n$. Chvátal [C8] gave the weakest possible sufficient condition of this kind, namely he determined all the *maximal* sequences which are not forcibly Hamiltonian. Independently of Chvátal, Dirac [D24] proved various best possible results in the same direction and Las Vergnas [L1] proved a theorem that contains Chvátal's result.

Bondy and Chvátal [BC2] observed that a result proved but not explicitly stated by Ore [O2], which was the driving force in the proofs of the results mentioned above, can be generalized to a principle that has a number of applications. Though the idea is very beautiful and can be used with great success, it is also very simple. In fact it is an immediate consequence of the definitions that follow.

Let P be a property of graphs of order n and let k be a natural number. We say that P is *k-stable* if whenever $G + xy$ has property P and $d_G(x)$

$+ d_G(y) \geqslant k$ then G also has P. Given a graph G (of order n) it is easily seen that there is a *unique maximal* graph H (of order n) containing G such that

$$d_H(x) + d_H(y) \leqslant k - 1 \quad \text{for } xy \notin E(H). \tag{1}$$

In fact H is the graph obtained from G by successively joining vertices whose degree sum is at least k. We call H the *k-closure* of G and denote it by $C_k(G)$.

THEOREM 4.1. *If P is a k-stable property and $C_k(G)$ has property P then so does G itself.*   ∎

If K^n has a k-stable property P then the most natural line of attack to prove that G has P is to show that $C_k(G) = K^n$. Bondy and Chvátal [BC2] proved the following technical lemma that will be used to do this in the case when P is the property of containing a Hamiltonian cycle or Hamiltonian path.

LEMMA 4.2. *Let k, n, t be natural numbers with $k \leqslant 2n - 4$ and $t < n$. Let G be a graph with $V(G) = \{x_1, \ldots, x_n\}$. Suppose there are no vertices x_i, x_j such that*

$$j \geqslant \max\{i + 1, 2n - k - i, n - t + 1\}, \quad x_i x_j \notin E(G)$$

$$d(x_i) \leqslant i + k - n, \qquad d(x_j) \leqslant j + k - n - 1, \tag{2}$$

$$d(x_i) + d(x_j) \leqslant k - 1.$$

Then $C_k(G)$ contains at least t vertices of degree $n - 1$.

Proof. Suppose that $H = C_k(G)$ contains at most $t - 1 \leqslant n - 2$ vertices of degree $n - 1$. Then H is not complete. Put

$$j = \max\{l : d_H(x_l) \neq n - 1\},$$

$$i = \max\{l : x_l x_j \notin E(H)\}.$$

Our aim is to show that, contrary to the assumption, the vertices x_i and x_j satisfy (2). Since $d(x_s) = n - 1$ for $s > j$, we have $n - j \leqslant t - 1$. By (1) we have

$$d_H(x_i) + d_H(x_j) \leqslant k - 1.$$

The choice of j and i gives

$$d_H(x_j) \geqslant n - i - 1$$

and

$$d_H(x_i) \geqslant n - j.$$

These three inequalities imply

$$d_G(x_i) \leqslant d_H(x_i) \leqslant (k - 1) - (n - i - 1) = i + k - n,$$

$$d_G(x_j) \leqslant d_H(x_j) \leqslant (k - 1) - (n - j) = j + k - n - 1,$$

$$i + j \geqslant (n - d_H(x_j) - 1) + (n - d_H(x_i)) \geqslant (2n - 1) - (k - 1) = 2n - k.$$

This completes the proof that x_i and x_j satisfy (2).   ∎

In order to apply Theorem 4.1 and Lemma 4.2 to obtain degree conditions for the existence of Hamiltonian cycles and paths we need the following result first proved by Ore [O2].

LEMMA 4.3. (i) *The property of being Hamiltonian is n-stable.*
(ii) *The property of containing a Hamiltonian path is $(n - 1)$-stable.*

Proof. (i) Suppose $d(x_1) + d(x_n) \geqslant n$ and $G + x_1x_2$ contains a Hamiltonian cycle C. If C is in G, we are home. Otherwise $C - x_1x_n$ is a Hamiltonian path, say $x_1x_2 \ldots x_n$. Since $d(x_1) + d(x_n) \geqslant n$, there is an index i, $2 < i < n$, such that $x_1x_i \in E(G)$ and $x_{i-1}x_n \in E(G)$. Then $x_ix_1x_2 \ldots x_{i-1}x_nx_{n-1} \ldots x_i$ is a Hamiltonian cycle.

(ii) The proof is similar to the proof of (i); we leave it to the reader (Ex. 25⁻).

Combining Theorem 4.1 with Lemmas 4.2 and 4.3(i) we obtain the theorem proved by Las Vergnas [L1]. A number of other results can be obtained by showing that a certain property is k-stable for some fixed k and then combining it with Theorem 4.1 and Lemma 4.2. Some of these possibilities are given as exercises (Exx. 25 and 26). In the next results $n \geqslant 3$.

THEOREM 4.4. *Let G be a graph with $V(G) = \{x_1, x_2, \ldots, x_n\}$. Suppose there* *are no indices i and j such that*

$$j \geqslant \max\{i + 1, n - i\}, \qquad x_i x_j \notin E(G),$$

$$d(x_i) \leqslant i, \qquad\qquad\qquad d(x_j) \leqslant j - 1,$$

$$d(x_i) + d(x_j) \leqslant n - 1.$$

Then G is Hamiltonian.   ∎

As a consequence of this theorem one gets Chvátal's [C8] description of all *maximal* degree sequences which are not forcibly Hamiltonian.

COROLLARY 4.5. (i) *Let* $d_1 \leqslant d_2 \ldots \leqslant d_n$ *be a graphic sequence such that*

$$d_k \leqslant k < \frac{n}{2} \quad \textit{implies} \quad d_{n-k} \geqslant n - k. \tag{3}$$

Then $(d_k)_1^n$ *is forcibly Hamiltonian.*

    (ii) *If a graphic sequence* $d_1 \leqslant d_2 \leqslant \ldots \leqslant d_k$ *does not satisfy* (3) *then it is majorized by a graphic sequence that is not forcibly Hamiltonian.*

Proof. Part (i) is an immediate consequence of Theorem 4.4. To see that (ii) also holds, suppose for some k we have

$$d_k \leqslant k < \frac{n}{2} \quad \text{and} \quad d_{n-k} \leqslant n - k - 1.$$

Then $(d_k)_1^n$ is majorized by the sequence $(d_k')_1^n$ whose first k terms are k, the next $n - 2k$ terms are $n - (k + 1)$ and the last k terms are $n - 1$.

    Put $\bar{H}(p, r, s) = (E^p \cup K^s) + K^r$. Thus the vertex set of $\bar{H}(p, r, s)$ can be partitioned into three subsets, say P, R and $S, |P| = p, |R| = r, |S| = s$, and two vertices are joined if and only if at least one of them is in R or both of them are in S. To complete the proof note that $G_1 = \bar{H}(k, k, n - 2k)$ is not Hamiltonian and its degree sequence is exactly $(d_k')_1^n$.   ∎

It is perhaps worth stating a particularly simple form of this corollary, first proved by Dirac [D7], which is the ancestor of all the results mentioned in the introduction.

COROLLARY 4.5'. *If $\delta(G) \geqslant n/2$ then G is Hamiltonian.* ∎

COROLLARY 4.6. *Let $d_1 \leqslant d_2 \leqslant \ldots \leqslant d_n$ be the degree sequence of G. Suppose $n \geqslant 2$ and*

$$d_k \leqslant k - 1 < \tfrac{1}{2}(n - 1) \quad implies \quad d_{n+1-k} \geqslant n - k. \tag{4}$$

Then G has a Hamiltonian path. Conversely, if $(d_k)_1^n$ does not satisfy (4) then there is a graph G' without a Hamiltonian path whose degree sequence majorizes $(d_k)_1^n$. Even more, if n is even and $(d_k)_1^n$ does not satisfy (4) then there is a graph G' without a 1-factor whose degree sequence majorizes $(d_k)_1^n$.

Proof. Though the first part follows from Theorem 4.1, Lemma 4.2 and Lemma 4.3(ii), we show how it can be deduced from Corollary 4.5. Suppose the monotone increasing degree sequence $(d_k)_1^n$ of G satisfies (4). Put $G^* = G + K^1$. Then it is easy to check that the degree sequence of G^* satisfies (3). Consequently G^* contains a Hamiltonian cycle C and $C - x$ is a Hamiltonian path of G.

If a monotone increasing degree sequence $(d_k)_1^n$ does not satisfy (4) then $d_k \leqslant k - 1 < \tfrac{1}{2}(n - 1)$ and $d_{n+1+k} \leqslant n - k - 1$ for some k. Consequently $(d_k)_1^n$ is majorized by the degree sequence of $G_2 = \bar{H}(k, k - 1, n - 2k + 1)$. The graph G_2 clearly does not contain a Hamiltonian path. In fact G_2 does not contain a set of independent edges such that every vertex in P $(|P| = k)$ is incident with one of the edges. ∎

A similar enlargement of the original graph enables one to deduce a result of Moon and Moser [MM1] on Hamiltonian cycles in bipartite graphs.

COROLLARY 4.7. *Let G be a bipartite graph with vertex classes $U = \{u_1, \ldots, u_n\}$ and $V = \{v_1, \ldots, v_n\}$. Suppose $d(u_1) \leqslant \ldots \leqslant d(u_n)$ and $d(v_1) \leqslant \ldots \leqslant d(v_n)$. If*

$$d(u_k) \leqslant k < n \quad implies \quad d(v_{n-k+1}) \geqslant n - k + 1 \tag{5}$$

then G is Hamiltonian. If (5) does not hold then there is a non-Hamiltonian bipartite graph with vertex classes $A = \{a_1, \ldots, a_n\}$ and $B = \{b_1, \ldots, b_n\}$ such that $d(a_1) \leqslant \ldots \leqslant d(a_n)$, $d(b_1) \leqslant \ldots \leqslant d(b_n)$, $d(u_k) \leqslant d(a_k)$ and $d(v_k) \leqslant d(b_k), k = 1, \ldots, n.$

Proof. Let G^* be the graph obtained from G by adding all the edges $v_i v_j$,

$1 \leqslant i < j \leqslant n$. If G satisfies (5) then G^* satisfies (3) so G^* has a Hamiltonian cycle C. As $|U| = |V|$, C is in G.

Suppose now that (5) is not satisfied, say it fails for k. Then $d(u_i)$ is majorized by $k, k, \ldots, k$ (k times), $n, n, \ldots, n$ ($n - k$ times) and $d(v_i)$ is majorized by $n - k, \ldots, n - k$ ($n - k$ times) and n (k times). These sequences are the degree sequences of the following non-Hamiltonian bipartite graph. Let the colour classes be $A = A_1 \cup A_2$ and $B = B_1 \cup B_2$, $|A_1| = |B_1| = k$ and $|A_2| = |B_2|$ $= n - k$. Join every vertex in A_2 to every vertex in B and every vertex in B_1 to every vertex in A.   ■

An obvious "obstruction" to Hamiltonian cycles is the existence of a set $S \subset V(G)$ for which $G - S$ has more than $|S|$ components. It is reasonable to expect that the degree conditions can be significantly slackened if there is no such obstruction. However, this is not the case. Jung [J4] proved that if $G - S$ has at most $|S|$ components for every S, $n \geqslant 11$ and $d(x) + d(y) \geqslant n - 4$ for every non-adjacent pair of vertices then G is Hamiltonian but there are infinitely many non-Hamiltonian graphs satisfying the conditions with $d(x) + d(y) \geqslant n - 5$.

There are very few results concerning Hamiltonian cycles which are not related to the fact that the property of being Hamiltonian is n-stable. One would like to hope that the flow of results in this area will quicken and will result in a cohesive theory. Answering a question of Szekeres, Erdös and Hobbs [EH6] proved that a 2-connected r-regular graph of order $2n$ is Hamiltonian if $r = n - 1$ or $n - 2$. Furthermore, they showed that there is a constant $c_1 > 0$ such that this holds for $n - c_1\sqrt{n} < r$ as well. This is particularly surprising in view of the above mentioned result of Jung. An essentially best possible result was proved by Bollobás and Hobbs [BH3], who showed that the condition above can be weakened to $n - \frac{1}{9}n \leqslant r$. Very recently W. Jackson further weakened this to $r \geqslant \frac{2}{3}n - 3$, which is best possible.

Let us investigate now the minimal number of edges needed to ensure the existence of certain long cycles. Let us start with another consequence of Corollary 4.4, observed by Erdös [E12].

COROLLARY 4.8. *Let $G = G(n, m)$, $n \geqslant 3$, and suppose that either $\delta = \delta(G)$* $\geqslant (n/2)$ *or*

$$m > \max\left\{\binom{n - \delta}{2} + \delta^2, \binom{\lfloor (n + 2)/2 \rfloor}{2} + \lfloor (n - 1)/2 \rfloor^2\right\}. \qquad (6)$$

Then G is Hamiltonian.

Proof. The maximal degree sequences which are not forcibly Hamiltonian are the degree sequences of $H_k = \overline{H}(k, k, n - 2k)$, $k = 1, \ldots, \lfloor (n - 1)/2 \rfloor$. Therefore $d_1 = \delta \geqslant n/2$ implies that G is Hamiltonian and if $\delta < (n/2)$ and G is not Hamiltonian then G has at most

$$\max\{e(H_k): \delta \leqslant k < (n/2)\}$$

edges. This maximum is exactly the right-hand side of (6).   ∎

Our next aim is to give conditions on G^n ensuring that $c(G^n) \geqslant c$, where c is a fixed integer, $3 \leqslant c \leqslant n$. Note first that if each block of a graph G^n has at most $c - 1$ vertices then $c(G^n) < c$. In particular, if G^n, $n = l(c - 2) + 1$, is a connected graph with l blocks such that each block is a K^{c-1} then G^n has $l\binom{c-1}{2} = (c - 1)(n - 1)/2$ edges and $c(G^n) = c - 1$. Erdös and Gallai [EG1] showed that this example is best possible. Though it would be natural to deduce this theorem from Theorem 4.11 we shall prove it here, independently of Theorem 4.11, since its proof is rather simple.

THEOREM 4.9. *If $G = G(n, m)$, $m \geqslant (c - 1)(n - 1)/2 + 1$, $3 \leqslant c \leqslant n$, then the circumference of G is at least c.*

Proof. The proof we present is due to Woodall [W23]. Fix c and apply induction on n. The result is true for $n = c - 1$ since there is no graph $G(n, m)$. Suppose now that $n \geqslant c$ and the theorem holds for smaller values of n.

Suppose that every cycle of $G = G(n, m)$ has length at most $c - 1$. Note that $\delta(G) > (c - 1)/2$ since otherwise we can pick a vertex x of degree $\delta(G)$ and then by the induction hypothesis $c(G - x) \geqslant c$.

Denote by $p - 1$ the maximal length of a path in G and let $P = x_1 x_2 \ldots x_p$ be a path of length $p - 1$ for which $d = d(x_1) \geqslant c/2$ is maximal. By the maximality of p the vertex x_1 is not joined to a vertex outside P. Furthermore, if $x_1 x_i \in E(G)$ $(i > 2)$ then $i \leqslant c - 1$ since $x_1 x_2 \ldots x_i$ is a cycle of G. Put

$$W = \Gamma'(x_1) = \{x_i : x_1 x_{i+1} \in E(G)\}.$$

For each $x_i \in W$ the path

$$x_i x_{i-1} \ldots x_1 x_{i+1} x_{i+2} \ldots x_p$$

has maximal length so $d(x_i) \leqslant d(x_1) = d$. Similarly to x_1, a vertex $x_i \in W$

can not be joined to x_j, $j \geqslant c$, since then

$$x_i x_{i-1} \ldots x_1 x_{i+1} x_{i+2} \ldots x_j$$

is a cycle of length $j \geqslant c$.

Let now $k = \min\{p, c - 1\}$, $Z = \{x_1, \ldots, x_k\}$. Then each vertex $x_i \in W$ is joined to at most d vertices, all of which are in Z. Thus the number of edges incident with at least one vertex in W is at most

$$|W||Z - W| + \tfrac{1}{2}|W|(d - |Z - W|) = d(k - d) + \tfrac{1}{2}d(2d - k) = \tfrac{1}{2}dk \leqslant \frac{c - 1}{2}d.$$

Consequently the graph $G^* = G - W$ has $n - |W| = n - d$ vertices and at least

$$m - \frac{c - 1}{2}d = \frac{c - 1}{2}(n - d - 1) + 1$$

edges. By the induction hypothesis $c(G^*) \geqslant c$. ∎

An immediate consequence of Theorem 4.9 is a theorem of Dirac [D7], stating that if $\delta(G) \geqslant 2$ then $c(G) \geqslant \delta(G) + 1$. This result can be strengthened considerably if we know that G is 2-connected. The first theorem in this direction is also due to Dirac [D7]. Later Pósa [P10], Bondy [B42], [B43] and Bermond [B11], [B12] proved various extensions of Dirac's result. Here we present a slightly sharper version of the result of Bondy [B43]. (It is also mentioned by Bermond [B12].)

THEOREM 4.10. *Suppose $G = G^n$ is 2-connected, $3 \leqslant c \leqslant n$,*

$$V = V(G) = \{v_1, v_2, \ldots, v_n\}, d_k = d(v_k) \text{ and } d_1 \leqslant d_2 \leqslant \ldots \leqslant d_n.$$

Suppose furthermore that

$$d_k \leqslant k < c/2, k \leqslant 1, d_{l+1} \leqslant l \quad and \quad v_k v_{l+1} \notin E(G) \quad imply \quad k + l \geqslant c.$$

Then $c(G) \geqslant c$.

Proof. Let us suppose that $G = G^n$ is a maximal 2-connected graph satisfying

(7) whose circumference is at most $c - 1$. Then every two nonadjacent vertices are joined by a path of length at least $c - 1$. Let $P = x_1 \ldots x_p (p \geqslant c)$ be a longest path in G such that $d(x_1) + d(x_p)$ is *maximal*. Note that $x_1 x_p \notin E(G)$.

Suppose first that $d(x_1) + d(x_p) \leqslant c - 1$. We can suppose without loss of generality that $d(x_1) = k < c/2$. Note that if $x_1 x_j \in E(G)$ then $x_{j-1} x_{j-2} \ldots x_1 x_j x_{j+1} \ldots x_p$ is also a longest path so by the maximality of $d(x_1) + d(x_p)$ we have $d(x_{j-1}) \leqslant d(x_1) = k$. Consequently $d_k \leqslant k < c/2$. Similarly, if $x_p x_j \in E(G)$ then the maximality of $d(x_1) + d(x_p)$ implies $d(x_{j+1}) \leqslant d(x_p) = l$. Therefore each vertex in

$$\{x_1\} \cup \{x_{j+1} : x_p x_j \in E(G)\} = \{x_1\} \cup \Gamma'(x_p)$$

has degree at most l. As this set has $l + 1$ elements, $d_{l+1} \leqslant l$, contradicting (7). Consequently $d(x_1) + d(x_p) \geqslant c$.

Suppose now that there are suffices i and j such that $1 < i < j < p$ and $x_1 x_j \in E(G)$, $x_i x_p \in E(G)$. Choose i and j in such a way that $j - i$ is *minimal*, say $j - i = t + 1$. Then $x_1 x_2 \ldots x_i x_p x_{p-1} \ldots x_{j+1} x_j$ is a cycle of length $p - t$ so $p - t \leqslant c - 1$. It is easy to check that the sets

$$\Gamma(x_1) = \{x_i : x_1 x_i \in E(G)\}$$

and

$$\Gamma'(x_p) = \{x_{j+1} : x_j x_p \in E(G)\}$$

are disjoint. Furthermore

$$\Gamma(x_1) \cup \Gamma'(x_p) \subset \{x_2, x_3, \ldots, x_{i+1}, x_j, x_{j+1}, \ldots, x_p\},$$

so

$$c \leqslant d(x_1) + d(x_p) = |\Gamma(x_1) \cup \Gamma'(x_p)| \leqslant p - t \leqslant c - 1.$$

This contradiction shows that $j < i$ whenever $x_1 x_j$ and $x_i x_p$ are edges of G.

Let $j_0 = \max\{j : x_1 x_j \in E(G)\}$ and $i_0 = \max\{i : x_p x_j \in E(G)\}$. As G is 2-connected, there is an $x_{f_1} - x_{l_1}$ path P_1 such that x_{f_1}, x_{l_1} are its only vertices on P and $f_1 < j_0 < l_1$. Choose a path P_1 for which l_1 is maximal. If $l_1 > i_0$, we stop here, otherwise we choose an $x_{f_2} - x_{l_2}$ path P_2 such that $f_2 < l_1 < l_2$ and l_2 is maximal. The existence of P_2 follows from the fact that G is 2-connected. Furthermore, as P_1 was chosen in such a way that l_1 was maximal, P_2 is vertex-disjoint from P_1. If $l_2 > i_0$ we stop, if not we repeat the process.

Thus we obtain $x_{f_r} - x_{l_r}$ paths P_r, $r = 1, \ldots, q$, say, such that

$$f_r < l_{r-1} < l_r, f_1 < j_0 < i_0 < l_q$$

and if $i < j$ then P_i and P_j are either vertex disjoint or $j = i + 2$ and $x_{l_i} = x_{f_{i+2}}$ is their only vertex in common.

Let now $j_1 > f_1$ be the minimal integer for which $x_1 x_{j_1} \in E(G)$. Similarly $i_1 < l_q$ is the maximal integer such that $x_{i_1} x_p \in E(G)$. It is easily checked that if q is odd then the cycle

$$x_1 P x_{f_1} P_1 x_{l_1} P x_{f_3} P_3 x_{l_3} \ldots P x_{f_q} P_q x_{l_q} P x_p x_{i_1} P x_{l_{q-1}} P_{q-1} x_{f_{q-1}} \cdots$$
$$P x_{l_2} P_2 x_{f_2} P x_{j_1} x_1,$$

illustrated in Fig. 4.1, has length at least $c + 2$ since it contains x_1 and $d(x_1)$ vertices joined to x_1 together with x_p and $d(x_p)$ vertices joined to x_p.

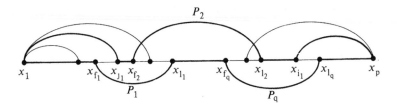

FIG. 4.1. Illustration to Theorem 4.10.

If q is even then the cycle

$$x_1 P x_{f_1} P_1 x_{l_1} P x_{f_3} P_3 x_{l_3} \ldots P x_{f_{q-1}} P_{q-1} x_{l_{q-1}} P x_{i_1} x_p P x_{l_q} P_q x_{f_q}$$
$$P x_{l_{q-2}} P_{q-2} x_{f_{q-2}} \ldots P x_{l_2} P_2 x_{f_2} P x_{j_1} x_1$$

has length $c + 2$. ∎

The condition in Theorem 4.9, unlike the condition in Corollary 4.5, seems to be very far from being best possible. It seems likely that essentially stronger assertions hold (cf. Conjecture 31). Note that if $\kappa(G^n) \geqslant 2$ and $2 \leqslant \delta(G^n) = \delta \leqslant n/2$ then the conditions of Theorem 4.10 are vacuously satisfied with $c = 2\delta$ so $c(G^n) \geqslant 2\delta$.

As we have already mentioned, Theorem 4.9 has been extended by Bondy [B43]. This extension may not seem very interesting in itself but it will play an important role in attacking the problems in the next section.

THEOREM 4.11. *Let C be a cycle of maximal length $c = c(G)$ in $G = G^n$. Then there are at most*

$$\tfrac{1}{2}c(n - c)$$

edges of G with at most one endvertex on C.

Remark. As there are at most $\binom{c}{2}$ edges joining vertices on the cycle C, Theorem 4.9 is indeed an immediate consequence of Theorem 4.11.

Proof. Note first that as C has length $c(G)$, a vertex not on c cannot be joined to vertices that are adjacent on C. In particular a vertex $x \in G - C$ is joined to at most $\lfloor c/2 \rfloor$ vertices of C. Furthermore if $c \geqslant 4$ and there is a path of length at least 2 from x to C then x can be joined to at most $(c - 2)/2$ vertices of C.

We apply induction on c and then on n. If $c = 3$ or $n = c$, the result is trivial. Suppose now that $3 < c < n$ and the result holds for all graphs with circumference less than c and for all graphs with circumference c and order less than n.

If $W \subset V(G) - V(C), |W| = w \geqslant 1$, then we may suppose that there are more then $(c/2)w$ edges incident with at least one vertex in W since otherwise the result follows by applying the induction hypothesis to $G - W$. In particular if $x \in G - C$ then $d(x) \geqslant (c + 1)/2$.

Suppose first that G is not 2-connected. Let $H = G^m$ be an end block of G, that is, a block containing only one cutvertex, say h. Naturally $(V(H) - \{h\}) \cap V(C) = \varnothing$. Every vertex of H different from h has degree at least $\delta(G) \geqslant (c + 1)/2$ so by Theorem 4.10 $\min(c + 1, m) \leqslant c(H) \leqslant c(G) = c$. Thus $m \leqslant c$ and H has at most $\binom{m}{2} \leqslant (c/2)(m - 1)$ edges, By induction $G - (V(H) - \{h\})$ has at most $(c/2)(n - m + 1)$ edges with at most one endvertex in C so G has at most $(c/2)(n - c)$ edges, as claimed. Therefore *we may assume that G is 2-connected.*

Let $B = G^{p+1}$ be an end block of $G^* = G - V(C)$. Putting $V(B) = \{b_1, \ldots, b_{p+1}\}$ we may assume that only b_{p+1} might be joined to vertices in $G^* - B$. Then $p \geqslant 1$ since otherwise $b_1 = b_{p+1}$ has to be joined to at least

$(c + 1)/2$ vertices of C. Similarly, $p = 1$ is also impossible since then b_1 is joined to at least $(c - 1)/2$ vertices of C and there is also a path from b_1 through b_2 to C. Thus $p \geqslant 2$ and so B contains a cycle. Put $d = c(B)$. Then B has at most $(d/2)p$ edges (either by the induction hypothesis or by Theorem 4.9). As there are at least $(c/2)p + \frac{1}{2}$ edges incident with at least one of the p vertices in $W = V(B) - \{b_{p+1}\}$, there are at least

$$\frac{c - d}{2} p + \tfrac{1}{2}$$

edges joining W to C. Therefore we may suppose without loss of generality that

at least $c - d + 1$ edges join b_1 and b_2 to C. \hfill (8)

Furthermore, as G is 2-connected we may assume without loss of generality that the notation is chosen in such a way that the following holds. There exist vertices $x_1 x_{p+1} \in C$, $x_1 \neq x_{p+1}$, such that $b_1 x_1 \in E(G)$ and there is a $b_{p+1} - x_{p+1}$ path P_{p+1} (possibly an edge) with only its endvertices in $V(B) \cup V(C)$.

We shall show that if $b_i, b_j \in B$, $i \neq j$, then there is a $b_i - b_j$ path $P_{i,j}$ of length at least $d/2 = c(B)/2$. For let D be a cycle of length d in B. As B is 2-connected there are vertex-disjoint paths P_i, P_j from b_i and b_j to D. The endvertices of these paths can be joined by two complementary arcs of G so at least one of these paths, say $P_{i,j}$, has length at least $d/2$. Then $P_i P_{i,j} P_j$ is a $b_i - b_j$-path of length at least $d/2$. Consequently if $x \in C$, $x \neq x_{p+1}$ and $b_i x \in E(G)$ then

$$d_C(x, x_{p+1}) \geqslant \left\lceil \frac{d}{2} \right\rceil + 2,$$ \hfill (9)

where $d_C(x, x_{p+1})$ denotes the distance on C. For if this does not hold then there is an $x - x_{p+1}$ path on C with length at least $c - 1 - \lceil d/2 \rceil$. This path can be continued to a cycle by $x b_i P_{i, p+1} b_{p+1} P_{p+1} x_{p+1}$ and this cycle has length at least $c - 1 - \lceil d/2 \rceil + 1 + \lceil d/2 \rceil + 1 = c + 1$. Similarly if $x, y \in C$, $x \neq y$ and $b_1 x, b_2 x \in E(G)$ then

$$d_C(x, y) \geqslant \left\lceil \frac{d}{2} \right\rceil + 2.$$

In particular, b_1 and b_2 are joined by at most 2 edges to 2 adjacent vertices on C and so b_1 and b_2 are joined

by at most $l + 1$ edges to l neighbouring vertices on C. \hfill (10)

Inequality (9) implies that

$$\left\lceil \frac{d}{2} \right\rceil + 2 \leqslant d_C(x_1, x_{p+1}) \leqslant \frac{c}{2}$$

and so

$$c \geqslant 2 \left\lceil \frac{d}{2} \right\rceil + 4.$$

Furthermore, by (9), b_1 and b_2 are joined by at most 2 edges to the $2\lceil d/2 \rceil + 3$ vertices on C within $\lceil d/2 \rceil + 1$ of x_{p+1}. Consequently, by (8), at least $c - d - 1$ edges join b_1 and b_2 to the remaining $c - 2\lceil d/2 \rceil - 3 \leqslant c - d - 3$ vertices of C, contradicting (10).   ■

To conclude this section we mention some results about Hamiltonian cycles in planar graphs and Hamiltonian decompositions of graphs. Some of the terms to be used here will be defined properly only in Chapter V, § 3, but most readers are likely to be familiar with them. In connection with the Four Colour Conjecture (4CC), to be discussed in Chapter V, § 3, Tait [T1] conjectured in 1884 that every cyclically 3-edge-connected cubic planar graph is Hamiltonian. A graph is said to be *cyclically k-edge connected* if the omission of fewer than k edges does not result in a graph having at least two components that contain cycles. If Tait's conjecture holds then so does the 4CC, as we shall point out in Chapter V, § 3. In fact, the 4CC is stronger than the conjecture that every cyclically 5-edge-connected cubic planar graph is Hamiltonian.

Tait's conjecture stood open for over sixty years and was *disproved* by Tutte [T7] in 1946. The graph constructed by Tutte is shown in Fig. 4.2. It is easily seen that the graph is 3-connected (so *a fortiori* cyclically 3-edge-connected) and it can be checked that it does not contain a Hamiltonian cycle. Extending a result of Whitney [W17], Tutte [T13] proved that every 4-connected planar graph is Hamiltonian, so in a way the counterexample above is best possible.

Tait's conjecture fails even in its weakest form: for $k = 3, 4$ and 5 there exist non-Hamiltonian cyclically k-edge-connected cubic planar graphs. (It is trivial that K^4 is the only cyclically 6-edge-connected cubic planar graph.) This was proved by Tutte [T13] for $k = 4$ and by Walther [W6] for $k = 5$. Later Walther [W8], [W9] (see also [W10] and Sachs [S4]) constructed non-Hamiltonian cyclically k-edge-connected r-regular planar graphs for $k = 5, 6$ and $r = 4, 5$. As Sachs [S4] proved that a 4- or 5-regular planar graph which is not 4-connected can not be cyclically 7-edge-connected, this shows that Tait's conjecture fails even in its weakest form.

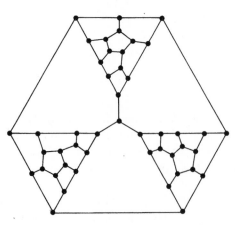

FIG. 4.2. Tutte's counterexample to Tait's conjecture.

A remarkable identity of Grinberg [G18] makes it rather easy to construct non-Hamiltonian planar graphs with various additional properties. Let G be a planar graph containing a Hamiltonian cycle C. Draw H on a sphere in such a way that the edges of C (and so all the vertices of G) are on the equator. The edges of G fall into three classes: the edges in C, the edges in the northern hemisphere and the edges in the southern hemisphere. Denote by n, n_1 and n_2 the number of edges in these classes (thus $|G| = n$). The n_1 diagonal edges decompose the northern hemisphere into $n_1 + 1$ faces. Denote by f_i' the number of i-sided faces in the northern hemisphere. The sum $\sum_{i=3} if_i'$ can be obtained by counting 1 for each edge of C and 2 for each edge in the northern hemisphere. Thus

$$\sum_{i=3}^{\infty} if_i' = n + 2n_1 = n - 2 + 2\sum_{i=3}^{\infty} f_i'$$

$$\sum_{i=3}^{\infty} if_i'' = n - 2 + 2\sum_{i=3}^{\infty} f_i'',$$

where f_i'' denotes the number of i-sided faces in the southern hemisphere. Consequently we arrive at the following identity, due to Grinberg [G18].

THEOREM 4.12. $\displaystyle\sum_{i=2}^{\infty} (i - 2)(f_i' - f_i'') = 0.$ (11)

If in a plane map there are f_i i-sided regions and relation (11) fails, no matter how we partition f_i into non-negative integers f_i' and f_i'', then the graph of the map is non-Hamiltonian. As pointed out by Grinberg and Kozyrev (see [S5]), this is the case if $f_i = 0$ whenever $i \not\equiv 2 \pmod 3$ with the single exception that $f_j \neq 0$ for one particular j with $j \not\equiv 2 \pmod 3$. This shows that the cyclically 5-edge-connected cubic planar graph in Fig. 4.3a, constructed by Grinberg and Kozyrev (see [S5]), is non-Hamiltonian. The

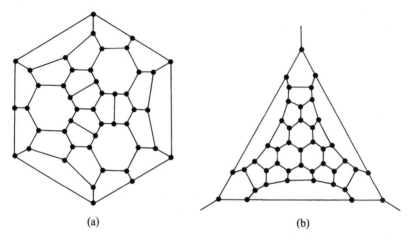

(a)　　　　　　　　　　　　　　　(b)

FIG. 4.3. Cyclically 5-edge-connected non-Hamiltonian cubic planar graphs.

slightly simpler example due to Tutte [T18] is shown in Fig. 4.3b. This graph has 24 faces. With the aid of computers Faulkner and Younger [FY1] proved that there is no example with 23 faces. Grünbaum [G20] constructed non-Hamiltonian cubic planar graphs of cyclic edge-connectivity 3, 4 and 5, respectively, with 21, 23 and 24 faces. Butler [B57] proved that for cyclic edge-connectivity 3 the example of Grünbaum has the minimum number of faces.

There are many problems related to Tait's conjecture but very few of them have positive answers. A positive result is due to Goodey [G15a]. He proved that if G is a 3-connected cubic plane graph each face of which is a triangle or a hexagon then G is Hamiltonian. Nash-Williams [N5] asked whether every 4-connected 4-regular graph is the edge-disjoint union of two Hamiltonian cycles. Meredith [M27] showed that this is not so. Answering a question of Grünbaum and Zaks [GZ1], Martin [M17] and Grünbaum and Malkevitch [GM3] proved that the answer is in the negative even for planar graphs. The strongest counterexample in this area is due to Zaks [Z2]. Answering a question raised in Grünbaum and Zaks [GZ1], Zaks proved that there exists a 5-connected 5-regular planar graph that does not contain two edge-disjoint Hamiltonian cycles.

Smith (see [T7]) proved that in a cubic graph every edge is contained in an even number of Hamiltonian cycles. Recently Thomason [T1b] extended this result considerably. Given a graph G, a vertex x_0 and a set W of vertices of G, denote by $\mathcal{L}(x_0, W)$ the set of Hamiltonian paths beginning at x_0 and ending in W. A graph consisting of a cycle and an $x_0 - x_1$ path, where x_1 is the only common vertex of the cycle and the path, is said to be a φ-*graph* (φ is an episemon, pronounced koppa). The vertex x_0 is the *root* of the φ-graph.

THEOREM 4.13. *Let W be the set of vertices of even degree in a graph G and let $x_0 \in G$. Then $|\mathcal{L}(x_0, W)|$ is even.*

Proof. Let H be the graph with vertex set $\mathcal{L}(x_0, V)$, in which $P_1 P_2$ is an edge iff there is a φ-graph with root x_0 that contains both P_1 and P_2. A φ-graph containing a path $P \in \mathcal{L}(x_0, V)$ clearly spans G and each spanning φ-graph with root x_0 contains exactly two paths from $\mathcal{L}(x_0, V)$. Furthermore, if $P \in \mathcal{L}(x_0, V)$ is an $x_0 - x_1$ path then the addition of every edge $x_1 y$ to P creates a φ-graph with root x_0. Hence the degree of P in H is exactly $d(x_1) - 1$. Therefore $\mathcal{L}(x_0, W)$ is the set of odd vertices of H and so $|\mathcal{L}(x_0, W)|$ is even. ∎

As a corollary of this theorem Thomason obtained an extension of the above mentioned theorem of Smith.

COROLLARY 4.14. *Let G be a graph without a vertex of even degree.*

(i) *Every edge of G is contained in an even number of Hamiltonian cycles.*
(ii) *If G does not contain a cycle of length $n - 1$ then the number of Hamiltonian cycles containing two adjacent edges is even.* ∎

In the vein of Corollary 4.14, Thomason [T1b] proved the following result about Hamiltonian decompositions. We call a pair of edge disjoint Hamiltonian cycles a *Hamiltonian pair*. Given a graph G and edges e and f, we denote by $P(e, f)$ the set of Hamiltonian pairs with e and f in the *same* cycle. $Q(e, f)$ is the set of Hamiltonian pairs with e and f in *different* cycles and P

is the set of *all* Hamiltonian pairs. Clearly if G is 4-regular then $P = P(e, f) \cup Q(e, f)$ for every $e, f \in E(G)$.

THEOREM 4.15. *Let G be a 4-regular multigraph with at least three vertices. Then $|P|$ is even and so are $|P(e, f)|$, $|Q(e, f)|$ for every $e, f \in E(G)$.*

Proof. We may assume that G does not contain a loop since otherwise $P = \varnothing$ and we have nothing to prove. We apply induction on $n = |G|$. If $n = 3$ then G is a triangle with double edges and so the assertions hold. Assume now that $n > 3$ and the theorem is true for smaller values of n.

Note that $Q(e, f) = P - P(e, f)$ and if e_1, e_2, e_3 and e_4 are the edges incident with a vertex x_0 then P is the disjoint union of $P(e_1, e_2)$, $P(e_1, e_3)$ and $P(e_1, e_4)$. This shows that it suffices to prove that $|P(e, f)|$ is even.

Assume first that e and f are adjacent, say $e = x_0 y_1, f = x_0 y_2$. Let the other edges incident with x_0 be $g = x_0 y_3$ and $h = x_0 y_4$, where the y_i need not be distinct. Let G' be the 4-regular multigraph obtained from $G - x_0$ by the addition of two edges: $e' = y_1 y_2$ and $f' = y_3 y_4$. Then $|P_G(e, f)| = |Q_G(e', f')|$, where the suffices indicate the graphs used in the definition of the sets. Hence $|P(e, f)|$ is even and so are $|Q(e, f)| = |P(e, g)| + |P(e, h)|$ and $|P| = |P(e, f)| + |Q(e, f)|$.

Assume now that $|P(e, f)|$ is even whenever e and f are at distance less than k. Let $e_0, e_1, \ldots, e_{k+1}$ be a sequence of edges such that e_i is adjacent to e_{i+1}. To complete the proof it suffices to show that $|P(e_0, e_{k+1})|$ is even. This is indeed so, since

$$|P(e_0, e_{k+1})| \equiv |P| - |Q(e_0, e_{k+1})| \equiv |Q(e_0, e_{k+1})|$$

$$\equiv |P(e_0, e_k) \Delta P(e_k, e_{k+1})| \equiv 0 \pmod 2. \quad \blacksquare$$

COROLLARY 4.16. *If a 4-regular multigraph with at least three vertices has a decomposition into two edge disjoint Hamiltonian cycles then it has at least 4 such decompositions.* $\blacksquare$

5. GRAPHS WITH CYCLES OF GIVEN LENGTHS

We have discussed conditions ensuring that a graph G^n contains short cycles, i.e. its girth is small, and also conditions ensuring that it contains long cycles, i.e. its circumference is large. In this section we are interested in conditions implying both. Furthermore we examine graphs containing a cycle of length r for each r in a given set. In particular we examine the graphs G^n, called *pancyclic* graphs, that contain cycles of all lengths r, $3 \leqslant r \leqslant n$.

In the first four results, due to Bondy [B43], [B44], we show that if a graph has sufficiently many edges and it contains a cycle of length l then it has cycles of all lengths r, $3 \leqslant r \leqslant l$. The first result is an exact form of this statement for the case $l = n$. The simple proof is due to C. Thomassen.

THEOREM 5.1. *Suppose* $G = G(n, m)$ *is Hamiltonian and* $m \geqslant n^2/4$. *Then either* G *is pancyclic or else* $G = K(n/2, n/2)$.

In particular, if $G = G(n, m)$ *is Hamiltonian and* $m > n^2/4$ *then* G *is pancyclic.*

Proof. Apply induction on n. For $n \leqslant 3$ there is nothing to prove so assume $n > 3$ and the result holds for smaller values of n.

Suppose first that G contains a cycle L of length $n - 1$. Let x be the vertex not on L. If $d(x) \geqslant n/2$ then it is immediate that for every r, $1 \leqslant r \leqslant (n-1)/2$, $\Gamma(x)$ contains two vertices at distance r on L. Hence G is pancyclic. If on the other hand $d(x) \leqslant (n-1)/2$ then by the induction hypothesis $G[L]$ is pancyclic since it has at least $n^2/4 - (n-1)/2 > (n-1)^2/4$ edges. Thus G is pancyclic in this case as well.

Assume now that G does not contain a cycle of length $n - 1$. Let $C = x_1 x_2 \ldots x_n$ be a Hamiltonian cycle in G. For the sake of convenience we use the suffices modulo n. Note that

$$\text{at most one of} \quad x_i x_k \quad \text{and} \quad x_{i+1} x_{k+2} \quad \text{is an edge of} \quad G, \tag{1}$$

since otherwise $L = x_k x_{k-1} \ldots x_{i+1} x_{k+2} x_{k+3} \ldots x_i$ is a cycle, and it has length $n - 1$ (see Fig. 5.1). For a fixed i each of the pairs $x_i x_j$ and $x_{i+1} x_j$ occurs exactly once among the $2n$ pairs enumerated in (1) so

$$d(x_i) + d(x_{i+1}) \leqslant n. \tag{2}$$

This implies that $m \leqslant n^2/4$. Hence $m = n^2/4$ and so n is even and equality holds in (2). In turn this implies that for every i and k

$$\text{exactly one of} \quad x_i x_k \quad \text{and} \quad x_{i+1} x_{k+2} \quad \text{is an edge of} \quad G. \tag{3}$$

If for each edge $x_i x_j$ of G we have $i \neq j \pmod 2$ then $G = K(n/2, n/2)$.

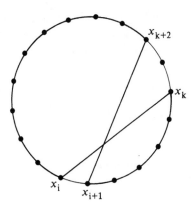

FIG. 5.1. Illustration to (1) and (2).

Supposing that this is not the case let s be the *minimal even* natural number for which G contains an edge of the form $x_{j+1}x_{j+1+s}$. Applying (1) (or (3)) with $i = j$ and $k = s - 1$ we see that $x_j x_{j-1+s} \notin E$; in particular $s \geq 4$. Putting $i = j - 1$ and $k = s - 2$, we see in turn from (3) that $x_{j-1}x_{j-3+s}$ is an edge (see Fig. 5.2). Since this contradicts the choice of s, the proof is complete. ∎

THEOREM 5.2. *Suppose $G = G(n, m)$, $c(G) = c$ and $m > (c/4)(2n - c)$. Then G contains a C^r for each r, $3 \leq r \leq c$.*

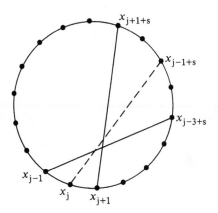

FIG. 5.2. The structure of G in cases (i) and (ii). Vertices joined by a dotted line are nonadjacent.

Proof. Let C be a cycle of length c in G. According to Theorem 4.11 there are at most $(c/2)(n - c)$ edges of G with at most one endvertex on C. Deleting these edges we obtain a graph H of order c with more than $(c/2)\{n - (c/2) - n + c\} = c^2/4$ edges. By Theorem 5.1 the graph H is pancyclic.   ∎

COROLLARY 5.3. *If $G = G(n, m)$ and $m > n^2/4$ then G contains a C^r for each r, $3 \leqslant r \leqslant c(G)$.*   ∎

COROLLARY 5.4. *If $G = G(n, m)$ and $m > n^2/4$ then G contains a C^r for each r, $3 \leqslant r \leqslant \lfloor \frac{1}{2}(n + 3) \rfloor$.*

Proof. $c(G) \geqslant \lfloor \frac{1}{2}(n + 3) \rfloor$ by Theorem 4.9.   ∎

This last Corollary considerably extends a result of Erdös [E13].

Let us turn now to the general problem: given n and l, $3 \leqslant l \leqslant n$, what is the *minimum of m such that every $G(n, m)$ contains a cycle of length r for each r, $3 \leqslant r \leqslant l$?* This minimum is denoted by $e(n, l) + 1$. In fact when investigating this problem we take into account the additional parameter $\delta = \delta(G)$. Denote by $e(n, l, \delta) + 1$ *the minimum of m such that whenever $G = G(n, m)$ satisfies $\delta(G) = \delta$ then G contains a C^r for each r, $3 \leqslant r \leqslant l$.* When examining $e(n, l, \delta)$ *we may assume that $\delta \leqslant n/2$* since by Corollary 4.5' and Theorem 5.1 the condition $\delta(G^n) > n/2$ in itself implies that G^n is pancyclic. Furthermore, for a number of values n, l and δ the results proved so far imply immediately the value of $e(n, l, \delta)$. We shall see that in each case there is an extremal graph of the form $K(p, r)$, $\bar{H}(p, r, s)$ or $H(p, r, s)$. The graph $\bar{H}(p, r, s)$ is the one defined in the proof of Corollary 4.5(ii), $\bar{H}(p, r, s) = (E^p \cup K^s) + K^r$, and $H(p, r, s)$ is obtained from $\bar{H}(p, r, s)$ by joining the vertices in the first class, i.e. $H(p, r, s) = (K^p \cup K^s) + K^r$ (see Fig. 5.3).

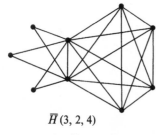   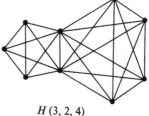

$\bar{H}(3, 2, 4)$ $H(3, 2, 4)$

FIG. 5.3. The extremal graphs $\bar{H}(3, 2, 4)$ and $H(3, 2, 4)$.

Theorem 5.5. (i) *If* $\delta = n/2$ *or* $l \leqslant \lfloor \frac{1}{2}(n + 3) \rfloor$ *(and* $\delta \leqslant n/2$*) then* $e(n, l, \delta) = \lfloor n^2/4 \rfloor$.

(ii) *If* $l \geqslant \frac{1}{2}(n + 3)$ *and* $\max\{\frac{1}{2}l, n - l + 1\} < \delta < n/2$ *then*

$$e(n, l, \delta) = \binom{n - \delta}{2} + \binom{\delta + 1}{2}.$$

Proof. (i) If $\delta = n/2$ then G^n is Hamiltonian by Corollary 4.5 so Theorem 5.1 implies that $e(n, l, n/2) \leqslant (n^2/4)$. The graph $K(n/2, n/2)$ shows that $n^2/4 \leqslant e(n, l, n/2)$.

Corollary 5.4 states exactly that $e(n, l, \delta) \leqslant \lfloor n^2/4 \rfloor$, if $(\delta \leqslant n/2$ and) $l \leqslant \lfloor \frac{1}{2}(n + 3) \rfloor$. The graph $K(\lfloor \frac{1}{2}n \rfloor, \lfloor \frac{1}{2}(n + 1) \rfloor)$ shows the converse inequality.

(ii) If a graph $G = G(n, m)$ is not 2-connected then it is the union of two subgraphs, say G_1 and G_2, $G_1 = G^{s+1}$ and $G_2 = G^{n-2}$ having exactly one vertex in common. Thus $\delta(G) \leqslant \min\{s, n - s + 1\}$ and $m \leqslant \binom{s+1}{2} + \binom{n-s}{2}$. Consequently if $G = G(n, m)$ is such that $\delta(G) = \delta < n/2$ and $m > \binom{\delta+1}{2} + \binom{n-\delta}{2}$ then G is 2-connected. Then Theorem 4.10 implies that $c(G) \geqslant 2\delta \geqslant l$ so by Corollary 5.3 G contains a cycle of length r for each r, $3 \leqslant r \leqslant l$. Thus $e(n, l, \delta) \leqslant \binom{\delta+1}{2} + \binom{n-\delta}{2}$.

The graph $H(\delta, l, n - \delta - 1)$ (i.e. the connected graph with blocks $K^{\delta+1}$ and $K^{n-\delta}$) shows the converse inequality. ■

The determination of $e(n, l, \delta)$ was completed by Woodall [23]. Naturally the result of Woodall includes the most difficult cases, namely the cases in which n is large compared to $n - l$. However, these cases are also easily dealt with since we can make use of a number of results proved so far. We need two more lemmas.

Lemma 5.6. *Suppose* $G = G(n, m)$ *is such that*

$$m \geqslant \binom{n - \delta}{2} + \binom{\delta + 1}{2} + 1,$$

$\delta(G) \geqslant \delta < (n/2)$ *and there is a path of length* l $(\geqslant 3)$ *in* G. *Then* G *contains a* C^r *for each* r, $3 \leqslant r \leqslant l$.

Proof. As $m > n^2/4$, by Corollary 5.3 it suffices to show that $c(G) \geqslant l$. This is done by an argument very similar to the proof of Theorem 4.10.

Suppose $c(G) \leqslant l - 1$. Let $P = x_1 x_2 \ldots x_p$ $(p \geqslant l + 1)$ be a longest path in G such that $d(x_1) + d(x_p)$ is maximal. Put $d_1 = d(x_1), d_p = d(x_p)$ Let

$$V_1 = \Gamma'(x_1) = \{x_{i-1} : x_1 x_i \in E(G)\},$$

$$V_p = \Gamma'(x_p) = \{x_{j+1} : x_p x_j \in E(G)\}.$$

Then $d(x_{i-1}) \leqslant d_1$ for every $x_{i-1} \in V_1$ since the path $x_{i-1} x_{i-2} \ldots x_1 x_i x_{i+1} \ldots x_p$ has the same length as P and also ends in x_p. Similarly $d(x_{j+1}) \leqslant d_p$ for every $x_{j+1} \in V_p$. Furthermore, V_1 and V_p are disjoint since if $x_k \in V_1 \cap V_p$ then $x_1 x_2 \ldots x_{k-1} x_p x_{p-1} \ldots x_{k+1}$ is a cycle of length $p - 1 \geqslant l$. Also, if x_{i_0-1} is the last vertex of P in V_1 then $x_{i_0} \notin V_1 \cup V_p$. Thus V_1 and V_p are disjoint subsets of $\{x_1, x_2, \ldots, x_{i_0-1}, x_{i_0+1}, \ldots, x_p\}$. By the definition of V_1 and V_p

$$|V_1| = d_1 \text{ and } |V_p| = d_p \text{ so } d_1 + d_p \leqslant p - 1 \leqslant n - 1.$$

The d_1 vertices in V_1 have degrees at most d_1 and the d_p vertices in V_p have degree at most d_p so

$$m \leqslant \tfrac{1}{2}\{d_1^2 + d_p^2 + (n - d_1 - d_p)(n - 1)\} = \binom{n}{2} - \tfrac{1}{2}\{d_1(n - 1 - d_1)$$

$$+ d_p(n - 1 - d_p)\}.$$

Note now that $\delta \leqslant d_1 \leqslant n - 1 - d_p \leqslant n - \delta - 1$ so $d_1(n - 1 - d_1) \leqslant \delta(n - 1 - \delta)$ and similarly $d_p(n - 1 - d_p) \leqslant \delta(n - 1 - \delta)$. Consequently

$$m \leqslant \binom{n}{2} - \delta(n - 1 - \delta) = \binom{n - \delta}{2} + \binom{\delta + 1}{2},$$

contradicting the assumption of the lemma.   ∎

LEMMA 5.7. *Let $G = G(n, m)$ be such that $\delta(G) \geqslant \delta$ and*

$$m \geqslant \binom{n - \delta}{2} + \binom{\delta + 1}{2} + 1. \tag{4}$$

Suppose furthermore that $(d_k)_1^n$, $d_1 \leqslant \ldots \leqslant d_n$, the degree sequence of G, is such that

$$d_{n-l+j} > j \quad \text{if} \quad j < \tfrac{1}{2}(l - 1) \tag{5}$$

and

$$d_{n-l+j+1} > j \quad if \quad j = \tfrac{1}{2}(l-1). \tag{6}$$

Then G contains a cycle of length r for each r, $3 \leqslant r \leqslant l$.

Proof. Let us use induction on $s = n - l$. For $s = 0$ the proposition holds by Corollary 4.5 and Theorem 5.1. Suppose now that $s > 0$ and the result holds for smaller values of s.

Put $G^* = G + K^1$. Then G^* satisfies the conditions of the proposition with $n + 1$, $\delta + 1$ and $s - 1$ in place of n, δ and s. Therefore $c(G^*) \geqslant n + 1 - (s - 1) = l + 2$ and so G contains a path of length at least l. The assertion follows from Lemma 5.6.   ∎

We are now ready to prove Woodall's theorem on graphs containing cycles of given lengths.

THEOREM 5.8. *Let $\tfrac{1}{2}(n + 3) \leqslant l \leqslant n, s = n - l$.*

(i) *If $n - l + 1 \leqslant \delta \leqslant l/2$, i.e. $s + 1 \leqslant \delta < \tfrac{1}{2}(n - s)$, then $e(n, l, \delta)$ is the maximum of the following three values:*

$$\binom{n - \delta}{2} + \binom{\delta + 1}{2}, \qquad \binom{l - \delta}{2} + \delta(\delta + n - l),$$

$$\binom{\lceil l/2 \rceil}{2} + \lfloor l/2 \rfloor (n - \lceil l/2 \rceil),$$

where the last value is considered only if l is odd.

(ii) *If $\delta \leqslant n - l + 1 = s + 1$, then*

$$e(n, l, \delta) = \binom{l - 1}{2} + \binom{n - l}{2}.$$

Proof. Denote by $e_0(n, l, \delta)$ the value of $e(n, l, \delta)$ asserted by the theorem.

(i) The graphs $H(\delta, 1, n - \delta - 1)$, $\bar{H}(n - 1 + \delta, \delta, l - 2\delta)$ and, if l is odd, $\bar{H}(n - \lceil l/2 \rceil, \lfloor l/2 \rfloor, 1)$, show that $e(n, l, r) \leqslant e_0(n, l, r)$ (Fig. 5.4).

To show the converse inequality we shall make use of Lemma 5.7. Let $G = G(n, m)$ be such that $\delta(G) \geqslant \delta$ and $m \geqslant e_0(n, l, r) + 1$. Then G satisfies

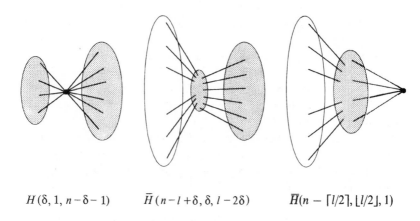

$$H(\delta, 1, n-\delta-1) \qquad \bar{H}(n-l+\delta, \delta, l-2\delta) \qquad \bar{H}(n - \lceil l/2 \rceil, \lfloor l/2 \rfloor, 1)$$

Fig. 5.4. Possible extremal graphs in Theorem 5.8(i).

(4) by the definition of $e_0(n, l, \delta)$. If G does not satisfy (5) then there is an index j $(\geqslant \delta)$ such that

$$d_{n-l+j} \leqslant j < \tfrac{1}{2}(l-1).$$

As $(d_k)_1^n$ is a graphic sequence, by Corollary II.3.5 we have

$$\sum_{k=n-l+j+1}^{n} d_k \leqslant (l-j)(l-j-1) + \sum_{k=1}^{n-l+j} d_k \leqslant (l-j)(l-j-1) + j(n-l+j),$$

that is G has at most as many edges as $H_j = \bar{H}(n-l+j, j, l-2j)$, $\delta \leqslant j < \tfrac{1}{2}(l-1)$.

Similarly one can check that if G does not satisfy (6) then l is odd and G has at most as many edges as $\bar{H}(n-\lceil l/2 \rceil, \lfloor l/2 \rfloor, 1)$. The result follows if we note that $e(j) = e(H_j)$ is quadratic in j, that is H_j with $j = \lfloor l/2 \rfloor$.

(ii) The graph $H(l-2, 1, n-l+1)$ (see Fig. 5.5) shows that $e(n, l, \delta)$ is at least as large as the value claimed. To show that it is at most as large we use induction on $s = n - l$. For $s = 0$ the result holds by Corollaries 4.8 and 5.3. Assume now that $s > 0$ and the result holds for smaller values of s. Let

$G = G(n, m)$ be such that $\delta(G) \geqslant \delta$ and $m \geqslant e_0(n, l, \delta)$. If $\delta(G) > s + 1$ the assertion follows from part (i) or Theorem 5.5. Thus we may suppose that $\delta(G) \leqslant s + 1$. Furthermore, as $e_0(n, l, \delta)$ is a non-increasing function of δ we may assume that $\delta(G) = \delta$. Let $x \in G$ be a vertex of minimal degree, $d(x) = \delta$. Then $G' = G - x$ satisfies the condition in (ii) with $n - 1$, $s - 1$ and $\delta - 1$ in place of n, s and δ. Therefore the induction hypothesis implies the assertion. ∎

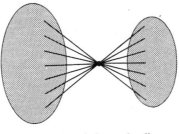

$$H(l - 2, 1, n - l + 1)$$

Fig. 5.5. An extremal graph of Theorem 5.8(ii).

The extremal graphs given in the proof of Theorem 5.8 show also that if $G = G(n, m)$ is such that $\delta(G) = \delta$ and $m = e(n, l, \delta) - 1$ then it does not necessarily follow that $c(G) \geqslant l$. Therefore the number of edges needed to ensure that G^n with $\delta(G^n) = \delta$ has circumference at least l also ensures that G^n contains a cycle of length r for each r, $3 \leqslant r \leqslant l$.

Recall that our original aim was to determine $e(n, l)$, the maximum of m such that not every $G(n, m)$ contains a cycle of length r for each r, $3 \leqslant r \leqslant l$. Since $e(n, l) = \max_{\delta} e(n, l, \delta)$, Corollary 4.8 (together with the graph $K(\lfloor n/2 \rfloor, \lfloor (n + 1)/2 \rfloor)$) and Theorem 5.8 have the following immediate consequence.

THEOREM 5.9.

$$e(n, l) = \lfloor n^2/4 \rfloor \qquad if \quad l < \tfrac{1}{2}(n + 3)$$

and

$$e(n, l) = \binom{l - 1}{2} + \binom{n - l + 2}{2} \qquad if \quad l \geqslant \tfrac{1}{2}(n + 3). \qquad ∎$$

The following reformulation of Theorem 5.9 seems to say even more about the set of cycles contained in a graph $G(n, m)$. Put

$$C(G(n, m)) = \{r: \ \forall G(n, m) \exists C^r \subset G(n, m)\}.$$

Then, as remarked after the proof of Theorem 5.8, the existence of the extremal graphs shown in Figs. 5.4 and 5.5 implies that if $m > n^2/4$ then

$$C(G(n, m)) = \{r: \ 3 \leqslant r \leqslant n, e(n, r) \leqslant m\}$$

and so Theorem 5.9 is equivalent to the following.

THEOREM 5.9'. *If* $m > n^2/4$ *then*

$$C(G(n, m)) = \left\{r: \ 3 \leqslant r \leqslant l, \lceil n/2 \rceil + 1 \leqslant l \ and \ \binom{l-1}{2} + \binom{n-l}{2} + 1 \leqslant m\right\}.$$

∎

It is not surprising that at present $C(G(n, m))$ is not known for every pair $n, m, n < m$. (If $m \leqslant n$ the problem is trivial.) Let us see first at most how large $C(G(n, m))$ can be. If $m \leqslant n^2/4$ then $C(G(n, m))$ consists of even numbers only since there exists a graph $G(n, m)$ which is bipartite. For the sake of convenience suppose now that n is even. Put $\delta = \lceil 2m/n \rceil$ and let g be the maximal natural number for which

$$2 \sum_0^{g-2} (\delta - 1)^i \leqslant n.$$

Then by Theorem 1.4' there exists a δ-regular graph G of order n with girth at least g. Clearly G contains a subgraph $G(n, m)$ so

$$C(G(n, m)) \subset \{g, g + 1, \ldots, n\}.$$

As we mentioned in § 1, Theorems 1.4 and 1.4' seem to be far from being best possible and it is likely that g can be chosen to be the maximal integer satisfying

$$C\delta^{g/2} \leqslant n$$

for some constant C. On the other hand if l and s are natural numbers,

$l(s - 1) = n - 1$ and $G = G(n, m) = K^1 + lK^{s-1}$ (i.e. G has l blocks and each block is a K^s) then

$$m = l\binom{s}{2}, \quad s = \frac{2m}{n-1},$$

and

$$c(G) = s.$$

Thus for this pair n, m we have

$$C(G(n, m)) \subset \{3, 4, \ldots, 2m/(n - 1)\}.$$

In particular, if $k \geqslant 2$ is a fixed integer, $\varepsilon > 0$, $m = \lfloor n^{1 + 1/k} \rfloor$ and n is sufficiently large then

$$C(G(n, m)) \subset \{2l: \ k/2 \leqslant l \leqslant (1 + \varepsilon)n^{1/k}\}.$$

and it is very likely that, in fact,

$$C(G(n, m)) \subset \{2l: \ k \leqslant l \leqslant (1 + \varepsilon)n^{1/k}\}.$$

It was conjectured and for $k = 2$ proved by Erdös [E17], [ES5] that essentially the converse of this last inclusion holds as well. More precisely, Erdös conjectured that for every k ($\geqslant 2$) there is a constant c_k such that if $m \geqslant c_k n^{1 + (1/k)}$ then

$$C(G(n, m)) \supset \{l: \ k \leqslant l \leqslant n^{1/k}\}.$$

This conjecture was proved by Bondy and Simonovits [BS3]. The proof of their result (Theorem 5.11) is based on a lemma about colourings (of the vertices) of a graph.

Though the lemma will be applied to a *proper* vertex colouring of a graph, we shall state and prove the lemma for arbitrary, not necessarily proper, colourings of the vertices. A (not necessarily proper) colouring of a graph is said to be *t-periodic* if the existence of a (simple) *x-y* path of length *t* implies that *x* and *y* have the same colour.

LEMMA 5.10. *Let* $G = G(n, m)$ *be a connected graph,* $m \geqslant (2t - 2)n, t \geqslant 1$.

Then a t-periodic (not necessarily proper) colouring of G is also 2-periodic. In particular, there are at most 2 colours in the colouring.

Proof. Note first that if $l > t$ then a t-periodic (not necessarily proper) colouring of C^l is also s-periodic where s is the greatest common divisor of t and l. Furthermore, if a cycle C^l ($l < t$) of G is s-periodic for some $s \leqslant t$ then the connectedness of G implies that G itself is s-periodic. Thus if t_0 denotes the minimal integer for which G is t_0-periodic and G contains a cycle of length $l > t \geqslant t_0$ then t_0 divides l.

Suppose now that G contains an edge xy and two disjoint $x - y$ paths of length at least p. We call the subgraph formed by these two $x - y$ paths and the edge xy a *θ-graph*. This θ-graph has three cycles, say of lengths l_1, l_2 and l_3 such that

$$l_1 + l_2 - l_3 = 2 \text{ and } p < l_1, l_2, l_3.$$

Since t_0 divides each l_i, t_0 divides 2 so $t_0 \leqslant 2$ as claimed. Thus to complete the proof it suffices to show that G does contain a θ-graph described above.

By (0.5) there is a subgraph H of G such that $\delta(H) \geqslant 2t - 1$. Let P be a maximal path in H, say $P = x_1 x_2 \ldots x_p$. Then every vertex adjacent to x_1 is on P. Denote by $x_{i_1}, x_{i_2}, \ldots, x_{i_s} (s \geqslant 2t - 1, i_j < i_{j+1})$ the vertices adjacent to x_1. Put $x = x_1, y = x_{i_s}, P_1 = x_1 x_2 \ldots x_{i_s}, P_2 = x_1 x_{i_s} x_{i_s - 1} \ldots x_{i_t}$. Then P_1, P_2 and the edge xy form a θ-graph with the required properties. ∎

THEOREM 5.11. *Let k, n and m be natural numbers such that*

$$m > 90kn^{1 + 1/k}.$$

Then $G = G(n, m)$ contains a cycle of length $2l$ for every integer l, $k \leqslant l \leqslant kn^{1/k}$.

Proof. Let l, n and m be natural numbers such that

$$m \geqslant \max\{90ln, 10ln^{1 + 1/l}\} = m_0.$$

To prove the theorem it suffices to show that if $G = G(n, m)$ then $C^{2l} \subset G$.

By (0.4) the graph G contains a bipartite subgraph with at least $m_0/2$ edges. In turn, by (0.5), that subgraph has a subgraph G_0 such that

$$\delta(G_0) \geqslant \left\lceil \frac{m_0}{2n} \right\rceil = \delta \geqslant \max\{45l, 5ln^{1/l}\}.$$

Suppose this bipartite graph G_0 contains no C^{2l}. Pick a vertex c of G_0 and put

$$V_i = \{x \in G_0: \quad d(x, c) = i\}.$$

We shall show that the assumption that G_0 has no C^{2l} implies

$$|V_{i+1}|/|V_i| \geqslant \frac{\delta}{5l}, \qquad i = 0, 1, \ldots, l - 1. \tag{7}$$

Since $\delta/5l \geqslant n^{1/l}$, (7) will give $|V_{l-1}| \geqslant n$ and this contradiction will complete the proof.

Let $H_i = G(n_i', m_i')$ be a component of the subgraph of G_0 spanned by $V_i \cup V_{i+1}$. Thus $n_i' = |H_i|$ and $m_i' = e(H_i)$. Our first aim is to show that

$$m_i' < 4ln_i', \qquad i = 1, \ldots, l - 1. \tag{8}$$

Put $W_1 = V(H_i) \cap V_i, W_2 = V(H_i) \cap V_{i+1}$. We may suppose that

$$|W_1| \geqslant 2 \quad \text{and} \quad |W_2| \geqslant 2$$

since otherwise (8) follows immediately.

We say that an x–y path P is *monotone* if $P = x_1 x_2 \ldots x_u$, $x = x_1$, $y = x_u$ and $x_j \in V_{s+j}$ $(j = 1, \ldots, u)$ for some s (i.e. if $d(x_j, c)$ is monotone increasing). Note that if $x \in V_r$, $Y \in V_s$, $r < s$, then the length of a monotone x–y path is exactly $s - r$. Let h be the *minimum* integer for which there are vertices $a \in V_h$, $x_1, x_2 \in W_1$ and two monotone paths P_1 and P_2 such that P_1 is an a–x_1 path, P_2 is an a–x_2 path and P_1 and P_2 are *independent*, i.e. have only the vertex a in common. Note that for every $x \in W_1$ there is a monotone a–x path since otherwise a could be replaced by the last common vertex of a monotone c–x path and a monotone c–x_1 path containing a.

Let us colour the vertices in W_1 with two colours, say red and blue, as follows. Colour $x \in W_1$ with red if there is a monotone a–x path P_x independent of P_1. Otherwise colour $y \in W_1$ with blue. Furthermore, colour the vertices in W_2 with green. Then x_1 is red, x_2 is blue and the vertices in W_2 are green so this colouring of H_i (which happens to be proper) uses exactly three colours. Let us show that this colouring of H_i is t-periodic with $t = 2(l - i + h)$. If a path of length t begins in a green vertex then it also ends in a green vertex since H_i is bipartite. Suppose now that there is an x–y path P of

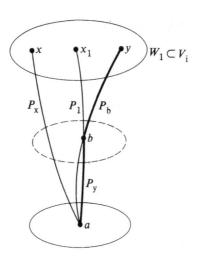

Fig. 5.6. Finding a path P_y independent of P_x.

length t such that x is red and y is blue. Let P_x be a monotone a–x path independent of P_1 and let b be the last vertex on the a–x_1 path P_1 for which there is a monotone b–y path P_b having only the vertex b in common with P_1(see Fig. 5.6). Note that P_b is disjoint from P_x for otherwise we could construct a monotone a–y path independent of P_1. Put $P_y = aP_1 bP_b y$. Then P_y is a monotone a–y path independent of P_x. Consequently P_y, P and P_x form a cycle of length

$$(i - h) + 2(l - i + h) + (i - h) = 2l,$$

contradicting the hypothesis. Therefore this colouring of H_i with three colours is indeed t-periodic with $t \leqslant 2(l - 1)$ and so Lemma 5.10 implies (8).

Denote by $G_i = G(n_i, m_i)$ the subgraph of G_0 spanned by $V_i \cup V_{i+1}$. Then, as (8) holds for every component H_i of G_i, we have

$$m_i < 4ln_i, \qquad i = 0, 1, \ldots, l - 1. \tag{8'}$$

Inequality (8') enables us to prove (7) by induction on i. For $i = 0$ inequality (7) holds since

$$d(c) \geqslant \delta(G_0) \geqslant \delta > \frac{\delta}{5l}.$$

Suppose now that $0 < i \leqslant l - 1$ and (7) holds for smaller values of i. As every vertex has degree at least δ,

$$\delta |V_i| \leqslant m_{i-1} + m_i.$$

Making use of inequality (8') to estimate m_{i-1} and m_i, and the induction hypothesis, we obtain

$$\delta |V_i| < 4l\{|V_{i-1}| + 2|V_i| + |V_{i+1}|\}$$

$$< 4l|V_i| (2 + 5l/\delta) + 4l|V_{i+1}|.$$

Since $\delta \geqslant 45l$, this gives

$$\frac{|V_{i+1}|}{|V_i|} > \frac{1}{4l}\left(\delta - 8l - \frac{20l^2}{\delta}\right) \geqslant \frac{1}{4l}(\delta - 9l) \geqslant \frac{\delta}{5l},$$

proving (7) and completing the proof of the theorem. ∎

6. EXERCISES, PROBLEMS AND CONJECTURES

1^-. Let G be a graph which is not a forest (i.e. contains at least one cycle). Prove that

$$g(G) \leqslant 2 \operatorname{diam} G + 1.$$

2. Let G^n be a δ-regular graph of girth at least g. Denote by N the number of cycles of length at most $2g - 1$ in G. Prove that if $p > N$ is a prime then there is a δ-regular graph G^{pn} of girth at most $2g$.

(Hint. Take the disjoint union of p copies of G^n, say $G_1, \ldots, G_p$. Denote by x_i the vertex of G_i corresponding to $x \in V(G)$. For a given function $\phi : E(G^n) \to \mathbb{Z}$ define G^{np}_ϕ on the vertex set $\bigcup_1^p V(G_i)$ as follows: if $ab \in E(G^n)$ join a_i to b_j for $j \equiv i + \phi(ab) \pmod{p}$. Prove that ϕ can be chosen in such a way that $g(G^{np}_\phi) \geqslant 2g$.)

3. Let $r \geqslant 3$, $m \geqslant 3$ and let $\alpha_3, \alpha_4, \ldots, \alpha_m$ be non-negative integers. Prove that there exists an r-regular graph containing exactly α_i cycles of length i, $3 \leqslant i \leqslant m$.

(Hint. Start with a set of disjoint cycles, α_i of which have length i, and imitate the proof of Theorem 1.4'.) (Sachs [S2], [S3].)

4. Let G be a δ-regular graph of girth at least g with minimal order. Prove that diam $G \leqslant g$.
[Hint. If $d(x, y) > g$, omit x and y and add k appropriate edges to G.]
(Note that by the result of Ex. 2 or Theorem 1.1 this implies

$$n(g, \delta) \leqslant 1 + \delta \frac{(\delta - 1)^g - 1}{q - 2}.)$$

(Erdös and Sachs [ES1].)

5. Show that Moore graphs of degree δ and girth 6 are in one to one correspondence with the finite abstract projective planes (i.e. planes not necessarily over a certain field) with δ points on a line. The Moore graph corresponding to a plane P is the bipartite graph with the set of points and the set of lines of P as the two colour classes. A "point vertex" is adjacent to a "line vertex" if and only if the point of the projective plane is incident with the line. Recalling that there are two different projective planes of order 9 (i.e. with $\delta = 10$ points on a line) conclude that Moore graphs are not necessarily unique.

Note that if p is a prime and s is a natural number,

$$n(6, p^s + 1) = 2(1 + p^s + p^{2s}).$$

(Kárteszi [K1], [K2], see also Singleton [S21] and Brown [B50].)

6⁻. Deduce from the result of the previous problem that

$$\lim_{\delta \to \infty} n(6, \delta)/\delta^2 = 2.$$

7. Based on the result of Ex. 5 note that $n(6, 3) = 14$ and produce another drawing of the Heawood graph (G_6 of Fig. 1.2).
Prove furthermore that the Heawood graph is the unique Moore graph of degree 3 and girth 6.

8. Show that the Petersen graph (G_5 of Fig. 1.1) is the only Moore graph of degree 3 and diameter 2. Construct a Moore graph of degree 7 and diameter 2.

9. Note that the graphs in Fig. 6.1 have girths 7 and 5. Deduce that $n(7, 3) = 24$ and $n(5, 4) = 19$.
(McGee [M25] and Robertson [R10].)

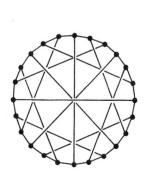
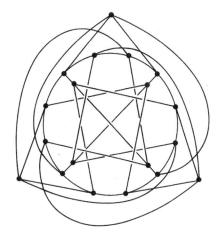

Fig. 6.1. Extremal regular graphs

10. PROBLEM. Denote by $\underset{\sim}{n}(g, \delta)$ the right-hand side of (1.1) (Theorem 1.2). Let $k \geqslant 2$ and $g \geqslant 4$ be integers and suppose $G = G(n, kn/2)$ has girth at least g. Is it true that

$$n \geqslant \underset{\sim}{n}(g, \delta)?$$

11. PROBLEM. (Cf. Problem 10.) Is it true that

$$n(g, \delta) - \underset{\sim}{n}(g, \delta)$$

cannot be arbitrarily large?

The answer is likely to be in the affirmative but we seem to be far from a proof. We cannot even answer the following.

Is there an $\varepsilon > 0$ such that if $\delta \geqslant 3$ is given and g is sufficiently large than

$$n(g, \delta) < (\underset{\sim}{n}(g, \delta))^{2-\varepsilon}?$$

In other words decide whether or not the exponents $g - 1$ and $g - 2$ in Theorem 1.4 can be replaced by $(1 - \varepsilon)g$ for some $\varepsilon > 0$.

12. PROBLEM. (This is the simplest case of Problem 11.) Prove or disprove

$$\lim_{g \to \infty} g^{-1} \log_2 n(g, 3) < 1.$$

13. CONJECTURE. There is a constant $C > 0$ such that

$$n(g, \delta) < C \, \underset{\tilde{}}{n}(g, \delta)$$

for every $g \geqslant 4$ and $\delta \geqslant 3$.

14[+]. Denote by $f(k)$ the minimal size of a k-regular *non-bipartite* graph of girth 4 and by $f(k, 2)$ the minimal size of a k-regular *non-bipartite* graph of girth 4 and diameter 2. Prove that

$$f(2k) = f(2k, 2) = 5k \quad \text{if} \quad k > 2$$

and

$$f(3(2k + 1), 3) = 8(2k + 1) \text{ if } k \geqslant 1.$$

(Sheehan [S13].)

15. PROBLEM. Let $g(G) \geqslant 4$. Can G be oriented in such a way that it contains no directed cycle and if we reverse the direction of any of its edges the resulting graph does not contain a directed cycle either? (P. Erdös.)

16. PROBLEM. Denote by $r(g, \delta)$ the minimum order of a δ-regular graph of girth at least g. Brown [B53] proved that if $\delta > 5$ then

$$\delta^2 + 3 \leqslant r(5, \delta) \leqslant 2(2\delta - 1)(\delta - 2).$$

Furthermore, given $\varepsilon > 0$ we have

$$r(5, \delta) \leqslant (2 + \varepsilon)\delta^2.$$

provided δ is sufficiently large. Prove the existence of the limit

$$\lim_{\delta \to \infty} r(5, \delta)/\delta^2.$$

17. Deduce from Kuratowski's theorem (Theorem V.3.1) that if $n \geqslant 6$ and $G = G(n, 3n - 5)$ then $G \in \Omega_2$, i.e. G contains two vertex disjoint cycles.

18. Deduce from Kuratowski's theorem that if G does not contain two vertex disjoint cycles then the cycles of G can be represented by three vertices. (Bollobás, see [EP2].)

19[-]. Prove that if $\delta(G) \geqslant 3k$ then $G \in \Omega'_k$.

20. PROBLEM. Determine the order of

$$\max\{k: \quad G(n, dn) \in \Omega'_k\}$$

as a function of the natural numbers n and d, $n > 2d$. To be precise find a lower bound $L(n, d)$ and an upper bound $U(n, d)$ such that

$$\sup_{n,d} U(n, d)/L(n, d) < \infty.$$

21. Suppose x is not a cut-vertex and its degree is at least $2k$. Prove that there are at least k edge-disjoint cycles containing x. (Dirac [D14].)

22. CONJECTURE. Let k and l be fixed natural numbers. Let $n = r(2k - 1)(l - 1) + 1 + 2k - 1$. Show that if r is sufficiently large and $G = G(n, m) \notin \Omega_k$ has girth at least $4l$ then

$$m \leqslant rl(2k - 1).$$

The graph obtained from $K(2k - 1, n - 2k + 1)$ by subdividing each edge with $l - 1$ vertices shows that the conjecture, if true, is best possible. (Cf. Theorems 2.5 and 2.9.)

23. Give detailed proofs of Theorem 3.2 and Corollary 3.3.

24. Suppose G^n ($n \geqslant 3$) is s-connected and contains at most t independent vertices. Prove the following assertions.

  (i) If $t = s$ then G^n contains a Hamiltonian cycle.
  (ii) If $t = s + 1$ then G contains a Hamiltonian path.
  (iii) If $t = s - 1$ then every two vertices of G^p are joined by a Hamiltonian path.

[Hint to (i). Consider a longest cycle C. If C does not contain a vertex x, there are s independent paths from x to C. Consider the set of successors of the endvertices in a fixed cyclic ordering of the vertices of C.] (Chvátal and Erdös [CE1].)

25⁻. Show that the property of containing a P^l is $(n - 1)$-stable.

26. Note that if a graph of order $n \geqslant l$ contains $\lfloor l/2 \rfloor$ vertices of degree $n - 1$ then it contains a P^l. State the result implied by Theorem 4.1, Lemma 4.2 and Ex. 24 concerning the existence of a P^l. What are the conditions for $l = n - 2$ and $l = 2$? (Bondy and Chvátal [BC2].)

27. Show that the property of containing a C^l is $(2n - l)$-stable $(3 \leqslant l \leqslant n)$ and if a graph contains $\lfloor (l + 1)/2 \rfloor$ vertices of degree $n - 1$ then it contains a C^l. As in Ex. 26, deduce from Theorem 4.1 and Lemma 4.2 some conditions implying the existence of a C^l. What are these conditions for $l = n - 1$ and $l = 3$? (Bondy and Chvátal [BC2]; see also Las Vergnas [L1].)

28$^-$. Prove that the maximal size of a non-Hamiltonian graph of order $n \geqslant 2$ is $\frac{1}{2}(n^2 - 3n + 4)$. Show furthermore that if $n \geqslant 6$ then there is a unique extremal graph, a K^{n-1} together with another vertex joined to K^{n-1} by an edge. (Ore [O3] and Bondy [B46].)

29. Let $(d_k)_1^n$ be the degree sequence of G^n, $d_1 \leqslant d_2 \leqslant \ldots \leqslant d_n$, $d_1 + d_2 < n$. Denote by l the maximal length of a path in G^n. Prove that

$$l \geqslant d_1 + d_2 + 1$$

unless $G^n = G^{d_1+1} + E^{n-d_1-1}$ for some graph G^{d_1+1} or $G = pK^{d_1} + K^1$ for some $p \geqslant 3$ $(n = pd_1 + 1)$. (Ore [O5].)

30. Let $d = (d_l)_1^n$, $n \geqslant 3$, be a maximal degree sequence which is not forcibly Hamiltonian. Prove that if G is a non-Hamiltonian graph with degree sequence $(d_l)_1^n$ then G is $\bar{H}(k, k, n - 2k)$ for some k, $1 \leqslant k < (n/2)$.

31. Call a graph G m-Hamiltonian if the removal of any l, $0 \leqslant l \leqslant m$, of the vertices results in a Hamiltonian graph. Prove that if the degree sequence $(d_k)_1^n$, $n \geqslant m + 3$, of G is such that

$$d_k \leqslant k + l < \tfrac{1}{2}(n + l) \text{ implies } d_{n-l-k} \geqslant n - k \quad (0 \leqslant l \leqslant m)$$

then G is m-Hamiltonian.

Show also that if $(d_k)_1^n$ does not satisfy the conditions above then it is majorized by the degree sequence of a graph which is not m-Hamiltonian. (Give two proofs for the first half, one based on Corollary 4.5 and another based directly on Theorem 4.1 and Lemma 4.2). (Chartrand, Kapoor and Lick [CKL1].)

32. CONJECTURE. Let G be a 2-connected graph with $V(G) = \{x_1, \ldots, x_n\}$ and let $3 \leqslant c \leqslant n$. Suppose there is no pair of vertices (x_j, x_k) such that

$$c - k \leqslant j < k, \qquad x_j x_k \notin E(G), \qquad d(x_j) \leqslant j,$$

$$d(x_k) \leqslant k - 1 \quad \text{and} \quad d(x_j) + d(x_k) \leqslant c - 1.$$

Then $c(G) \geqslant c$. (Bermond [B12].)

33. PROBLEM. Let $d_1 \leqslant d_2 \leqslant \ldots \leqslant d_n$ be the degree sequence of a 2-connected graph G. Suppose $3 \leqslant c \leqslant n$ and if $d_k \leqslant k < c/2$ then $d_{n-k} \geqslant c - k$. Does it follow that $c(G) \geqslant c$?

34. CONJECTURE. Let $V(G) = \{x_1, x_2, \ldots, x_n\}$ and $4 \leqslant c \leqslant n$. Suppose there is no pair of vertices (x_j, x_k) such that

$$j < k, \quad x_j x_k \notin E(G), \quad d(x_j) \leqslant j - 1, \quad d(x_k) \leqslant k - 2$$

and

$$d(x_j) + d(x_k) \leqslant \min\{2c - 4, n - 1\}.$$

Then $c(G) \geqslant c$. (Bermond [B12].)

35. PROBLEM. Given n let m be the minimal integer for which

$$C(n) = \bigcap_{G(n, m)} C(G(n, m)) \neq \varnothing,$$

where the intersection is taken over all graphs $G(n, m)$. Give bounds on the elements of the set $C(n)$. (If $k \in C(n)$ then one may say that C^k is a cycle whose existence in $G(n, m)$ we can *first* guarantee as m increases.)

36. CONJECTURE. Let G be a k-connected r-regular graph of order n. If $n \leqslant (k + 1)(r + 1)$ then G is Hamiltonian.

37. PROBLEM. Give sufficient conditions on a sequence $d = (d_k)_1^n$ implying that d is a forcibly Hamiltonian degree sequence but *not* every degree sequence majorizing d is forcibly Hamiltonian.

38. PROBLEM. Let $d = (d_k)_1^n$ be a degree sequence $\kappa \geqslant 0$ and $3 \leqslant c \leqslant n$. Give conditions on d implying that if d is the degree sequence of G and $\kappa(G) \geqslant \kappa$ then $c(G) \geqslant c$. Prove similar results for the girth.

39. PROBLEM. Given n and $l, 3 \leqslant l \leqslant n$, estimate the minimal size of a C^l-saturated graph of order n (see p. 103): $c_l(n) = \min\{m: \text{there is a maximal } G(n, m) \text{ without a } C^l\}$.

40. (Cf. Problem 39.) Note that $c_3(n) = n - 1$ and prove that $c_4(n) = \lceil 3n/2 \rceil - 2$ if $n \geqslant 2$.
[Hint. The graphs $K^1 + mK^2$ and $K^1 + (mK^2 \cup K^1)$ are C^4-saturated.

To prove the converse inequality investigate cutvertices and vertices of degree 2. Cf. Ex. 16 Ch. IV.]

41⁻. (Cf. Problem 39.) Show that $c_n(n) \geqslant 3n/2$. Probably equality holds for infinitely many n.

[Hint. Let $G = G^n$ be maximal non-Hamiltonian. Show that if $\kappa(G) \geqslant 2$ then a vertex of degree 2 cannot be adjacent to a vertex of degree at most 3. Note also that if $x, y \in \Gamma(z)$ and $d(x) = d(y) = 2$ then $G - \{x, y\}$ is complete.] (Bondy [B46].)

42⁻. Call a graph G *hypo-Hamiltonian* if $G - x$ is Hamiltonian for every $x \in G$. Show that if $G = G(n, m)$ is hypo-Hamiltonian then $\delta(G) \geqslant 3$ and so $\frac{3}{2}n$.

43. The *Coxeter graph* C_{28} is a cubic graph of order 28 and girth 7. It consists of 3 independent 7-cycles, say $a_1 a_2 \ldots a_7, b_1 b_2 \ldots b_7, c_1 c_2 \ldots c_7$ together with 7 other vertices $d_1, d_2, \ldots, d_7$ and 21 edges of the form $a_i d_i, b_i d_i, c_i d_i, i = 1, \ldots, 7$. Prove that

   (i) all oriented paths of length 3 in C_{28} are equivalent under the automorphism group of the graph,

   (ii) C_{28} is hypo-Hamiltonian.
     (Bondy [B46], see also Tutte [T12].)

44. Define a generalized Petersen graph $G_{k,l}$ by

$$V(G_{k,l}) = \{a_i, b_i: \quad i \in \mathbb{Z}_k\}, \quad E(G_{k,l}) = \{a_i a_{i+1}, b_i b_{i+l}, a_i b_i: \quad i \in \mathbb{Z}_k\}.$$

Show that if $k = 3m + 2$ and m is odd then $G_{k,2}$ is a cubic hypo-Hamiltonian graph (and so a hypo-Hamiltonian graph of order $2k$ and minimal size). (Bondy [B46], see also the constructions of Sousselier [HDV1] and Lindgren [L9].)

45.⁺ Suppose $\delta(G) = \delta > (n + 1)/3$ where $|G| = n$. Prove that either G is Hamiltonian or it is separable or else it contains $\delta + 1$ independent vertices. (Nash-Williams [N6].)

46. PROBLEM. Consider a $G(n, kn)$ and let L be the set of integers l for which our graph contains a C^l. Determine or estimate

$$\min \sum_{l \in L} 1/l,$$

where the minimum is extended over all $G(n, kn)$. It seems likely that this minimum is $(\frac{1}{2} + o(1)) \log k$ but it has not even been proved that it tends to ∞ with k. (Erdös and Hajnal, see [E28].)

Chapter IV

The Diameter

When talking about the applications of graph theory to non-mathematical problems, Menger's theorem, the max-flow min-cut theorem and Philip Hall's theorem are invariably mentioned. Results concerning the diameter of a graph are also often said to have applications. Though it is dubious how genuine these applications have been so far, it cannot be denied that a number of papers that started new areas of research concerning the diameter claimed to have been motivated by the possibility of applications. Furthermore, it seems very likely that the present results form a poor beginning of a theory with many important applications, in particular to multi-processor systems of constrained connectibility.

The following "real life problem" was considered by Erdös and Rényi [ER4]. Each of the n towns of a country has one airport. Each airport is equipped to handle at most k direct flights to other airports. The flights must be arranged in such a way that from each town it is possible to fly to any other town by landing at fewer than d ($d \geqslant 2$) airports on the way. At least how many direct flights must be set up? In order words: determine the minimum size $e_d(n, k)$ of a graph of order n whose diameter is at most d and whose maximal degree is at most k. In §1 we present bounds on $e_d(n, k)$, with special attention to $e_2(n, k)$ and $e_3(n, k)$. It seems that for a given value of d the nature of the problem depends on the relation between n and k. Though there are non-trivial results, the estimates for $e_d(n, k)$ are very poor at present.

Murty and Vijayan [MV2] investigated the following very general problem. A communication network has n centres and a certain number of two-way communication lines joining the centres. From each centre it is possible to send information to any other centre by making use of at most d of the lines. The network satisfies the following reliability condition: if any s of the *centres* are out of order then in the remainder of the system it is still possible to send information from each centre to any other centre by using at most d' of the lines. Determine $f(n, d, d', s)$, the *minimum* of the number of lines needed to construct such a system. A similar problem arises if we wish to insure ourselves against the breakdown of *lines*. In §§2 and 3 we attack the graph theoretic form of this problem. It turns out that there are two essentially different problems. If d' is much larger than d then all that matters is that $d' \leqslant n - 1$, i.e. the graph remains connected after the omission of s vertices or edges. Though the exact value of $f(n, d, n - 1, s)$ is known only for small values of d and s, as the main result of the chapter, first proved in [B31], [B33], we determine the asymptotic value of $f(n, d, n - 1, s)$ as $n \to \infty$ and d and s remain fixed. The results concerning the case when d' is not much larger than d are rather embryonic. The aim would be the determination of $f(n, d, d, s)$. Though in §3 we present some exact results for small values of d and s, the estimates of $f(n, d, d, s)$ in the general case are rather bad.

In §4 we investigate the possibility of factoring a complete graph into k factors of small diameter. Due mostly to a refinement of a construction of Sauer [S8], the results of the section go a long way towards a complete solution of the problem.

As we have mentioned, the results of §1 answer only a small fraction of the problems arising in the area. It is obvious from the results of the section that the method applied to estimate $e_d(n, k)$ has to depend on the relation of the parameters. It might not be hopelessly difficult to give a good estimate of $e_d(n, k)$ when n is about a given power of k (Problem 13).

Looking through the results of this chapter it will be clear that the problem to be discussed in §3 is also in great need of new ideas. Put $\hat{\phi}(n, d, s) = f(n, d, d, s - 1) - sn/2$. For fixed values of d and s we clearly have $\hat{\phi}(n, d, s) = O(n)$. However, the known general results on $\hat{\phi}(n, d, s)$ are very shallow; both the lower and upper bounds are essentially trivial. Not surprisingly the gap between these bounds is very large. There is little doubt that the determination of $\lim_{n \to \infty} n^{-1} \hat{\phi}(n, d, s)$ would contribute a great deal to our understanding of the role of the diameter (cf. Problem 24).

1. DIAMETER, MAXIMAL DEGREE AND SIZE

What is the minimal size of a graph of order n and diameter d? What is the maximal size of a graph of order n and diameter d? It is not surprising that these straightforward questions can be answered without the slightest effort (Exx. 1–6) just as the similar questions concerning the connectivity or the chromatic number of a graph. The answers to these questions and their variants are not very enlightening. In particular, the class of maximal graphs of order n and diameter d is easy to describe and reduces every question concerning maximal graphs to a (not necessarily easy) question about binomial coefficients, as in [O6], [W11], [HO1], [OSH1] and [HS5]. Therefore we shall investigate the *minimal* size of a graph of order n and diameter d under various additional conditions. The first additional condition we impose on the graph is an upper bound on the maximal degree.

Let d, k and n be natural numbers, $d < n$ and $k < n$. Denote by $\mathscr{H}_d(n, k)$ the set of all graphs of order n with maximal degree k and diameter at most d. Put

$$e_d(n, k) = \min\{e(G): G \in \mathscr{H}_d(n, k)\}.$$

If $\mathscr{H}_d(n, k)$ is empty then, following the usual convention, we shall write $e_d(n, k) = \infty$. The study of the function $e_d(n, k)$ was begun by Erdös and Rényi [ER4]. The aim of this section is to present some of the results concerning $e_d(n, k)$, especially $e_2(n, k)$ and $e_3(n, k)$.

Clearly $\mathscr{H}_d(n, 1) = \varnothing$ unless $n \leqslant 2$. Furthermore, $\mathscr{H}_d(n, 2) = \varnothing$ unless $d \geqslant \lfloor n/2 \rfloor$. If $d = n - 1$ then $e_d(n, 2) = n - 1$ and if $\lfloor n/2 \rfloor \leqslant d < n - 1$ then $e_d(n, 2) = n$. To avoid these trivial cases in the sequel we suppose that $k \geqslant 3$.

Consider a graph $G \in \mathscr{H}_d(n, k)$. Pick a vertex $x_0 \in G$. Since $\Delta(G) \leqslant k$, there are at most k vertices at distance 1 from x_0 (i.e. joined to x_0), there are at most $k(k - 1)$ vertices at distance 2 from x_0 and in general there are at most $k(k - 1)^{r-1}$ vertices at distance r from x_0. As diam $G \leqslant d$ this implies

$$n \leqslant 1 + k \sum_{r=1}^{d} (k - 1)^{r-1} = 1 + k\frac{(k - 1)^d - 1}{k - 2}. \tag{1}$$

If both n and k are odd then G cannot be k-regular. Consequently we can choose an x_0 with degree at most $k - 1$ and then (1) can be improved to

$$n \leqslant 1 + (k - 1)\frac{(k - 1)^d - 1}{k - 2}. \tag{2}$$

We shall see that for $d = 2$ inequalities (1) and (2) are essentially sufficient to ensure that $\mathscr{H}_2(n, k) \neq \varnothing$. However, for $d = 3$ *and large* k we know little about the maximum order of a graph of diameter d and maximum degree k (cf. Problem 7 and Conjecture 8).

Let us note next that if for a given pair (d, k) the order n is sufficiently small then it is again trivial to determine $e_d(n, k)$. For if d is even and

$$n \leqslant 1 + k \sum_{r=1}^{d/2} (k - 1)^{r-1}, \tag{3'}$$

or if d is odd and

$$n \leqslant 2 \sum_{r=1}^{(d+1)/2} (k - 1)^{r-1} \tag{3''}$$

then there is a tree of order n, maximal degree k and diameter at most d. Consequently if (3') or (3'') is satisfied then $e_d(n, k) = n - 1$. In the sequel we shall always suppose that inequalities (1) and (2) hold and inequalities (3') and (3'') do not hold.

Following Erdös and Rényi [ER4] one can easily get a lower bound on $e_d(n, k)$.

THEOREM 1.1.

$$e_d(n, k) \geqslant \frac{n(n - 1)(k - 2)}{2((k - 1)^d - 1)} \tag{4}$$

Proof. Let $G = G(n, m) \in \mathscr{H}_d(n, k)$. Then G contains m paths of length 1. Each edge is contained in at most $2(k - 1)$ paths of length 2. In this way each path of length 2 is counted twice so there are at most $m(k - 1)$ paths of length 2. Similarly each edge is contained in at most $r(k - 1)^{r-1}$ paths of length r. Each of these paths is counted r times so there are at most $m(k-1)^{r-1}$ paths of length r. Since there are $\binom{n}{2}$ pairs of vertices and for each pair there is a path of length at most d joining them,

$$\binom{n}{2} \leqslant m \sum_{r=1}^{d} (k - 1)^{r-1},$$

implying the required inequality.   ∎

The argument above implies that equality holds in (4) iff there is a Moore graph of order n, degree k and diameter d. As we mentioned in Chapter III §1 (p. 106) there are very few Moore graphs. In particular, unless $d = 2, k = 3, 7$ or 57 *and equality holds in* (1), inequality (4) is strict.

Let us examine now the function $e_2(n, k)$ in detail. Note that because of (1) and (3') we assume $\sqrt{n - 1} \leqslant k \leqslant n - 2$. The following two theorems are due to Erdös and Rényi [ER4] and Erdös, Rényi and Sós [ERS1].

THEOREM 1.2. (i) *If* $k = n - 2$ *or* $n - 3$ *then*

$$e_2(n, k) = n + k - 2.$$

Furthermore,

$$e_2(n, n - 4) = 2n - 5.$$

(ii) *If* $(2n - 2)/3 \leqslant k \leqslant n - 5$ *then*

$$e_2(n, k) = 2n - 4.$$

Proof. Let us show first that $e_2(n, k)$ is at most as large as we claim. We do this by exhibiting some graphs that will be shown to be extremal. The graph $K(2, n - 2)$ shows $e_2(n, n - 2) \leqslant 2n - 4$. The graphs G_1 and G_2 of Fig. 1.1 show $e_2(n, k) \leqslant 2n - 5$ if $n - 4 \leqslant k \leqslant n - 3$. Finally the graph G_3 exhibited in Fig. 1.1 shows $e_2(n, k) \leqslant 2n - 4$ if $(2n - 2)/3 \leqslant k \leqslant n - 5$. Note that in this case $(2n - 2)/3 \leqslant n - 5$ so $n \geqslant 13$. To simplify the notation in the representation of G_3 we put $k = n - j$ where $5 \leqslant j \leqslant (n + 2)/3$.

Let us turn now to the proof of the converse inequality which is the essential assertion of the theorem.

Let G be a graph of order n with diameter 2 and maximal degree k.

(i) Let $n - 4 \leqslant k \leqslant n - 2$ and let $x_0 \in G, d(x_0) = k$. Then $\{x_0\} \cup \Gamma(x_0) \neq V$ so there is a vertex $x_1 \neq x_0$ not joined to x_0. If $x \in V - \{x_0, x_1\}$ then the path of length at most 2 joining x_1 to x cannot contain x_0. Hence $G - x_0$ is connected and so $e(G) \geqslant d(x_0) + e(G - x_0) \geqslant k + n - 2$, which is the required inequality if $n - 3 \leqslant k \leqslant n - 2$. To complete the proof we have to show only that if $k = n - 4$ then $G - x_0$ is not a tree. Let us assume that $G - x_0$ *is a tree* T. Let $V - \Gamma(x_0) \cup \{x_0\} = \{x_1, x_2, x_3\}$ and suppose x_2 separates x_1 from x_3 on T. Let $x_1' \in \Gamma(x_0) \cap \Gamma(x_1)$. Then $d(x_3, x_1') \geqslant 3$: contradicting diam $G \leqslant 2$.

(ii) Let $(2n - 2)/3 \leqslant k \leqslant n - 5$ and suppose that $e(G) \leqslant 2n - 5$. Clearly $\delta(G) \leqslant 3$. Since $\Delta(G) < n - 1$ and diam $G \leqslant 2$, we must have $2 \leqslant \delta(G) \leqslant 3$.

Suppose first that $\delta(G) = 3$ and, say, $d(x_0) = 3$. Put $\Gamma(x_0) = \{x_1, x_2, x_3\}$. Then $d(x_1) + d(x_2) + d(x_3) \geqslant n - 1$, with equality iff x_i is not joined to $x_j, 1 \leqslant i < j \leqslant 3$, and x_0 is the only vertex joined to at least 2 of x_1, x_2 and x_3. Hence, counting the sum of the degrees,

$$2n - 5 = \tfrac{1}{2}\{3(n - 3) + n - 1\} \leqslant e(G) \leqslant 2n - 5.$$

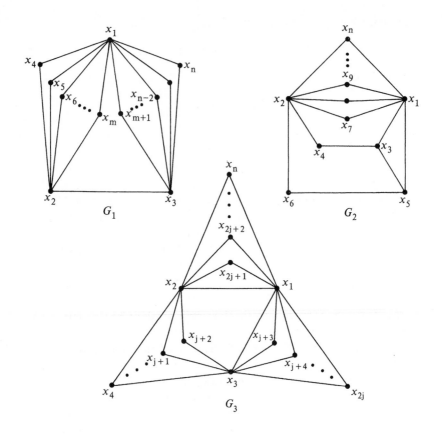

FIG. 1.1. Some of the extremal graphs in Theorem 1.2.

As we must have equalities above, $d(x_1) + d(x_2) + d(x_3) = n - 1$ and every vertex $x \notin \{x_1, x_2, x_3\}$ has degree 3. Suppose $d(x_1) \leqslant d(x_2) \leqslant d(x_3)$, so $d(x_1) \leqslant (n - 1)/3$. Let $x_1' \in \Gamma(x_1) - \{x_0, x_2, x_3\}$ and put $\Gamma(x_1') = \{x_1', x_2', x_3'\}$. Note that $\{x_2', x_3'\} \cap \{x_2, x_3\} = \varnothing$ and, as before, $d(x_1) + d(x_2') + d(x_3') = n - 1$. Since $d(x_2') = d(x_3') = 3$, this gives $n - 7 = d(x_1) \leqslant (n - 1)/3$, contradicting $n \geqslant 13$.

Assume now that $\delta(G) = 2$ and, say, $d(x_0) = 2, \Gamma(x_0) = \{x_1, x_2\}$. Put

$$X_0 = \Gamma(x_1) \cap \Gamma(x_2), \quad |X_0| \geqslant 1, \quad X_1 = \Gamma(x_1) - X_0 \quad \text{and} \quad X_2 = \Gamma(x_2) - X_0.$$

Note that $|X_1| = n - d(x_2) - 2 \geqslant 3$ and similarly $|X_2| \geqslant 3$. Let X_0' be the set of vertices of X_0 joined to at least one vertex of $X_1 \cup X_2$. Since the path of length at most 2 joining a vertex $y_1 \in X_1$ to a vertex $y_2 \in X_2$ cannot contain x_1 or x_2, the graph $G[X_1 \cup X_2 \cup X_0']$ is connected. Hence

$$2n - 5 \geqslant e(G) \geqslant n + |X_0| - 2 + e(G[X_1 \cup X_2 \cup X_0'])$$

$$\geqslant n + |X_0| + |X_1| + |X_2| + |X_0'| - 3 \geqslant 2n - 5.$$

This implies that $X_0' = \varnothing$ and $T = G[X_1 \cup X_2]$ is a tree. If $y_1 \in X_1$ then $d(y_1, x_2) = 2$ implies that y_1 is joined to a vertex in X_2. Similarly every $y_2 \in X_2$ is joined to a vertex in X_1. Let $y_1 \in X_1$ be an endvertex of T. If y_1 is joined to $y_2 \in X_2$ then y_2 must be joined to every other vertex of X_2. This implies that each endvertex of T belongs to X_1. Since $|X_2| > 1$, there is an endvertex $z_1 \in X_1$ joined to a vertex $z_2 \in X_2$, $z_2 \neq y_2$. Then z_2 is also joined to every other vertex of X_2. If follows from $|X_2| \geqslant 3$ that $T[X_2]$ contains a triangle, contradicting the fact that T is a tree.   ∎

We shall show by examples that $e_2(n, k)$ is at most as large as claimed in the next theorem. The converse inequality is proved by examination of cases, on the lines of the proof of Theorem 1.2. We leave the proof to the reader.

THEOREM 1.3.

$$e_2(n, k) = \begin{cases} 3n - k - 6 & \text{if } \dfrac{3n - 3}{5} \leqslant k < \dfrac{2n - 2}{3}, \\[2ex] 5n - 4k - 10 & \text{if } \dfrac{5n - 3}{9} \leqslant k < \dfrac{3n - 3}{5}, \\[2ex] 4n - 2k - 11 & \text{if } \dfrac{n + 1}{2} \leqslant k < \dfrac{5n - 3}{9}. \end{cases}$$

Some of the extremal graphs are shown in Fig. 1.2.   ∎

For the sake of convenience in the sequel we use a slightly modified definition of $\mathscr{H}_d(n, k)$ and $e_d(n, k)$: instead of requiring that $\Delta(G) = k$ we shall require only that $\Delta(G) \leqslant k$. This difference is rather inessential but, among others, enables us to talk about $e_d(n, cn)$ instead of $e_d(n, \lfloor cn \rfloor)$. Though in the

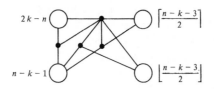

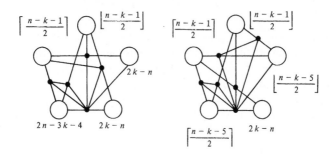

FIG. 1.2. One extremal graph for each of the three cases of Theorem 1.3. The number next to a circle is the number of vertices in that group.

range $\sqrt{n-1} \leqslant k \leqslant n/2$ we do not know the exact value of $e_2(n, k)$, we shall prove asymptotically exact estimates. If q is a prime power, one can easily construct a graph G of order $q^2 + q + 1$ such that diam $G = 2, \Delta(G) = q + 1$ and $e(G) = \frac{1}{2}q(q + 1)^2$. To define G let $V(G)$ be the set of points of the projective plane $PG(2, q)$ and join a point (x, y, z) to a point (x', y', z') if $xx' + yy' + zz' = 0$, that is if (x', y', z') lies on the line $[x, y, z]$. Then clearly diam $G = 2$, $\Delta(G) = q + 1$ and $\delta(G) = q$. This shows, in particular, that if $n_k = k^2 - k + 1$ then

$$\liminf_{k \to \infty} \frac{e_2(n_k, k)k}{n_k(n_k - 1)} = \frac{1}{2},$$

i.e. Theorem 1.1 is asymptotically best possible in this case. Our next aim is to improve the lower bound given by Theorem 1.1 when $d = 2$ and k is substantially larger than $n^{\frac{1}{2}}$. We present a slightly weaker result than the one proved in [B26] but this makes it possible to give a simpler proof.

THEOREM 1.4. *If $n - 1 < k^2$ then*

$$e_2(n, k) \geqslant \frac{n(n-1)}{k}\left(1 + \frac{n-1}{k^2}\right)^{-1}. \tag{5}$$

Proof. Let $G = G(n, m)$ be such that $\Delta(G) \leqslant k$ and diam $G \leqslant 2$. Then $\delta = \delta(G) \geqslant (n - 1)/k$. Let $d_1, d_2, \ldots, d_n$ be the degree sequence of G. Any two vertices of G are joined by an edge or a path of length 2. As there are $\sum_1^n \binom{d_i}{2}$ paths of length 2, the following inequality holds:

$$\sum_1^n \binom{d_i}{2} \geqslant \binom{n}{2} - m.$$

Define p by the relation

$$pk + (n - p)\frac{n-1}{k} = 2m. \tag{6}$$

Note that p is well defined (it is not necessarily an integer) and $0 < p < n$. Then

$$p\binom{k}{2} + (n - p)\binom{(n-1)/k}{2} \geqslant \sum_1^n \binom{d_i}{2} \geqslant \binom{n}{2} - \tfrac{1}{2}\left(pk + (n - p)\frac{n-1}{k}\right).$$

Hence

$$p \geqslant \frac{n(n-1)}{k^2}\left(1 + \frac{n-1}{k^2}\right)^{-1}$$

and so (6) gives

$$m \geqslant \frac{n(n-1)}{k}\left(1 + \frac{n-1}{k^2}\right)^{-1}. \quad\blacksquare$$

As in [B26] let us construct graphs of diameter 2 showing that (5) is almost best possible.

THEOREM 1.5. *Let q be a power of an odd prime and let m be an arbitrary natural number. Put $n = (q^2 + q + 1)(m + 1)$ and $k = (q + 1)(m + 1)$. Then*

$$e_2(n, k) \leqslant (q + 1)(q^2 + q + 1)m + \frac{q(q+1)^2}{2}. \tag{7}$$

Proof: We shall construct a graph G of order n, showing (7). Take two copies of $PG(2, q)$, the projective plane over the field with q elements (cf. §0). Put one vertex of G in each of the points of the first copy and put m vertices in each point of the second. Join the vertex in a *point* (x, y, z) of the first copy of $PG(2, q)$ to those vertices which are in the points of the *line* $[x, y, z]$ in *either* copy (except, of course, the vertex itself if $x^2 + y^2 + z^2 = 0$). It is immediate that $\Delta(G) = k$, diam $G = 2$ and $e(G)$ is exactly the right-hand side of (7).

We may use the fact that there is a prime between x and $x + x^{2/3}$ if x is sufficiently large (cf (0.7)) to extend (7) to other pairs (n, k). Combining it with (5) we obtain the following results.

COROLLARY 1.6.

$$e_2(n, k)kn^{-2} \to 1 \quad if \quad \frac{n}{k} \to \infty \quad and \quad \frac{k^2}{n} \to \infty.$$

In particular, if $c > 0$ and $\frac{1}{2} < \alpha < 1$ are constants then

$$\lim_{n \to \infty} \frac{e_2(n, cn^\alpha)c}{n^{2-\alpha}} = 1. \qquad \blacksquare$$

COROLLARY 1.7. *If $c > 0$ is sufficiently small and n is sufficiently large (depending on c) then*

$$\frac{n}{c} - 2c^{-3} \leqslant e_2(n, cn) \leqslant \frac{n}{c}(1 + c^3). \qquad \blacksquare$$

By refining slightly the proof of Theorem 1.1 one can obtain the following lower bound on $e_d(n, k)$.

THEOREM 1.8. *If $k \leqslant n - 1$ and $d \geqslant 3$ then*

$$e_d(n, k) \geqslant \frac{n^2}{k^{d-1}}\left[1 - 4\left(\frac{n}{k^d}\right)^{\frac{1}{3}}\right]. \tag{8}$$

Proof. Put $\gamma = 4(n/k^d)^{\frac{1}{3}}$. We may assume that $\gamma < 1$ for otherwise (8) is trivial. Furthermore, since $e_d(n, k) \geqslant n - 1$ always holds, the right-hand side of (8) can be supposed to be greater than $n - 1$. Thus we may assume that

$$(64\,n)^{1/d} < k < n^{1/(d-1)},$$

and so $n > 64^{d-1}$ and $k > 64$. Put $r = 4n/(k^{d-1}\gamma)$.

Let $G = G(n, m)$ be a graph with diam $G \leqslant d$ and $\Delta(G) \leqslant k$. Our aim is to show that G is at least as large as the right-hand side of (8), i.e. $m \geqslant n^2 k^{1-d}(1-\gamma)$. Put $X = \{x \in G : d(x) < r\}$ and let $Y = V - X$. If $|X| \leqslant n(1 - [\gamma(1-\gamma)/2])$ then

$$m \geqslant \tfrac{1}{2}|Y|r \geqslant n^2 k^{1-d}(1 - \gamma),$$

so we may suppose that $|X| > n\{1 - \gamma(1 - \gamma)/2\}$.

If each $x \in X$ is joined to at least $(1 - (\gamma/2))n\,k^{1-d}$ vertices of Y then

$$m \geqslant |X|\left(1 - \frac{\gamma}{2}\right)nk^{1-d} \geqslant n^2 k^{1-d}(1 - \gamma).$$

Thus we may assume that there is a vertex $x_0 \in X$ joined to less than $(1 - (\gamma/2))\,nk^{1-d}$ vertices of Y. We shall prove that this is impossible by showing that the number of non-trivial paths of length at most d starting at x_0 is less than $n - 1$, contradicting diam $G \leqslant d$.

The number of paths $x_0 y \ldots, y \in Y$ of length at most d is at most

$$|\Gamma(x_0) \cap Y| \sum_{s=0}^{d-2} (k - 1)^s \leqslant \left(1 - \frac{\gamma}{2}\right)nk^{1-d} \sum_{s=0}^{d-2} (k - 1)^s < \left(1 - \frac{\gamma}{2}\right)n.$$

Similarly the number of paths $x_0 x \ldots, x \in X$, of length at most d is at most

$$r\left(1 + r \sum_{s=0}^{d-2} (k - 1)^s\right) < \frac{\gamma n}{3}.$$

These two inequalities show the required contradiction diam $G > d$. ∎

In a certain range Theorem 1.8 is essentially best possible, as shown by the following result of [B26].

THEOREM 1.9. *If c is a sufficiently small positive constant and n is sufficiently large (depending on c) then*

$$c^{-2}(1 - n^{-\frac{1}{4}})n < e_3(n, cn^{\frac{1}{2}}) < 7c^{-2}n.$$

Proof. Let $t = \lceil 2/c \rceil$ and let q be the smallest prime satisfying $[(q + 1)/t] + t < cq$. Note that $q(c^2/4) \to 1$ as $c \to 0$. Let $A_i, B_j (i, j = 1, 2, \ldots, q^2 + q + 1)$ be disjoint sets such that $A_i = \{a_l^i : l = 1, 2, \ldots, tm\}$ and $|B_j| = m^2$, where m is an arbitrary natural number. We construct a graph G on the union of these sets as follows. Take first the complete graphs with vertex sets $A_1, A_2, \ldots, A_{q^2+q+1}$, respectively, and then add the complete graphs with vertex sets

$$\{a_l^i : \quad i = 1, 2, \ldots, q^2 + q + 1\}, \qquad l = 1, 2, \ldots, m.$$

Consider the sets A_i as *lines* and the sets B_j as *points* of the projective plane $PG(2, q)$. Let B_i' be the union of the sets B_j which, as points of the plane, are on the line A_i. To complete the construction of G join each vertex of B_i' to a vertex of $A_i, i = 1, 2, \ldots, q^2 + q + 1$, in such a way that each vertex of A_i is joined to $\lfloor (q + 1) m/t \rfloor$ or $\lceil (q + 1) m/t \rceil$ vertices of B_i'.

It is easily seen that the diameter of G is 3,

$$n = |G| = (q^2 + q + 1)(m^2 + tm), \quad \Delta(G) = \lceil \frac{q + 1}{t} m \rceil + tm + q^2 + q - 1$$

and

$$e(G) = (q^2 + q + 1)(q + 1)m^2 + (q^2 + q + 1)\binom{tm}{2} + \binom{q^2 + q + 1}{2} tm.$$

If c is sufficiently small and m is sufficiently large, the existence of this graph G implies the required inequality. ∎

If $k \neq O(n^{\frac{1}{2}})$ then both (4) and (8) become trivial. In view of the next construction this is not very surprising. Take $K^t \cup E^{n-t}$ and join each vertex of E^{n-t} to exactly one vertex of K^t in such a way that each vertex of K^t has degree $\lfloor (n - t)/t \rfloor + t - 1$ or $\lfloor (n - t)/t \rfloor + t$. The resulting graph G_t has diameter 3 if $t > 2$ and $n - t > 2$; furthermore $\Delta(G) = \lceil n/t \rceil + t - 2$ and

$$e(G_t) = \binom{t}{2} + n - t = n + \binom{t - 1}{2} - 1.$$

It is likely that as $kn^{-\frac{1}{2}} \to \infty$ this construction is essentially best possible. This expectation is supported by the following result of Erdős, Rényi and Sós [ERS1]. The proof is based on the observation that if diam $G \leqslant 3$ and $\Gamma(x_i) = \{y_i\}, i = 1, 2, y_1 \neq y_2$ then y_1 must be joined to y_2.

We leave the details of the proof to the reader.

THEOREM 1.10. *If* $\lceil n/(s+1) \rceil + s - 1 \leqslant k < \lfloor n/s \rfloor + s - 2$ *and* $1 \leqslant s$ $\leqslant \lfloor (n/2)^{\frac{1}{2}} \rfloor$ *then*

$$e_3(n, k) = n + \binom{s}{2} - 1.$$

The graph G_{s+1} *constructed above is one of the extremal graphs.* ∎

At present there is no nontrivial upper bound of $e_d(n, k)$ for every triple (d, n, k). However, given an upper bound for $e_d(n, k)$ we may obtain an upper bound for $e_{d+2}(n', k)$ as follows.

If diam $G \leqslant d$ then adding to it a set of vertices and joining each to some vertex of G the resulting graph H has diameter at most $d + 2$. In particular, if $\Delta(G) \leqslant k$ and a vertex x of G is joined to at most $k - d(x)$ new vertices then $\Delta(H) \leqslant k$. Hence

$$e_{d+2}(n + p, k) \leqslant e_d(n, k) + p,$$

whenever $0 \leqslant p \leqslant kn - 2e_d(n, k)$.

2. DIAMETER AND CONNECTIVITY

Let n, d, d' and s be natural numbers, $s < n$ and $d \leqslant d' \leqslant n - 1$. Denote by $\mathscr{V}(n, d, d', s)$ the set of graphs of order n and diameter at most d such that the deletion of any s of the vertices results in a graph of diameter at most d'. Similarly $\mathscr{E}(n, d, d', s)$ denotes the corresponding class if instead of vertices we delete edges. Put

$$f(n, d, d', s) = \min\{e(G): G \in \mathscr{V}(n, d, d', s)\},$$

$$g(n, d, d', s) = \min\{e(G): G \in \mathscr{E}(n, d, d', s)\}.$$

The study of the functions f and g was initiated by Murty and Vijayan [MV2]. A number of other results have been obtained by Murty [M36], [M37], [M38], Bollobás [B24], [B25], Bondy and Murty [BM1], Bollobás and Eldridge [BE4], Caccetta [C1] and Bollobás and Erdös [BE9]. All these results concern the cases $d = 2, s \geqslant 1$ or $d \leqslant 4, s = 1$.

If $d_1 \leqslant d_2, d_1' \leqslant d_2'$ and $s_1 \geqslant s_2$ then by definition $\mathscr{V}(n, d_1, d_1', s_1) \subset \mathscr{V}(n, d_2, d_2', s_2)$ and $\mathscr{E}(n, d_1, d_1', s_1) \subset \mathscr{E}(n, d_2, d_2', s_2)$. Therefore $f(n, d_1, d_1', s_1) \geqslant f(n, d_2, d_2', s_2)$ and $g(n, d_1, d_1', s_1) \geqslant g(n, d_2, d_2', s_2)$. The results referred to above indicate that for any given triple (d, d', s) the order of $f(n, d, d', s)$ and $g(n, d, d', s)$ depends either on d and s or on d' and s and not on both d and d'. Namely in the known cases either

$$\lim_{n \to \infty} f(n, d, d', s)/f(n, d, n - 1, s) = 1$$

or else

$$\lim_{n \to \infty} f(n, d, d', s)/f(n, d', d', s) = 1,$$

and the function g satisfies similar relations. Thus it is natural to concentrate on $f(n, d, n - 1, s)$, $f(n, d', d', s)$, $g(n, d, n - 1, s)$ and $g(n, d', d', s)$. In this section we present some results concerning $f(n, d, n - 1, s)$ and $g(n, d, n - 1, s)$, together with the rather obvious implications of these results about the general functions, and the next section is devoted to the other case.

A superficial observer might think that the problems mentioned above are similar to questions about the minimal diameter of *strongly connected* directed graphs of given order and size. (A directed graph is said to be *strongly connected* if for every pair (x, y) of its vertices it contains a path directed from x to y. In other words, a directed graph $\vec{G}$ is strongly connected iff diam $\vec{G} < \infty$.) This is certainly not the case, the various results proved in [G12], [G13], [G14], [V4] about the diameter and radius of strongly connected directed graphs have no relation to the results concerning 2-connected graphs. The problems in this area happen to be substantially easier than the ones we are going to discuss in §§2 and 3. Incidentally, we shall see that the general problem is very far from being completely solved even today, contrary to the claim in [B10, p. 73].

To simplify the notation we put $f(n, d, s + 1) = f(n, d, n - 1, s)$ and $g(n, d, s + 1) = g(n, d, n - 1, s)$. We have replaced s by $s + 1$ in the new functions since this way $f(n, d, k)$ (resp. $g(n, d, k)$) is the *minimal size* of a k-connected (resp. k-edge-connected) graph whose order is n and whose diameter is at most d. As in [B31] and [B33] we shall determine the order of these functions for every fixed pair $(d, k), d \geqslant 2, k \geqslant 2$. It will turn out that $f(n, d, k)n^{-1}$ and $g(n, d, k)n^{-1}$ tend to the same finite positive limit as $n \to \infty$.

Given $d \geqslant 2$, $k \geqslant 3$ and $n > k$, following Bollobás and Harary [BH²] one can easily construct a graph of order n, diameter at most d and connectivity k whose size will be almost minimal as $n \to \infty$. To shorten the discussion we construct this graph only for $d = 2m$ and use the trivial inequalities $f(n, 2m + 1, k) \leqslant f(n, 2m, k)$ and $g(n, 2m + 1, k) \leqslant g(n, 2m, k)$ to obtain upper bounds on $f(n, 2m + 1, k)$ and $g(n, 2m + 1, k)$.

Throughout this section we shall find it convenient to use the notation

$$S(r) = S(k, r) = \sum_{i=0}^{r-1} (k - 1)^i.$$

Let $d = 2m$, $m \geqslant 1$, and put $a = \lceil (n - 1)/S(m) \rceil$. In other words a is the minimal natural number for which there is a tree T_0 of order at least n with radius at most m whose centre c has degree a and in which every other vertex that is not an endvertex has degree k. If $a \leqslant k$ then there is a tree T_1 of order n, radius at most m (and so diameter at most $d = 2m$), whose maximal degree is at most k. Starting with an appropriate tree of this kind we may add to it edges in such a way that the resulting graph is k-connected, its maximal degree is k or $k + 1$ and in the latter case there is exactly one vertex of degree $k + 1$. This shows that if $a \leqslant k$, that is if $kS(m) \geqslant n - 1$ then

$$f(n, 2m + 1, k) = f(n, 2m, k) = g(n, 2m + 1, k) = g(n, 2m, k) = \left\lceil \frac{kn}{2} \right\rceil. \qquad (1)$$

Suppose now that $a \geqslant k$ and consider the tree T_0. By omitting groups of $k - 1$ endvertices joined to the same vertex, from T_0 we can obtain a tree T_1 of order n',

$$n - (k - 1) < n' \leqslant n.$$

Add $p = n - n'$ new vertices to T_1. (Clearly $p \equiv n - 1 \bmod(k - 1)$ and $0 \leqslant p < k - 1$). Join each of these vertices to c and denote by T the resulting tree. Let W be the set of endvertices of T. Since $|W| \geqslant k$ we may add to T a $(k - 1)$-connected graph with vertex set W and with minimal size $\lceil \frac{1}{2}(k - 1)|W| \rceil$. The resulting graph G is k-connected and has diameter at most $2m$. Therefore

$$f(n, 2m + 1, k) \leqslant f(n, 2m, k) \leqslant e(G) = \lceil \tfrac{1}{2}\{(n - 1) k + a + p\} \rceil$$

$$< \tfrac{1}{2}\{(kn + a - 1\} < \frac{n}{2}\{ k + S(m)^{-1}\}. \qquad (2)$$

Almost the same construction gives

$$g(n, 2m + 1, k) \leqslant g(n, 2m, k) \leqslant \lceil \tfrac{1}{2}\{(n -)k + a\} \rceil < \frac{n}{2}\{k + S(m)^{-1}\}. \qquad (2')$$

In fact, to obtain a slightly weaker form of (2'), it suffices to add to T a graph with vertex set W, size $\lceil \frac{1}{2}(k - 1)|W| \rceil$ and minimal degree $k - 1$, which has at

most one edge in the subgraph spanned by any two components of $T - c$.
(Note that a component has at most $S(k, m)$ vertices.) If $n \geqslant k^m S(k, m)$ then
this can clearly be done. Incidentally, the same graph G also shows that for
$n \geqslant k^m S(k, m)$ we have

$$g(n, 2m + 1, 4m, k - 1) \leqslant g(n, 2m, 4m, k - 1) \leqslant \frac{n}{2}(k + S(m)^{-1}), \qquad (2'')$$

since every vertex of G is joined to c by a path of length at most m and by k
edge disjoint paths of length at most $2m$.

In order to obtain upper bounds on $f(n, d, 2)$ and $g(n, d, 2)$ we need slightly
different constructions. Join two vertices by at least 2 disjoint paths such that
a longest path has length $d + 1$ or d and the other paths have length at most
d (Fig. 2.1(i)). If the resulting graph G, has order n then

$$G \in \mathscr{V}(n, d, 2d - 2, 1) \subset \mathscr{V}(n, d, n - 1, 1).$$

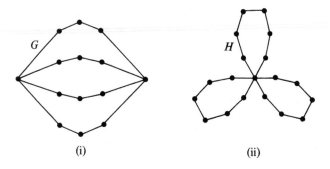

FIG. 2.1. $G \in \mathscr{V}(14, 4, 6, 1)$ and $H \in \mathscr{E}(19, 6, 9, 1)$.

With an appropriate choice of the paths we see that

$$f(n, d, 2) \leqslant f(n, d, 2d - 2, 1) \leqslant \left\lfloor \frac{d}{d - 1} n - \frac{d + 3}{d - 1} \right\rfloor \qquad (3)$$

if $n \geqslant 2d + 1$. Similarly, let H be the union of cycles of length at most l with
one vertex in common (Fig. 2.1(ii)). If H has order n then

$$H \in \mathscr{E}(n, 2\lfloor l/2 \rfloor, \lfloor 3l/2 \rfloor - 1, 1) \subset \mathscr{E}(n, 2\lfloor l/2 \rfloor, n - 1, 1).$$

Choosing $l = 2\lfloor d/2 \rfloor + 1$ and taking as few cycles as possible we see that

$$g(n, d, 2) \leqslant g(n, f, 3\lfloor d/2 \rfloor, 1) \leqslant \frac{2\lceil d/2 \rceil + 1}{2\lfloor d/2 \rfloor} (n + 1) \qquad (4)$$

if $n \geqslant 2\lfloor d/2 \rfloor + 1$.

The aim of this section is to show that all these constructions are essentially best possible. The proofs of the main results are based on two lemmas.

LEMMA 2.1. *Let H be a graph of order h containing a set $X = \{x_1, x_2, \ldots, x_t\}$ of t vertices such that every vertex of H is within distance r of X. Let $d_1, d_2, \ldots, d_h$ be the degree sequence of H. Then for $k \geqslant 2$ we have*

$$\sum_{i=1}^{h} \max\{d_i - k, 0\} \geqslant \frac{h - t}{S(r)} - tk. \qquad (5)$$

Proof. When proving (5) we may suppose that each vertex y of H, $y \notin X$, is joined to exactly one vertex that is nearer to X than y. If $d(y) = l > k$, take $l - k$ of the vertices joined to y and at distance at least $d(y, X)$ from X, delete the edges joining them to y and join them to $x_1 \in X$ instead. This operation does not change the left hand side of (5). By repeated applications of this operation we arrive at a graph H' of order h in which every vertex is at distance at most r from X and a vertex $y \notin X$ has degree at most k. Note now that if there are q vertices at distance 1 from X then there are at most

$$t + q \sum_{0}^{p} (k - 1)^i$$

vertices within distance $p + 1$ of X, so

$$t + q \sum_{0}^{r-1} (k - 1)^i = t + qS(r) \geqslant h.$$

The sum of the degrees of the vertices in X is clearly at least q, so

$$\sum_{1}^{h} \max\{d_i - k, 0\} \geqslant q - tk \geqslant \frac{h - t}{S(r)} - tk. \qquad \blacksquare$$

In certain cases it is more convenient to apply the following reformulation of Lemma 2.1.

LEMMA 2.1'. *Suppose* $|G| = n$, $\delta(G) \geq k$ *and* G *contains a set* X *of at most* t *vertices such that* n' *of the vertices of* G *are within distance* r *of* X.

Then

$$2e(G) \geq \frac{n' - t}{S(r)} + k(n - t).$$ ∎

LEMMA 2.2. *Suppose* $|G| = n$, $\delta(G) \geq k \geq 2$. *Assume furthermore that* $\gamma \geq k$ *and for every vertex* $y \in G$ *there is a vertex* $x \in G$ *such that*

$$d(x) \geq \gamma \quad and \quad d(x, y) \leq r.$$

Then

$$2e(G) \geq \left(k + \frac{\gamma - k}{1 + \gamma S(r)} \right) n.$$

Proof. Let $X = \{x_1, x_2, \ldots, x_t\} = \{x \in G : d(x) \geq \gamma\}$.

Then

$$2e(G) \geq (n - t)k + t\gamma,$$

so we may assume that

$$(\gamma - k)t \leq n\frac{\gamma - k}{1 + \gamma S(r)},$$

i.e.

$$t \leq \frac{n}{1 + \gamma S(r)} \tag{6}$$

Let $d_1, d_2, \ldots, d_n$ be the degree sequence of G. Then

$$2e(G) = \sum_{i=1}^{n} d_i = nk + \sum_{i=1}^{n} \max\{d_i - k, 0\}.$$

Applying Lemma 2.1 to the graph G and the set X of vertices, by (6) we obtain the required inequality.   ∎

Armed with these lemmas we can give reasonably good lower bounds on the functions $f(n, d, k)$ and $g(n, d, k)$. In most cases it will be sufficient to investigate the minimal size of a graph G^n provided diam $G^n \leq d$ and $\delta(G^n) \geq k$ (instead of $\kappa(G^n) \geq k$ or $\lambda(G^n) \geq k$). It will turn out that this bound gives the proper order of the functions $f(n, d, k)$ and $g(n, d, k)$. The only exception is the case d even and $k = 2$ which we shall treat separately.

THEOREM 2.3. *Suppose* $|G| = n$, $\delta(G) \geqslant k \geqslant 2$, $m \geqslant 1$ *and* diam $G \leqslant 2m + 1$. *Then*

$$e(G) \geqslant \frac{n}{2}(k + S(m)^{-1}) - (k + 1) n^{(m+1)/(m+2)} \equiv e(n, 2m + 1, k),$$

where

$$S(m) = \sum_{0}^{m-1} (k - 1)^i.$$

Proof. Let γ be the positive real number defined by

$$n = \gamma^{m+1}(1 + \gamma S(m)). \tag{7}$$

Suppose first that for every vertex $y \in G$ there exists a vertex $x \in G$ such that

$$d(x) \geqslant \gamma \quad \text{and} \quad d(x, y) \leqslant m.$$

Then, by Lemma 2.2,

$$e(G) \geqslant L_1 = \frac{n}{2}\left(k + \frac{\gamma - k}{1 + \gamma S(m)}\right).$$

Note that

$$n \left\{ S(m)^{-1} - \frac{\gamma - k}{1 + \gamma S(m)} \right\} = n \frac{1 + kS(m)}{S(m)(1 + \gamma S(m))}$$

$$< n \frac{k + 1}{1 + \gamma S(m)} = (k + 1) \gamma^{m+1} < (k + 1) n^{(m+1)/(m+2)}.$$

Consequently

$$e(G) \geqslant L_1 > e(n, 2m + 1, k).$$

Assume now that there is a vertex $y \in G$ such that if $x \in G$ and $d(x, y) \leqslant m$ then $d(x) < \gamma$. Set

$$X = \{x \in G: \quad d(x, y) = m + 1\}.$$

Then $|X| \leqslant \gamma(\gamma - 1)^m < \gamma^{m+1} - 1$ and every vertex of G, with the exception of y, is within distance m of X. Therefore Lemma 2.1' implies that

$$e(G) \geqslant L_2 = \frac{k}{2}n + \frac{n - \gamma^{m+1}}{S(m)} - k\gamma^{m+1}.$$

It is easily checked that (7) is equivalent to the fact that $L_1 = L_2$. Hence, as before,

$$e(G) \geqslant L_2 \geqslant e(n, 2m + 1, k). \qquad \blacksquare$$

Inequalities (2), (2'), (3), (4) and Theorem 2.3 give the order of $f(n, 2m + 1, k)$ and $g(n, 2m + 1, k)$.

COROLLARY 2.4. *Let $k \geqslant 2$ and $m \geqslant 1$. Then*

$$\lim_{n \to \infty} f(n, 2m + 1, k) n^{-1} = \lim_{n \to \infty} g(n, 2m + 1, k) n^{-1} = \tfrac{1}{2}(k + S(m)^{-1}).$$

If, furthermore, $d' \geqslant 4m$ and $d'' \geqslant 3m$ then

$$\lim_{n \to \infty} \frac{f(n, 2m + 1, d', 1)}{n} = \lim_{n \to \infty} \frac{g(n, 2m + 1, d'', 1)}{n} = 1 + \frac{1}{2m}. \qquad \blacksquare$$

It is slightly more troublesome to determine the order of $f(n, 2m, k)$ and $g(n, 2m, k)$, mostly because, as we mentioned earlier, the case $f(n, 2m, 2)$ needs separate treatment.

THEOREM 2.5. *Suppose $|G| = n$, $\delta(G) \geqslant k \geqslant 2$, $m \geqslant 1$ and diam $G \leqslant 2m$. Then*

$$e(G) \geqslant \frac{n}{2}(k + S(k, m)^{-1}) - C(k, m),$$

where

$$C(2, m) = (2m + 1)(2m - 1)^{m-1},$$

$$C(3, m) = 4^{m+1}$$

and

$$C(k, m) = k^{m+1} \quad if \quad k \geqslant 4.$$

Proof. Put $\gamma = (1 + kS(k, m))(k - 1)^{-m+1}$. Then

$$\frac{\gamma - k}{1 + \gamma S(k, m - 1)} = S(k, m)^{-1}.$$

Therefore by Lemma 2.2 we may assume that there is a vertex $y \in G$ such that if $d(x, y) \leqslant m - 1$ then $d(x) \leqslant \gamma_0 = \lfloor \gamma \rfloor$. Putting

$$X = \{x \in G : d(x, y) = m\},$$

we see that $t = |X| \leqslant \gamma_0(\gamma_0 - 1)^{m-1}$. Since diam $G \leqslant 2m$, every vertex of G is within distance m of X. Therefore by Lemma 2.1'

$$2e(G) \geqslant kn + \frac{n - t}{S(k, m)} - tk.$$

Note now that if $k = 2$ then $\gamma_0 = 2m, t = 2m(2m - 1)^{m-1}, S(2, m) = m$ and so

$$e(G) \geqslant n \left\{ 1 + \frac{1}{2m} \right\} - (2m + 1)(2m - 1)^{m-1}.$$

If $k = 3$ then $\gamma_0 = 5, t = 5 \cdot 4^{m-1}$ and $S(3, m) = 2^m - 1$, giving

$$e(G) \geqslant \frac{n}{2} \left\{ 3 + \frac{1}{2^m - 1} \right\} - 4^{m+1}.$$

If $k \geqslant 4$ then $\gamma_0 = k + 1, t = (k + 1)k^{m-1}, S(k, m) = (k - 1)^m - 1/k - 2$ and so

$$e(G) \geqslant \frac{n}{2} \left\{ k + \frac{k - 2}{(k - 1)^m - 1} \right\} - k^{m+1}. \qquad \blacksquare$$

COROLLARY 2.6. *If $m \geqslant 1$ and $k \geqslant 3$ then*

$$\frac{n}{2} \left\{ k + \frac{k - 2}{(k - 1)^m - 1} \right\} - C(k, m) \leqslant g(n, 2m, k) \leqslant f(n, 2m, k)$$

$$\leqslant \frac{n}{2} \left\{ k + \frac{k - 2}{(k - 1)^m - 1} \right\}.$$

In particular,

$$\lim_{n \to \infty} f(n, 2m, k) \, n^{-1} = \lim_{n \to \infty} g(n, 2m, k) \, n^{-1} = \frac{1}{2} \left\{ k + \frac{k - 2}{(k - 1)^m - 1} \right\}.$$

Furthermore, if $d' \geqslant 3m$ then

$$\lim_{n \to \infty} \frac{g(n, 2m, 2)}{n} = \lim_{n \to \infty} \frac{g(n, 2m, d', 1)}{n} = 1 + \frac{1}{2m}.$$

Proof. Inequalities (2), (2′), and Theorem 2.5 imply the assertions.   ■

Let us note also that inequality (2″) and Theorems 2.3 and 2.5 have the following immediate consequence.

COROLLARY 2.7. *If $m \geqslant 1$, $k \geqslant 2$ and $d' \geqslant 4m$ then*

$$\lim_{n \to \infty} g(n, 2m + 1, d', k)\, n^{-1} = \lim_{n \to \infty} g(n, 2m, d', k) n^{-1} = \frac{1}{2}\left(k + 1 + \frac{k - 1}{k^m - 1}\right). \quad ■$$

This corollary shows that d' does not have an essential effect on the function $g(n, d, d', k)$, $k \geqslant 2$, unless $d' < 4\lfloor d/2 \rfloor$.

THEOREM 2.8. *Let G be a 2-connected graph of order n with diameter at most $2m$. Then for $m = 1$ we have*

$$e(G) \geqslant 2n - 5$$

and if $m \geqslant 2$ then

$$e(G) \geqslant \left(1 + \frac{1}{2m - 1}\right) n - 4(4m - 2)^{m - 1}.$$

Proof. Suppose first that for every vertex $a \in G$ there exists a vertex $b \in G$ such that

$$d(a, b) \leqslant m - 1 \quad \text{and} \quad \deg b \geqslant 4m.$$

Then, by Lemma 2.2

$$e(G) \geqslant \left(1 + \frac{4m - 2}{2(1 + 4m(m - 1))}\right)n = \left(1 + \frac{1}{2m - 1}\right)n.$$

Therefore we may assume without loss of generality that there is a vertex $a \in G$ such that if $d(a, b) \leqslant m - 1$ then $d(b) \leqslant 4m - 1$. If $m = 1$ then let a be a vertex of minimal degree.

Put

$$A = \{x \in G : d(x, a) \leqslant m\}, \quad p = |A|.$$

Then

$$p \leqslant 1 + (4m - 1)\frac{(4m - 2)^m - 1}{4m - 3}. \tag{8}$$

If $m = 1$ then $p = 1 + \delta(G)$, $2 \leqslant \delta(G) \leqslant 3$, and every vertex of G is within distance 1 from the set $A' = \{x \in G : d(x, a) = 1\}$. By Lemma 2.1' we have

$$2e(G) \geqslant n - \delta(G) + \delta(G)(n - \delta(G)).$$

Therefore if $m = 1$ we may assume that $\delta(G) = 2$ and so

$$p = 3. \tag{8'}$$

The graph $H = G[A]$ is connected, implying

$$e(H) \geqslant p - 1. \tag{9}$$

Let K be a maximal subgraph of G that contains H and is such that

$$e(K) - e(H) \geqslant (q - p)\left(1 + \frac{1}{2m - 1}\right), \tag{10}$$

where $q = |K|$. Put $A^* = V(K)$ and

$$B_i = \{b \in G - A^* : d(b, A^*) = i\}, \quad i = 0, \ldots, m.$$

For $x \in G - A^*$ we have

$$2m \geqslant d(x, a) \geqslant d(x, A^*) + m,$$

so

$$A^* \cup \bigcup_1^m B_i = A^* \cup B^*$$

is a partition of $V(G)$.

Suppose now that $r \leqslant m - 1$ and

$$x \in B_r,\, xx_1,\, xx_2 \in E(G) \quad \text{and} \quad x_1, x_2 \in \bigcup_0^r B_i. \tag{11}$$

Let P_i be a shortest path from x_i to $A^* = \dot{B}_0$, $i = 1, 2$. Put

$$Z = V(P_1) \cup V(P_2) \cup \{x\} - A^*.$$

Then

$$e(G[A^* \cup Z]) - e(K) \geq |Z| + 1$$

and

$$|Z| \leq 2r + 1 \leq 2m - 1,$$

since $P_i (i = 1, 2)$ has length at most r and an endvertex of P_i belongs to A^*. Note now that the existence of the graph $G[A^* \cup Z]$ contradicts the maximality of the graph K. Consequently (11) does not hold.

A similar argument shows that

$$x \in B_m, \ xx_1, xx_2 \in E(G) \quad \text{and} \quad x_1, x_2 \in B_{m-1} \tag{12}$$

cannot hold either.

If $B^* \neq \varnothing$ let C_0 be a component of $G - A^* = G[B^*]$. Let C be the subgraph of G obtained from C_0 by adding to it the $C_0 - A^*$ edges together with the endvertices of these edges.

As neither (11) nor (12) holds, C has the following structure. For each vertex $x \in C \cap B_m$ there is a *unique* path P_x of length m joining x to A^*. The graph C is the union of $C_0 \cap G[B_m]$ and these paths. As $\delta(G) \geq 2$, for each vertex $x \in B_m \cap C$ there is an edge $xy \in E(C)$, $y \in B_m$.

There is an edge $x_0 y_0 \in E(C_0)$, $x_0, y_0 \in B_m$, such that P_{x_0} is an $x_0 - x_0'$ path, P_{y_0} is a $y_0 - y_0'$ path and $x_0' \neq y_0'$. For if there is no such edge the structure of C implies that there is a vertex $z \in A^*$ that is an endvertex of *every* path P_x, $x \in B_m \cap C$. However, then z separates G, contradicting $\kappa(G) \geq 2$.

Note that the existence of this edge $x_0 y_0 \in E(C)$ implies $C_0 = G - A^*$. For otherwise we can find vertices $u_1, u_2 \in B_m$ belonging to *different* components of $G - A^*$ and vertices $v_1, v_2 \in A^*$, $v_1 \neq v_2$, such that for each i, $i = 1, 2$, v_i is the *unique* vertex of A^* at distance m from u_i. Then

$$d(u_1, u_2) \geq d(u_1, A^*) + 1 + d(u_2, A^*) = 2m + 1,$$

contradicting diam $G \leq 2m$.

Let us say that a graph G_2 is obtained from a graph G_1 *adjoining a path P* if $G_2 = G_1 \cup P$ and $V(G_2) \cap V(P)$ consists of the endvertices of P. Denote by L the graph obtained from K by adjoining to it the $x_0' - y_0'$ path $x_0' P_{x_0} x_0 y_0 P_{y_0} y_0'$.

Then

$$|L - K| = 2m \quad \text{and} \quad e(L) - e(K) = 2m - 1. \tag{13}$$

It follows from the above description of the structure of C and from $C_0 = G - A^*$ that G can be obtained from L by successively adjoining to it paths of length at most $2m$. If one adjoins a path of length l to a graph, the order is increased by $l - 1$ and the size by l. Consequently if $r = |L| = q + 2m$ then

$$e(G) - e(L) \geqslant (n - r)\left(1 + \frac{1}{2m - 1}\right). \tag{14}$$

The proof of Theorem 2.8 is easily completed by putting together the inequalities above. Suppose first that $B^* = \varnothing$, that is $q = n$. Then (9) and (10) imply

$$e(G) \geqslant (n - p)\left(1 + \frac{1}{2m - 1}\right) + p - 1 = \left(1 + \frac{1}{2m - 1}\right)n - \frac{p}{2m - 1} - 1.$$

If $m = 1$ then (8′) gives

$$e(G) \geqslant 2n - 4$$

and if $m \geqslant 2$ then (8) implies

$$e(G) \geqslant \left(1 + \frac{1}{2m - 1}\right)n - 4(4m - 2)^{m-1}.$$

Suppose now that $B^* \neq \varnothing$. Then (9), (10), (13) and (14) imply

$$e(G) = e(H) + (e(K) - e(H)) + (e(L) - e(K)) + (e(G) - e(L))$$

$$\geqslant p - 1 + (q - p)\left(1 + \frac{1}{2m - 1}\right)$$

$$+ 2m + 1 + (n - q - 2m)\left(1 + \frac{1}{2m - 1}\right)$$

$$= \left(1 + \frac{1}{2m - 1}\right)n - \frac{p}{2m - 1} - \frac{2m}{2m - 1}.$$

Applying (8) and (8′) we see that if $m = 1$ then

$$e(G) \geqslant 2n - 5$$

and if $m \geqslant 2$ then

$$e(G) \geqslant \left(1 + \frac{1}{2m - 1}\right) n - 4(4m - 2)^{m-1}. \qquad \blacksquare$$

COROLLARY 2.9. *If* $n \geqslant 5$ *then* $f(n, 2, 2) = 2n - 5$. *If* $m \geqslant 1$ *and* $d' \geqslant 4m - 2$ *then*

$$\lim_{n \to \infty} \frac{f(n, 2m, 2)}{n} = \lim_{n \to \infty} \frac{f(n, 2m, d', 1)}{n} = 1 + \frac{1}{2m - 1}.$$

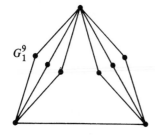 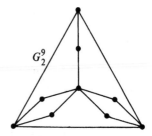

FIG. 2.2. $G_j^9 \in \mathscr{V}(9, 2, 3, 1), j = 1, 2.$

Proof. Let $j = 1, 2$ and $n \geqslant 2j + 1$. Let G_j^n be a graph of order n obtained by joining a vertex to the vertices of a K^{j+1}, with independent paths of length 2 (Fig. 2.2). It is easily seen that $G_j^n \in \mathscr{V}(n, 2, 3, 2)$, proving $f(n, 2, 2) \geqslant 2n - 5$.

Together with Theorem 2.8 this implies the first assertion. The second assertion follows from Theorem 2.8 and inequality (3). $\qquad \blacksquare$

3. GRAPHS WITH LARGE SUBGRAPHS OF SMALL DIAMETER

The aim of this section is to investigate the functions $f(n, d, d', k)$ and $g(n, d, d', k)$, introduced in the preceding section, in the case when d' is not much larger than d. (In particular, $d' < 4\lfloor d/2 \rfloor$ in the second case, by Corollary

2.7.) As we remarked there, it seems that in this case the functions are essentially independent of d, in particular

$$\lim_{n \to \infty} f(n, d, d', k)/f(n, d', d', k) = 1$$

and

$$\lim_{n \to \infty} g(n, d, d', k)/g(n, d', d', k) = 1.$$

Analogously to the functions defined in the preceding section, put

$$\hat{f}(n, d, k) = f(n, d, d, k - 1) \quad \text{and} \quad \hat{g}(n, d, k) = g(n, d, d, k - 1).$$

Thus $\hat{f}(n, d, k)$ is the minimal size of a graph of order n provided *any subgraph obtained by omitting any $k - 1$ of the vertices has diameter at most d.* Replacing "vertices" by "edges" in the definition above we obtain the definition of $\hat{g}(n, d, k)$.

At present rather little is known about these functions $\hat{f}$ and $\hat{g}$. The case $d = 2$ is well understood but in the case $d > 2$ there are only a few scattered results and rather weak estimates.

The graph $K(k, n - k)$ belongs to $V(n, 2, 2, k - 1)$ if and only if $n \geqslant 2k$. Murty and Vijayan [MV2] proved that $\hat{f}(n, 2, k) = k(n - k)$ and $K(k, n - k)$ is the only extremal graph if $n > (2 + \sqrt{2})k - \sqrt{2}$. Bollobás and Eldridge [BE4] extended this result by showing that whenever $K(k, n - k) \in \mathscr{V}(n, 2, 2, k - 1)$ the graph $K(k, n - k)$ is the unique extremal graph of $\mathscr{V}(n, 2, 2, k - 1)$.

THEOREM 3.1. *Let $2 \leqslant 2k \leqslant n$. Then $K(k, n - k)$ is the only extremal graph in the class $\mathscr{V}(n, 2, 2, k - 1)$. In particular, $\hat{f}(n, 2, k) = k(n - k)$.*

Proof. Since $K(k, n - k) \in \mathscr{V}(n, 2, 2, k - 1)$, $\hat{f}(n, 2, k) \leqslant k(n - k)$. Furthermore, $K(n, n - k)$ is a minimal graph in the class $\mathscr{V}(n, 2, 2, k - 1)$. Consequently to prove the theorem it suffices to show that if $G \in \mathscr{V}(n, 2, 2, k - 1)$ and $e(G) = k(n - k)$ then $G = K(k, n - k)$.

The class $\mathscr{V}(n, 2, 2, k - 1)$ consists of graphs in which every two non-adjacent vertices have at least k common neighbours. Consequently $\delta(G) \geqslant k$. Furthermore, if $\delta(G) = k$, say $x \in G$ and $d(x) = k$, then every vertex of $G - \Gamma(x)$ is joined to every vertex in $\Gamma(x)$. Therefore in this case G contains a $K(k, n - k)$ and as $e(G) = k(n - k) = e(K(k, n - k))$, we have $G = K(k, n - k)$, as required.

Suppose now that $\delta = \delta(G) \geqslant k + 1$. To prove the result we shall show that this leads to contradiction.

Pick a vertex y_1 with $d(y_1) = \delta$. Let

$$X = \Gamma(y_1) = \{x_1, x_2, \ldots, x_\delta\} \quad \text{and} \quad Y = V(G) - \Gamma(y_1) = \{y_1, y_2, \ldots, y_{n-\delta}\}.$$

Put $d_i = |\Gamma(x_i) \cap X|$, $e_i = |\Gamma(x_i) \cap Y|$, $f_j = |\Gamma(y_j) \cap X|$ and $h_j = |\Gamma(y_j) \cap Y|$. Set furthermore

$$D = \sum_1^\delta d_i,$$

$$E = \sum_1^\delta e_i = \sum_1^{n-\delta} f_i,$$

and

$$H = \sum_1^{n-\delta} h_i.$$

Note that

$$E + \frac{D}{2} + \frac{H}{2} = e(G) = k(n - k). \tag{1}$$

Since $\delta = \delta(G)$ we have

$$k(n - k) = e(G) \geqslant \frac{\delta n}{2} \tag{2}$$

and

$$f_i + h_i \geqslant \delta,$$

implying

$$E + H = \sum_1^{n-\delta} (f_i + h_i) \geqslant \delta(n - \delta). \tag{3}$$

As $\delta \geqslant k + 1$, inequality (2) implies $n \geqslant 2k + 2$. For $j \geqslant 2$ the vertex y_j is not adjacent to y_1 so y_j and y_1 have at least k common neighbours. Consequently

$$k \leqslant f_j \leqslant \delta \tag{4}$$

and

$$E \geqslant \delta + (n - \delta - 1) k. \tag{5}$$

If x_i is not adjacent to y_j then there are at least k vertices adjacent to both of them, say k_1 from X and k_2 from Y. Then $k_1 \leqslant d_i$ and $k_2 \leqslant h_j$, implying $k \leqslant d_i + h_j$. Summing these inequalities for every pair (x_i, y_j) not joined by an edge, we obtain

$$k(\delta(n - \delta) - E) \leqslant \sum_1^\delta d_i(n - \delta - e_i) + \sum_1^{n-\delta} h_j(\delta - f_j) \leqslant (n - \delta)D + (\delta - k)H,$$

where the second inequality is a consequence of (4). Thus

$$E + \frac{n - \delta}{k} D + \frac{\delta - k}{k} H \geqslant \delta(n - \delta). \tag{6}$$

To complete the proof we show that the inequalities (1), (2), (3), (5) and (6) cannot all hold.

Notice first that (1), (3) and (5) imply

$$2k(n - k) = 2E + D + H \geqslant 2E + H \geqslant \delta(n - \delta) + \delta + (n - \delta - 1)k.$$

Hence

$$n(\delta - k) \leqslant (\delta - k)(2k + \delta - 1),$$

giving

$$n \leqslant 2k + \delta - 1.$$

From (2) and the last inequality we have

$$\delta \leqslant \frac{2k(n - k)}{n} \leqslant \frac{2k(k + \delta - 1)}{2k + \delta - 1}, \tag{8}$$

implying

$$\delta \leqslant \sqrt{2}k < \tfrac{3}{2}k. \tag{9}$$

(1) and (3) imply

$$\frac{H}{2} \geqslant (\delta - k)(n - \delta - k) + \frac{D}{2}, \tag{10}$$

and (1) and (6) imply

$$\frac{2n - 2\delta - k}{2k} D - \frac{3k - 2\delta}{2k} H \geqslant (\delta - k)(n - \delta - k).$$

By (9) and (10) this inequality gives

$$\frac{2n - 2\delta - k}{2k} D \geq (\delta - k)(n - \delta - k)\left(1 + \frac{3k - 2\delta}{k}\right) + \frac{3k - 2\delta}{2k} D,$$

and so

$$\frac{D}{2} \geq (\delta - k)(n - \delta - k)(2k - \delta)/(n - 2k). \tag{11}$$

Putting this into (10) we have

$$\frac{H}{2} \geq (\delta - k)(n - \delta - k)\left(1 + \frac{2k - \delta}{n - 2k}\right)$$

$$= (\delta - k)(n - \delta - k)(n - \delta)/(n - 2\delta). \tag{12}$$

Summing (5), (11) and (12), and taking (1) into account we obtain

$$k(n - k) \geq \delta + k(n - \delta - 1) + (\delta - k)(n + 2k - 2\delta)/n - 2k,$$

i.e.

$$(n - \delta - k)(n + 2k - 2\delta)/(n - 2k) \leq k - 1.$$

Finally, rearranging this inequality, taking into account (8) and then $n \geq 2k + 2$, we arrive at the required contradiction

$$0 \geq n^2 - 3\delta n + n + 2\delta^2 - 2k$$

$$\geq n^2 - 6k(n - k) + n + 8k^2(n - 1)^2/p^2 - 2k > 0. \qquad \blacksquare$$

In the range $0 < n < 2k$ the function $\hat{f}(n, 2, k)$ is not too interesting. Suppose $G \in \mathscr{V}(n, 2, k - 1)$. If $n \leq k$, then $G = K^n$. If $k < n < 2k$ then $\delta(G) \geq k$ so $\hat{f}(n, 2, k) \geq kn/2$. This trivial estimate is fairly accurate. For certain pairs (n, k) it is easy to determine the exact value of $\hat{f}(n, 2, k)$. E.g. if $r \geq 3, l \geq 1, n = rl$ and $k = (r - 1)l$ then $\hat{f}(n, 2, k) = kn/2$ (see Ex. 17).

Let us turn now to the function $\hat{g}(n, 2, k)$. It turns out that for a given k the range of n divides naturally into two parts. For large values of n the function $\hat{g}(n, 2, k)$ was determined by Murty [M36]. For the other pairs (n, k) the exact value of $\hat{g}(n, 2, k)$ is not known but Bollobás and Erdös [BE9] determined the order of the function.

Note that $\mathscr{E}(n, 2, 2, k - 1)$ is the set of all graphs of order n in which every two vertices are joined by at least k independent paths of length at most 2. In particular, if $G \in \mathscr{E}(n, 2, k - 1)$ then $\delta(G) \leqslant k$ and so $|G| \leqslant k + 1$. Furthermore, if $n \geqslant k + 1$, then $K^k + E^{n-k} \in \mathscr{E}(n, 2, 2, k - 1)$.

THEOREM 3.2. *If* $n \geqslant \left[(3 + \sqrt{5})/2\right] k$ *then* $K^k + E^{n-k}$ *is the unique extremal graph of the class* $\mathscr{E}(n, 2, 2, k - 1)$. *In particular,* $\hat{g}(n, 2, k) = \binom{k}{2} + k(n - k)$.

Proof. Let $G \in \mathscr{E}(n, 2, 2, k - 1)$ and $e(G) \leqslant \binom{k}{2} + k(n - k)$. If $\delta(G) = k$, say $x \in G$ and $d(x) = k$, then every vertex of G is adjacent to every vertex in $\Gamma(x)$. Thus $G \supset K^k + E^{n-k}$ and so $G = K^k + E^{n-k}$. To complete the proof we shall show that $\delta(G) \geqslant k + 1$ cannot hold.

Suppose $\delta = \delta(G) \geqslant k + 1$, say $y \in G$, $d(y) = \delta$. Put $X = \Gamma(y)$, $Y = V(G) - X$. Then $\delta(G[X]) \geqslant k - 1$ so $e(G[X]) \geqslant \delta(k - 1)/2$. Every vertex in Y is joined to at least k vertices in X so there are at least $(n - \delta - 1)k$ edges joining $Y - \{y\}$ to X. Furthermore, $\delta(G) = \delta \geqslant k + 1$ implies that there are at least $(n - \delta - 1)k + (n - \delta - 1)(\delta - k)/2$ edges incident with at least one vertex of $Y - \{y\}$. Consequently

$$\frac{\delta(k - 1)}{2} + \delta + (n - \delta - 1)\frac{\delta + k}{2} \leqslant e(G) \leqslant \binom{k}{2} + k(n - k).$$

This is easily seen to imply

$$\delta = n - k. \tag{13}$$

Trivially

$$\frac{n\delta}{2} \leqslant e(G) \leqslant \binom{k}{2} + k(n - k),$$

giving

$$\delta \leqslant \frac{k(2n - k - 1)}{n}.$$

Combining this with (13) we find that

$$n - k \leqslant \frac{k(2 - k - 1)}{n},$$

i.e.

$$n^2 - 3kn + k^2 + k \leqslant 0.$$

This inequality cannot hold since $n \geqslant [(3 + \sqrt{5})/2] k$ implies that $n^2 + k^2 \geqslant 3kn$.

When $n < [(3 + \sqrt{5})/2] k$ then $K^k + E^{n+k}$ is essentially never an extremal graph of $\mathscr{E}(n, 2, 2, k - 1)$, as shown by the next result.

THEOREM 3.3. *Let* $1 < c < (3 + \sqrt{5})/2$, $n = [ck]$. *Then*

$$\hat{g}(n, 2, k) = \tfrac{1}{2}c^{\frac{3}{2}}k^2 + o(k^2).$$

Proof. (i) Let $G \in \mathscr{E}(n, 2, 2, k)$, $e(G) = \hat{g}(n, 2, k)$. As in the proof of the previous theorem, pick a vertex $y \in G$ having minimal degree $\delta = \delta(G)$. Put $X = \Gamma(y)$, $Y = V(G) - X$ and denote by E the number of $X - Y$ edges. Clearly

$$E \geqslant k(n - \delta)$$

and

$$e(G[X]) \geqslant \frac{(k - 1)\delta}{2}.$$

Furthermore, $\delta(G) = \delta$ implies that

$$e(G[Y]) \geqslant \frac{(n - \delta)\delta - E}{2}.$$

Consequently

$$e(G) \geqslant \tfrac{1}{2}(n(\delta + k) - \delta^2) - \frac{\delta}{2} = \frac{nk}{2} + \tfrac{1}{2}(n - 1 - \delta)\delta.$$

Taking into account the trivial relation $e(G) \geqslant \delta n/2$, it follows from this inequality that

$$e(G) = \hat{g}(n, 2, k) \geqslant \tfrac{1}{2}c^{\frac{3}{2}}k^2 + O(k).$$

(ii) To prove the required upper bound we apply a probability argument to construct a graph $G \in \mathscr{E}(n, 2, 2, k)$.

Let $\varepsilon > 0$. Take $n = \lfloor ck \rfloor$ vertices and choose each edge with probability $d = c^{-\frac{3}{2}} + \varepsilon$. Let G be a graph obtained in this way. By the law of large numbers $e(G) = \binom{n}{2}(d + o(1))$ with probability tending to 1 as $k \to \infty$. Appealing again to the law of large numbers we see that with probability tending to 1 every two vertices have $(d^2 + o(1))n$ common neighbours. Thus, as $n \to \infty$, with probability tending to 1 we have $G \in \mathscr{E}(n, 2, 2, k)$ and

$$e(G) \leqslant \tfrac{1}{2}(d + \varepsilon) \, n^2 \leqslant \tfrac{1}{2}c^{\frac{3}{2}}k^2 + \varepsilon c^2 k^2. \qquad \blacksquare$$

Let us turn to some results concerning the functions $\hat{f}(n, d, 2)$ and $\hat{g}(n, d, 2)$. We start with a reformulation of a conjecture of Bollobás and Harary [BH2] which is a slight modification of an incorrect conjecture in [B25].

CONJECTURE 3.4. *If $d \geqslant 2$ then*

$$\lim_{n \to \infty} \frac{\hat{f}(n, d, 2)}{n} = \frac{\lfloor d/2 \rfloor + 1}{\lfloor d/2 \rfloor} \tag{14}$$

and

$$\lim_{n \to \infty} \frac{\hat{g}(n, d, 2)}{n} = \frac{\lfloor 2d/3 \rfloor + 1}{\lfloor 2d/3 \rfloor}. \tag{15}$$

The conjecture is motivated by almost the same graphs that prove inequalities (2.3) and (2.4). Let G^n be obtained by joining two vertices with s independent paths of length $l = \lfloor d/2 \rfloor + 1$ and one pair of length at most l. Then $G^n \in \mathscr{V}(n, d, d, 1)$ and

$$\lim_{n \to \infty} \frac{e(G^n)}{n} = \frac{\lfloor d/2 \rfloor + 1}{\lfloor d/2 \rfloor}.$$

Let H^n be the union of s cycles of length $m = \lfloor 2d/3 \rfloor$ and at most two cycles of length at most m such that the cycles have exactly one vertex in common. Then

$$H^n \in \mathscr{E}(n, d, d, 1) \quad \text{and} \quad \lim_{n \to \infty} \frac{e(H^n)}{n} = \frac{\lfloor 2d/3 \rfloor + 1}{\lfloor 2d/3 \rfloor}.$$

These examples show that

$$\limsup_{n \to \infty} \frac{\hat{f}(n, d, 2)}{n} \leqslant \frac{\lfloor d/2 \rfloor + 1}{\lfloor d/2 \rfloor} \tag{14'}$$

and

$$\limsup_{n \to \infty} \frac{\hat{g}(n, d, 2)}{n} \leqslant \frac{\lfloor 2d/3 \rfloor + 1}{\lfloor 2d/3 \rfloor}. \tag{15'}$$

It is easily seen (cf. Ex. 20) that if Conjecture 3.4 is true then for every pair of natural numbers (d, d'), $2 \leqslant d \leqslant d'$, we have

$$\lim_{n \to \infty} \frac{f(n, d, d', 1)}{n} = \max\left\{\frac{d}{d-1}, \frac{\lfloor d'/2 \rfloor + 1}{\lfloor d'/2 \rfloor}\right\} \tag{14''}$$

and

$$\lim_{n \to \infty} \frac{g(n, d, d', 1)}{n} = \max\left\{\frac{2\lfloor d/2 \rfloor + 1}{2\lfloor d/2 \rfloor}, \frac{\lfloor 2d'/3 \rfloor + 1}{\lfloor 2d'/3 \rfloor}\right\}. \tag{15''}$$

Thus, in order to determine the order of $f(n, d, d', 1)$ and $g(n, d, d', 1)$ for all values of d and d' it suffices to prove Conjecture 3.4. Though at present we seem to be rather far from a proof of this conjecture, Theorems 3.1 and 3.2 imply its validity for $d = 2$ and the results below, mostly from [B31], show that it also holds for $d \leqslant 4$. The following lemma, in the vein of Lemma 2.1, will be useful in the sequel.

LEMMA 3.5. *Suppose* $a > 0$ *and* G *is such that for every* $x \in G$ *we have*

$$d(x) + \sum_{\substack{y \in \Gamma(x) \\ d(y) > a}} \frac{d(y) - a}{d(y)} \geqslant a. \tag{16}$$

Then

$$e(G) \geqslant \frac{a}{2}|G|. \tag{17}$$

If there is a vertex x *for which inequality* (16) *is sharp then so is* (17).

Proof. If $xy \in E(G)$ and $d(y) \geqslant a$ let $w(x, y) = (d(y) - a)/d(y)$. If $x \in G$ and $d(x) < a$ put

$$d'(x) = d(x) + \sum_{\substack{y \in \Gamma(x) \\ d(y) > a}} w(x, y).$$

If $y \in G$ and $d(y) \geqslant a$, set

$$d'(y) = d(y) - \sum_{x \in \Gamma(y)} w(x, y) = a.$$

By assumption $d'(x) \geqslant a$ for every vertex $x \in G$. Therefore

$$2e(G) = \sum_{x \in G} d(x) \geqslant \sum_{x \in G} d'(x) \geqslant a|G|. \qquad \blacksquare$$

THEOREM 3.6. *If* $n \geqslant 3$ *then*

$$\hat{f}(n, 3, 2) \geqslant 2n - 2(2n)^{\frac{1}{2}}$$

and

$$\hat{f}(n, 4, 2) \geqslant \tfrac{3}{2}n - 2n^{\frac{2}{3}}.$$

In particular, (14) *holds for* $d \leqslant 4$.

Proof. The assertions are trivial for $n = 3$ and 4 so $n \geqslant 5$. Let $G \in \mathscr{V}(n, 3, 3, 1)$. Our aim is to show that $e(G) \geqslant 2n - 2(2n)^{\frac{1}{2}}$.

Suppose first that each vertex $x \in G$ of degree 2 is adjacent to two vertices of degree at least $\lfloor (2n)^{\frac{1}{2}} - 1 \rfloor$ and each vertex $y \in G$ of degree 3 is adjacent to at least one vertex of degree at least $\lfloor (2n)^{\frac{1}{2}} - 1 \rfloor$. In this case routine calculations show that G satisfies the conditions of Lemma 3.5 with $a = 4 - (4\sqrt{2}/n^{\frac{1}{2}})$ and so $e(G) \geqslant 2n - 2(2n)^{\frac{1}{2}}$.

Therefore we may assume that there exists a vertex $x_0 \in G$ of degree at most 3 such that at most one vertex adjacent to x_0, say x_1, has degree at least $\lfloor (2n)^{\frac{1}{2}} - 1 \rfloor$. Put $X = \Gamma(x_0)$, $X' = X - \{x_1\}$, $A = \Gamma(X') - X$, $A^* = A \cup X \cup \{x_0\}$ and $B = \Gamma(x_1) - A^*$. Let furthermore $p = |A^*|$, $q = |B|$, $H = G[A^*]$ and $K = G[A^* \cup B]$. Then

$$p \leqslant 2(2n)^{\frac{1}{2}} - 2$$

and by construction

$$e(H) \geqslant p - 1.$$

Let $b \in B$. As b is at distance at most 3 from x_0 in the graph $G - \{x_1\}$, it is joined to a vertex in A. Consequently

$$e(K) \geqslant e(H) + 2q \geqslant p - 1 + 2q.$$

Let now $c \in G - A^* \cup B$ and suppose c is adjacent to *at most one vertex in* $A^* \cup B$. This means that for some vertex $d \in A^* \cup B$ we have $\Gamma(c) \cap (A^* \cup B - \{d\}) = \varnothing$. However, this is impossible since then the distance between c and x in $G - d$ is at least 4. Consequently every vertex in $G - A^* \cup B$ is adjacent to at least 2 vertices in $A^* \cup B$. Therefore

$$e(G) \geqslant e(K) = 2(n - p - q) \geqslant 2n - p - 1$$

$$\geqslant 2n - 2(2n)^{\frac{1}{2}} + 1,$$

proving the first inequality.

The proof of the second inequality is essentially a simpler version of the proof of Theorem 2.7. We leave the details to the reader (Ex. 21).

Inequality (14') and the inequalities of Theorem 3.6 imply that (14) holds for $d \leqslant 4$. ∎

THEOREM 3.7. *If $n \geqslant 3$ then*

$$\hat{g}(n, 4, 2) \geqslant \tfrac{3}{2}n - (\tfrac{3}{2}n)^{\frac{1}{2}} - \tfrac{1}{2}.$$

In particular, (15) holds for $d \leqslant 4$.

Proof. Let $G \in \mathscr{V}(n, 4, 4, 1)$. We have to show that $e(G) \geqslant \tfrac{3}{2}n - (\tfrac{3}{2}n)^{\frac{1}{2}}$. Suppose first that every vertex of degree 2 is adjacent to a vertex of degree at least $\lfloor (\tfrac{3}{2}n)^{\frac{1}{2}} \rfloor$. It is easily checked that G satisfies the conditions of Lemma 3.5 with $a = 3 - (6n)^{\frac{1}{2}}$ and so it does have the required number of edges.

Assume now that there exists a vertex $x_0 \in G$ of degree 2 whose neighbours have degree at most $(\tfrac{3}{2}n)^{\frac{1}{2}} - 1$. Put $X = \Gamma(x_0)$, $A = \Gamma(X)$, $A^* = A \cup X$, $p = |A^*|$ and $H = G[A^*]$. Then

$$p \leqslant 2(\tfrac{3}{2}n)^{\frac{1}{2}} - 1 \tag{81}$$

and

$$e(H) \geqslant p - 1. \tag{19}$$

As in the proof of Theorem 2.7, let K be a maximal subgraph of G such that $K \supset H$ and

$$e(K) - e(H) \geqslant \tfrac{3}{2}(|K| - |H|) = \tfrac{3}{2}(|K| - p). \tag{20}$$

Our next aim is to show that $K = G$. Suppose this is not so. Let b be a vertex not in K that is adjacent to a vertex $c \in K$. Since $G \in \mathscr{E}(n, 4, 4, 1)$, the graph $G - c$ contains a b–x path of length at most 4. The vertices within distance 2 from x belong to $A^* \subset V(K)$ so there is a path P of length at most 2 from b to K in the graph $G - c$. Then $G[V(K) \cup V(P)]$ strictly contains K and also satisfies the conditions, contradicting the maximality of K. Thus $K = G$, is claimed.

The inequality in the theorem is an immediate consequence of inequalities (19), (18) and (20):

$$e(G) = (e(G) - e(H)) + e(H) \geqslant \tfrac{3}{2}(n - p) + p - 1$$
$$= \tfrac{3}{2}n - \tfrac{1}{2}p - 1 \geqslant \tfrac{3}{2}n - (\tfrac{3}{2}n)^{\frac{1}{2}} - \tfrac{1}{2}.$$

Appealing to (15') we see that (15) holds for $d \leqslant 4$. ∎

With slightly more discussion one can show that $\hat{g}(n, 3, 2) = \lfloor \frac{3}{2}n \rfloor - 1$ if $n \geqslant 5$, as proved in [B25] (cf. Ex. 22).

To conclude the section we turn to another problem concerning the reliability of networks. The problem, proposed by van Lint [V3], [V4], originated in computer designs but we state it in terms of bipartite graphs. Let $G = G_2(2n, m)$ be a bipartite graph with vertex classes $A = \{a_1, a_2, \ldots, a_{2n}\}$ and $B = \{b_1, b_2, \ldots, b_m\}$. Let G be such that for any partition $\{a_{i_1}, a_{j_1}\}$, $\{a_{i_2}, a_{j_2}\}, \ldots, \{a_{i_n}, a_{j_n}\}$ of the vertices of A into n pairs. there are distinct vertices $b_{k_1}, \ldots, b_{k_n}$ of B such that b_{k_s} is joined to a_{i_s} and a_{j_s} for every s, $1 \leqslant s \leqslant n$. Let $\mathscr{S}(n, m)$ be the set of all such graphs G. Thus $\mathscr{S}(n, m)$ consists of graphs $G_2(2n, m)$ in which any n disjoint pairs of vertices belonging to the first class can be joined by vertex disjoint paths of length 2. Denote by $\Delta_A(G)$ the maximal degree of the vertices in A. Then define

$$\lambda(n, m) = \min\{\Delta_A(G): G \in \mathscr{S}(n, m)\},$$

$$\lambda(n) \quad = \min\{\lambda(n, m): m \geqslant n\},$$

$$l(n) \quad = \min\{m: \lambda(n) = \lambda(n, m)\}.$$

The problem is to determine the functions $\lambda(n, m)$, $\lambda(n)$ and $l(n)$ together with the graphs for which the bounds are attained and simple algorithms for finding the vertices $b_{k_1}, \ldots, b_{k_n}$ for every possible partition of A.

The formulation of this problem is reminiscent of the problems about graphs of diameter 2, but the similarity is very superficial. Suppose $G \in \mathscr{S}(n, m)$ and $\Delta_A(G) = \lambda(n, m) = r$, say. Then we may assume without loss of generality that all the sets $\Gamma(a_i)$ have order r. If $\{a_{i_1}, a_{j_1}\}, \ldots \{a_{i_n}, a_{j_n}\}$ is any partition of A into n pairs then the sets $B_k = \Gamma(a_{i_k}) \cap \Gamma(a_{j_k})$, $k = 1, 2, \ldots, n$, have a set of distinct representatives. Thus, by Hall's theorem (Theorem I.2.3, see also Theorem II.1,1'), $G \in \mathscr{S}(n, m)$ iff

$$\left| \bigcup_{k \in L} B_k \right| \geqslant |L|$$

for every subset L of $\{1, 2, \ldots, n\}$.

In order to make the left-hand side above large, it seems desirable to choose the $\Gamma(x_i)$ to be independent r-sets of a relatively small set B, that is subsets of r elements each that represent independent events in the probability space B in which each point b_i has probability $1/m$. The difficulty is that if n is large and r is small then it is impossible to choose the r-sets independently and so one has to find graphs in which these sets are approximately independent and have large pairwise intersections. The upper bounds in the next result, due to Erdős and Spencer (see [V3]), are proved exactly this way. Though this is the essential part of the result, we leave the details to the reader (Ex. 26).

Theorem 3.8. $\lambda(n, n) \geq \lfloor n/2 \rfloor$ *and* $\lim n^{-1}\lambda(n, n) = \frac{1}{2}$. *Furthermore,* $\lambda(n)$ $= O(n^{\frac{1}{2}}\log n)$ *and* $\lambda(n) \geq ((2n - 1)!!)^{1/(2n)} \sim (2/e)^{\frac{1}{2}}n^{\frac{1}{2}}$.

Proof of the lower bounds. Let $G \in \mathcal{S}(n, n)$. Given $b \in B$ and a partition of A into pairs, there is a pair such that b is joined to both vertices of the pair. Consequently $d(b) \geq n$ for every $b \in B$ and so $e(G) \geq n^2$ and $\Delta_A(G) \geq \lfloor n^2/2n \rfloor$ $=\lfloor n/2 \rfloor$.

In order to obtain a lower bound for $\lambda(n)$, let $G \in \bigcup_{m \geq n} \mathcal{S}(n, m), \Delta_A(G) = \lambda(n)$.
There are $(2n)!/(2^n n!) = (2n - 1)!!$ partitions of A into pairs. For each partition $\{a_{i_1}, a_{j_1}\}, \ldots, \{a_{i_n}, a_{j_n}\}$ there is a set of choices $b(a_k) \in \Gamma(a_k)$ such that $b(a_{i_s}) = b(a_{j_s})$ for $s = 1, \ldots, n$, and $b(a_{i_s}) \neq b(a_{i_t})$ if $s \neq t$. Clearly each set of choices is suitable for exactly one partition. Since there are at most $(\Delta_A(G))^{2n} = \lambda(n)^{2n}$ sets of choices, we have

$$(2n - 1)!! \leq \lambda(n)^{2n},$$

as required.   ∎

As proposed by van Lint [V4], it would be of interest to find a constructive proof of Theorem 3.8.

4. FACTORS OF SMALL DIAMETER

Given a sequence $d_1, d_2, \ldots, d_k$ of natural numbers, is there a complete graph $K^n, n \geq 2$, that can be factored into k factors, say $F_1, F_2, \ldots, F_k$, such that diam $F_i \leq d_i, i = 1, 2, \ldots, k$? If $k \geq 2$ then in order to expect an affirmative answer each d_i must be at least 2 since a graph of diameter 1 is complete. It is not entirely obvious (though not far from it) that if $d_i \geq 2$ for every i then such a complete graph K^n does exist. Denote by $f(d_1, d_2, \ldots, d_k)$ the minimal integer that may be the order of such a complete graph. In order to simplify the notation we put $f_k(d) = f(d_1, d_2, \ldots, d_k)$ if $d = d_1 = d_2$

$= \ldots = d_k$. The investigation of the function $f(d_1, d_2, \ldots, d_k)$ was initiated by Bosák, Rosa and Znám [BRZ1] and a number of results concerning this function and other related problems were obtained by Bosák, Erdös, Rosa [BER1], Sauer [S8], Palumbiny [P1], [P2], Sauer and Schaer [SS1], Bosák [B48] and Erdös, Sauer, Schaer and Spencer [ESSS1]. The main aim of this section is to give bounds on $f_k(2)$, though we start by showing a pleasant property of $f(d_1, d_2, \ldots, d_k)$, observed by Bosák, Rosa and Znám [BRZ1].

THEOREM 4.1. *If $m \geqslant f(d_1, d_2, \ldots, d_k) \geqslant 2$ then K^m can be factorized into k factors such that the diameter of the ith factor is at most d_i, $1 \leqslant i \leqslant k$.*

Proof. Let K^n be the complete graph with vertex set $\{x_1, x_2, \ldots, x_n\}$ and let K^{n+1} be the complete graph with vertex set $\{x_1, x_2, \ldots, x_{n+1}\}$. Suppose K^n has a factorization into k factors, say $F_1, F_2, \ldots, F_k$ such that diam $F_i \leqslant d_i$. To prove the theorem it suffices to show that K^{n+1} has an analogous factorization.

Let F_i' be the subgraph of K^{n+1} obtained from F_i by adding to it the set of edges

$$\{x_{n+1}x_j : x_n x_j \in E(F_i)\}.$$

Then each F_i', $1 \leqslant i \leqslant k$, is a factor of K^{n+1} and each edge of K^{n+1} belongs to exactly one factor F_i', with the exception of the edge $x_n x_{n+1}$. Add this edge to F_1' and denote by F_1'' the resulting graph. Then $F_1'' \cup F_2' \cup \ldots \cup F_k'$ is a factorization of K^{n+1}, $d(F_1'') \leqslant d_1$ and $d(F_i') \leqslant d_i$, $2 \leqslant i \leqslant k$. ∎

When $d \geqslant 3$ the function $f_k(d)$ is easily determined.

THEOREM 4.2. *if $d \geqslant 3$ then $f_k(d) = 2k$.*

Proof. By definition $f_k(3) \geqslant f_k(4) \geqslant \ldots$. A factor of K^n having finite diameter is connected so has at least $n - 1$ edges. Therefore if K^n has a factorization into k factors of finite diameter then $k(n - 1) \leqslant e(K^n) = \frac{1}{2}n(n - 1)$, implying $n \geqslant 2k$. Thus $f_k(d) \geqslant 2k$ for every $d \geqslant 2$.

To complete the proof it suffices to give a factorization of K^{2k} into k factors of diameter 3. Let $V(K^{2k}) = \{x_1, x_2, \ldots, x_{2k}\}$. For $1 \leqslant i \leqslant k$ define a factor F_i of K^{2k} by

$$E(F_i) = \{x_{2i-1}x_{2h}, x_{2i-1}x_{2j-1}, x_{2i}x_{2h-1}, x_{2i}x_{2j} : h \leqslant i < j\}.$$

Then $\bigcup_1^k F_i$ is a factorization of K^{2k} into factors of diameter 3. ∎

We turn now to the non-trivial results of this section namely to those concerning the function $f_k(2)$. In order to emphasize that k is the variable, from now on we write $d_2(k)$ instead of $f_k(2)$. The first bounds on $d_2(k)$ were obtained by Bosák, Erdös and Rosa [BER1]. Their main result was that

$$d_2(k) \leqslant (\tfrac{49}{10})^2 k^2 \log k,$$

provided k is sufficiently large. A breakthrough was achieved by Sauer [S8] who proved

$$d_2(k) \leqslant 7k.$$

Using the method of Sauer, Bosák [B48] improved this to $d_2(k) \leqslant 6k$ and showed that in the other direction we have $6k - 52 \leqslant d_2(k)$. Bosák also proved various better lower bounds for $d_2(k)$ if $k \leqslant 370$. As the main result of this section we present a slight sharpening of these bounds for $d_2(k)$, noted in [B30].

THEOREM 4.3. $d_2(2) = 5$ and

$$6k \geqslant d_2(k) \geqslant g(k) = \begin{cases} \lceil \tfrac{9}{2}(k-1) \rceil + 1 & \text{if} \quad k = 3 \quad \text{or} \quad 4 \\ 20 & \text{if} \quad k = 5 \\ 6k - 9 & \text{if} \quad k \geqslant 6. \end{cases}$$

Proof. (i) There is only one way to factor K^4 into 2 connected factors and these factors have diameter 3. On the other hand K^5 is the edge disjoint union of 2 pentagons. Thus $d_2(2) = 5$.

(ii) Let $k \geqslant 2$ and let $K = K^{6k}$ be the complete graph with vertex set $V(K) = \{a_{i,s} : 1 \leqslant i \leqslant k, 1 \leqslant s \leqslant 6\}$. In order to prove $d_2(k) \leqslant 6k$ we explicitly factorize K into k factors of diameter 2. This factorization is based on the following system of subsets of $\{1, 2, \ldots, 6\}$:

$$A_1 = B_2 = \{1, 3, 4\}, \qquad A_2 = B_1 = \{2, 3, 4\},$$

$$A_3 = B_4 = \{3, 5, 6\}, \qquad A_4 = B_3 = \{4, 5, 6\},$$

$$A_5 = B_6 = \{5, 1, 2\}, \qquad A_6 = B_5 = \{6, 1, 2\}.$$

For $1 \leqslant i \leqslant k$ define a factor F_i of K by joining each vertex $a_{i,s}, 1 \leqslant s \leqslant 6$, to each vertex in the set

$$\{a_{i,t} : s < t \leqslant 6\} \cup \{a_{j,t} : i < j \leqslant k, t \in A_s\} \cup \{a_{j,t} : 1 \leqslant j < i, t \in B_s\}.$$

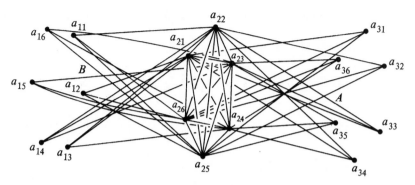

FIG. 4.1. One factor, F_2, of K^{18}.

It is easy to check that $F_1, F_2, \ldots, F_k$ are edge disjoint factors of diameter 2 (cf. Fig. 4.1).

(iii) There remains to prove that $g(k)$ is a lower bound of $d_2(k)$. The proof will be based on Lemma 3.5 and though there is no theoretical difficulty, there are a number of details to be taken care of.

Let $k \geqslant 3$, $n \geqslant d_2(k)$ and let $F_1, F_2, \ldots, F_k$ be a factorization of K^n into factors of diameter 2. Put $F = F_1$.

Let $x_0 \in F$ and put $X = \Gamma(x_0) = \{x_1, \ldots, x_s\}$. Since diam $F = 2$, $\Gamma(X) \cup X = V(G)$ and so

$$\sum_1^s d(x_i) \geqslant n - 1. \tag{1}$$

Furthermore, $\bar{F} = \bigcup_2^k F_i \in \mathscr{E}(n, 2, 2, k - 2)$, i.e. any 2 vertices of $\bar{F}$ are joined by at least $k - 1$ paths of length at most 2. In particular, if $1 \leqslant i < j \leqslant s$ then there are at least $k - 2$ vertices *not joined to either of* x_i and x_j in F. Hence

$$\Sigma' d(x_i) \geqslant k - 3 + s - 2 = k + s - 5, \tag{2}$$

where $\Sigma' d(x_i)$ denotes the sum of any $s - 2$ of the numbers $d(x_1), d(x_2), \ldots, d(x_s)$. An immediate consequence of this inequality is that $\delta(F) \geqslant 3$. Furthermore, since

$$\Delta(F) + \sum_2^k \delta(F_i) \leqslant n - 1 = \Delta(K^n),$$

we have

$$\Delta(F) \leqslant n + 2 - 3k. \tag{3}$$

Choose the notation in such a way that $\delta(F) = \min\limits_{1 \le i \le k} \delta(F_i)$. Let $x_0 \in F$, $d(x_0) = \delta$. Since $\bigcup\limits_{1}^{k} F_i$ is a factorization of K^n,

$$n - 1 = \Delta(K^n) \ge \Delta(F) + \sum_{2}^{k} \delta(F_i) \ge \Delta(F) + (k-1)\delta.$$

Hence if $\delta \ge 5$ we have $n \ge 5k + 1$. If $\delta = 4$ then (1) implies $\Delta(F) \ge (n-1)/4$ and so

$$n - 1 \ge \frac{n-1}{4} + 4(k-1),$$

giving

$$n \ge \frac{16}{3}(k-1) + 1.$$

Finally, if $\delta = 3$ then

$$\Delta(F) \ge \frac{n-1}{3}$$

from (1) and so

$$n - 1 \ge \frac{n-1}{3} + 3(k-1),$$

implying

$$n \ge \lceil \tfrac{9}{2}(k-1) \rceil + 1$$

This shows that the last inequality holds for every $k \ge 3$.

In order to sharpen this inequality for $k \ge 5$, we proceed as follows. Suppose that, contrary to the assertion, $d_2(k) < g_2(k)$ for some k, $k \ge 5$. Put

$$n = g_2(k) - 1$$

and

$$e = \left\lfloor \frac{n(n-1)}{2k} \right\rfloor.$$

Choose the indices of the factors $F_1, F_2, \ldots, F_k$ in such a way that

$$e(F) = \min_{1 \le i \le k} e(F_i).$$

Since

$$e(K^n) = \binom{n}{2} = \sum_{1}^{k} e(F_i) \ge k e(F),$$

we have

$$e(F) \leqslant e.$$

Put $a = 2e/n < 6$. Our aim is to show that the conditions of Lemma 3.5 are satisfied with $G = F$ and (3.16) is sharp for every vertex x_0 of degree at most $a < 6$. Having done this Lemma 3.5 will give $e(F) > e$, contradicting the assumption.

Let $x_0 \in F$ and suppose $d(x_0) = 3$, $d(x_1) \leqslant d(x_2) \leqslant d(x_3)$. Then (2), (3) and (1) imply

$$k - 2 \leqslant d(x_1) \leqslant d(x_2) \leqslant d(x_3) \leqslant n + 2 - 3k$$

and

$$\sum_1^3 d(x_i) \geqslant n - 1.$$

If $1 < p < q$ then

$$\frac{1}{p} + \frac{1}{q} < \frac{1}{p-1} + \frac{1}{q+1}.$$

Therefore

$$3 + \sum_1^3 \frac{d(x_i) - a}{d(x_i)} = 6 - a \sum_1^3 \frac{1}{d(x_i)}$$

$$\geqslant 6 - a \left(\frac{1}{k-2} + \frac{1}{n+2-3k} + \frac{1}{2k-1} \right) = 6 - ar_3 = a_3. \qquad (4)$$

Similarly, if $d(x_0) = 4 \leqslant a$ then (1) and (3) imply

$$4 + \sum_{d(x_i) > a} \frac{d(x_i) - a}{d(x_i)} \geqslant 6 - a \left(\frac{1}{n+2-3k} + \frac{1}{3k-2a-3} \right) = 6 - ar_4 = a_4.$$

$$\qquad (5)$$

Note that if $d(x_1) = d(x_2) = a, d(x_4) = n + 2 - 3k$ and $d(x_3) = n - 1 - 2a - (n + 2 - 3k) = 3k - 2a - 3$, then we have equality in (5).

Finally, if $d(x_0) = 5 \leqslant a$ and $k \geqslant 10$ then it follows from (1) and (3) that

$$5 + \sum_{d(x_i) > a} \frac{d(x_i) - a}{d(x_i)} \geqslant 7 - a \left(\frac{1}{n+2-3k} + \frac{1}{3k-3a-3} \right) = 7 - ar_5 = a_5.$$

$$\qquad (6)$$

Thus, to complete the proof of the theorem it suffices to show that $a_3 > a$ always holds, $a_4 > a$ provided $a \geqslant 4$ and $a_5 > a$ provided $a \geqslant 5$. This can be done by routine calculations. For $k \leqslant 13$ we omit all the details. If $14 \leqslant k \leqslant 18$ then $e = 18k - 60$, if $19 \leqslant k \leqslant 27$ then $e = 18k - 61$, if $28 \leqslant k \leqslant 55$ then $e = 18k - 62$ and, finally, if $56 \leqslant k$ then $e = 18k - 63$. The required inequalities can be checked in each of these four cases. We give the details in the last case.

Let $k \geqslant 56$. Then $n = 6k - 10$, $e = 18k - 63 = 9(2k - 7)$ so $a = 9(2k - 7)/3k - 5)$. Consequently

$$ar_3 = \frac{9(2k - 7)}{3k - 5}\left(\frac{1}{k - 2} + \frac{1}{3k - 8} + \frac{1}{2k - 1}\right)$$

$$< \frac{9(2k - 7)}{3k - 5}\frac{1}{k - \frac{7}{2}}\left(1 + \frac{1}{3} + \frac{1}{2}\right) = \frac{33}{3k - 5} = 6 - a.$$

Hence $6 - ar_3 = a_3 > a$, as required. Similarly

$$ar_4 < \frac{9(2k - 7)}{3k - 5}\left(\frac{1}{3k - 8} + \frac{1}{3k - 15}\right) < \frac{33}{3k - 5} = 6 - a$$

and

$$ar_5 < \frac{9(2k - 7)}{3k - 5}\left(\frac{1}{3k - 8} + \frac{1}{3k - 21}\right) < 1 + \frac{33}{3k - 5} = 7 - a.$$

This completes the proof of the inequality $d_2(k) \geqslant 6k - 9$ if $k \geqslant 56$. ∎

COROLLARY 4.4. $\lim\limits_{k \to \infty} d_2(k)/k = 6$. ∎

We close the section with the conjecture that the upper bound in Theorem 4.3 is exact for large values of k (Conjecture 31).

It is unlikely that this conjecture can be proved by refining the proof of $g(k) \leqslant d_2(k)$. However, one might be able to show that in a proper factorization of $K^n, n = d_2(k)$, each factor has at least 6 vertices of degree at least $(n - 3k + 6)/2$. It is easily seen that this would imply $n \geqslant 6k$.

5. EXERCISES, PROBLEMS AND CONJECTURES

1^-. Characterize the set of *maximal* graphs of order n and diameter d.

2^-. Let $d \geqslant 2$ and $k \geqslant 1$. Prove that

$$\min\{|G|: \operatorname{diam} G = d, \kappa(G) = k\}$$

$$= \min\{|G|: \operatorname{diam} G = d, \kappa(G) \geqslant k\} = k(d - 1) + 2.$$

(Watkins [W11].)

3^-. Let $2 \leqslant d < n$. Prove that

$$\max\{e(G^n): \operatorname{diam} G^n = d\} = \max\{e(G^n): \operatorname{diam} G^n \geqslant d\} = \binom{n - d + 2}{2} + d - 3.$$

(Ore [O6].)

4. (Ex. 3 ctd.) Determine the number of non-isomorphic extremal graphs. (Homenko and Ostroverhiĭ [HO1].)

5^-. Show that if G is connected then

$$\chi(G) \leqslant |G| - \operatorname{diam} G + 1.$$

6^+. Determine the *minimal* size of a *maximal* graph of order n and diameter d. (Homenko and Ostroverhiĭ [HO1].)

7. PROBLEM. Given $d \geqslant 3$ and $k \geqslant 3$, estimate the order of

$$\max\{n: \mathcal{H}_d(n, k) \neq \varnothing\}$$

as $k \to \infty$. (Note that by (1.1) and (1.2) this maximum is at most k^d.)

8. CONJECTURE. (Problem 7 ctd.) If $\varepsilon > 0$, $d \geqslant 3$ and k is sufficiently large then the maximum is at least $(1 - \varepsilon)k^d$.

9. Prove that $e_2(n, k)$ is at least as large as claimed in Theorem 1.3.

10. PROBLEM. Estimate $n^{-1}e_3(n, cn^{\frac{1}{2}}) - 1$ if c is large (and fixed) and $n \to \infty$. How near is it to $\frac{1}{2}c^{-2}$?

(Cf. Theorems 1.9 and 1.10 and the construction given between these two theorems.)

11. Prove that

$$k^2 + \tfrac{1}{2}k^{\frac{1}{2}} \leqslant e_4(k^2 + 2, k) \leqslant k^2 + k.$$

(Bollobás [B26].)

12. PROBLEM. Is there an absolute constant $c, 0 < c < 1$, such that

$$e_{2^d}(k^d + 2, k) \geqslant k^d + ck?$$

If not, is there a constant depending on d with this property? In particular, is there a constant $c_4 > 0$ such that

$$e_4(k^2 + 2, k) \geqslant k^2 + c_4 k?$$

(Bollobás [B26].)

13. PROBLEM. Let $c > 0$ be a real number, let $1 \leqslant l \leqslant d < k$ be integers and put $n = \lfloor ck^l \rfloor$. Determine the order of $e_d(n, k)$ as $k \to \infty$.

14. CONJECTURE. $\lim\limits_{n \to \infty} n^{-1} e_2(n, cn)$ exists for each $c, 0 < c < 1$. (Cf. Corollary 1.7.)

15. Deduce the following assertions from the proof of Theorem 2.5. If $n \geqslant 4$ then $g(n, 2, 2) = \lfloor 3n/2 \rfloor - 1$. If $n = 2p + 1$ is odd then $K^1 + pK^2$ is the unique extremal graph. If $n = 2p + 2$ is even, $K^1 + (P_3 \cup (p-1)K^2)$ is the unique extremal graph (see Fig. 5.1). (Murty [M38].)

FIG. 5.1. Extremal graphs of the function $g(n, 2, 2)$.

16. By elaborating the proof of the first assertion of Theorem 2.7, prove that for $n \geqslant 5$ the extremal graphs of the function $f(n, 2, 2)$ are the graphs described in the proof of Corollary 2.8 (cf. Fig. 2.2) and the Petersen graph (cf. Fig. III. 1.2). (Murty [M37].)

17. Prove that $f(n, 3, 4, 1) = \lceil 3n/2 \rceil - 3$ if $n \geqslant 5$ and $f(n, 3, d', 1) = \lfloor 3n/2 \rfloor - 3$ if $d' \geqslant 5$ and $n \geqslant 6$. (Bollobás [B24].)

18. Suppose $r \geqslant 3, l \geqslant 1, n = rl$ and $k = (r - 1)l$. Prove by induction that

$T_r(n)$ is the unique extremal graph in the class $\mathscr{V}(n, 2, 2, k - 1)$. (Bollobás and Eldridge [BE4].)

19. Rewrite the proof of Theorem 3.3 to obtain an entirely combinatorial proof. (Bollobás and Erdös [BE9].)

20. Assuming the truth of Conjecture 3.4, deduce relations (3.14'') and (3.15'').

21. Give a proof of the second inequality of Theorem 3.6, modelled on the proof of Theorem 2.7.

22. Prove that $\hat{g}(n, 3, 2) = \lfloor \frac{3}{2}n \rfloor - 1$ for $n \geqslant 5$.

[Hint. Let $G \in \mathscr{E}(n, 3, 3, 1)$ be an extremal graph. Note that $\delta(G) = 2$, say $d(x_0) = 2, x_0 \in G$. Put $H = G[\{x_0\} \cup \Gamma(x_0)]$ and go through the second part of the proof of Theorem 3.7.] (Bollobás [B25].)

23. Put $\hat{\phi}(n, d, k) = \hat{f}(n, d, k) - kn/2$. Prove that for $d \geqslant 2$ and $k \geqslant 3$ we have

$$S(k, \lfloor d/2 \rfloor)^{-1} \leqslant \liminf_{n \to \infty} n^{-1} \hat{\phi}(n, d, k) \leqslant \limsup_{n \to \infty} n^{-1} \hat{\phi}(n, d, k)$$

$$\leqslant \tfrac{1}{2} S(k, \lfloor d/4 \rfloor + 1)^{-1}$$

24. CONJECTURE. (Ex. 23 ctd.) $\lim_{n \to \infty} n^{-1} \hat{\phi}(n, d, k)$ exists.

25. PROBLEM. Determine the limit in Conjecture 24. As a partial result, determine the *order* of the limit as a function of d and k, i.e. improve the bounds given in Ex. 23 until the upper bound is at most an absolute constant multiple of the lower bound.

26. Prove the lower bounds given in Theorem 3.8.

[Hint. Estimate the number of choices for the r-sets $\Gamma(x_i)$ of B that do not result in a graph belonging to $S(2n, m)$ and compare it with the number of all possible choices.] (Erdös and Spencer, see [V3].)

27. Prove that $d_2(3) = 12$ or 13. (Bosák, Rosa and Znam [BRZ1].)

28. Suppose G and $\bar{G}$ are both connected. Prove that diam $G \geqslant 3$ implies diam $\bar{G} \leqslant 3$ and diam $G \geqslant 4$ implies diam $\bar{G} = 2$. Deduce that if G is self-complementary then $2 \leqslant$ diam $G \leqslant 3$. (Sachs [S1] and Ringel [R8a].)

29⁻. Determine the maximum and the minimum of

$$\text{diam } G + \text{diam } \bar{G}$$

as G runs over all connected graphs of order n for which $\bar{G}$ is connected.

30. Let $d_1, d_2, \ldots, d_k$ be natural numbers, $d_i \geqslant 2, i = 1, 2, \ldots, k$. Prove that there exists a natural number n_0 such that K^{n_0} can be factored into k factors, say $F_1, F_2, \ldots, F_k$, such that diam $F_i = d_i, i = 1, 2, \ldots, k$. Prove furthermore that if $n \geqslant n_0$ then K^n has a similar factorization. (Bosák, Erdös and Rosa [BER1].)

31. CONJECTURE. If k is sufficiently large then $d_k(2) = 6k$, that is K^{6k-1} can not be factored into k factors of diameter 2.

32^-. Show that a tree of order n and radius r has at least $\lceil (n-1)/r \rceil$ endvertices. Note that the result is best possible for $n - 1 > r$.

33^-. A directed graph is said to be *strongly connected* if for any two vertices x and y there is a directed path from x to y. Show that if G is a strongly connected graph of order n with m edges then

$$\text{rad } G \geqslant \left\lceil \frac{n-1}{m-n+1} \right\rceil.$$

Show also that equality can be attained for every pair $(n, m), m \geqslant n$. (Goldberg [G14].)

34^+. Show that if G is a strongly connected directed graph of order n with $m \geqslant n + 1$ edges then

$$\text{diam } G \geqslant \left\lceil \frac{2(n-1)}{m-n+1} \right\rceil.$$

Show also that equality can be attained for every pair $(n, m), m \geqslant n + 1$.
[Hint. Let $x_0 x_1 \ldots x_l$ be a longest directed path in G such that for each $i, 0 \leqslant i < l$, only the edge $x_i x_{i+1}$ is directed away from x_i. Let T be a tree in G which is such that each vertex of G can be reached from x_0 in T. Show first that T has at most $m - n$ endvertices.] (Goldberg [G13].)

35. Determine

$$\max\{e(G^n): \text{rad } G^n = r\}.$$

(Vizing [V4])

36. A *polarity* α of a projective plane is a bijection from the set of points onto the set of lines satisfying $P \in \alpha(Q)$ iff $Q \in \alpha(P)$. Baer [B1] proved that if α is a polarity of a finite projective plane then there is at least one point P satisfying $P \in \alpha(P)$. (In fact, a plane of order q has at least $q + 1$ such points.) Use this result to prove the following "friendship theorem".

Suppose $|\Gamma(x) \cap \Gamma(y)| = 1$ for every pair of distinct points x, y of a graph G. Then $G = rK^2 + K^1$.

[Hint. Suppose that $\Delta(G) < |G| - 1$. Show that then G is regular and the structure of G contradicts Baer's theorem.] (Erdös, Rényi and Sós [ERS1].)

37. CONJECTURE. (Cf. Ex. 35.) If $k > 2$ there is no graph G that for every pair of distinct vertices x, $y \in G$ contains a *unique* x–y path of length k. (A. Kotzig)

38⁺. Each of $n \geqslant 4$ elderly dons knows some item of gossip not known to the others. They communicate by telephone and in each conversation they tell each other all that they know at that time. Prove that the minimum number of calls required before each don knows everything is $2n - 4$.

[Hint. It is true for $n = 4$. Suppose $2n - 4$ calls are not always needed. Let m be the minimal n for which $2m - 5$ calls will do in a certain arrangement S of the calls. Show first that no don can hear his information from another in S. Deduce that a call is either a first call for both parties or for neither. Look at the graph formed by the $m - 5$ intermediate calls.]

(The origin of the problem is unclear, but various solutions were given by R. T. Bumby, Hajnal, Milner and Szemerédi [HMS1], Baker and Shostak [BS1] and Bermond [B13]. The hint above is from [BS1].)

39. PROBLEM. Give bounds on

$$\min\{e(G): |G| = n, \text{diam } G \leqslant d, \delta(G) \geqslant \delta \quad \text{and} \quad \Delta(G) \leqslant \Delta\}.$$

(This problem is a common extension of the problems discussed in §§ 1 and 2. However, in an appropriate range of n, d, δ and Δ it may be accessible.)

Chapter V

Colourings

This is the most popular area of graph theory. It would be impossible to cover all the results in the literature, but we do try to present all the main results in an appropriate setting.

In §1 after some preliminary observations we present the fundamental colouring results, including the theorem of Brooks [B49], Gallai's theorem [G10], a theorem of Erdös and Hajnal [EH3] and some results from [B32] about uniquely colourable graphs. In the section we also prove the only substantial result about edge colouring, Vizing's theorem [V3], stating that the edge chromatic number of a graph G, $\chi'(G)$, is either $\Delta(G)$ or $\Delta(G) + 1$. In some way this theorem is too strong to allow the development of a rich theory of edge colourings of graphs; as $\chi'(G)$ is almost determined by $\Delta(G)$, there is not much room for estimates based on other properties of the graphs.

At the beginning of Chapter I we remarked that, in order to describe k-connected graphs, it is considerably more important to obtain information about minimally k-connected graphs than about critical k-connected graphs. There is no such difference between the corresponding classes of k-chromatic graphs. If $\chi(G) = k$ and $\chi(G - x) = k - 1$ for each vertex $x \in G$ then G is a *critical k-chromatic* graph. One might call a graph minimally k-chromatic if $\chi(G - e) = k - 1$ for each edge $e \in E(G)$. However, as there is a rather obvious 1–1 correspondence between the results concerning these two classes, we follow the custom of using the term *edge-critical k-chromatic* instead of minimally k-chromatic. The investigation of critical k-chromatic graphs was started by Dirac [D6] and a theorem of his [D9], extending the theorem of Brooks, is the main result of §2. Ideally the aim of the theory of critical k-chromatic graphs is to prove structure theorems and rather restrictive inequalities. At present this goal seems to be very far indeed; most of the results show how rich the class of critical graphs is.

The chromatic number of graphs drawn on orientable surfaces is discussed in §3. Though the best known problem of graph theory, the Four Colour Conjecture, also belongs to this section, we present only a small fraction of the relevant results. Nevertheless, the results show why the problem of colouring planar graphs is so different from the problem of colouring graphs on orientable surfaces. Given an orientable surface F, Euler's formula gives a bound $p = p(F)$ on the chromatic number of the graphs drawn on F. If F is the plane, this bound is 5 though K^5 is not planar. However, if F is not the plane, the bound p can be attained iff K^p can be drawn on F. It so happens that this is the case for every such F. Thus for a surface different from the plane the problem is whether a certain *complete* graph can be drawn on it or not. This problem has clearly very little to do with the problem whether or not there is a *large* planar graph, which is very far from being a complete graph, that needs 5 colours.

Perhaps the most important problem concerning graph colourings is the connection between the chromatic number and contractions onto complete subgraphs asserted by Hadwiger's conjecture [H1]. The results in this area are embryonic. We shall present them in Chapter VIII, where we discuss contractions.

If $K^k \subset G$ then $\chi(G) \geqslant k$; thus $\chi(G) \geqslant \text{cl}(G)$. It is trivial that here we need not have equality but it is less trivial that a graph with small clique number (say, without triangles) may have large chromatic number. In §4 we present results mostly due to Erdős showing that much more is true: a graph with large chromatic number need not contain subgraphs of small order and relatively large size. The rudimentary results of this kind were obtained by Tutte [D3], Zykov [Z8] and Mycielski [M39]. The breakthrough was achieved by Erdős [E7] who realized that the proofs of the results should be based on the properties of random graphs. Among others we show that a graph of large girth and large chromatic number may even be uniquely colourable.

In §5 we investigate the *perfect* graphs introduced by Berge [B7]. These are the graphs for which $\chi(H) = \text{cl}(H)$ for every induced subgraph H. (Thus perfect graphs are as far from the graphs of §4 as possible.) The main result of the section is that a graph is perfect if its complement is. This was conjectured by Berge and was proved by Fulkerson [F5] and Lovász [L18]. A stronger conjecture of Berge concerning perfect graphs is still open (Conjecture 33).

The last section (§6) of the chapter is on Ramsey type theorems, though the connection between Ramsey theorems and genuine colouring results is rather superficial. This somewhat artificial arrangement is forced on us by the size of the book. We start the section with the fundamental and well known theorem of Ramsey [R6], stating that if for a finite r the r-sets of an infinite set S are coloured with 2 colours then there is an infinite set $S' \subset S$ whose r-sets are coloured with the same colour. An almost immediate consequence of this is that if k and l are natural numbers then there is a minimal integer $R(k, l)$ such that if $G = G^n$ and $n > R(k, l)$ then either $K^k \subset G$ or else $K^l \subset \bar{G}$. Clearly the estimation of the numbers $R(k, l)$ is a very natural extremal problem in graph theory.

There are two, clearly distinguishable, types of Ramsey theorems. One might say that a *qualitative* Ramsey theorem guarantees the existence of certain objects (e.g. it guarantees the existence of $R(k, l)$) and a *quantitative* Ramsey theorem estimates the *size* of that object. Most Ramsey type results in graph theory are of the quantitative kind. The existence of the constants they estimate is usually guaranteed by Ramsey's original theorem. As most proofs contain long case by case examinations, they are not very suitable for a book of this kind. The proofs of the quantitative results are very different; they tend to rely on ingenious constructions.

The main aim of § 6 is to present a deep result of Nešetřil and Rödl [NR2], which is the most important qualitative Ramsey theorem of finite graph theory known at present.

Not surprisingly, there are a great number of basic open problems concerning colourings. Some of these problems might not even be too difficult. The state of the theory of edge colourings is rather poor. Most of the results proved so far can be read out almost immediately from the proof of Vizing's theorem. In particular, we do not even know whether for a given $\Delta > 0$ which of the two equalities $\chi'(G) = \Delta$ and $\chi'(G) = \Delta + 1$ is satisfied by most graphs $G = G^n$ with $\Delta(G) = \Delta$, provided n is sufficiently large (Conjecture 16).

Though there are only few positive results about critical k-chromatic graphs, it might not be impossible to determine the *minimal* size of a critical k-chromatic graph of order n (Problem 21). Naturally a result of this kind would be a considerable extension of the theorems of Brooks and Dirac. It is less likely that one could determine the *maximal* size of a critical k-chromatic graph of order n, but for fixed k and large n one should obtain good estimates (Problem 25).

Erdös and Hajnal conjectured that each graph of large chromatic number contains a subgraph of large chromatic number and large girth (Conjecture 26). This conjecture is particularly fascinating since it seems to withstand the attack of the customary method of random graphs.

1. GENERAL COLOURING THEOREMS

Let us begin with some trivial observations. Suppose G is connected and does not contain an odd cycle. Pick a vertex $x \in G$ and colour a vertex y with 0 or 1 according to $d(x, y)$ is even or odd. This gives a 2-colouring of G and shows that a graph is 2-colourable if and only if it does not contain an odd cycle. It is perhaps regrettable but very natural that there is no similar characterization of k-chromatic graphs for $k \geqslant 3$.

Let V_1 be a colour class of maximal order in a k-colouring of G. Then $|V_1| \geqslant n/k$. Furthermore, V_1 consists of independent vertices, so $\beta_0(G) \geqslant |V_1|$. As $\bar{G}$, the complement of G, contains a complete graph of order $\beta_0(G)$, we have $\beta_0(G) \leqslant \chi(\bar{G})$. Note also that if $V_1 \subset V(G)$ is a set of $\beta_0(G)$ independent vertices, we may colour them with one colour and use one colour for each additional vertex. Therefore the following inequalities hold:

$$|G|/\chi(\bar{G}) \leqslant |G|/\beta_0(G) \leqslant \chi(G) \leqslant |G| + 1 - \beta_0(G). \tag{1}$$

We shall see shortly (Corollary 1.4) that (1) can be strengthened by replacing the right-hand side by $|G| + 1 - \chi(\bar{G}) \leqslant |G| + 1 - \beta_0(G)$.

We call the next observation a theorem, since it has some interesting consequences.

THEOREM 1.1. *Let G be a graph with vertex set $V(G) = \{x_1, x_2, \ldots, x_n\}$ and let $H = G[x_1, x_2, \ldots, x_h]$. Suppose H is coloured with some of the colours $1, 2, \ldots, k + 1$ and each $x_j, h + 1 \leqslant j \leqslant n$, is such that if its neighbours in H are coloured with c_j different colours then x_j is adjacent to at most $k - c_j$ vertices x_i with $h + 1 \leqslant i \leqslant j$. Extend the colouring of H to a colouring of G with colours $1, 2, \ldots$ as follows. Having coloured $x_1, x_2, \ldots, x_l, h \leqslant l < n$, colour x_{l+1} with the minimal natural number that is not used to colour any of the neighbours of x_{l+1} we have already coloured.*

This colouring uses at most $k + 1$ colours. In particular, $\chi(G) \leqslant k + 1$. ∎

The observation above was formulated by Halin [H8] (see also Matula [M18], Szekeres and Wilf [SW1] and Nordhaus and Gaddum [NG1]) in a slightly different form.

COROLLARY 1.2. *Given a graph G let $k = \max \delta(G')$ where the maximum is taken over all spanned subgraphs G' of G. Then $\chi(G) \leqslant k + 1$.*

Proof. Define a sequence $H_n \supset H_{n-1} \supset \ldots$ of spanned subgraphs of G, $|G| = n$, as follows. Put $H_n = G$. Having defined $H_n \supset H_{n-1} \supset \ldots \supset H_l$, $l \geqslant 2$, pick a vertex $x_l \in V(H_l)$ with $d_{H_l}(x_l) \leqslant k$ and put $H_{l-1} = H_l - x_l$. Then $V(G) = \{x_1, x_2, \ldots, x_n\}$, $H_1 = (\{x_1\}, \varnothing)$ is 1-colourable and each x_j, $1 \leqslant j \leqslant n$, is joined to at most k vertices x_i, $1 \leqslant i < j$. Thus $\chi(G) \leqslant k + 1$. ∎

The next corollary of Theorem 1.1 was noted by Bondy [B40] which, in its turn, implies the promised sharpening of (1), due to Nordhaus and Gaddum [NG1].

COROLLARY 1.3. *Let $d_1 \geqslant d_2 \geqslant \ldots \geqslant d_n$ be the degree sequence of G. Then*

$$\chi(G) \leqslant \max_{1 \leqslant i \leqslant n} \min\{d_i + 1, i\}. \qquad \blacksquare$$

In other words, if k is the maximal natural number such that $k \leqslant d_k + 1$ then $\chi(G) \leqslant k$.

COROLLARY 1.4.

$$\chi(G) \leqslant |G| + 1 - \chi(\overline{G}).$$

Proof. Let $d_1 \geqslant d_2 \geqslant \ldots \geqslant d_n$ be the degree sequence of G. Then $\bar{d}_1 \geqslant \bar{d}_2 \geqslant \ldots \geqslant \bar{d}_n$ is the degree sequence of $\overline{G}$ where

$$\bar{d}_i = n - d_{n+1-i}.$$

Denote by k the maximal natural number satisfying $k \leqslant d_k + 1$. Then, by Corollary 1.3,

$$\chi(G) + \chi(\overline{G}) \leqslant k + \max_i \min\{n - d_{n+1-i}, i\}$$

$$= k + n + 1 - \min_i \max\{d_{n+1-i} + 1, n + 1 - i\}$$

$$= n + 1 + k - \min_i \max\{d_i + 1, i\} \leqslant n + 1. \qquad \blacksquare$$

Either of Corollaries 1.2 and 1.3 implies the next simple fact.

COROLLARY 1.5. $\chi(G) \leqslant \Delta(G) + 1.$ ∎

Let us digress for a moment to remark on some generalizations of the chromatic number. A graph G is said to be *k-degenerate* if $s(G) = \max \delta(G') \leqslant k$, where the maximum is taken over all (induced) subgraphs, as in Corollary 1.2. The lth *vertex partition number* of a graph G, denoted by $\chi_l(G)$, is the *minimum* number of subsets V_i in a partition $V(G) = \bigcup_1^k V_i$ such that each $G[V_i]$ is l-degenerate, $i = 1, 2, \ldots, k$. Then $\chi_0(G)$ is exactly the chromatic number and $\chi_1(G)$ is the so called *vertex-arboricity*. (Note that $s(H) \leqslant 1$ iff H is a forest, so $\chi_1(G)$ is the minimum number of classes in a partition of $V(G)$ such that each class spans a forest.) Many questions can be asked about k-degenerate graphs and vertex partition numbers, and numerous papers have been written posing and answering these questions. Most of the questions are trivial or well-known for $\chi_0 = \chi$ and can be answered easily for χ_l, $l \geqslant 1$; we state a number of these as easy exercises (Exx. 1–7). In spite of the existence of some non-trivial and rather curious results, the theory of vertex partition numbers is in great need of some substantial theorems and applications that would justify the generalization of $\chi_0 = \chi$ to χ_l and would change the present collection of easy exercises into an essential part of graph theory.

Let us return to the study of the chromatic number. If G is an odd cycle then $\Delta(G) = 2$ and $\chi(G) = 3$. Conversely, if $\Delta(G) \leqslant 2$ and $\chi(G) = 3$ then G is not bipartite so it contains an odd cycle and this odd cycle must be a component of G. Also if $G = K^{k+1}$ $(k \geqslant 3)$ then $\Delta(G) = k$ and $\chi(G) = k + 1$. These examples show that the trivial inequality in Corollary 1.5 is best possible for $\Delta(G) \geqslant 3$ as well. Brooks [B49] proved that the complete graphs are essentially the only graphs for which the chromatic number is larger than the maximal degree, provided the maximal degree is at least 3. Different proofs of this fundamental theorem of Brooks were given by Gerencsér [G11], Melnikov and Vizing [MV1], Ore [O4] and Ponstein [P7]. Szekeres and Wilf [SW1] claimed that Corollary 1.2 implies Brook's theorem. Though strictly speaking this is true only if the graph is not regular, Lovász [L23] noticed that Brooks' theorem does follow from a simple form of Theorem 1.1. First we give this short and direct proof. We present another simple proof, due to Melnikov and Vizing [MV1], since it illustrates the device of switching colours in a subgraph spanned by the vertices of certain colours. Suppose a graph H is k-coloured, say with $1, 2, \ldots, k$, and L is a *component* of the subgraph of H spanned by the vertices of colour $1, 2, \ldots, l$. Keeping the colours of the vertices in $V[H - L]$ and permuting the colours in L in any way we obtain another k-colouring of H.

THEOREM 1.6. *Let* $\Delta \geqslant 3$ *and let* G *be a graph such that* $\Delta(G) \leqslant \Delta$ *and* $K^{\Delta+1} \nsubseteq G$. *Then* $\chi(G) \leqslant \Delta$.

1st proof. We may assume without loss of generality that $\Delta(G) = \Delta$, $|G| \geqslant \Delta + 2$ and G is *2-connected*.

Suppose G contains a vertex x with $d(x) < \Delta$. The connectedness of G implies that the vertices of G can be arranged in a sequence $x_n = x, x_{n-1}, \ldots, x_1$ such that each x_k, $k > n$. is adjacent to a vertex x_{k+1}, $l \geqslant 1$, i.e. to a vertex appearing earlier in the sequence. Then $\chi(G) \leqslant \Delta$ follows from Corollary 1.2. Therefore we may assume that G is Δ-regular.

Let us show that G contains two vertices, a and b, say, such that $d(a, b) = 2$ and $G - a - b$ is connected. This is clear if $G - x$ is 2-connected for some $x \in G$. If $G - x$ is separable, take two endblocks of $G - x$, say B_1 and B_2. Since G is 2-connected there are vertices $a \in B_1 \cap \Gamma(x)$ and $b \in B_2 \cap \Gamma(x)$ that are not cutvertices of $G - x$. Then $G - a - b$ is connected and as a and b are joined to x, $d(a, b) = 2$.

Let x be a vertex joined to both a and b. As earlier, the connectedness of $G - a - b$ implies that its vertices can be arranged in a sequence $x_n = x$, $x_{n-1}, \ldots, x_3$ such that each x_k, $k < n$, is adjacent to a vertex x_{k+1}, $l \geqslant 1$. Then $V(G) = \{a, b, x_3, x_4, \ldots, x_n\}$, $H = G[a, b]$ and the colouring of H with the single colour 1 satisfy the conditions of Theorem 1.1 with $k = \Delta - 1$. Hence $\chi(G) \leqslant \Delta$. ∎

2nd proof. Suppose the theorem fails and let G be a graph of minimal order showing this. Pick a vertex $x \in G$ and put $\Gamma(x) = \{x_1, x_2, \ldots, x_d\}$, $d \leqslant \Delta$. As G is a minimal counterexample, for $H = G - x$ we have $\chi(H) \leqslant \Delta$. Consider any Δ-colouring of H, say with colours $1, 2, \ldots, \Delta$. If one of these colours, say colour i, is not used to colour at least one of the vertices in $\Gamma(x)$ then the Δ-colouring of H extends to a Δ-colouring of G if we assign x the colour i. This shows the following fact.

(i) $d = \Delta$ *and the vertices* $x_1, x_2, \ldots, x_\Delta$ *are coloured with different colours in every* Δ-*colouring of* H. For simplicity we assume that x_i is coloured with colour i, $1 \leqslant i \leqslant \Delta$. Denote by H_{ij} the subgraph of H induced by the vertices coloured i and j.

(ii) *The vertices* x_i *and* x_j, $1 \leqslant i < j \leqslant \Delta$, *belong to the same component* C_{ij} *of* H_{ij}. For if this is not so then by interchanging the colours i and j in the component of H_{ij} containing x_j we obtain a Δ-colouring of H in which both x_i and x_j have colour i. This contradicts (i).

(iii) C_{ij} *is an* x_i–x_j *path*, $1 \leqslant i < j \leqslant \Delta$. Note first that $\Gamma_H(x_i)$ contains exactly one vertex of colour j since otherwise in H we could recolour x_i with a colour different from i, again contradicting (i). Similarly C_{ij} can not contain a vertex y with degree at least 3 in C_{ij}. For if y is the *first* such vertex on an x_i–x_j path in C_{ij} then we could recolour y with a colour different from i and j and in this new colouring x_i and x_j would belong to different components of H_{ij}.

(iv) $C_{ij} \cap C_{ik} = \{x_i\}$ *for* $j \neq k$. For if y is another vertex in the intersection then y has two neighbours of colour j and two of colour k. Therefore we can recolour y with a colour different from i, j and k. In this new colouring x_i and x_j once again belong to different components of H_{ij}, contradicting (i).

The information we have gathered about the structure of H enables one to complete the proof very easily. Since $K^{\Delta+1} \nsubseteq G$ we may assume that $x_1 x_2 \notin E(G)$. Then the path C_{12} contains a vertex $y \neq x_2$ adjacent to x_1. Naturally y is coloured with 2. Interchange now colours 1 and 3 in the path C_{13} and denote the new paths by C'_{ij}. Then $y \in C'_{23}$ since y is coloured with 2 and it is adjacent to x_1 whose new colour is 3. Furthermore, $y \in C_{12} - \{x_1\} \subset C'_{12}$ (see Fig. 1.1). Thus $y \in C'_{12} \cap C'_{23}$, contradicting (iv).   ∎

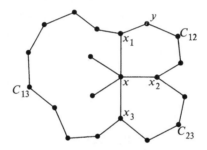

FIG. 1.1. The final step in the second proof of Theorem 1.6.

Gallai [G10] and Roy [R12] proved independently the following result connecting the chromatic number of a graph with the maximal length of a directed path in any orientation of the graph. An *orientation* of a graph $G = (V, E)$ is a way of directing the edges of G. With an orientation G clearly becomes an oriented graph $\vec{G} = (V, \vec{E})$. A *directed cycle* in $\vec{G}$ is a cycle of G in which the edges are cyclically directed; a *directed path* is defined analogously (cf. § 0). Denote by $l(\vec{G})$ the maximal number of *vertices* in a directed path in $\vec{G}$.

THEOREM 1.7. $\chi(G) = \min_{\rightarrow} l(\vec{G})$, *where* $\min_{\rightarrow}$ *means that the minimum is taken over all orientations of* G.

Proof. Given a colouring of G with colours $1, 2, \ldots, k$, direct an edge $xy \in E(G)$ from x to y if the colour of x is smaller than the colour of y. For the obtained directed graph $\vec{G}$ clearly $l(\vec{G}) \leqslant k$. Thus $\chi(G) \geqslant \min_{\rightarrow} l(\vec{G})$.

To prove the real assertion of the theorem let $\vec{G} = (V, \vec{E})$ be an orientation of G. We have to show that G is k-colourable, where $k = l(\vec{G})$.

Let $\vec{E}_0$ ($\subset \vec{E}$) be a minimal set of directed edges such that $\vec{H} - (V, \vec{E} - \vec{E}_0)$ does not contain a directed cycle. For $x \in V$ let $c(x)$ be the maximal number of vertices on a directed path in $\vec{G}$ beginning at x. Then $1 \leqslant c(x) \leqslant k$ for every $x \in V$. Colour x with $c(x)$. Let us check that this colouring of V is a proper vertex colouring of G.

If $\vec{xy} \in \vec{E} - \vec{E}_0$ then $c(x) = c(y) + 1$, so $c(x) \neq c(y)$. If $\vec{xy} \in \vec{E}_0$ then in H there is a directed path from y to x. Denoting its length by p ($p \geqslant 1$) we have $c(y) = c(x) + p$ so $c(x) \neq c(y)$. ∎

The following extension of Theorem 1.7 was conjectured by Las Vergnas [L2] and proved recently by Bondy [B47]. Denote by $c(\vec{G})$ the maximum length of a directed cycle in an orientation $\vec{G}$ of the graph G. Then $\chi(G) = \min c(\vec{G})$, where min means that the minimum is taken over all orientations of G. Chvátal and Komlós (see [R5], [CK2] and [C9]) noticed that Theorem 1.7 implies the following extension of a result of Busolini [B55].

THEOREM 1.8. *Let G be a graph and let $\vec{G} = (V, \vec{E})$ be some orientation of G. Let $E = \bigcup_{1}^{k} E_i$ and suppose*

$$\chi(G) > \prod_{1}^{k} n_i,$$

where $n_1, n_2, \ldots, n_k$ are natural numbers. Then there is an index i, $1 \leqslant i \leqslant k$, such that $\vec{G}_i = (V, \vec{E}_i) \subset \vec{G}$ contains a directed path with n_i edges.

Proof. Put $G_i = (V, E_i)$. Then

$$\chi(G) \leqslant \prod_{1}^{k} \chi(G_i),$$

since if $f_i: V \to C_i$ is a colouring of G_i then $f: V \to \underset{i=1}{\overset{k}{\times}} C_i$, where $f(a) = (f_1(a), \ldots, f_k(a))$, is a colouring of G. Therefore $\chi(G_i) \geqslant n_i + 1$ for some i, so the result follows from Theorem 1.7. ∎

If $\chi(G) \geqslant 3$ then G must contain an odd cycle. The graph $G = K^{2k}$ shows that a $2k$-chromatic graph need not contain odd cycles longer than $2k - 1$. Erdös and Hajnal [EH3] proved that this example is extremal in the sense that every $(2k + 1)$-chromatic graph must contain an odd cycle of length at least $2k + 1$. The proof makes use of the following simple technical lemma.

LEMMA 1.9. *Let G be a 2-connected graph containing an odd cycle. Let C be an even cycle in G. Then there exist vertices $x, y \in C$ and an x–y path P independent of C such that $d_C(x, y) \not\equiv d_P(x, y)$ (mod 2), where $d_H(x, y)$ denotes the distance from x to y in H.*

Proof. The conditions imply immediately that there is an odd cycle, say $D = x_1 x_2 \ldots x_{2l+1}$, that intersects C in at least two vertices. Put

$$V(C) \cap V(D) = \{x_{i_1}, x_{i_2}, \ldots, x_{i_k} : 1 \leqslant i_1 < \ldots < i_k \leqslant 2l+1\}, \qquad k \geqslant 2.$$

Assume that the assertion fails. Then

$$d_C(x_{i_j}, x_{i_{j+1}}) \equiv i_{j+1} - i_j \pmod 2, \qquad 1 \leqslant j < k,$$

and

$$d_C(x_{i_k}, x_{i_1}) \equiv i_1 - i_k + 1 \pmod 2.$$

Summing these congruences the left-hand side is 0 since C is even and the right-hand side is 1. ∎

THEOREM 1.10. *If every odd cycle in G has length at most $2k - 1$ $(k \geqslant 1)$ then G is 2k-colourable.*

Proof. Apply induction on the order of G. Assume that $\chi(G) > 2k$ but the result holds for graphs of order less than $|G|$. Then $\delta(G) \geqslant 2k$ since $\chi(G - x) = 2k$ for every $x \in G$. Hence, by Theorem III.4.10, $c(G) \geqslant \min\{4k, |G|\} = 4k$. Instead of invoking this unnecessarily strong result we can see this in the following simple way as well. Let $Q = x_1 x_2 \ldots x_q$ be a maximal path in G. Then $\Gamma(x_1) \subset \{x_2, \ldots, x_q\}$. Put

$$J = \{i : x_i \in \Gamma(x_1)\},$$

$$j = \max J.$$

Then $|J| \geqslant 2k$ and for $i \geqslant k$ we have $2i + 1 \notin J$ and $j - 2i + 1 \notin J$. Hence $j \geqslant 4k$. The graph G contains the cycle $C = x_1 x_2 \ldots x_j$ and so j is even.

By Lemma 1.9 we may assume that there is a path P of length p independent of C that joins x_1 to x_l and $p \equiv l$ (mod 2). Then $x_1 x_2 \ldots x_l P x$ and $x_l x_{l+1} \ldots x_1 P x_l$ are odd cycles. As the sum of their lengths is at least $4k + 2$, at least one of them has length greater than $2k$, contradicting the original assumption. ∎

A graph is called *uniquely k-colourable* if there is only one partition of its vertex set into k colour classes. The totally disconnected (i.e. empty) graphs are the only uniquely 1-colourable graphs and the uniquely 2-colourable graphs are exactly the connected bipartite (i.e. 2-chromatic) graphs. Let G be a uniquely k-colourable graph of order n. Clearly $n \leqslant k$ iff G is complete. To avoid this trivial situation we assume in the sequel that $n \geqslant k \geqslant 3$. In this case $k = \chi(G)$ and $\delta(G) \geqslant k - 1$. The subgraph induced by the union of any $l \geqslant 2$ colour classes is a uniquely l-colourable graph since any l-colouring of it extends to a k-colouring of G. In particular, the subgraph induced by the union of any two colour classes is connected, as remarked by Cartwright and Harary [CH1]. It is almost equally trivial to see, as pointed out by Chartrand and Geller [CG1], that G is $(k - 1)$-connected and so the subgraph induced by any l of the colour classes is $(l - 1)$-connected.

Adding a new vertex to a uniquely k-colourable graph G and joining it to $k - 1$ vertices of G, one from every colour class of a k-colouring of G, we again obtain a uniquely k-colourable graph. Starting with K^k this way we can construct a uniquely k-colourable graph of order n for every $n \geqslant k$ with $\binom{k}{2} + (n - k)\,(k - 1) = (k - 1)\,n - (k/2)$ edges. It is not known whether there are uniquely k-colourable graphs with significantly fewer edges (cf. Problem 21).

Given $k \geqslant 2$ and $l \geqslant 1$, let $G_2 = 2K^{l,l}$ and let G_k be the join of G_2 and the complete $(k - 2)$-partite graph with $3l$ vertices in each class: $G_k = G_2 + K_{k-2}(3l)$. Then $|G_k| = n = (3k - 2)l$ and clearly $\chi(G_k) = k$ and $\delta(G_k) = n(3k - 5)/(3k - 2)$. Furthermore, as G_2 is disconnected, G_k is not uniquely k-colourable. The following result, proved in [B32], shows that G_k has the largest minimal degree among all non-uniquely k-colourable graphs of order n.

THEOREM 1.11. *Let G be a k-colourable $(k \geqslant 2)$ graph of order n such that*

$$\delta(G) > \frac{3k - 5}{3k - 2}n.$$

Then G is uniquely k-colourable.

Proof. The assertion is almost immediate by induction on n. If $k = 2$ and G is disconnected then it contains a component H of order $m \leqslant n/2$. Hence $\delta(H) \geqslant \delta(G) > n/4 \geqslant m/2$ and this implies that H contains a triangle, contradicting $\chi(G) = 2$. Thus the assertion holds for $k = 2$.

To prove the induction step, for $x \in G$ put $G_x = G[\Gamma(x)]$ and $n_x = |G_x|$. Then

$$n_x > \frac{3k-5}{3k-2} n.$$

and each vertex $y \in G_x$ has degree at least

$$\frac{3k-5}{3k-2} n - (n - n_x) = n_x - \frac{3}{3k-2} n > \frac{3(k-1)-5}{3(k-1)-2} n_x.$$

Therefore by the induction hypothesis G_x is uniquely $(k-1)$-colourable.

To complete the proof, consider vertices $u_1, u_2 \in G$. Since

$$d(u_i) \geqslant \delta(G) > \frac{3k-5}{3k-2} n \geqslant \frac{4}{7} n > \frac{1}{2} n,$$

there is a vertex x adjacent to both u_1 and u_2, i.e. $u_1, u_2 \in G_x$. Now a k-colouring of G always restricts to a $(k-1)$-colouring of G_x. As this $(k-1)$-colouring is unique, either u_1 and u_2 get the same colour or they get different colours, independently of the k-colouring of G. Thus G is uniquely k-colourable. ∎

In order to improve the bound in Theorem 1.11 let us impose on G the trivial necessary condition that it has a k-colouring in which the subgraph induced by the union of any two colour classes is connected. The example given before Theorem 1.11 clearly does not satisfy this condition. To construct an appropriate graph let H_3 be the graph obtained from the graph of the triangular prism (Fig. 1.2) by replacing each vertex with $l \, (l \geqslant 1)$ independent vertices. Thus two vertices of H_3 are joined iff they correspond to two adjacent vertices of the prism. Put $H_k = H_3 + K_{k-3}(3l)$. Then $|H_k| = n = 3(k-1)l$, $\chi(H_k) = k$ and $\delta(H_k) = 3(k-2)l = n(k-2)/(k-1)$. Clearly H_k

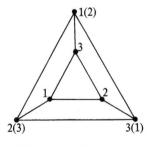

FIG. 1.2. The prism and its two 3-colourings.

is not uniquely k-colourable since H_3 is not uniquely 3-colourable (cf. Fig. 1.2). The subgraph induced by the union of any l colour classes is $(l - 1)$-connected. The next result shows that under these conditions the minimal degree of H_k is as large as possible. The proof consists of the examination of several cases. The interested reader will find it in [B32].

THEOREM 1.12. *Let G be a graph of order $n \geqslant k$ having a k-colouring in which the union of any two colour classes is connected. If*

$$\delta(G) > \left(1 - \frac{1}{k-1}\right)n$$

then G is uniquely k-colourable. ■

An *edge colouring* of a graph G is an assignment of colours to the edges in such a way that no two adjacent edges have the same colour. The minimal number of colours needed in an edge colouring of G is the *edge chromatic number* of G, denoted by $\chi'(G)$. Thus by definition $\chi'(G) = \chi(L(G))$. If $\Delta(G) = 1$ then $L(G)$ is totally disconnected and so $\chi'(G) = 1$. Let us note some bounds on $\chi'(G)$. The edges incident with a vertex in G span a complete subgraph in $L(G)$ so

$$\Delta(G) \leqslant \mathrm{cl}(L(G)) \leqslant \chi'(G). \tag{2}$$

As the edges coloured with the same colour are independent, a colour class contains at most $\beta_1(G)$ edges and so

$$e(G)/\beta_1(G) \leqslant \chi'(G).$$

Furthermore we have

$$\chi'(G) = \chi(L(G)) \leqslant \Delta(L(G)) + 1 \leqslant 2\Delta(G) - 1.$$

If $\Delta(G) \geqslant 3$ then, invoking Brooks' theorem, we see that the right-hand side can be decreased by 1.

Vizing [V3] proved the beautiful and deep result that though (2) is entirely trivial, $\chi'(G)$ is always very near to $\Delta(G)$. We present here the original proof of Vizing, based on switching colours in certain subgraphs.

THEOREM 1.13. $\Delta(G) \leqslant \chi'(G) \leqslant \Delta(G) + 1$ *for every graph G.*

Proof. As we have remarked already, the first inequality is trivial. Suppose that the second inequality, which is the real assertion of the theorem, *fails*. Then there is a graph G with $\Delta(G) \leqslant \Delta$ whose edges cannot be coloured with colours $1, 2, \ldots, \Delta + 1$. Our aim is to arrive at a contradiction. We may assume that there is an edge $ab \in E(G)$ such that $H = G - ab$ has got an edge colouring $\mu : E(H) \rightarrow S = \{1, 2, \ldots, \Delta + 1\}$.

Before starting the actual proof, let us make some preliminary remarks. For $x \in G$ denote by $M(x)$ (resp. $N(x)$) the colours appearing (resp. not appearing) at x:

$$M(x) = \{\mu(xy) : xy \in E(H)\} \subset S,$$

$$N(x) = S - M(x).$$

If we want to emphasize the dependence of $M(x)$ and $N(x)$ on the colouring μ, we write $M_\mu(x)$ and $N_\mu(x)$. Note that $|M(x)| \leqslant \Delta < \Delta + 1 = |S|$ so $N(x) \neq \varnothing$. For $\{s, t\} \subset S, s \neq t$, denote by $H_{s,t}$ the factor of G whose edges are the edges of H coloured with s and t: $E(H_{s,t}) = \mu^{-1}\{s, t\}$. Denote by $P_{s,t}(x)$ the component of $H_{s,t}$ containing $x \in V(G)$. By the definition of an edge colouring, $P_{s,t}(x)$ is a path or a cycle whose edges are alternatively coloured with s and t. If $\{s, t\} \not\subset M(x)$ then $P_{s,t}(x)$ is a *path* and x is an end-vertex of this path. This implies that if x_1, x_2, x_3 are different vertices and $\{s, t\} \not\subset M(x_i), 1 \leqslant i \leqslant 3$, then $P_{s,t}(x_1) = P_{s,t}(x_2) = P_{s,t}(x_3)$ cannot hold. Therefore there is an x_i such that

$$V(P_{s,t}(x_i)) \cap \{x_1, x_2, x_3\} = \{x_i\}. \tag{3}$$

If $N(a) \cap N(b) \neq \varnothing$ then the edge ab can be coloured with any of the colours in $N(a) \cap N(b)$, extending μ to an edge colouring of G with at most $\Delta + 1$ colours. Consequently for *any colouring* μ of H we have

$$N_\mu(a) \cap N_\mu(b) = \varnothing, \quad \text{i.e.} \quad N_\mu(b) \subset M_\mu(a). \tag{4}$$

This implies that if $t \in N(a)$ and $s \in N(b)$ then

$$P_{s,t}(a) = P_{s,t}(b). \tag{5}$$

For if this were not so, by interchanging the two colours in $P_{s,t}(a)$ we would get a colouring μ' of H for which $s \in N_{\mu'}(a) \cap N_{\mu'}(b)$.

For every $x \in \Gamma_G(a) - \{b\} = \Gamma_H(a)$ choose a colour $v(x) \in N(x)$. We will use this function v to replace μ by a colouring μ' of H for which (4) does not hold. Pick a colour $s_0 \in N(b) \subset M(a)$. Construct a sequence $s_0, s_1, \ldots$ of colours and a sequence $x_1, x_2, \ldots$ of vertices as follows. Suppose $s_0, s_1, \ldots, s_l$ and $x_1, x_2, \ldots, x_l$ ($l \geq 0$) have been chosen. If $s_l \in N(a)$ stop both sequences. If $s_l \in M(a)$ let $x_{l+1} \in \Gamma_H(a)$ be the unique vertex with

$$\mu(ax_{l+1}) = s_l \text{ and put } s_{l+1} = \nu(x_{l+1}). \tag{6}$$

The definition of ν implies that $s_{i+1} \neq s_i, x_{i+1} \neq x_i, i \geq 0$.

(a) Suppose first that the sequences end in s_l and x_l ($l \geq 1$). Note that then $s_0 \neq s_i$ for some $i > 0$. Change μ to a colouring μ' by colouring each edge ax_i with $\nu(x_i), 1 \leq i \leq l$, instead of $\mu(ax_i)$. It is clear that μ' is a good edge colouring of H and

$$s_0 \in N_{\mu'}(a) \cap N_{\mu'}(b).$$

As this contradicts (4), this case cannot occur.

(b) The sequence $s_0, s_1, \ldots$ of colours in $M(a)$ is infinite. Let $k > 0$ be the minimal index such that $s_k = s_i$ for some $i, 0 \leq i < k$. We distinguish two cases according to $i = 0$ or $i \neq 0$.

1. The case $s_k = s_0$. Choose a colour $t \in N(a)$. As $s_0 \in N(b)$ and $s_0 \in N(x_k)$, the three vertices a, b, x_k and two colours s_0, t satisfy the conditions of (3). By (5) we have $P_{s_0, t}(a) = P_{s_0, t}(b)$ so (3) implies that $P_{s_0, t}(x_k)$ does not contain either of a and b. Interchanging the two colours in $P_{s_0, t}(x_k)$ we obtain a colouring μ' of H for which $t \in N_{\mu'}(x_k)$. Then we may change ν to ν' by putting $\nu'(x_k) = t$ instead of $\nu(x_k) = s_k$. For this pair μ', ν' the sequence $s_0, s_1, \ldots$ becomes finite: $s_0, s_1, \ldots, s_k, t$ and so we are home by (a).

2. The case $s_k = s_i, 1 \leq i \leq k - 1$. As $s_k \neq s_{k-1}$ we have, in fact, $1 \leq i \leq k - 2$. Also by the construction of the sequences $x_1, x_2, \ldots, x_k$ are all different. Pick a colour $t \in N(a)$. As $s_i \in N(x_i)$ and $s_i \in N(x_k)$, the three vertices a, x_i, x_k and two colours s_i, t satisfy the conditions of (3). Therefore at least one of the following three cases must occur.

(i) $P_{s_i, t}(x_k)$ contains neither a nor x_i. Interchanging the colours in this path we obtain a colouring μ' such that $t \in N_{\mu'}(x_k)$. Then we change ν to ν' by putting $\nu'(x_k) = t$ instead of $\nu(x_k) = s_k$. As in (1), the sequence $s_0, s_1, \ldots$ is finite for this pair μ', ν' so we are home by (a).

(ii) $P_{s_i, t}(x_i)$ contains neither a nor x_k. This case is analogous to (i), only k is replaced by i.

(iii) $P_{s_i, t}(a)$ contains neither x_i nor x_k. Then $P_{s_i, t}(a)$ begins with the edge ax_{i+1} of colouring s_i. Interchanging the two colours in this path we obtain a colouring μ' such that $s_i \in N_{\mu'}(a)$. For this colouring μ' the sequence $s_0, s_1, \ldots$ ends in s_i, so by (a) we once again arrive at a contradiction. ∎

The bounds in Theorem 1.13 can not be improved except for $\Delta(G) = 0, 1$ when $\chi'(G) = \Delta(G)$. If $G = K^{1,\Delta}(\Delta \geqslant 0)$ then $\chi'(G) = \Delta(G) = \Delta$. Furthermore, for $\Delta \geqslant 2$ there is a graph G such that $\Delta(G) = \Delta$ and $\chi'(G) = \Delta + 1$. If $\Delta = 2k$ then $G = K^{2k+1}$ is like that: $\chi'(K^{2k+1}) \geqslant \binom{2k+1}{2}/\beta_1(G) = \binom{2k+1}{2}/k = 2k + 1$. Similarly, if G is a graph of order $2\Delta + 1$ with $\Delta(G) = \Delta$ and $e(G) \geqslant \Delta^2 + 1$ (e.g. if G is obtained from $K^{\Delta, \Delta}$ by dividing an edge into two new edges) then $\chi'(G) \geqslant e(G)/\beta_1(G) \geqslant e(G)/\Delta > \Delta$ and so $\chi'(G) \geqslant \Delta + 1$.

A graph is *uniquely edge colourable* if the partition of $E(G)$ defined by an edge colouring of G with $\chi'(G)$ colours is unique; if $\chi'(G) = k$ we say G is *uniquely k-edge colourable*. Any path or even cycle is uniquely 2-edge colourable. The stars $K^{1,k}$ are uniquely k-edge colourable, and K^4 is uniquely 3-edge colourable. Suppose that G is cubic and that there is a partition of V into sets V_1 and V_2 such that $G[V_1]$ is joined to $G[V_2]$ by exactly 3 edges. The graphs G_i are formed by contracting $G[V_i]$ to a single vertex. Then G is uniquely 3-edge colourable if and only if both G_1 and G_2 are uniquely edge colourable. Thus for instance we may construct new uniquely 3-edge colourable graphs from old ones by "expanding" a vertex of degree 3 into a triangle. Apart from graphs described so far, the only known example of a uniquely 3-edge colourable graph is the graph shown in Fig. 1.3, constructed by Tutte [T20].

If $k \geqslant 4$, there are no other uniquely k-edge colourable graphs. This was conjectured by Wilson [W21], and proved by Thomason [T24].

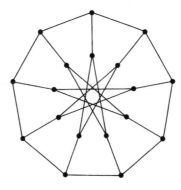

FIG. 1.3. Tutte's uniquely edge colourable graph.

THEOREM 1.14. *For $k \geq 4$, the only uniquely k-edge colourable graphs are the stars, $K^{1,k} = K^1 + E^k$.*

Proof. Suppose G is uniquely k-edge colourable and $k' \leq k$. Then the subgraph of G spanned by the edges of any k' colours is uniquely k'-edge colourable; hence it suffices to prove the theorem for the case $k = 4$.

Assume that G is a uniquely 4-edge colourable graph and $G \neq K^{1,4}$. Consider an edge colouring of G with red, blue, green and yellow. We may assume that *either* G is 4-regular *or* two vertices do not have degree 4, both these vertices have degree 2 and are incident with, say, red and green edges (Ex. 19).

A subgraph spanned by the edges of any two colours is a uniquely 2-edge colourable graph, and so it is a path or an even cycle. If G is 4-regular then each of the three partitions of the colours into pairs gives a decomposition of G into Hamiltonian cycles, for example the red and blue edges form a Hamiltonian cycle and the green and yellow edges form another Hamiltonian cycle. By Corollary III.4.16 the graph G contains a fourth pair of Hamiltonian cycles. But any pair of edge disjoint Hamiltonian cycles induces an edge colouring of G; hence G cannot be 4-regular.

Therefore G contains two vertices, say u and v, of degree two as described above; let u be adjacent to u_i and v to v_i, $i = 1, 2$. The red and blue edges form a Hamiltonian cycle C_1, and the green and yellow edges form a cycle C_2 of length $|G| - 2$.

Construct *a multigraph* G' by removing u and v and adding edges $u_1 u_2$ and $v_1 v_2$. Then C_1 and C_2 give rise to a pair of edge disjoint Hamiltonian cycles C_1' and C_2' in G', where both $u_1 u_2$ and $v_1 v_2$ are in C_1'. By Corollary III.4.16, there is another such pair D_1' and D_2', where $u_1 u_2$ and $v_1 v_2$ are in D_1'. D_1' induces a Hamiltonian cycle D_1 in G such that the remaining edges form a cycle D_2 of length $|G| - 2$. By colouring D_1 with red and blue and D_2 with green and yellow we achieve a different colouring of G, contradicting our assumption. ∎

2. CRITICAL *k*-CHROMATIC GRAPHS

A graph G is said to be *critical k-chromatic* if $\chi(G) = k$ and $\chi(G - x) = k - 1$ for every vertex x of G. Similarly G is *edge-critical k-chromatic* if $\chi(G) = k$ and $\chi(G - \alpha) = k - 1$ for every edge α of G. An edge-critical k-chromatic graph is also critical k-chromatic but it is trivial to construct critical k-

chromatic graphs that are not edge-critical. Clearly every k-chromatic graph contains an edge-critical k-chromatic subgraph: every minimal k-chromatic subgraph is such. Furthermore, if P is a monotone property (cf. p. 91), e.g. if P is the property of not containing a subgraph belonging to a given family of graphs, then there exists a k-chromatic graph possessing P iff there exists a critical (or edge-critical) graph possessing P. The study of critical k-chromatic graphs was started by Dirac [D6], [D8] and Gallai [G4], [G5]. Clearly K^2 is the only critical 2-chromatic graph and the critical 3-chromatic graphs are exactly the odd cycles. For $k \geqslant 4$ the situation is rather different; the characterization of critical k-chromatic $(k \geqslant 4)$ graphs seems hopeless. Instead of pointing towards a possible characterization, the results of this section will show the great variety of critical k-chromatic graphs. Since the critical 2- and 3-chromatic graphs are entirely trivial, we always suppose that $k \geqslant 4$ unless we state otherwise. Furthermore, to simplify the terminology, instead of saying that a graph is critical (edge-critical) k-chromatic we say sometimes that it is *k-critical (k-edge-critical)*.

If G is k-critical then $\delta(G) \geqslant k - 1$. For if $x \in G$ and $d(x) \leqslant k - 2$ then a $(k - 1)$-colouring of $G - x$ can be extended to a $(k - 1)$-colouring of G (cf. Theorem 1.1). Furthermore, if G is k-critical then it cannot be separated by a uniquely $(k - 1)$-colourable subgraph. For if G is the union of two proper subgraphs, say G_1 and G_2, such that $H = G_1 \cap G_2$ is uniquely $(k - 1)$-colourable then a $(k - 1)$-colouring of G_1 and a $(k - 1)$-colouring of G_2 can be put together to give a $(k - 1)$-colouring of G since they can be made to coincide on H. In particular, a k-critical graph cannot be separated by a complete graph. In turn this shows that the k-critical graphs are 2-connected since they cannot be separated by a vertex, i.e. a complete graph of order 1. Hence by Theorem III.4.10 we have the following theorem of Dirac [D7].

THEOREM 2.1. *If G is k-critical then either G is Hamiltonian or its circumference is at least $2k - 2$.* ∎

Let G be a k-critical graph of order n. Since $\delta(G) \geqslant k - 1, e(G) \geqslant \frac{1}{2}(k - 1)n$. Furthermore, if $n > k$ then G does not contain a K^k since $\chi(K^k) = k$. Therefore Brooks' theorem (Theorem 1.6) implies that in this case $e(G) \geqslant \frac{1}{2}\{(k - 1)n + 1\}$; in fact Brooks' theorem is equivalent to this inequality. This result was extended by Dirac [D9]. The simple proof we present is due to Kronk and Mitchem [KM1]; another simple proof was given by Weinstein [W14a]. In order to simplify the notation we formulate it for $(k + 1)$-critical graphs.

THEOREM 2.2. *If G is a $(k + 1)$-critical $(k \geqslant 3)$ graph of order $n > k + 1$ then*

$$2e(G) \geqslant kn + k - 2.$$

Proof. Assume that, contrary to the claim, G is $(k + 1)$-critical, $|G| = n > k + 1$ and $2e(G) \leqslant kn + k - 3$. Then $G \nsupseteq K^{k+1}$. Since $\delta(G) \geqslant k$, at most $k - 3$ vertices of G have degree at least $k + 1$. Let x be a vertex of degree k, say $\Gamma(x) = \{x_1, x_2, \dots, x_k\}$, and put $H = G - x$. Then $\chi(H) = k$, for G is $(k + 1)$-critical. Consider a k-colouring of H with colours $1, 2, \dots, k$. Exactly as in the second proof of Theorem 1.6, we can make the following two observations.

(i) *The vertices* $x_1, \dots, x_k$ *are coloured with different colours in every k-colouring of H.* We may assume that i is the colour of x_i, $1 \leqslant i \leqslant k$. For $1 \leqslant i < j \leqslant k$ the vertex x_i is adjacent to a vertex of colour j.

(ii) *The vertices x_i and x_j, $1 \leqslant i < j \leqslant k$, belong to the same component C_{ij} of the subgraph H_{ij} induced by the vertices coloured i and j.*

Denote by q the number of colours assigned to the vertices of degree at least $k + 1$ and put $p = k - q$. We may assume without loss of generality that the colours $1, 2, \dots, p$ are assigned only to vertices of degree k. Again recalling the second proof of Theorem 1.6, we see that the structure of H is as follows.

(iii) *The component C_{ij}, $1 \leqslant i < j \leqslant p$, is an x_i–x_j path.*

(iv) $C_{ij} \cap C_{il} = \{x_i\}$ *for* $i, j, l = 1, 2, \dots, p, i \neq j \neq l \neq i$.

(v) *If* $1 \leqslant i < j \leqslant p$ *then* $x_i x_j \in E(H)$.

Since $G \nsupseteq K^{k+1}$, there exist non-adjacent vertices x_u and x_v, $1 \leqslant u < v \leqslant k$. Let $1', 2', \dots, r'$ be the colours of the $k - 2$ colour classes not containing x_u and x_v that have at most one vertex adjacent to x_u and at most one vertex adjacent to x_v. Then (iv) implies that

$$d(x_u) + d(x_v) \geqslant 2k + k - 2 - r$$

so $2e(G) \geqslant k(n - 2) + 3k - 2 - r = kn + k - 2 - r$. Consequently $r \geqslant 1$.

(vi) $\{1', 2', \dots, r'\} \cap \{1, 2, \dots, p\} \neq \varnothing$. Suppose this is not so, i.e. for each i, $1 \leqslant i \leqslant r$, there is a vertex of degree at least $k + 1$ coloured with i'. Then

$$2e(G) = \sum_{x \in V - \{x_u, x_v\}} d(x) + d(x_u) + d(x_v) \geqslant k(n - 2) + r + 3k - 2 - r$$

$$= kn + k - 2,$$

contradicting our initial assumption.

To simplify the notation suppose that $1 \in \{1', 2', \dots, r'\}$.

(vii) *The vertex x_1 is adjacent to at most one of x_u and x_v.* Otherwise $V(C_{1u}) = \{x_1, x_u\}$ and $V(C_{1v}) = \{x_1, x_v\}$. Interchanging the colours in C_{1u} we obtain a colouring in which x_v is not adjacent to a vertex of colour 1. Let us assume that x_1 and x_v are not adjacent.

(viii) *There is at least one colour i, $2 \leqslant i \leqslant p$, such that C_{vi} is a path.* Otherwise each C_{vi} contains a vertex coloured v whose neighbours are coloured with all the $k - 1$ other colours and which has at least two neighbours coloured i. Consequently if there are t such vertices then the sum of their degrees is at least

$$tk + p - 1.$$

Taking into account that there are at least $q - 1$ vertices of degree at least $k + 1$ that are not coloured v, we see that

$$2e(G) \geqslant kn + p - 1 + q - 1 = kn + k - 2.$$

Assume that $2, 3, \ldots, s$ are the colours satisfying (viii). By (vii) there is a vertex y (coloured v) in C_{1v} adjacent to x_1 and different from x_v.

(ix) $y \notin \bigcap_{i=2}^{s} V(C_{iv})$ For otherwise

$$d(y) \geqslant 2s + (k - 1 - s) = k + s - 1.$$

Furthermore, analogously to the argument in (viii) the sum of the degrees of the other, say t', vertices of colour v is at least

$$kt' + p - s.$$

There are at least $q - 1$ vertices of degree at least $k + 1$ not coloured with v. Thus

$$2e(G) \geqslant kn + s - 1 + p - s + q - 1 = kn + k - 2,$$

a contradiction. Thus we may assume that $y \notin V(C_{2v})$.

We are ready to deduce our final contradiction. By (iii) and (v) the component C_{12} contains only two vertices, x_1 and x_2. Thus if we interchange the colours in C_{2v} we obtain a colouring in which no neighbour of x_1 is assigned the colour 2, contradicting (i).   ∎

Dirac [D25] determined all the graphs for which equality holds in Theorem 2.2. Equality can be attained for every $k \geqslant 3$ but each extremal graph has order $2k + 1$. It is likely that for large n the critical graphs *must have* many more edges than the number guaranteed by Theorem 2.2 (cf. Problem 23).

One might expect that a k-critical graph of order n *can not have* too many edges. We shall show that this is not the case. Put

$$f_k(n) = \max \{e(G): |G| = n, \ G \text{ is } k\text{-edge-critical}\}.$$

Note first that if $f_k(n) > cn^2$ for every $n \geq k$, where $c > 0$, then $f_{k+1}(n) > cn^2$ for every $n \geq k + 1$ as well. For if G is k-edge-critical then $G + K^1$ is $(k + 1)$-edge-critical and has $|G|$ more edges than G. Hence

$$f_{k+1}(n) \geq f_k(n - 1) + n - 1 > c(n - 1)^2 + n + 1 > cn^2$$

since the example of $G = K^k$ shows that $ck^2 < \binom{k}{2}$, i.e. $c < (k - 1)/2k$.

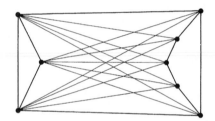

FIG. 2.1. The edge-critical 6-chromatic graph $C^3 + C^5$.

Dirac [D6] noticed that $G = C^{2l+1} + C^{2m+1}$, $l, m \geq 1$, is edge-critical 6-chromatic (see Fig. 2.1). Analogously to this example it is trivial that the sum of s disjoint odd cycles is critical $3s$-chromatic. These examples imply that if $k = 3s \geq 6$ then

$$\frac{s - 1}{2s} n^2 < f_k(n)$$

for infinitely many values of n. Furthermore, for $k \geq 6$

$$n^2/4 < f_k(n)$$

holds for infinitely many values of n. It is more complicated to construct critical 4- and 5-chromatic graphs with many, say cn^2, edges. Zeidl [Z4] was the first to do so: he constructed critical 4-chromatic graphs of order

$n = 6m + 4$ and size $(n^2 + 5n)/6$. Though some years later, but independently of Zeidl, Toft [T2] managed to alter the above construction of Dirac to give 4- and 5-chromatic graphs with many edges. Here we present only some of the graphs constructed by Toft. Let $H_l = 2C^{2l+1} \cup K(2l + 1, 2l + 1)$. Add to H_l a 1-factor such that each edge of this 1-factor joins either the first C^{2l+1} to the first vertex class of $K(2l + 1, 2l + 1)$ or it joins the second C^{2l+1} to the second vertex class of $K(2l + 1, 2l + 1)$. It is almost immediate to see that the obtained graph G_l is edge-critical 4-chromatic (cf. Fig. 2.2). Clearly $|G_l| = 4(2l + 1)$ and $e(G_l) = (2l + 1)(2l + 5)$.

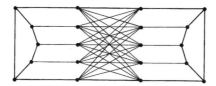

FIG. 2.2. The edge-critical 4-chromatic graph G_2.

In order to construct an edge-critical 5-chromatic graph let $l \geqslant 1$, $m = 8l + 4$ and let $n \geqslant 2m + 3$ be odd. Take $G_l \cup (E^m + C^{n-2m})$ and add to it a 1-factor of $G_l \cup E^m$, each edge of which joins a vertex of G_l to a vertex of E^m. It is again trivial that the obtained graph $G_{l,n}$ is edge-critical 5-chromatic (cf. Fig. 2.3). Clearly $|G_{l,n}| = n$ and $e(G_{l,n}) = (8l + 5)(n - 16l - 8) + (2l + 1)(2l + 9)$. With an appropriate choice of parameters these examples imply that

$$n^2/16 < f_4(n)$$

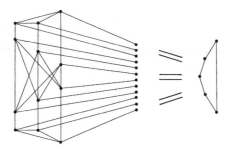

FIG. 2.3. The edge-critical 5-chromatic graph $G_{1,29}$.

holds for infinitely many values of n and

$$\frac{4}{31} n^2 < f_5(n)$$

for infinitely many values of n. In fact Toft proved that these inequalities hold for all values of n. Furthermore, by elaborating the examples above he obtained good lower bounds on $f_k(n)$. We state these together with the inequality of Dirac mentioned earlier.

Let $k = 3s + r \geqslant 4, 0 \leqslant r \leqslant 2$. Then

$$\lim_{n \to \infty} \sup f_k(n)n^{-2} \geqslant \begin{cases} \dfrac{s - 1}{2s} & \text{if} \quad r = 0, \\[3ex] \dfrac{7s - 6}{14s + 2} & \text{if} \quad r = 1, \\[3ex] \dfrac{23s - 15}{46s + 16} & \text{if} \quad r = 2. \end{cases}$$

While attempting to construct critical graphs of large size, Toft [T4a] proved the following results about the structure of critical graphs. For the rather involved proof we refer the reader to the original paper.

THEOREM 2.3. *A graph G is vertex-critical if and only if each block of $\bar{G}$, the complement of G, is the complement of a vertex-critical graph.* ∎

Clearly, Theorem 2.3 is eminently suitable for constructing large critical graphs; for example, by making use of the graph of Zeidl and an odd cycle, one can construct critical 6-chromatic graphs with n vertices and more than $\frac{3}{10}n^2$ edges.

It is rather natural to expect that a k-chromatic graph cannot contain too many independent vertices, i.e. the edges cannot be covered with too few vertices. Once again the k-critical graphs are too unruly to live up to this expectation. Note that

$$\min\{\alpha_0(G): |G| = n \text{ and } G \text{ is } k\text{-critical}\}$$

$$= \min\{\alpha_0(G): |G| = n \text{ and } G \text{ is } k\text{-edge-critical}\},$$

so for simplicity we consider k-critical graphs. Brown and Moon [BM2] proved that for $k = 4$ the minimum above is $o(n^{1/2})$. This result was extended by Simonovits [S20] and Toft [T4]. The following theorem of Lovász [L22] gives the correct order of magnitude of $\min\{\alpha_0(G) : |G| = n$ and G is k-critical$\}$.

THEOREM 2.4. (i) *Let* $G = G^n$ *be a critical k-chromatic graph, where* $k \geqslant 4$. *Then the vertex covering number* $\alpha_0(G)$ *satisfies*

$$\alpha_0(G) \geqslant \frac{k-2}{3} n^{1/(k-2)} \geqslant \frac{k}{6} n^{1/(k-2)}.$$

(ii) *For every* $k \geqslant 4$ *there are infinitely many critical k-chromatic* $G = G^n$ *such that*

$$\alpha_0(G) \leqslant k n^{1/(k-2)}.$$

Proof. (i) Let T be a set of vertices covering the edges of G. Put $t = |T|$. We have to show that

$$n \leqslant \left(\frac{3t}{k-2}\right)^{k-2}. \tag{1}$$

If there exists a vertex $x \in V - T$ with $d = d(x) \geqslant k$ then we replace x by $\binom{d}{k-1}$ new vertices and join each of them to a different $(k-1)$-set in $\Gamma(x)$. The resulting graph is k-chromatic so omitting some of the new vertices we obtain a k-critical graph. This graph has at least n vertices and the set T covers all its edges. Repeating this operation a number of times we arrive at a critical k-chromatic graph G' of order $n' \geqslant n$ in which T covers all the edges and in which every vertex not in T has degree $k - 1$. Thus, when proving (1) we may assume that every vertex in $S = V - T$ has degree $k - 1$. Consider the following system of $(k-1)$-sets in T:

$$\Sigma = \{\Gamma(x) : \quad x \in S\}.$$

Though we shall not use it in the sequel, note that a set $\Gamma_1 = \Gamma(x)$ has multiplicity 1 in Σ since $\Gamma(x) = \Gamma(y)$, $x \neq y$, cannot hold in a critical graph.

(a) *If* $x_1, x_2, \ldots, x_s$ *are distinct vertices of S then there exists a $(k-2)$-set that is contained in an odd number of* $\Gamma(x_i)$'s.

To see this, colour $G - x_1$ with $1, 2, \ldots, k - 1$. This restricts to a colouring of T with $1, 2, \ldots, k - 1$ in which $\Gamma(x_1)$ meets every colour but $\Gamma(x_2), \Gamma(x_3), \ldots,$ $\Gamma(x_s)$ do not. Consider the $(k-2)$-sets in T coloured with $1, 2, \ldots, k - 2$. The set $\Gamma(x_1)$ contains exactly 1 such $(k-2)$-set and each $\Gamma(x_i)$ $(2 \leqslant i \leqslant s)$ contains 0 or 2. Consequently at least one of these $(k-2)$-sets is contained in an odd number of $\Gamma(x_i)$ sets.

In order to estimate $|\Sigma| = |S|$ we reformulate (a) in terms of a vector space over $\mathbb{Z}_2$, the field of order 2. Let $p = \binom{t}{k-2}$ and let $\Delta_1, \Delta_2, \ldots, \Delta_p$ be the $(k-2)$-sets in T. For $\Gamma \in \Sigma$ define

$$v_\Gamma = (v_1, v_2, \ldots, v_p),$$

where

$$v_i = \begin{cases} 1 & \text{if} \quad \Delta_i \subset \Gamma, \\ 0 & \text{if} \quad \Delta_i \not\subset \Gamma. \end{cases}$$

Then (a) states that the *vectors* v_Γ, $\Gamma \in \Sigma$, *are independent over* $\mathbb{Z}_2$. In particular,

$$|\Sigma| = |S| \leqslant p,$$

so

$$n \leqslant \binom{t}{k-2} + t \leqslant \left(\frac{3t}{k-2}\right)^{k-2}.$$

(ii) An odd cycle C^{2l+1} is 3-critical and so $C^{2l+1} + K^{m-3}$ $(m \geqslant 3)$ is m-critical. In particular, for every $p \geqslant m \geqslant 3$ there is an m-critical graph of order p or $p + 1$. Let $p \geqslant k \geqslant 4$ and let G_m be disjoint m-critical graphs with vertex sets V_m, $|V_m| = n_m = p$ or $p + 1$, $m = 3, 4, \ldots, k - 1$. Let G_2 be obtained from a copy G'_{k-1} of G_{k-1}, disjoint from the G_is, by adding to it a set V_2 of $|G_{k-1}|$ independent vertices and joining each new vertex to one vertex of G'_{k-1}, different vertices to different vertices. Note that in a colouring of G_2 the set V_2 behaves as a 2-critical graph in the following sense: in any k-colouring of G_2 at least 2 colours must occur in V_2 but omitting any edge of G_2 there is a k-colouring of G_2 in which every vertex of V_2 gets the same colour. Finally let G_1 be the trivial graph with vertex set $V_1 = \{v_1\}$ such that v_1 is not in any of the V_is or $V(G_2)$. Let G be obtained from $\bigcup_1^{k-1} G_i$ as follows. Consider all $(k-1)$-sets meeting each V_i, $i = 1, 2, \ldots, k-1$, in exactly one element. Add one new vertex for each $(k-1)$-set and join it to the vertices in that set (Fig. 2.4).

We claim that G is critical k-chromatic. Consider a colouring of G with $1, 2, \ldots, k-1$. We may assume that the vertex x_1 of V_1 is coloured with 1. As each vertex of V_2 can not be coloured with 1 (since $\chi(G'_{k-1}) = k-1$), we may

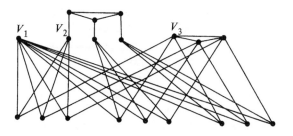

FIG. 2.4. A 4-chromatic graph with $p = 3$.

assume that V_2 contains a vertex x_2 coloured 2. Continuing this way, by $\chi(G_i) = i$ we may assume that G_i contains a vertex x_i coloured i, $i = 3, 4, \ldots$, $k - 1$. There is a vertex v_k joined to each of $v_1, v_2, \ldots, v_{k-1}$ and so v_k must be coloured with a colour different from $1, 2, \ldots, k - 1$. This shows that $\chi(G) \geqslant k$. In fact it is trivial that $\chi(G) = k$. Though it is not immediate that G is edge critical, we leave the routine proof to the reader.

The edges of G are covered by the set $T = \bigcup_1^{k-1} V(G_i)$. By construction

$$|T| \leqslant (k - 1)(p + 1) + 1.$$

Furthermore,

$$n = |G| \geqslant p^{k-2},$$

so

$$\alpha_0(G) \leqslant (k - 1)n^{1/(k-2)} + k < kn^{1/(k-2)},$$

if n is sufficiently large. ∎

3. COLOURING GRAPHS ON SURFACES

Given a graph $G = (V, E)$, $V = \{x_1, x_2, \ldots, x_n\}$ choose n points in $\mathbb{R}^3$, say $p_1, p_2, \ldots, p_n$, in such a way that no four of them are in a plane. Denote by (p_i, p_j) the open straight line segment joining p_i to p_j. Then the topological space

$$T(G) = \bigcup_1^n p_i \cup \{(p_i, p_j) : x_i x_j \in E(G)\} \subset \mathbb{R}^3$$

is said to be a *realization* of G. We say that the graph G can be *embedded* in a topological space X if $T(G)$ is homeomorphic to a subspace of X. Recall that a graph G is said to be *homeomorphic to H* if they have isomorphic subdivisions. Clearly G is homeomorphic to H iff the topological spaces $T(G)$ and $T(H)$ are homeomorphic. Naturally a graph G is said to be *planar* if it can be embedded in $\mathbb{R}^2$, the plane. If ϕ is a homeomorphism of $T(G)$ into $\mathbb{R}^2$ then $\phi(T(G))$ is said to be a *drawing* of G. We may identify G and $\phi(T(G))$; in this case we call G a *plane* graph. It must be emphasized that planarity involves absolutely no topological difficulty. It is trivial that a planar graph can always be drawn in the plane in such a way that the edges are broken lines. With a little effort one can prove that the edges can even be made straight line segments, as shown by Fáry [F1]. When talking of plane graphs, we shall always assume that the edges are particularly pleasant curves. Planar graphs were characterized by Kuratowski [K15] in 1930.

THEOREM 3.1. *A graph G is planar if and only if it does not contain a subgraph homeomorphic to K^5 or $K^{3,3}$.* ∎

As we shall not make serious use of this theorem, we do not prove it here. Beside the original proof of Kuratowski a simpler proof was given by Dirac and Schuster [DS8] (see also Harary [H19]). Tutte [T16] gave an algorithm for drawing a planar graph in the plane and as a corollary obtained another proof of Kuratowski's theorem. As was pointed out by Wagner [W1], [W2], and rediscovered by Harary and Tutte [HT1], Theorem 3.1 is equivalent to the following.

THEOREM 3.1'. *A graph is planar if and only if it does not have a subgraph contractible to K^5 or $K^{3,3}$.* ∎

Other characterizations of planar graphs were given by Whitney [W19], [W20] and MacLane [M1].

Instead of the plane we may wish to embed a graph in a surface of genus γ. Given a graph G, the minimal number $\gamma \geqslant 0$ such that G can be embedded in an orientable surface of genus γ is called the *genus of G* and is denoted by $\gamma(G)$. Euler's polyhedron theorem states that a polyhedron of genus γ with n vertices, e edges and f faces satisfies

$$n - e + f = 2 - 2\gamma.$$

An immediate consequence of this is that if $\gamma(G) \leqslant \gamma$, i.e. if G is embeddable in a surface of genus γ then

$$e(G) \leqslant 3n + 6(\gamma - 1). \tag{1}$$

Though Kuratowski's theorem has not been carried over to other surfaces, Vollmerhaus [V9] proved that for every $\gamma \geqslant 0$ there is a finite family of graphs $\mathcal{F}_\gamma$ such that $\gamma(G) \leqslant \gamma$ iff G does not contain a subgraph homeomorphic to a member of $\mathcal{F}_\gamma$. Related results are discussed in Milgram [M28], [M29].

In this section we present some not too deep results about the chromatic number of graphs on a surface S. Naturally the dominant problem in this area is the Four Colour Conjecture (usually referred to as the 4CC), which is by far the most famous problem in graph theory. For the long and fascinating early history of the problem we refer to the reader to the recent work of Biggs, Lloyd and Wilson [BLW1].

CONJECTURE 3.2. (4CC) *Every planar graph is 4-colourable.*   ∎

The graph K^4 is one of an infinite family of trivial examples showing that a planar graph may be 4-chromatic.

Basing their work on ideas of Heesch [H24] and making extensive use of a computer, Appel and Haken [AH1] and Appel, Haken and Koch [AHK1] *have recently settled the* 4CC *in the affirmative* and so turned it into the 4CT (Four Colour Theorem). We shall say a few words about their long proof at the end of the section. Nevertheless, in the rest of the section we keep the name in preference to 4CT, in order to stress that we only prove pre-4CT type results, results that can be obtained without the 4CT.

Attacks on the 4CC have produced a good number of interesting results. We have already discussed the history of Tait's conjecture (Chapter III, §4), a proof of which would have implied a proof of the 4CC. Here we shall prove the equivalence of the 4CC to a number of other assertions. We prove the simple result that a planar graph is 5-colourable and present the similarly easy upper bound on the chromatic number of a graph of genus $\gamma \geqslant 1$, due to Heawood [H22]. It is a rather different matter that this straightforward bound happens to be best possible; this is a hard theorem of Ringel and Youngs [RY1]. In this section we do not even sketch their long and complicated proof; we refer the interested reader to the book of Ringel [R9] written about this problem (cf. also Ore [O4]).

The first colouring result we present is the simple 5-colour theorem, proved by Kempe [K5] in his unsuccessful attempt at a proof of the 4CC.

THEOREM 3.3. *Every planar graph is 5-colourable.*

Proof. The proof is a simpler version of the proof of Theorem 1.3. Suppose the assertion is false and let G be a minimal counter example, drawn in the plane. Pick a vertex $x \in G$ of degree $\delta(G) \leqslant 5$. The existence of such a vertex is guaranteed by (1), since $\gamma = 0$ for the plane. Put $H = G - x$. Then H is 5-colourable, say with colours $1, 2, \ldots, 5$, and each of these colours must occur as the colour of a vertex adjacent to x. Thus we may assume that $\Gamma(x) = \{x_1, \ldots, x_5\}$, in the drawing of G the vertices $x_1, x_2, \ldots, x_5$ are arranged cyclically about x and for each $i, 1 \leqslant i \leqslant 5, x_i$ is coloured with colour i. As in the proof of Theorem 1.3, it is easily seen that for $1 \leqslant i < j \leqslant 5$ the graph H contains an $x_i - x_j$ path P_{ij} whose vertices are coloured alternatively with i and j. The cycle $C = xx_1 P_{13} x_3$ of G separates x_2 from x_4 in the plane. Consequently the path P_{24} intersects the cycle C. This is impossible. ∎

Let now S be an orientable surface of genus $\gamma > 0$. Suppose that S contains a graph G with $\chi(G) > k$. Then, deleting some vertices of G if necessary, we may assume that G is critical $(k + 1)$-chromatic. Then $\delta(G) \geqslant k$ and $n = |G| \geqslant k + 1$, so (1) implies

$$\tfrac{1}{2}kn \leqslant e(G) \leqslant 3n + 6(\gamma - 1).$$

Hence if $k \geqslant 6$ then

$$\frac{k+1}{2}(k - 6) \leqslant 6(\gamma - 1),$$

and so

$$k \leqslant \tfrac{1}{2}\{5 + \sqrt{1 + 48\gamma}\}.$$

Thus we have the following result, due to Heawood [H22].

THEOREM 3.4. *If a graph G can be drawn on a surface of genus $\gamma > 0$ then*

$$\chi(G) \leqslant H(\gamma) = \lfloor \tfrac{1}{2}\{7 + \sqrt{1 + 48\gamma}\}\rfloor.$$ ∎

Dirac [D10], [D11] proved the following extension of this result.

THEOREM 3.5. *Suppose G is drawn on a surface of genus $\gamma > 0$ and $\chi(G) = H(\gamma) = k$. Then G contains a K^k.*

Proof. Assume that the assertion fails. Then we may suppose that G is critical k-chromatic and $G \not\supseteq K^k$ ($k \geqslant 7$). Then $|G| = n \geqslant k + 2$ and, by Theorem

2.2, we have $2e(G) \geqslant (k - 1)n + k - 3$. Therefore

$$(k - 1)n + k - 3 \leqslant 6n + 12(\gamma - 1),$$

so

$$(k - 7)(k + 2) \leqslant (k - 7)n \leqslant 12(\gamma - 1) - (k - 3).$$

Hence

$$k^2 - 4k - (5 + 12\gamma) \leqslant 0,$$

$$k \leqslant \tfrac{1}{2}(4 + \sqrt{36 + 48\gamma}) < H(\gamma). \qquad \blacksquare$$

When investigating the chromatic number of a graph drawn on a surface S it is natural to define the *chromatic number of S*, $\chi(S)$, as the maximal chromatic number of a graph drawn on S. With this notation Theorems 3.4 and 3.5 can be reformulated as follows.

THEOREM 3.5'. *If S_γ is an orientable surface of genus $\gamma > 0$ then*

$$\chi(S_\gamma) \leqslant H(\gamma) = \lfloor \tfrac{1}{2}(7 + \sqrt{1 + 48\gamma}) \rfloor, \qquad (2)$$

with equality iff $K^{H(\gamma)}$ can be drawn on S_γ. $\blacksquare$

As we have mentioned already, equality *does hold* in (2) for every $\gamma > 0$, that is $K^{H(\gamma)}$ can be drawn on a surface of genus γ for every γ. This is easily seen for $\gamma = 1$; Fig. 3.1. illustrates a drawing of K^7 on the torus, proving $\chi(T) = \chi(S \times S) = 7$, due to Heawood [H22].

Note that $H(\gamma + 1) \leqslant H(\gamma) + 1$ $(\gamma > 0)$ so every natural number $p \geqslant 7$ is the value of $H(\gamma)$ for some $\gamma > 0$. Clearly $K^{H(\gamma)}$ can be drawn on an orientable surface S_γ of genus $\gamma > 0$ iff

$$\gamma(K^p) \leqslant \min\{\gamma > 0 : H(\gamma) = p\}, \qquad p \geqslant 7. \qquad (3)$$

In fact, inequality (3) is exactly the converse of (2). Straightforward calculations give

$$\min\{\gamma > 0 : H(\gamma) = p\} = \left\lceil \frac{(p - 3)(p - 4)}{12} \right\rceil, \qquad p \geqslant 4.$$

Therefore equality holds in (2) iff K^p $(p \geqslant 7)$ can be drawn on an orientable

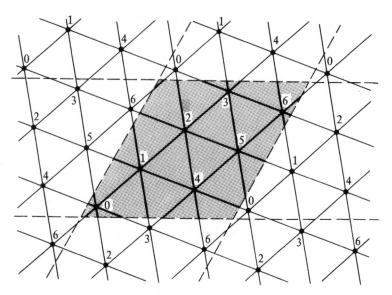

Fig. 3.1. A drawing of K^7 on the torus. As usual, the opposite side of the rhombus are identified to give a representation of the torus.

surface of genus $\lceil (p-3)(p-4)/12 \rceil$, i.e. iff

$$\gamma(K^p) = \left\lceil \frac{(p-3)(p-4)}{12} \right\rceil, \qquad p \geqslant 7. \tag{3'}$$

(It is trivial to see that (3') holds for $1 \leqslant p \leqslant 6$.) Heawood [H22] gave a faulty proof of (3') in 1890 but, after various partial results by Heffter [H26], Ringel [R8], and Youngs [Y4], a correct proof for all $p \geqslant 7$, with the exception of 3 values, was given only in 1967 by Ringel and Youngs [RY1]. The 3 exceptional values of p, 18, 20 and 23, were taken care of by Mayer [22]. Thus, the following so-called Heawood colouring theorem is due mostly to Ringel and Youngs.

THEOREM 3.6. $\gamma(K^p) = \lceil (p-3)(p-4)/12 \rceil$ for all $p \geqslant 1$ or, equivalently, if S_γ is an orientable surface of genus $\gamma > 0$ then

$$\chi(S_\gamma) = H(\gamma) = \lfloor \tfrac{1}{2}(7 + \sqrt{1 + 48\gamma}) \rfloor. \qquad \blacksquare$$

If G is a plane graph, $\mathbb{R}^2 - G$ falls into a finite number of connected components, say $R_1, R_2, \ldots, R_p$. The graph G together with the set $\{R_1, R_2, \ldots, R_p\}$ of regions it determines is called a *plane map*, $R_1, R_2, \ldots, R_p$ are the *countries* or *faces* of this map. Two countries are *neighbouring* or *adjacent* if their boundaries have an edge in common. Clearly each edge is contained in the boundary of at most two countries. With a slight abuse of notation we denote by G the map determined by a plane graph G. Given a plane map G it is natural to define a plane multigraph G^* from G as follows. Pick a point $r_i \in R_i$, $i = 1, 2, \ldots, p$. Let $r_1, r_2, \ldots, r_p$ be the vertices of G^*. If $\alpha \in E(G)$ is on the boundary of R_i and R_j $(i \neq j)$ then join r_i to r_j by a simple curve $d(\alpha)$ intersecting α and no other edge of G. These curves can be chosen to be such that only their endpoints may be in common. Considering these curves as the edges of G^* we obtain a multigraph, said to be a *geometric dual* of G, already drawn in the plane. If G is connected and bridgeless then G can be taken as a dual of the plane multigraph G^*; if the boundaries of any two countries of the map of G contain at most one edge then G^* is a *graph*. It is clear from the definition that G^* is determined by the *map* G, i.e. by the drawing of G, so a *graph* G may have a good number of geometric duals.

It is just as natural to colour maps as graphs, in fact the 4CC is usually formulated in terms of maps. A colouring of a map G is an assignment of colours to the countries of G in such a way that any two neighbouring countries get different colours. Thus a colouring of a map G is exactly a colouring of G^*. This implies immediately that the 4CC is true iff every plane map is 4-colourable. The next result is only slightly less trivial.

THEOREM 3.7. *The 4CC is true iff every cubic bridgeless plane map is 4-colourable.*

Proof. Suppose every cubic bridgeless plane map is 4-colourable. Given a plane map G we would like to colour with 4 colours it is trivial that we may assume that $\delta(G) \geqslant 3$ and G is bridgeless. Replace each vertex $x \in G$, $d(x) \geqslant 4$, by a small new country C_x, i.e. let C_x be a small disc with centre x (cf. Fig. 3.2). A 4-colouring of this new cubic map induces a 4-colouring of the original map G. ∎

THEOREM 3.8. *Let G be a bridgeless cubic plane graph and let G^* be its geometric dual drawn in the plane.*

Then $\chi'(G) = 3$ iff G^* is 4-colourable, i.e. iff the map of G is 4-colourable.

In particular, the 4CC holds iff $\chi'(G) = 3$ for every bridgeless cubic planar graph G.

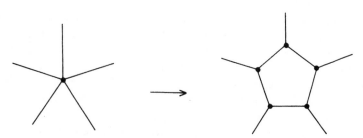

FIG. 3.2. Replacement of a vertex by a small country.

Proof. (i) Suppose G^* is 4-colourable. Take a colouring ϕ of G^* with the elements of the Klein four-group $F = \mathbb{Z}_2 \times \mathbb{Z}_2$; $\phi: V(G^*) \to F$. Let $ab \in E(G)$. If $xy = d(ab) \in E(G^*)$ is the edge of G^* intersecting ab, colour ab with $\psi(ab) = \phi(x) + \phi(y) \in F$. Then, as $\phi(x) \neq \phi(y)$, $\psi(ab) \neq 0$. Hence ψ maps $E(G)$ into the 3-element set $F - \{0\}$. It is immediate that ψ is a proper colouring so $\chi'(G) = 3$.

(ii) Let ψ be a colouring of the edges of G with the non-zero elements of the Klein four-group F, $\psi: E(G) \to F - \{0\}$. Given $ab \in E(G^*)$ let $ab = d(xy)$, i.e. let xy be the unique edge of G intersecting ab. Put $\psi(ab) = \psi(xy)$. If $y_0 y_1 y_2$ is a triangle in G^* containing exactly one vertex of G then $\psi(y_0 y_1) + \psi(y_1 y_2) + \psi(y_2 y_0) = 0$. If $C = x_0 x_1 \ldots x_k$ is a cycle in G^* then, taking a triangulation of this cycle, we see that

$$\sum_{i=0}^{k} \psi(x_i x_{i+1}) = 0, \tag{4}$$

where we have put $x_{k+1} = x_0$ (cf. Fig. 3.3).

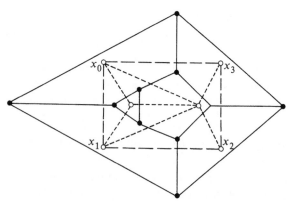

FIG. 3.3. A cycle $x_0 x_1 x_2 x_3$ and its triangulation.

Pick a vertex $x_0 \in G^*$ and colour x_0 with $0 \in F$. Given another vertex $y \in G^*$, choose a path $x_0 x_1 \ldots x_{k+1}$, $x_{k+1} = y$, in G^* and colour $y = x_{k+1}$ with $\sum_{i=0}^{k} \psi(x_i x_{i+1})$. Relation (4) implies that this colouring of $V(G^*)$ is well defined. Neighbouring countries do not get the same colour since the difference of their colours is the colour of an edge of G so it is a non-zero element of F. ■

As an immediate consequence of Theorem 3.8, we see that the truth of Tait's conjecture (Chapter III, §4) would have indeed implied the truth of the 4CC.

The results presented so far fall very short of a proof of the 4CC. Nevertheless, we turn now to two conjectures that are even stronger than the 4CC. Recall that a *contraction* of a graph G is the identification of the vertices of a connected subgraph. If H is the result of a sequence of contractions then H is a *contraction* of G and we write $G \succ H$. If H is a contraction of a subgraph of G then *H is a subcontraction of G*: $G > H$. Thus $G > H$ with $V(H) = \{y_1, y_2, \ldots, y_m\}$ iff G contains vertex disjoint connected subgraphs G_1, $G_2, \ldots, G_m$ such that $y_i y_j \in E(H)$, $i \neq j$, iff G contains a $G_i - G_j$ edge. Clearly $G \succ H$ implies $G > H$, and both $\succ$ and $>$ are transitive relations. If G contains a subgraph homeomorphic to H (i.e. a topological H, a TH) then $G > H$, but the converse need not hold.

In 1943 Hadwiger [H1], [H2] made the following conjecture.

CONJECTURE 3.9. *If $\chi(G) = k$ then $G > K^k$.*   ■

The conjecture is trivial for $k = 1, 2$ and 3. Dirac [D6] proved it for $k = 4$; we shall present his result in Chapter VII when discussing contractions. Clearly the truth of the conjecture for $k = p$ implies its truth for $k < p$ since $\chi(H + K^1) = \chi(H) + 1$. Knowing the truth of the conjecture for $k = p$ does not seem to help in the proof of the conjecture for $k = p + 1$. Furthermore, by Kuratowski's theorem (more precisely, by Theorem 3.1') Hadwiger's conjecture for $k = 5$ implies the 4CC. Thus Hadwiger's general conjecture is considerably stronger than the 4CC.

Hajós [H3] conjectured that every k-chromatic graph contains a TK^k, a topological complete graph of order k. For $k \leqslant 4$ this is clearly equivalent to Hadwiger's conjecture, but for $k \geqslant 5$ it is even stronger. Wagner [W1] proved that the 4CC is, in fact, equivalent to Hadwiger's conjecture for $k = 5$. Other proofs were given by Halin [H6], [H7] and Ore [O4]. The simple proof of this result we are going to present is due to Young [Y3] and is based on the following lemma, first proved by Halin [H6].

LEMMA 3.10. *If G is 4-connected and non-planar then* $G > K^5$.

Proof. By Kuratowski's theorem we may assume that G contains a subgraph H homeomorphic to $K^{3,3}$. Let the two classes of branchvertices of H be $\{a, b, c\}$ and $\{d, e, f\}$. If a path in H corresponding to an edge of $K^{3,3}$ is a path of length at least 2, then contract the subgraph spanned by the interior vertices to a single vertex. With a slight abuse of notation we keep the same notation for the images of G and H. Note that H is now a subgraph homeomorphic to $K^{3,3}$ and each path corresponding to an edge of $K^{3,3}$ has length at most 2, i.e. has at most one interior vertex. We call such a path an H *line*. Clearly an H line joins two branchvertices, one of which belongs to $\{a, b, c\}$ and the other to $\{d, e, f\}$. The new graph G need not be 4-connected but it can not be disconnected by omitting any 3 of the original (i.e. unchanged by the contractions) vertices of G. In particular, no 3 branchvertices of H disconnect G.

Let P be a path from a to $\{b, c\}$ in $G - \{d, e, f\}$. Let a' be the last vertex on P (from a) that is on an H line from a. Let b' be the first vertex on P after a' that is on an H line. We may assume that b' is on an H line from b and neither a' nor b' is the midvertex of an H line from f. Take now a path Q from f to $\{d, e\}$ in $G - \{a, b, c\}$. Let f' be the last vertex on Q (from f) that is on an H line from f. Let x be the first vertex on Q after f' that belongs to $H \cup P_{a'b'}$.

If $x \notin P_{a'b'}$ we may assume that x is on an H line from e. Identifying a with a', b with b', f with f' and x with e the graph becomes homeomorphic to L_1, shown in Fig. 3.4. Identifying c with d in L_1 we obtain K_5. Thus $G > K^5$.

If $x \in P_{a'b'}$ we distinguish two cases according to the positions of a' and b'.

1. *Not both* $a' = a$ *and* $b' = b$. By symmetry we may assume that $a' \neq a$. If $x \neq b$ contract $P_{a'x}$ and identify b with b' and f with f'. This gives a sub-contraction of G homeomorphic to L_2, shown in Fig. 3.4. As $L_2 > K^5$ we have $G > K^5$. If $x = b'$ then, identifying a with a' and f with f', we see that G subcontracts to an isomorphic copy of L_2. Hence $G > K^5$.

 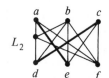 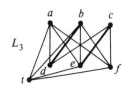

FIG. 3.4. The graphs L_i all become K^5 on contracting the thick edges.

2. $a' = a$ and $b' = b$. Let e' be the first vertex of Q (traversing it from f) after x in H. We may assume that e' is on an H line from e. Contract the connected subgraph spanned by the interior vertices of $P_{a'b'}$ and $Q_{f'e'}$ to a single vertex t. Identifying f with f' and e with e' we see that $G > L_3$, shown in Fig. 3.4. As $L_3 > K^5$, $G > K^5$ in this case as well.   ∎

Armed with this lemma we can easily prove Wagner's equivalence theorem [W1].

THEOREM 3.11. *The 4CC is true iff* $\chi(G) = 5$ *implies* $G > K^5$.

Proof. As we have mentioned already if $G > K^5$ then by Theorem 3.1′ G is nonplanar. Thus to prove the theorem we have to show that if the 4CC is true then $\chi(G) = 5$ implies $G > K^5$. Suppose therefore that the 4CC is true, $\chi(G) = 5$ but $G \not> K^5$. Then G is non-planar so by Lemma 3.10 we may assume that G is not 4-connected. We may also assume that every proper subcontraction of G is 4-colourable.

Let $S \subset V$ be a minimal set disconnecting G, $|S| \leqslant 3$. Let $C_1, C_2, \ldots, C_k$ be the components of $G - S$ $(k \geqslant 2)$ and put $G_i = G[V(C_i) \cup S]$, $1 \leqslant i \leqslant k$. Let T be a maximal independent set in S. As $|S| \leqslant 3$ it is easily seen that either $T = S$ or $G[S - T]$ is a complete graph of order 1 or 2. Pick a graph G_i and let $C_j \neq C_i$. Contract C_j to a vertex x_j in G. Then x_j is joined to every vertex in S since S is a minimal set disconnecting G. Identify the vertices in $T \cup \{x_j\}$ and denote by y_j the new vertex. The graph H_j we obtain this way is a contraction of G so it is 4-colourable. Furthermore, y_j and the vertices in $S - T$ span a complete subgraph of H_j. Consequently the 4-colouring of H_j induces a 4-colouring of G_i such that the vertices in T get one colour and the vertices in $S - T$ get $|S - T|$ other colours. Hence these 4-colourings of the G_is can be put together to give a 4-colouring of G, contradicting $\chi(G) = 5$.
                                                                                      ∎

There are a very large number of results stating that the plane graphs of small order are 4-colourable. Supposing that the 4CC is false, denote by N the minimal order of a 5-chromatic planar graph. Among others the following bounds have been obtained: $N \geqslant 40$ by Ore and Stemple [OS1], $N \geqslant 48$ by Mayer [M23] and $N \geqslant 52$ by Stromquist [S24], [S28]. These bounds were obtained by finding various so called *reducible* configurations in small plane graphs, i.e. subgraphs that cannot exist in a minimal 5-chromatic plane graph. Reducible configurations have been discovered by Heesch [H24], [H25], who has also made extensive use of computers, Bernhart [B16],

Allaire and Swart [AS1]. Before the proof of the 4CC the record, $N \geqslant 96$, was held by Mayer [M24], a professor of French literature at Montpellier.

Appel and Haken [AH1] used the methods of Heesch [H25] to find a finite unavoidable set of configurations, i.e. a set of subgraphs such that a 5-chromatic planar triangulation must contain one of them. With Koch, in [AHK1] they completed the proof of the 4CC by showing that each of these configurations is reducible. Their set contained 1936 configurations and it is is possible that this number cannot be reduced substantially (Heesch's reduction method was extended recently by Dürre, Heesch and Miche [DHM1], who thereby found 2669 reducible configurations. Using these it may be easier to find an unavoidable set of reducible configurations.) In spirit, this proof is similar to the celebrated false proof of the 4CC given by Kempe [K5], using Kempe chains, i.e. subgraphs spanned by the vertices of two colour classes (cf. the second proof of Theorem 1.6, and the proofs of Theorems 1.13, 2.2 and 3.3).:Several papers have been devoted to repairing the flaw in Kempe's reasoning by investigating Kempe-chain arguments. The first comprehensive paper of this kind was written by Birkhoff [B17] on reducibility. This was followed by A. Bernhart [B14], [B15], Errera [E3], Ore and Stemple [OS1], F. Bernhart [B16], Allaire and Swart [AS1], Mayer [M24] and others. More detailed reducibility algorithms were given by Heesch [H24] and Tutte and Whitney [TW1]. Most of the recent work on these algorithms has been carried out by computers.

To show that a set of configurations is unavoidable, Heesch [H25] invented a so called discharging procedure. Again such procedures can become very complicated as in Stromquist [S24] and Osgood [O7] and it seems to be necessary to use computers. An example of a discharging programme is that of Appel and Haken [AH3].

4. SPARSE GRAPHS OF LARGE CHROMATIC NUMBER

The simplest way of ensuring that a graph has large chromatic number is to make it contain large complete subgraphs. It is not at all obvious that a graph without a triangle can have an arbitrarily large chromatic number. Publishing under a pseudonym in the mathematical journal of Cambridge undergraduates, Tutte [D3] (see also [D4]) was the first to prove that this is possible. Tutte's theorem was rediscovered independently by Zykov [Z8] and Mycielski [M39]. Kelly and Kelly [KK1] constructed graphs of girth 6 with arbitrarily high chromatic number. Erdős [E7] (see also [E8]) proved a strong generalization of these results. He showed that for every g and k there are graphs of girth at least g and chromatic number at least k. Using the counting method he proved the existence of these graphs but did not construct them. Such graphs were constructed by Nešetřil [N8] for $g = 8$. Lovász [L16] (see also [L14]) gave a rather complicated constructive proof of the above theorem of Erdős. A simpler constructive proof was given by Nešetřil and Rödl [NR5]. Here we start with a proof based on the original proof of Erdős, since it is not only simple but it also gives an almost best possible relation between the order, the girth and the chromatic number of a graph, which has no constructive proof at present.

In the proofs in this section we shall frequently make use of certain estimates of binomial coefficients. According to Stirling's formula

$$a! = \left(\frac{a}{e}\right)^a \sqrt{2\pi a} \; e^{\alpha/12a},$$

where $0 < \alpha < 1$. Therefore if $0 < b < a$ then

$$\frac{1}{2\sqrt{a}} \left(\frac{a}{b}\right)^b < \frac{1}{2\sqrt{a}} \left(\frac{a}{a-b}\right)^{a-b} \left(\frac{a}{b}\right)^b < \binom{a}{b} < \left(\frac{a}{a-b}\right)^{a-b} \left(\frac{a}{b}\right)^b < \left(\frac{ea}{b}\right)^b. \quad (1)$$

The expansion of the binomial coefficients into quotients of products of factorials gives that if $0 \leqslant x < b$ and $b + x < a$ then

$$\left(\frac{a-b-x}{a-x}\right)^x \leqslant \binom{a-x}{b}\binom{a}{b}^{-1} \leqslant \left(\frac{a-b}{a}\right)^x \leqslant e^{-(b/a)x} \quad (2)$$

and

$$\left(\frac{b-x}{a-x}\right)^x \leqslant \binom{a-x}{b-x}\binom{a}{b}^{-1} \leqslant \left(\frac{b}{a}\right)^x. \quad (3)$$

Let us make one more remark before stating and proving the theorem of Erdős. In the proofs we shall use the language of probability. This is purely for *convenience* since we do nothing more than estimate the number of elements in certain subsets of a set. We shall take a finite set $\mathscr{G}$ of graphs and consider $\mathscr{G}$ as a probability space whose elements have equal probability. All the graph invariants, e.g. chromatic number, vertex independence number, clique number, girth, the number of cycles C^l, etc., are then random variables on $\mathscr{G}$.

For $2 \leqslant k \leqslant n$ denote by $g(n; \chi \geqslant k)$ the *maximal girth* of a graph of order n with chromatic number at least k. Clearly $g(n; \chi \geqslant 2) = \infty$ and $g(n; \chi \geqslant 3) = 2\lceil n/2 \rceil - 1$ $(n \geqslant 3)$ so to avoid these trivial cases in the sequel we shall always suppose that $k \geqslant 4$.

THEOREM 4.1.

$$g(n; \chi \geqslant k) \geqslant \left\lfloor \tfrac{1}{4} \frac{\log n}{\log k} \right\rfloor.$$

Proof. Since $g(G) \geqslant 3$ for every graph G, we may assume that the right-hand side of the inequality at hand is at least 4 and so $n \geqslant k^{16} \geqslant 2^{32}$. This shows that even the simplest non-trivial case of this theorem concerns rather large graphs. Put

$$g_0 = \left\lfloor \tfrac{1}{4} \frac{\log n}{\log k} \right\rfloor \geqslant 4.$$

Denote by $\mathscr{G}$ the set of all graphs with a given set V of n vertices and with $m = k^3 n$ edges. Note that $|\mathscr{G}| = \binom{\binom{n}{2}}{m}$. Denote by $N(l)$ the *expected number* of cycles of length l, where $3 \leqslant l < g_0$. There are

$$\binom{n}{l} \frac{l!}{2l} < \frac{1}{2l} n^l$$

ways of choosing a cycle C^l with $V(C^l) \subset V$. Each of these cycles is contained in

$$\binom{\binom{n}{2} - l}{m - l}$$

of the graphs, so

$$N(l) < \frac{1}{2l} n^l \binom{\binom{n}{2} - l}{m - l} \binom{\binom{n}{2}}{m}^{-1} \leqslant \frac{1}{l} n^l \left(\frac{2m}{n^2} \right)^l = \frac{1}{l} 2^l k^{3l},$$

where we have made use of (3). Hence

$$N = \sum_{3 \leqslant l < g_0} N(l) < \frac{1}{g_0} 2^{g_0} k^{3g_0} \leqslant \frac{1}{g_0} n^{\frac{1}{4}} n^{\frac{3}{4}} = \frac{1}{g_0} n^{\frac{7}{8}} = f.$$

Denote by $M(q)$ the expected number of sets $W \subset V, |W| = p = \lfloor n/(k-1) \rfloor$ such that $G[W]$ has exactly q edges. Clearly

$$M(q) = \binom{n}{p}\binom{\binom{p}{2}}{q}\binom{\binom{n}{2}-\binom{p}{2}}{m-q}\binom{\binom{n}{2}}{m}^{-1} < (ek)^{n/(k-1)}p^{2q}\, e^{-m\binom{p}{2}\binom{n}{2}^{-1}} < n^{2q}k^{n/3}\, e^{-kn}$$

where we have applied (1) and (2). Consequently

$$M = \sum_{q \leqslant 3f} M(q) < n^{6f}k^{n/2}e^{-kn}.$$

Taking the logarithm we see that

$$\log M < \tfrac{3^{n718}}{2}\log n + \tfrac{1}{2}n\log k - kn < -n,$$

so

$$M < e^{-n}.$$

Denote by $\mathscr{G}_1 \subset \mathscr{G}$ the set of all graphs containing at most $3f$ cycles of length less than g_0. As $N < f$, at least two thirds of the graphs in $\mathscr{G}$ belong to $\mathscr{G}_1$, i.e.

$$|\mathscr{G}_1| \geqslant \tfrac{2}{3}|\mathscr{G}| = \tfrac{2}{3}\binom{\binom{n}{2}}{m}.$$

Denote by $\mathscr{G}_2$ the set of all graphs that do not contain a set W of $p = \lfloor n/(k-1) \rfloor$ vertices such that $G[W]$ contains at most $3f$ edges. Since $M < e^{-n} < \tfrac{1}{3}$, at least two thirds of the graphs belong to $\mathscr{G}_2$, i.e.

$$|\mathscr{G}_2| \geqslant \tfrac{2}{3}|\mathscr{G}|.$$

Therefore

$$|\mathscr{G}_1 \cap \mathscr{G}_2| \geqslant \tfrac{1}{3}|\mathscr{G}|,$$

in particular $\mathscr{G}_1 \cap \mathscr{G}_2 \neq \varnothing$. Now we are home since, though the graphs in $\mathscr{G}_1 \cap \mathscr{G}_2$ need not show the truth of the theorem, starting with each graph $G \in \mathscr{G}_1 \cap \mathscr{G}_1$ we can omit at most f of its edges in such a way that the resulting graph, G^* say, satisfies $g(G^*) \geqslant g$ and $\chi(G^*) \geqslant k$. For if $G \in \mathscr{G}_1 \cap \mathscr{G}_2 \subset \mathscr{G}_1$ then one can find f edges in G whose omission results in a graph G^* of girth at least

$$g_0 = \left\lfloor \tfrac{1}{4}\frac{\log n}{\log k} \right\rfloor.$$

Furthermore, $G \in \mathscr{G}_2$ implies that $\beta_0(G^*) < \lfloor n/(k-1) \rfloor$. Hence $\chi(G^*) \geqslant n/\beta_0(G^*) \geqslant k$. ∎

The bound given in Theorem 4.1 is essentially best possible, as pointed out by Erdös [E10]. For if $\chi(G) \geqslant k$ and $|G| = n$ then G contains a k-critical subgraph H of order at most n. Since $\delta(H) \geqslant k - 1$, Theorem III.1.2 has the following consequence. (Recall that $4 \leqslant k \leqslant n$.)

THEOREM 4.2.

$$g(n; \chi \geqslant k) \leqslant 2 \frac{\log n}{\log (k-2)} + 1.$$ ∎

We say that a graph G has property $P(\beta_0, c)$ if

$$\beta_0(H) \geqslant c|H|$$

for *every* subgraph H of G. (Naturally, we may consider only induced subgraphs.) Clearly if G has property $P(\beta_0, \frac{1}{2})$ then G does not contain an odd cycle so $\chi(G) \leqslant 2$. Furthermore, if G has property $P(\beta_0, c)$ and G contains an odd cycle C^{2l+1} then $l/(2l+1) \geqslant c$ since $\beta_0(C^{2l+1}) = l$. In particular, if $c > \frac{1}{3}$ then $g(G) \geqslant 4$. It would not be unreasonable to expect that the chromatic number of a graph with property $P(\beta_0, c)$ cannot be too large if c is only slightly smaller than $\frac{1}{2}$. However, Erdös and Hajnal [EH5] proved that this is not so. They based their proof on the following theorem of Borsuk (see [D27, p. 349]): *the k-dimensional unit sphere S^k cannot be covered by $k + 1$ closed (resp. open) sets none of which contains a pair of antipodal points.*

THEOREM 4.3. *For every $c < \frac{1}{2}$ and natural number k there is a graph G having property $P(\beta_0, c)$ such that $\chi(G) > k$.*

Proof. It is convenient to construct such a graph G with an infinite set of vertices. The vertices of G will be the points of the k-dimensional unit sphere S^k. Choose $\varepsilon > 0$ so small that if F denotes the spherical cap of diameter $2 - \varepsilon$ and $m(S)$ denotes the area of a surface S then

$$cm(S^k) < m(F).$$

For $x, y \in S^k$ let $xy \in E(G)$ iff $|x - y| > 2 - \varepsilon$. A colouring of G then corresponds to a partition of S^k into sets of diameter at most $2 - \varepsilon$. Borsuk's theorem implies that $\chi(G) > k + 1$.

To see that every finite spanned subgraph of G has property $P(\beta_0, c)$ take n points on S^k, say $x_1, x_2, \ldots, x_n$. Denote by F_i the spherical cap of diameter $2 - \varepsilon$ and centre x_i, $i = 1, 2, \ldots, n$. As $\sum_1^n m(F_i) > cn\, m(S^k)$, at least one point of S^k is contained in at least cn of the F_i's, say in $F_{i_1}, F_{i_2}, \ldots, F_{i_t}$, $t \geqslant cn$. Then $x_{i_1}, x_{i_2}, \ldots, x_{i_t}$ are independent vertices of G.

It is very easy to extract a finite example from the infinite one we have just constructed. One way is to invoke a simple theorem of de Bruijn and Erdős [BE12], stating that if $\chi(G) > k$ (k finite) then G contains a finite subgraph H with $\chi(H) > k$. Another theoretically very simple but slightly laborious way is to replace S^k by a large number of uniformly distributed points on it. ■

Another sparse graph of high chromatic number was constructed by Kneser [K7]. Let $T = \{1, 2, \ldots, t\}$ and let r be a natural number. The *Kneser graph* $K_t^{(r)}$ has vertex set $T^{(r)}$ and two r-sets of T are joined in $K_t^{(r)}$ iff they are disjoint. It is clear that $K_{2r+k}^{(r)}$ has a $(k + 2)$-colouring given by the sets $A_1, A_2, \ldots, A_{k+2}$, where $A_i = \{\sigma \in T^{(r)} : \min \sigma = i\}$ for $1 \leqslant i \leqslant k + 1$ and $A_{k+2} = \{k + 2, k + 3, \ldots, k + 2r\}^{(r)}$. Kneser [K7] conjectured that this colouring is best possible, that is $\chi(K_{2r+k}^{(r)}) = k + 2$. The conjecture was open for over twenty years; $K_{2r+k}^{(r)}$ was a fascinating example of a simply constructed graph whose chromatic number could not be determined. It is easily seen that a Kneser graph does not contain short odd cycles but is very rich in even cycles. The property made M. Simonovits wonder whether the Kneser graphs were not closely related to the graphs in Theorem 4.3, constructed by Erdős and Hajnal, in particular whether the conjecture could not be proved by using Borsuk's theorem. This belief was justified when Lovász [L26] gave a proof of Kneser's conjecture, based on Borsuk's theorem and some homotopy theory. Shortly afterwards Bárány [B1a] gave a charmingly simple proof, the proof we are going to give, based on Borsuk's theorem and a theorem of Gale [G1a]. This result of Gale asserts that one can select $2r + k$ open hemispheres of S^k in such a way that every point of S^k is in at least r hemispheres. Here we remark only that this can be proved by induction on k. An r-fold cover of the equator of S^k by $2r + k - 1$ hemispheres gives an r-fold cover of S^k, except for the north and south poles. Move the first hemisphere up a little, the second down a little, the third up, and so on. It is easily shown that these moves can be arranged in such a way that after $2r - 1$ moves we obtain an r-fold cover, except in a small neighbourhood of the south pole, where the cover is only $(r - 1)$-fold. By adding the southern hemisphere we obtain an r-fold cover.

THEOREM 4.4. *The Kneser graph $K_{2r+k}^{(r)}$ has chromatic number $k + 2$.*

Proof. Let $H_1, H_2, \ldots, H_{2r+k}$ be open hemispheres forming an r-fold cover of S^k. Let $V = \{1, 2, \ldots, 2r + k\}^{(r)}$ be the vertex set of $K_{2r+k}^{(r)}$ and suppose c is a $(k + 1)$-colouring of $K_{2r+k}^{(r)}$ with colours $1, 2, \ldots, k + 1$. Put

$$U_i = \bigcup_{c(\sigma) = i} \bigcap_{j \in \sigma} H_j, \qquad i = 1, 2, \ldots, k + 1.$$

Then the choice of the H_j implies that $U_1, U_2, \ldots, U_{k+1}$ is an open cover of S^k. No U_i contains a pair of antipodal points since otherwise V contains disjoint r-sets of colour i. This contradicts Borsuk's theorem. ∎

In the original proof of Theorem 4.4, Lovász [L26] discovered a fascinating connection between the topology of the neighbourhood complex of a graph and its chromatic number. The result is of great interest on its own, especially since it may be followed by other deep results in the same vein. In order to present this theorem of Lovász we need some definitions; for the topological concepts we refer the reader to Dugundji [D27].

The neighbourhood complex $N(G)$ of a graph G is the abstract simplicial complex whose simplices are sets of vertices which are joined to a common vertex of G. Thus the 0-simplices are the vertices of G, provided G has no isolated vertices, and the maximal simplices are the sets $\Gamma(x)$ for $x \in G$. The *first derived complex* $N'(G)$ has, as its 0-simplices, sets of vertices which are joined to a common vertex of G, and, as its r-simplices, the nested sequences of $(r + 1)$ such sets. A simplicial map

$$\Gamma : N'(G) \to N'(G)$$

may be defined by sending a 0-simplex X of $N'(G)$ to the 0-simplex $\Gamma(X) = \bigcap_{x \in X} \Gamma(x)$. The image of Γ is a subcomplex $M(G)$ and Γ^2 gives the identity map on $M(G)$. Hence the corresponding map of the underlying polyhedra

$$\Gamma^2 : |N(G)| \to |M(G)|$$

is a retraction.

THEOREM 4.5. *If $|N(G)|$ is $(k - 1)$-connected then $\chi(G) \geqslant k + 2$.*

Proof. We shall first show that there is a continuous map

$$\beta_k : S^k \to |M(G)|$$

with

$$\beta_k(-x) = \Gamma \circ \beta_k(x) \quad \text{for all} \quad x \in S^k. \tag{4}$$

If $k = 0$ we may simply choose $v \in |M(G)|$ and define

$$\beta_0(1) = v, \qquad \beta_0(-1) = \Gamma(v).$$

By induction we may assume that a map

$$\beta_{k-1}: S^{k-1} \to |M(G)|$$

has been constructed. Since $|N(G)|$ is $(k-1)$-connected, so is its retract $|M(G)|$. Thus, if we regard β_{k-1} as being defined on the equator of S^k, then there is a continuous extension of it over the upper hemisphere. This defines β_k on the upper hemisphere of S^k and (4) may be used to extend it to the entire sphere.

Assume that G is coloured with $\{1, \ldots, k+1\}$ and let M_i be the subcomplex of $M(G)$ spanned by the 0-simplices $\Gamma(X)$ where X contains a vertex with colour i. Since adjacent vertices have different colours, the set $\Gamma(X)$ can contain no vertex with colour i and so

$$|M_i| \cap |\Gamma(M_i)| = \varnothing. \tag{5}$$

Now the sets $|M_i|$ form a closed cover of $|M(G)|$, so the sets $\beta_k^{-1}|M_i|$ form a closed cover of S^k. Moreover, conditions (4) and (5) show that no one of these sets contains a pair of antipodal points. Since this contradicts Borsuk's theorem, the proof is complete. ∎

Harary, Hedetniemi and Robinson [HHR1] showed that for every $k \geqslant 3$ there is a uniquely k-colourable graph that does not contain a complete graph of order k. In fact, considerably more is true: a uniquely k-colourable graph may have arbitrarily large girth. This was sketched by Erdös [E27] and Müller [M39] and proved in detail by Bollobás and Sauer [BS2]. Extending this result, Bollobás and Thomason [BT1] studied sparse uniquely partitionable graphs with respect to some property P. Given a graph G and a property P, a (k, P)-*partition* of G is a partition of $V(G)$ into k classes, called the colour classes, each of which induces a subgraph with property P. The minimum k for which there is such a partition is denoted by $\chi_P(G)$. If $\chi_P(G) = k$ and G has exactly one (k, P)-partition then G is said to be *uniquely* (k, P)-*partitionable*. If P is the property of being l-degenerate then in the notation above we simply replace P by l. Thus $\chi_l(G)$ is the lth vertex partition number of G. The rather long proof of the next result is based on the counting method, whose rather simple form was applied in the proof of Theorem 4.1, and makes use of the relations (1), (2) and (3). The graphs under consideration are obtained from the disjoint union of k given graphs by joining them to each other with a given number of "random" edges. For the details of the proof the reader is referred to [BT1].

THEOREM 4.6. *Let P be a property of graphs and let $g \geqslant 3$ be a natural number. Let*

$$m = \max\{e(G^n): g(G^n) \geqslant g \text{ and } G^n \text{ has } P\}.$$

Suppose that if n is sufficiently large then there is a graph G^n with property P for which $g(G^n) \geqslant g$, $e(G^n) \geqslant m - u$ and $\Delta(G^n) \leqslant \Delta$, where $u = O(n^\beta)$ and $\Delta = O(n^{\beta/2g^2})$ for some β, $0 < \beta < \frac{1}{2}g$. Then for every sufficiently large n there is a uniquely (k, P)-partitionable graph of order n with girth at least g. ∎

There are a number of properties P of graphs G that trivially satisfy the conditions of the theorem above, including the following properties: (i) G is planar, (ii) G contains at most l edge disjoint cycles, (iii) $e(G) < l$, (iv) $\Delta(G) \leqslant \Delta$, (v) G is a forest of size at most l. By Theorem 4.6 for these properties P there are uniquely (k, P)-partitionable graphs of large girth. Another property that does satisfy the conditions of Theorem 4.6 is that of being l-degenerate. It is trivial (see Ex. 2(iii)) that an l-degenerate graph of order n has at most $ln - \binom{l+1}{2}$ edges. On the other hand it is not too easy to prove that this maximum can almost be attained on graphs of large girth and bounded maximum degree, as was proved in [BT1].

THEOREM 4.7. *Given natural numbers l and g, there is a constant c such that for every $n \geqslant 1$ there exists an l-degenerate graph $G = G^n$ with $\Delta(G) \leqslant 6l$, $g(G) \geqslant g$ and $e(G) \geqslant (n - c)l$.* ∎

Theorems 4.6 and 4.7 have the following immediate consequence.

THEOREM 4.8. *Given natural numbers k and g, there is a constant n_0 such that for every $n \geqslant n_0$ there is a uniquely (k, l)-partitionable graph of order n and girth at least g.* ∎

Call a graph G *critically uniquely* (k, l)-*partitionable* if it is uniquely (k, l)-partitionable but no proper subgraph of it is. It is somewhat surprising that it is not too easy to show that for given $k \geqslant 3$ and $l \geqslant 0$ there are critically uniquely (k, l)-partitionable graphs of arbitrarily large order. Bollobás and Sauer [BS1] showed the existence of critically uniquely k-colourable (that is, $(k, 0)$-partitionable) graphs of arbitrarily large order and the general case was settled in [BT1]. Indeed, Theorem 4.8 guarantees the existence of a uniquely (k, l)-partitionable graph G with $g(G) \geqslant g > k$. Let H be a critically uniquely (k, l)-partitionable subgraph of G. Since $g(H) \geqslant g$, for $k \geqslant 3$ we have $|H| \geqslant g$, implying the following corollary.

COROLLARY 4.9. *For every* $k \geqslant 3$, $l \geqslant 0$ *and* n *there is a critically uniquely* (k, l)-*partitionable graph of order at least* n. ∎

5. PERFECT GRAPHS

In the previous section we examined graphs with high chromatic number but without small subgraphs of high chromatic number. The aim of this section is to investigate graphs with a more or less converse property: graphs in which the chromatic number of each induced subgraph is only as large as its clique number. These so called perfect graphs were introduced by Berge [B7], [B8], [B9]. Thus a graph G is *perfect* if

$$\chi(H) = \text{cl}(H)$$

for every induced subgraph $H \subset G$. Bipartite graphs and unions of complete k-partite graphs are trivial examples of perfect graphs. Slightly less trivial examples are the complements of bipartite graphs. Berge conjectured that a graph is perfect iff neither itself nor its complement contains a cycle C^{2l+1} ($l \geqslant 2$) without diagonals. Another (clearly weaker) conjecture of Berge was that a graph is perfect iff its complement is. After partial results of Berge [B10], [B8], Berge and Las Vergnas [BL1] and Sachs [S6], Fulkerson [F5] reduced this second conjecture to a simple case of Lemma 5.1. A complete proof of the conjecture was given by Lovász [L18], proving Lemma 5.1 on the way. Later Lovász [L19] gave a simpler proof of the entire conjecture and this is the proof we shall present. In the second part of the section we discuss some results related to the main conjecture of Berge.

Let $G = (V, E)$ and $G_x = (V_x, E_x)$, $x \in V$, be vertex disjoint graphs. Define a graph G^* with vertex set $V^* = \bigcup\limits_{x \in V} V_x$ by joining u to v if $uv \in E_x$ or $u \in V_x$, $v \in V_y$ and $xy \in E$. We say that G^* is obtained from G by *substituting* G_x *for* x, $x \in V$. If for each $x \in V$, G_x consists of $h(x)$ independent vertices then we say that G^* is the result of *multiplying* G *by* h, and we put $G^* = h \cdot G$. Furthermore, if $h(x_0) > 1$ and $h(x) = 1$ for $x \in V - \{x_0\}$ then we may say that $h \cdot G$ is obtained from G by multiplying the vertex x_0 by $h(x_0)$. The definition implies that if $G^* = h \cdot G$ then

$$\mathrm{cl}(G^*) = \mathrm{cl}(G), \qquad \chi(G^*) = \chi(G)$$

and

$$\mathrm{cl}(\overline{G^*}) = \beta_0(G^*) = \max_Z \sum_{x \in Z} h(x), \qquad (1)$$

where $\max\limits_Z$ means that the maximum is taken over all independent sets Z in G. An induced subgraph of G^* can be obtained by multiplying an induced subgraph of G. In particular, the relations above imply that G is perfect iff $G^* = h \cdot G$ is perfect for some h, where $h(x) \geqslant 1$ for all $x \in V$.

In the proof of the next theorem we need an analogue of this last remark. We state and prove it in greater generality than necessary.

LEMMA 5.1. *Let G^* be obtained from a perfect graph by substituting perfect graphs for some of the vertices. Then G^* is perfect.*

Proof. It suffices to prove that if G^* is obtained from a perfect graph G by substituting the perfect graph H for the vertex $x \in V$ then G^* is perfect. We prove this by induction on $p = \mathrm{cl}(G^*)$. For $p = 1$ the assertion is trivial so let us proceed to the induction step. Put $q = \mathrm{cl}(G), r = \mathrm{cl}(H)$ and denote by s the maximal order of a clique in G containing x. Then

$$p = \max\{q, r + s - 1\}.$$

Take a q-colouring of G and let V_1 be the colour class containing x. Similarly let W_1 be a colour class in an r-colouring of H. Note that every q-clique of G meets V_1 and every r-clique of H meets W_1. Put $Z = (V_1 - \{x\}) \cup W_1$. Then Z is a set of independent vertices of G^*. Let $K = K^p$ be a p-clique of G^*. If $V(K) \cap V(H) \neq \varnothing$ then K contains an r-clique of H so $Z \cap V(K) \neq \varnothing$. If $V(K) \cap V(H) = \varnothing$ then K is a q-clique of G so $Z \cap V(K) \neq \varnothing$. Therefore Z intersects every p-clique of G^*, i.e. $\mathrm{cl}(G^* - Z) \leqslant p - 1$. By the induction hypothesis $G^* - Z$ is $(p - 1)$ colourable. Assigning a new colour to the vertices in Z we obtain a p-colouring of G^*.   ∎

Lemma 5.1 implies in particular that if $\bar{G}$ is perfect and $G_0 = h \cdot G$ then $\bar{G}_0$ is perfect. The following characterisation of perfect graphs was conjectured by A. Hajnal and proved by Lovász [L19].

THEOREM 5.2. *A graph G is perfect if and only if*

$$\text{cl}(G')\beta_0(G') = \text{cl}(G')\,\text{cl}(\bar{G}') \geqslant |G'| \tag{2}$$

for every induced subgraph G' of G. In particular, G is perfect iff $\bar{G}$ is.

Proof. One of the implications is trivial. If G is perfect then

$$\text{cl}(G') = \chi(G') \geqslant |G'|/\beta_0(G')$$

for every induced subgraph G' of G.

To prove the converse implication we use induction on $|G|$. For $|G| = 1$ there is nothing to prove so we proceed to the induction step. Let G be a graph satisfying (2). The induction hypothesis implies that if H is an induced subgraph of G then both H and $\bar{H}$ are perfect. In particular we have to prove only that $\chi(G) = \text{cl}(G)$.

LEMMA 5.3. *If G satisfies (2) and* $G_0 = h \cdot G$ $(h \geqslant 0)$ *then*

$$\text{cl}(G_0)\,\text{cl}(\bar{G}_0) \geqslant |G_0|.$$

Proof. Suppose the lemma is false and let $G_0 = h \cdot G$ a graph of minimal order showing this. Pick a vertex $x \in V$ with $h(x) \geqslant 2$ and let $V_x = \{x_1, x_2, \ldots, x_h\}$. Then

$$\text{cl}(G_0 - x_h)\,\beta_0(G_0 - x_h) \geqslant |G_0| - 1$$

by the minimality of G_0. Consequently

$$\text{cl}(G_0 - x_h) = \text{cl}(G_0) = p, \qquad \text{cl}(\bar{G}_0 - x_h) = \beta_0(\bar{G}_0) = q$$

and

$$|G_0| = pq + 1.$$

G_0 is obtained from $G_1 = G_0 - \{x_2, x_3, \ldots, x_h\}$ by multiplying x_1 by h. Therefore (1) implies that if Z is any set of independent vertices in G_1 *containing* x_1 then

$$|Z| + h - 1 \leqslant \mathrm{cl}(\bar{G}_0) = q. \tag{3}$$

In particular, $h \leqslant q$.

The induction hypothesis implies that $\bar{G} - x$ is perfect. The graph $G_2 = G_0 - V_x$ is obtained from $G - x$ by multiplication so by Lemma 5.1 $\bar{G}_2$ is perfect, implying $\chi(\bar{G}_2) = \mathrm{cl}(\bar{G}_2) \leqslant \mathrm{cl}(\bar{G}_0) = q$. Let $V_1, V_2, \ldots, V_q$ be the colour classes in a q-colouring of $\bar{G}_2$, $|V_1| \geqslant |V_2| \geqslant \ldots \geqslant |V_q|$.

Since $|G_2| = |G_0| - h = pq + 1 - h$,

$$|V_1| = |V_2| = \ldots = |V_{q-h+1}| = p.$$

Put $G_3 = G_0[V_1 \cup \ldots \cup V_{q-h+1} \cup \{x_1\}] \subset G_1 \subset G_0$. Then G_3 is a proper induced subgraph of G_0 so by the induction hypothesis

$$\mathrm{cl}(G_3)\ \mathrm{cl}\,(\bar{G}_3) \geqslant |G_3| = (q - h + 1)p + 1.$$

Since $\mathrm{cl}(G_3) \leqslant \mathrm{cl}(G_0) = p$ this implies that G_3 contains a set Z of $q - h + 2$ independent vertices. As $|Z \cap V_i| \leqslant 1$ for each $1 \leqslant i \leqslant q - h + 1$, we see that $x_1 \in Z$, contradicting (3). ∎

Let us return to the proof of Theorem 5.2. As in the proof of Lemma 5.1, to prove $\chi(G) = \mathrm{cl}(G)$ it suffices to find an independent set Z in G such that $\mathrm{cl}(G - Z) < \mathrm{cl}(G) = p$.

Suppose there is no such independent set Z, i.e. whenever Z is an independent set, $G - Z$ contains a p-clique K_Z. For $x \in V$ denote by $h(x)$ the number of p-cliques K_Z containing x and denote by k the number of independent sets. Putting $G_0 = h \cdot G$ we have

$$|G_0| = \sum_{x \in V} h(x) = \sum_Z |K_Z| = kp,$$

$$\mathrm{cl}(G_0) = \mathrm{cl}(G) = p,$$

and, by (1),

$$\mathrm{cl}(\bar{G}_0) = \max_Z \sum_{x \in Z} h(x) = \max_Z \sum_{Z'} |Z \cap K_{Z'}| \leqslant \max_Z \sum_{Z' \neq Z} 1 = k - 1.$$

Therefore Lemma 5.3 gives the following contradiction:

$$kp = |G_0| \leqslant \mathrm{cl}(G_0)\,\mathrm{cl}(\bar{G}_0) \leqslant p(k-1). \qquad \blacksquare$$

Though it is not known that if neither G nor $\bar{G}$ contains an odd cycle without a diagonal then G is perfect, there are some similar conditions (non-symmetric in G and $\bar{G}$) implying that G is perfect.

Let $C = x_1 x_2 \ldots x_k$ be a cycle in a graph G. If $x_i x_k$ and $x_j x_l$ are diagonals of C (i.e. $x_i x_k, x_j x_l \in E(G) - E(C)$), we say that they *cross* iff $i < j < k < l$ or $j < i < l < k$. The diagonals joining vertices at distance 2 in C are called *shortest* diagonals. A cycle C^l is said to be *triangulated* if there are $l - 3$ diagonals, no two of which cross. If a cycle of length at least 5 is triangulated then it contains at least 2 non-crossing shortest diagonals.

Hajnal and Surányi [HS2] proved that if every cycle of length at least 4 in a graph G contains a diagonal (i.e. if every cycle of G can be triangulated) then G is perfect. Gallai [G3] extended this theorem by showing that a graph is perfect if every *odd* cycle in it can be triangulated. He also pointed out that it is not sufficient to require that every odd cycle has a diagonal (see Fig. 5.1).

FIG. 5.1. A graph G with $\chi(G) = 4$ and $\mathrm{cl}(G) = 3$ in which every odd cycle has a diagonal.

The elegant proof of this theorem of Gallai we shall present is due to Surányi [S26]. It is more convenient to formulate the theorem in terms of the chromatic number and clique number of the complement. After the proof we reformulate the result as a statement about perfect graphs.

THEOREM 5.4. *If every odd cycle of G is triangulated then $\chi(\bar{G}) = \alpha_0(G) = cl(\bar{G})$.*

Proof. We apply induction on $|G|$. As for $|G| \leqslant 3$ the assertion is trivial we proceed to the induction step.

Suppose therefore that $\chi(\bar{G}) > \mathrm{cl}(\bar{G}) = r$. Let $A \subset V$ be a set of r independent vertices in G and let $b_0 \in V - A$. Put $H = G - b_0$. Then $\chi(\bar{H}) = \mathrm{cl}(\bar{H}) = r$.

By a *partition* P we mean a partition of V into $r + 1$ sets: into $\{b_0\}$ and r colour classes of an r-colouring of $\bar{H}$. The colour class containing a vertex x is denoted by $P(x)$ and $\mathscr{P}$ denotes the set of all partitions. If $x \neq b_0$ then b_0 is not joined to every vertex of $P(x)$ ($P \in \mathscr{P}$) since then we could add b_0 to the class $P(x)$ and obtain an r-colouring of $\bar{G}$.

We shall define inductively a sequence of integers $k_{-1} = -1 < k_0 = 0 < k_1 < \ldots < k_p$ a sequence of sets $V_0 = \{b_0\}, V_1, \ldots, V_{k_l} \subset V$, a sequence of sets of partitions $\mathscr{P}_0 = \mathscr{P} \supset \mathscr{P}_1 \supset \ldots \supset \mathscr{P}_l \neq \varnothing$, two sequences of vertices $a_1, a_2, \ldots, a_{k_l} \in A$ and $b_0, b_1, \ldots, b_{k_l} \in V - A$ together with a function $\sigma : \{1, 2, \ldots, k_l\} \to \{0, 1, \ldots, k_{l-1}\}$ with the following properties (see Fig. 5.2).

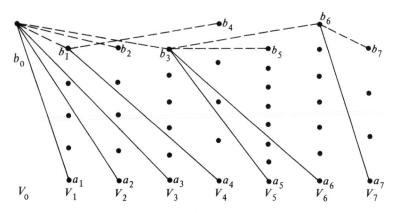

FIG. 5.2. The construction in the proof of Theorem 5.4. A dotted line indicates that the corresponding edge is missing.

(i) $P_l(a_i) = P_l(b_i) = V_i, P_l(b_0) = V_0$ for $P_l \in \mathscr{P}_l$ and $1 \leqslant i \leqslant k_l$,

(ii) $a_i b_i \in E(G)$, $a_i b_{\sigma(i)} \in E(G)$ and $b_i b_{\sigma(i)} \notin E(G)$ for $1 \leqslant i \leqslant k_l$,

(iii) $b_i x \notin E(G)$ for $i \leqslant k_{l-1}$ and $x \in V - \bigcup_1^{k_l} V_j$,

(iv) if V_s is a class of $P_l \in \mathscr{P}_l$ different from $V_0, V_1, \ldots, V_{k_l}$ then no vertex b_i, $0 \leqslant i \leqslant k_p$ is joined to every vertex of V_s.

Suppose the objects have been constructed up to k_m, $m \geqslant 0$. Take those vertices of A which are joined to at least one b_i, $i \leqslant k_m$, and which have not yet been enumerated. If there is no such vertex, finish the construction. Otherwise index these vertices to give the next block: $a_{k_m+1}, a_{k_m+2}, \ldots, a_{k_{m+1}}$. Then for each j, $k_m < j \leqslant k_{m+1}$, there is a vertex $b_{\sigma(j)}$, $\sigma(j) \leqslant k_m$, that justifies the selection of a_j, that is for which $a_j b_{\sigma(j)}$ is an edge. Let $P_{m+1} \in \mathscr{P}_m$ be such that

$$\sum_{i>k_m}^{k_{m+1}} |P_{m+1}(a_i)| \quad \text{is minimal.} \tag{4}$$

Put $\mathscr{P}_{m+1} = \{P \in \mathscr{P}_m : P(a_i) = P_{m+1}(a_i), 1 \leqslant i \leqslant k_{m+1}\}$ and $V_i = P_{m+1}(a_i), k_m < i \leqslant k_{m+1}$. By (iv) the vertex $b_{\sigma(i)}$ is not joined to every vertex of V_i ($k_m < i \leqslant k_{m+1}$). Pick a vertex $b_i \in V_i$ such that $b_i b_{\sigma(i)} \notin E(G)$.

Conditions (i), (ii) and (iii) are automatically satisfied up to k_{m+1} as well. Suppose (iv) is not satisfied, i.e. a vertex b_i ($k_m < i \leqslant k_{m+1}$) is joined to every vertex of a class V_s of P_{m+1}, where V_s is disjoint from $V_0, V_1, \ldots, V_{k_{m+1}}$. Let P'_{m+1} be the partition obtained from P_{m+1} by replacing V_i with $V_i - \{b_i\}$ and V_s with $V_s \cup \{b_i\}$. The existence of this partition P'_{m+1} contradicts (4), so (iv) is also satisfied.

The sequence $k_{-1} < k_0 < k_1 < \ldots$ must come to an end since G is finite. Denote its last term by k_l. (In fact $k_l \leqslant r$.) Put $A^* = \{a_1, a_2, \ldots, a_{k_l}\}$, $B^* = \{b_0, b_1, \ldots, b_{k_l}\}$, $E^* = \{a_i b_i, a_i b_{\sigma(i)} : 1 \leqslant i \leqslant k_l\}$. The subgraph $T = (A^* \cup B^*, E^*)$ of G is a tree and it is also a bipartite graph with vertex classes A^*, B^*. Suppose G contains an edge $b_{i_1} b_{j_1}$, $1 \leqslant i_1 < j_1 \leqslant k_l$. Then T contains a $b_{i_1} - b_{j_1}$ path P which together with $b_{i_1} b_{j_1}$ gives a cycle C in G. The path P is the union of two independent paths of the form $P_1 = b_{i_1} a_{i_1} b_{i_2} a_{i_2} \ldots b_{i_s}$ and $P_2 = a_{j_1} b_{j_1} a_{j_2} b_{j_2} \ldots b_{j_t}$, where $i_s = j_t$, $\sigma(i_m) = i_{m+1}$ and $\sigma(j_m) = j_{m+1}$ (see Fig. 5.3.) This cycle C has length at least 5 and it does *not* contain the

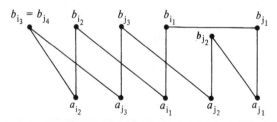

FIG. 5.3. The cycle C in G.

shortest diagonals joining vertices at distance 2 on P_1 and P_2. Hence the only shortest diagonals C might contain are $a_{i_1} b_{j_1}$ and $a_{j_1} b_{i_1}$. These two diagonals cross so C does not contain two non-crossing shortest diagonals. Thus C cannot be triangulated, contradicting the assumption of the theorem. Consequently B^* is a set of independent vertices. As there is no edge joining B^* to

$A - A^*$, so is $A' = (A - A^*) \cup B^*$. Since $|B^*| = |A^*| + 1$ we have $|A'| \geqslant |A| + 1 = r + 1$, and this contradicts $\beta_0(G) = \chi(\overline{G}) = r$. ∎

Recalling Theorem 5.2 we can reformulate Theorem 5.4 as follows.

THEOREM 5.4'. *If every odd cycle in a graph can be triangulated then the graph is perfect.* ∎

6. RAMSEY TYPE THEOREMS

In 1928 Ramsey [R6] proved two remarkable theorems which have found a great number of applications in various branches of mathematics. Restricted to graphs, the second of these theorems states that if k and l are natural numbers, n is sufficiently large (depending on k and l) and $G = G^n$, then either G contains a complete graph of order k or else the complement of G contains a complete graph of order l. Putting it another way,

$$R(k, l) = \inf\{n: \text{ if } |G| = n \text{ and } cl(G) < k \text{ then } cl(\overline{G}) \geqslant l\} \qquad (1)$$

is finite. The theorems themselves have been generalized and extended in numerous directions and, especially in set theory, have given rise to deep results. The extensive literature of Ramsey type theorems could hardly be covered in a book, let alone in a short section. However, since the determination of $R(k, l)$ is a natural extremal problem in graph theory, any book on extremal graph theory must contain some Ramsey type theorems.

This section, rather artificially included in the chapter on colourings, is devoted to the most basic and simplest Ramsey theorems and to a powerful and deep theorem of Nešetřil and Rödl [NR2] on the existence of certain graphs with the so called Ramsey property. Most Ramsey type theorems in graph theory concern the estimation of certain constants (the simplest of which is $R(k, l)$) whose existence is a trivial consequence of Ramsey's original theorem. We do not attempt to give an account of these estimates. The theorem of Nešetřil and Rödl is of an entirely different nature. The existence of the graphs constructed by Nešetřil and Rödl does not follow from any other Ramsey type result. The estimates of the minimal orders of these graphs are hopelessly bad at present but it is remarkable that such graphs exist at all.

Let us begin with the two theorems due to Ramsey [R6]. Recall that if X is a set, $X^{(r)}$ denotes the set of all r-sets of X (i.e. subsets of X with r elements).

If in a colouring of $X^{(r)}$ the elements of $Y^{(r)}$, $Y \subset X$, have the same colour (say c_i) then with a slight abuse of terminology we say that Y is *monochromatic* (with colour c_i).

THEOREM 6.1. *Let r and k be natural numbers, let A be an infinite set and let c be a k-colouring of $A^{(r)}$ with colours $c_1, c_2, \ldots, c_k$. Then A contains a monochromatic infinite set.*

Proof. Though it is trivial to see that it suffices to prove the theorem for $k = 2$ we do not bother to make this simplification.

The theorem is trivial when $r = 1$ and we prove it for all values of r by induction. Let us assume it, therefore, when $r = t - 1$ and deduce it for $r = t$. We shall define an infinite subset $X = \{x_1, x_2, \ldots\}$ of A and a nested sequence $B_0 \supset B_1 \supset \ldots$ of infinite subsets of A, together with a k-colouring $\tilde{c}$ of X with colours $c_1, c_2, \ldots, c_k$. We put $B_0 = A$ and pick $x_1 \in B_0$ arbitrarily. Suppose we have defined $x_1, x_2, \ldots, x_l$ and $B_0 \supset B_1 \supset \ldots \supset B_{l-1}$ such that $x_i \in B_{i-1}$, $1 \leqslant i \leqslant l$. Set $C_{l-1} = B_{l-1} - \{x_l\}$ and define a k-colouring c' of $C_{l-1}^{(r-1)}$ by putting $c'(\sigma) = c(\sigma \cup \{x_l\})$, $\sigma \in C_{l-1}^{(r-1)}$. By the induction hypothesis C_{l-1} contains a monochromatic infinite subset B_l, say with colour c_i. We put $\tilde{c}(x_l) = c_i$. Then B_l is an infinite subset of $C_{l-1} \subset B_{l-1}$ such that all r-sets obtained by adding x_l to an $(r-1)$-subset of B_{l-1} have the same colour $\tilde{c}(x_l)$.

Having constructed the infinite set X and its k-colouring $\tilde{c}$ the proof is easily completed. Put $X_i = \{x \in X : \tilde{c}(x) = c_i\}$, $i = 1, 2, \ldots, k$. Then $X = \bigcup_1^k X_i$ so for some index i the set X_i is infinite. By construction each r-subset of X_i has colour c_i. ∎

THEOREM 6.2. *Let r, k and s be natural numbers. There exists a natural number R such that if A is a set with R elements then for any k-colouring of $A^{(r)}$ there is a monochromatic s-subset S of A.*

Proof. Assume the theorem fails, i.e. for every n there is a k-colouring $c^{(n)}$ of $[1, n]^{(r)}$, $[1, n] = \{1, 2, \ldots, n\}$, in which there is no monochromatic s-subset. For convenience we assume that $c_1, c_2, \ldots, c_k$ are the colours in each k-colouring $c^{(n)}$.

Let $\rho_1, \rho_2, \ldots$ be an enumeration of the r-sets of $\mathbb{N} = \{1, 2, \ldots\}$. We define a nested sequence $I_0 = \mathbb{N} \supset I_1 \supset \ldots$ of infinite sets and a colouring $\tilde{c}$ of $\mathbb{N}^{(r)}$ as follows. Suppose we have defined $I_0 \supset I_1 \supset \ldots \supset I_m$ and have specified the values of $\tilde{c}$ on $\{\rho_1, \rho_2, \ldots, \rho_m\}$, $m \geqslant 0$. Let I_{m+1} be be an infinite subset of I_m such that each colouring $c^{(i)}$, $i \in I_{m+1}$, gives the same colour to ρ_{m+1}, say c_j. Put $\tilde{c}(\rho_{m+1}) = c_j$.

Then for every l-subset L of $\mathbb{N}$ there is a colouring $c^{(n)}$ that coincides with $\tilde{c}$ on $L^{(r)}$. Consequently in the colouring $\tilde{c}$ of $\mathbb{N}^{(r)}$ there is no monochromatic l-subset, contradicting Theorem 6.1.   ∎

Denote by $R_k^{(r)}(s)$ the minimal value of R that will do in Theorem 6.2. The number $R_k^{(r)}(s)$ is said to be the *Ramsey number with parameters r, k and s*. The proof above does not give any bound on $R_k^{(r)}(s)$ though it is straightforward to rewrite it in such a way that it does. In order to obtain a better bound on $R_k^{(r)}(s)$ it turns out to be more convenient to introduce Ramsey numbers depending on more parameters. Namely let $R_k^{(r)}(s_1, s_2, \ldots, s_k)$ be the *minimal* integer R for which every k-colouring of $[1, R]^{(r)}$ with colours $c_1, c_2, \ldots, c_k$ contains a monochromatic s_i-subset of colour c_i for some i. Clearly $R_k^{(r)}(s_1, s_2, \ldots, s_k)$ is invariant under a permutation of the variables $s_1, s_2, \ldots, s_k$. To simplify the notation we omit those of the indices r and k that are equal to 2, i.e.

$$R^{(r)}(s_1, s_2) = R_2^{(r)}(s_1, s_2),$$

$$R_k(s_1, s_2, \ldots, s_k) = R_k^{(2)}(s_1, \ldots, s_k)$$

and

$$R(s_1, s_2) = R_2^{(2)}(s_1, s_2),$$

as defined in (1). Theorem 6.2 implies that $R_k^{(r)}(s_1, s_2, \ldots, s_k)$ exists for each choice of $r, k, s_1, s_2, \ldots, s_k$. However, the existence (i.e. finiteness) of these numbers will also follow from the next result whose simple proof is independent of Theorems 6.1 and 6.2. Note that for $k \geqslant 3$ we have

$$R_k^{(r)}(s_1, \ldots, s_k) \leqslant R_{k-1}^{(r)}(s_1, s_2, \ldots, s_{k-2}, R^{(r)}(s_{k-1}, s_k)). \tag{2}$$

To see this, replace a k-colouring of $[1, N]^{(r)}$ with colour classes $C_1, C_2, \ldots, C_k$ with the $(k - 1)$-colouring whose colour classes are $C_1, \ldots, C_{k-2}, C_{k-1} \cup C_k$. Because of inequality (2) it would be sufficient to give bounds on $R^{(r)}(s_1, s_2)$. Furthermore, the definition of $R_k^{(r)}(s_1, \ldots, s_k)$ implies immediately that if $p < r \leqslant s_i$ then

$$R_k^{(r)}(p, s_2, \ldots, s_k) = p$$

$$R_k^{(r)}(r, s_2, s_3, \ldots, s_k) = R_{k-1}^{(r)}(s_2, s_3, \ldots, s_k), \tag{3}$$

$$R^{(r)}(r, s_2) = R_1^{(r)}(s_2) = s_2 \tag{4}$$

$$R^{(1)}(s_1, s_2) = s_1 + s_2 - 1. \tag{5}$$

In view of inequalities (3), (4) and (5) the following fundamental inequality, due to Erdös and Szekeres [ES6] and rediscovered by Greenwood and Gleason [GG2], gives an upper bound on any Ramsey number $R_k^{(r)}(s_1, \ldots, s_k)$ and so also proves its existence.

THEOREM 6.3. *If* $s_i > r$, $i = 1, 2, \ldots, k$, *then*

$$R_k^{(r)}(s_1, s_2, \ldots, s_k) \leqslant R_k^{(r-1)}\{R_k^{(r)}(s_1 - 1, s_2, \ldots, s_k),$$

$$R_k^{(r)}(s_1, s_2 - 1, s_3, \ldots, s_k) \ldots, R_k^{(r)}(s_1, \ldots, s_{k-1}, s_k - 1)\} + 1. \tag{6}$$

Furthermore, if $s_1 > 2$ *and* $s_2 > 2$ *then*

$$R(s_1, s_2) \leqslant R(s_1 - 1, s_2) + R(s_1, s_2 - 1), \tag{7}$$

and

$$R(s_1, s_2) \leqslant \binom{s_1 + s_2 - 2}{s_1 - 1}. \tag{8}$$

Proof. Denote by $R + 1$ the right-hand side of (6). Let c be a k-colouring of $A^{(r)}$ with colours $c_1, \ldots, c_k$ where $|A| > R$. Pick an element $a \in A$ and put $B = A - \{a\}$. Define a k-colouring $\tilde{c}$ of $B^{(r-1)}$ by putting $\tilde{c}(\sigma) = c(\sigma \cup \{a\})$, $\sigma \in B^{(r-1)}$. Since $|B| \geqslant R$, we may assume without loss of generality that B contains a monochromatic s-set C_1 with colour c_1, where $s = R_k^{(r)}(s_1 - 1, s_2, \ldots, s_k)$. The k-colouring c restricts to a k-colouring of $C_1^{(r)}$. The definition of $R_k^{(r)}(s_1 - 1, s_2, \ldots, s_k)$ implies that C_1 contains a monochromatic set D such that either $|D| = s_1 - 1$ and D is of colour c_1 or else $|D| = s_i$ for some i, $2 \leqslant i \leqslant k$, and D is of colour c_i. In the latter case we are home since D is a monochromatic s_i-set of A with colour c_i. To complete the proof of (6) we note that in the first case the set $D \cup \{a\}$ is a monochromatic s_1-set of A with colour c_1.

Inequality (7) follows from (5) and (6) and inequality (8) can be proved by induction if one makes use of (7). ∎

In general the bounds obtainable from Theorem 6.3 are very far from the actual value of $R_k^{(r)}(s_1, s_2, \ldots, s_k)$ and the determination or even the estimation of the Ramsey numbers is rather difficult. There are a great many results giving bounds on various Ramsey numbers, especially $R(s_1, s_2)$, but most proofs are long case by case examination proofs and since the space available is rather limited, we prefer to omit them. Here we note only two important relations. The first is

$$c_1 \frac{s^2}{(\log s)^2} \leqslant R(3, s) \leqslant c_2 s^2 \frac{\log \log s}{\log s},$$

where c_1 and c_2 are positive absolute constants. The lower bound is due to Erdös [E8] (see Ex. 44) and the upper bound was proved by Graver and Yackel [GY1]. In the same note Graver and Yackel estimated various specific Ramsey numbers. Furthermore, Yackel [Y1], [Y2] proved the existence of a constant c such that for every given s_1 and sufficiently large s_2 (depending on s_1) we have

$$R(s_1, s_2) \leqslant c \left(\frac{\log \log s_2}{\log s_1} \right)^{s_1 - 2} s_2^{s_1 - 2}.$$

In order to formulate some Ramsey type problems and results in which some of the parameters are graphs instead of natural numbers, we introduce the following notations. (The original version of the arrow notation was introduced by Erdös and Rado [ER1] in the study of Ramsey problems concerning large cardinals.) Let $H, G_1, G_2, \ldots, G_k$ be graphs. Denote by $H \to (G_1, \ldots, G_k)$ the following statement: given any k-colouring of the *edges* of H with colours $c_1, c_2, \ldots, c_k$ there is an index i such that H contains a subgraph isomorphic to G_i whose edges have colour c_i (shortly H contains a monochromatic G_i of colour c_i or H contains a c_i-chromatic G_i). We denote by $H \overset{v}{\to} (G_1, \ldots, G_k)$ the statement obtained from the one above if instead of edges we colour the vertices. As usual, if $G_1 = G_2 = \ldots = G_k = G$ then instead of $H \to (G_1, \ldots, G_k)$ and $H \overset{v}{\to} (G_1, \ldots, G_k)$ we write $H \to (G)_k$ and $H \overset{v}{\to} (G)_k$. We say that a k-colouring c of the *edges* of G is *linear* if we can find an order $x_1 < x_2 < \ldots < x_n$ of the vertices together with a k-colouring c' of the vertices (depending on c and the order) such that $c(x_i x_j) = c'(x_j)$ if $x_i x_j \in E(G)$ and $i < j$. Finally we write $H \overset{\text{lin}}{\longrightarrow} (G)_k$ to denote the statement that for any k-colouring of the *edges* of H there is a *linearly coloured* subgraph G' of H isomorphic to G.

Theorem 6.2 restricted to graphs states that if $s_1, s_2, \ldots, s_k$ are natural numbers and n is sufficiently large (depending on $s_1, s_2, \ldots, s_k$) then $K^n \to$

$(K^{s_1}, K^{s_2}, \ldots, K^{s_k})$. In fact

$$R_k(s_1, s_2, \ldots, s_k) = \min\{n : K^n \to (K^{s_1}, K^{s_2}, \ldots, K^{s_k})\}.$$

Replacing the complete graphs $K^{s_1}, K^{s_2}, \ldots, K^{s_k}$ by arbitrary graphs G_1, $G_3, \ldots, G_k$ in the formula above, we obtain the so called *generalized Ramsey numbers*:

$$r(G_1, G_2) = \min\{n : K^n \to (G_1, G_2)\},$$

$$r_k(G_1, G_2, \ldots, G_k) = \min\{n : K^n \to (G_1, G_2, \ldots, G_k)\}.$$

The existence of these generalized Ramsey numbers is an *immediate* consequence of Theorem 6.2. In fact clearly

$$\max_{1 \leqslant i \leqslant k} |G_i| \leqslant r_k(G_1, G_2, \ldots, G_k) \leqslant R_k(|G_1|, |G_2|, \ldots | |G_k|).$$

The first non-trivial result concerning generalized Ramsey numbers is due to Gerencsér and Gyárfás [GG1] who proved that if $m \geqslant n \geqslant 2$ then

$$r(P^m, P^n) = m + \left\lfloor \frac{n}{2} \right\rfloor - 1.$$

(Recall that P^k is the path of order k, i.e. length $k - 1$). However, the systematic study of generalized Ramsey numbers was proposed only some years later by Cockayne [C10a]. The problem was rediscovered and publicised by Chvátal and Harary [CH5], [CH6]. Since then numerous, and in some cases quite surprising, results have been obtained about generalized Ramsey number. Burr gives an account of some of these results in a recent survey article [B54].

The other natural generalization of Ramsey's theorem has a completely different nature. It does not concern itself with the size of the constant appearing in the result but the *existence* part of Ramsey's theorem. Quite often the arising objects have considerably more structure than graphs or set systems. A classical example of such a Ramsey type theorem is Van der Waerden's theorem [V2] stating that if $N \geqslant N(r)$ then in any 2-colouring of the integers $1, 2, \ldots, N$ there is a monochromatic arithmetic progression of length r. These generalizations of Ramsey's theorem are said to form the *partition theory*. Most of the problems in partition theory concerning finite graphs are of the following type. Let $G_1, G_2, \ldots, G_k$ be given graphs and let $\mathcal{H}$ be a family of graphs. Is there a graph $H \in \mathcal{H}$ such that $H \to (G_1, G_2, \ldots, G_k)$? E.g., given an integer $r, r \geqslant 3$, is there a graph H with $\text{cl}(H) = r$ such that $H \to (K^r)_2$? This question was raised by Galvin (see [EH4]) and is highly nontrivial even for $r = 3$. In fact it is very easy to find a graph H with $\text{cl}(H) = 5$ such that $H \to (K^3)_2$ (cf. Ex. 41) but already the condition $\text{cl}(H) = 4$ is not easily satsified.

We end this section by presenting a proof of a deep theorem of Nešetřil and Rödl [NR2] answering in the affirmative a generalization of the question above. The theorem, extending a number of weaker results of Folkman [F2], Erdös, Hajnal and Pósa [EHP1], Rödl [R12], Nešetřil and Rödl [NR1], and Deuber [D2], states that for every G and r there exists a graph H such that $H \to (G)_r$ and $\text{cl}(H) = \text{cl}(G)$. This theorem will be an easy consequence of analogous results concerning $\overset{v}{\to}$ and $\overset{\text{lin}}{\longrightarrow}$. To start with we note a consequence of Theorem 6.2.

LEMMA 6.4. *Let* $q_1, q_2, s_1, s_2,$ *and* k *be natural numbers. Then there are integers* R_1 *and* R_2 *such that if* $[1, R_1]^{(q_1)} \times [1, R_2]^{(q_2)}$ *is* k*-coloured then there are sets* $S_1 \subset [1, R_1]$, $S_2 \subset [1, R_2]$, $|S_1| = s_1$, $|S_2| = s_2$ *for which* $S_1^{(q_1)} \times S_2^{(q_2)}$ *is monochromatic.*

Proof. Note that $R_1 = R_k^{(q_1)}(s_1)$ and $R_2 = R_r^{(q_2)}(s_2)$ have the required properties where $r = k^{\binom{R_1}{q_1}}$. ∎

The main idea of the construction is the use of *interlace graphs*, which are generalizations of the edge graphs of Erdös and Hajnal [EH2]. Let X be a linearly ordered set. Usually X will be a subset of the natural numbers. Let M_1, M_2, N_1, N_2 be p-sets of X. We say that M_1 and M_2 have the same *type of interlacing* as N_1 and N_2 if there is an order preserving map ϕ: $M_1 \cup M_2 \to N_1 \cup N_2$ such that for each $i = 1, 2$ we have $\phi(M_i) = N_{j(i)}$ where $\{j(1), j(2)\} = \{1, 2\}$. Denote by $t[M_1, M_2]$ the type of M_1 and M_2, and denote by $\mathcal{L}_{(p)}$ the set of all types of interlacing of p-sets (of ordered sets). If $L \subset \mathcal{L}_{(p)}$ let $\langle X, L, p \rangle$ be the graph with vertex set $X^{(p)}$ in which $M_1 M_2$ is an edge $(M_1, M_2 \in X^{(p)})$ iff $t[M_1, M_2] \in L$. The graph $\langle X, L, p \rangle$ is said to be an *interlace graph*. To simplify the notation we put $\langle n, L, p \rangle = \langle [1, n], L, p \rangle$. Define furthermore $\text{cl}(L) = \sup_n \text{cl}(\langle n, L, p \rangle)$.

Let G be a graph, X an ordered set and $\phi: V(G) \to X^{(p)}$ a map. With a slight abuse of notation put

$$\phi(G) = \{t[\phi(x), \phi(y)]: xy \in E(G)\} \subset \mathcal{L}_{(p)}.$$

Note that ϕ induces a homomorphism $\bar{\phi}: G \to \langle X, \phi(G), p \rangle$. If ϕ is an injection than $\bar{\phi}$ is an embedding.

If $M \in X^{(p)}$ and $1 \leqslant i \leqslant p$ then $M(i)$ denotes the ith element of M (in the natural order, induced by X). For $Q \subset [1, p]$ we put $M(Q) = \{M(i): i \in Q\}$. Furthermore, if $L \subset \mathscr{L}_{(p)}, |Q| = q$, then define

$$L(Q) = \{t[M(Q), N(Q)]: \; t[M, N] \in L\} \subset \mathscr{L}_{(q)}.$$

We are ready to prove the remaining three lemmas needed in the proofs of the theorems.

LEMMA 6.5. *Suppose* $L \subset \mathscr{L}_{(p)}, Q \subset [1, p]$ *and if* $t[N, M] \in L$ *then* $N(Q) \neq M(Q)$. *Then*

$$\mathrm{cl}(L(Q)) \geqslant \mathrm{cl}(L).$$

Proof. Suppose $M_1, M_2, \ldots, M_r$ are the vertices of a K^r in $\langle n, L, p \rangle$. Then $M_1(Q), M_2(Q), \ldots, M_r(Q)$ are the vertices of a K^r in $\langle n, L(Q), q \rangle, q = |Q|$. ∎

LEMMA 6.6. *For every graph G there are n, p and an injection* $\phi: V(G) \to [1, n]^{(p)}$ *such that*

$$\mathrm{cl}(\phi(G)) = \mathrm{cl}(G).$$

Proof. The lemma is trivial when $|G| = 1$ and we prove it for every graph G by induction on $|G|$. Let us assume it, therefore, for graphs of order less than $|G| (|G| > 1)$. Pick any vertex $x \in G$. Put $G_1 = G - x$ and $G_2 = G[\Gamma(x)]$. By the induction hypothesis we have injections $\phi_1: V(G_1) \to [1, n_1]^{(p_1)}$ and $\phi_2: V(G_2) \to [n_1 + 1, n_2]^{(p_2)}$ such that

$$\mathrm{cl}(\phi_i(G_i)) = \mathrm{cl}(G_i), i = 1, 2.$$

Let $V(G) - \{x\} \cup \Gamma(x) = \{z_1, z_2, \ldots, z_l\}$ and set $n = n_2 + l(p_2 + 1) + 1 + p_1 + p_2$ and $p = p_1 + p_2 + 1$. For $y \in G$ define

$$\phi(y) = \begin{cases} \phi_1(y) \cup \phi_2(y) \cup \{n_2 + l(p_2 + 1) + 1\} & \text{if } y \in \Gamma(x), \\ \phi_1(y) \cup [(j-1)(p_2+1) + 1, j(p_2+1)] & \text{if } y = z_j, \\ [n_2 + l(p_2 + 1) + 1, n] & \text{if } y = x. \end{cases}$$

It can be checked that the injection $\phi\colon V(G) \to [1,n]^{(p)}$ satisfies $\mathrm{cl}(\phi(G)) = \mathrm{cl}(G)$. ∎

LEMMA 6.7. *Let G_0 be a graph and let $G_1, G_2, \ldots, G_s$ be some induced subgraphs of order at least 2. Then there are n, p and an injection $\phi\colon V(G_0) \to [1,n]^{(p)}$ such that*

(i) $\mathrm{cl}(G_i) = \mathrm{cl}(\phi(G_i))$, $i = 0, 1, \ldots, s$,
(ii) *for $1 \leqslant i \leqslant s$ we have*

$$t[M, N] \in \phi(G_i) \quad \textit{iff} \quad t[M, N] \in \phi(G_0) \quad \textit{and} \quad M(p - s + i) = N(p - s + i).$$

Proof. Put $n_{-1} = 0$ and let $\phi_i\colon V(G_i) \to [n_{i-1} + 1, n_i]^{(p_i)}$ be the injection guaranteed by Lemma 6.6. We would like to fit them together and use new elements to "index" the images of the vertices so that (ii) is also satisfied. To do this let $\alpha_i\colon V(G_0) \to [n_{i-1} + 1, n_i]^{(p_i)}$ be an arbitrary extension of ϕ_i.

Put $n = n_s + sm$, where $m = |G_0|$ and $p = \sum_0^s p_i + s$. For $1 \leqslant i \leqslant s$ let $\beta_i\colon V(G_0) \to [n_s + (i-1)m + 1, n_s + im]$ be a map such that if $x \neq y$ then $\beta_i(x) = \beta_i(y)$ iff $x, y \in G_i$.

Define $\phi\colon V(G) \to [1,n]^{(p)}$ by

$$\phi(x) = \bigcup_0^s \alpha_i(x) \cup \bigcup_1^s \beta_i(x), \quad x \in V(G).$$

By the construction of β_i this injection ϕ satifies (ii). To see that (i) holds as well, put $L_i = \phi(G_i)$ and

$$Q_i = \left[\sum_0^{i-1} p_j + 1, \sum_0^i p_j \right].$$

Then $L_i(Q_i) = \phi_i(G_i)$ so by Lemma 6.5 we have

$$\mathrm{cl}(G_i) \leqslant \mathrm{cl}(\phi(G_i)) \leqslant \mathrm{cl}(L_i(Q_i)) = \mathrm{cl}(\phi_i(G_i)) = \mathrm{cl}(G_i). \qquad ∎$$

THEOREM 6.8. *For every graph G and natural number k there is a graph H such that $H \overset{v}{\to} (G)_k$ and $\mathrm{cl}(H) = \mathrm{cl}(G)$.*

Proof. Let $\bar{\phi}\colon G \to \langle n, L, p \rangle$ be an embedding guaranteed by Lemma 6.6 with $\mathrm{cl}(L) = \mathrm{cl}(G)$. We claim that $H = \langle N, L, p \rangle$ has the required properties if $N > R_k^{(p)}(n)$. For, given a k-colouring of $[1, N]^{(p)}$, there exists a set $X \subset [1, N]$,

$|X| = n$, such that all its p-subsets have the same colour. The subgraph of H spanned by these p-subsets is exactly $\langle n, L, p \rangle$ so contains a subgraph isomorphic to G. ∎

THEOREM 6.9. *For every graph G and natural number k there is a graph H such that $H \xrightarrow{\text{lin}} (G)_k$ and* $\text{cl}(H) = \text{cl}(G)$.

Proof. We use induction on $|G|$. The result is trivial for $|G| = 1$ so assume $|G| > 1$ and the theorem holds for graphs of smaller order. Pick a vertex $x \in G$ and put $G_0 = G - x$, $G_1 = G[\Gamma(x)]$. By the induction hypothesis there is a graph H_0 such that $H_0 \xrightarrow{\text{lin}} (G_0)_k$ and $\text{cl}(H_0) = \text{cl}(G_0)$. We may clearly assume that $|G_1| \geqslant 2$ otherwise the assertion of the theorem follows trivially.

Let $\bar{\phi} : H_0 \to \langle n, L, p \rangle$ be the embedding guaranteed by Lemma 6.7 where the induced subgraphs of H_0, say $H_1, H_2, \ldots, H_s$ are all those with $\text{cl}(H_i) < \text{cl}(G), |H_i| \geqslant 2$.

For $N \geqslant n$ we define a graph $H(N)$ as follows:

$$V(H(N)) = [1, N]^{(p)} \cup \{x_{ij} : 1 \leqslant i \leqslant N, 1 \leqslant j \leqslant s\}$$

$$E(H(N)) = E(\langle n, L, p \rangle) \cup \{x_{ij}M : M \in [1, N]^{(p)} \text{ and } M(p - s + j) = i\}.$$

We shall show that for N sufficiently large the graph $H(N)$ has the required properties.

Let us show first that $\text{cl}(H(N)) \leqslant \text{cl}(G)$. Because of Lemma 6.5 we have to check only that if $x_{ij}, M_1, M_2, \ldots, M_l$ are the vertices of a complete subgraph of $H(N)$ then $\text{cl}(G) \geqslant l + 1$. If $1 \leqslant u < v \leqslant l$ then $M_u(p - s + j) = M_v(p - s + j) = i$ and $t[M_u, M_v] \in L$ so by Lemma 6.7 we have $t[M_u, M_v] \in \phi(H_j)$. Thus $l \leqslant \text{cl}(H_j)$ and so $l + 1 \leqslant \text{cl}(G)$.

Suppose now that c is a k-colouring of the edges of $H = H(N)$. Let $Y = \{y_1, \ldots, y_n\} \subset [1, N]$ and let $\alpha_Y : \langle n, L, p \rangle \to \langle Y, L, p \rangle \subset H$ be the embedding induced by the order preserving map $[1, n] \to Y$. This embedding induces a colouring of $\langle n, L, p \rangle$. There are at most $w = k^{\binom{p}{2}}$ different k-colourings of $\langle n, L, p \rangle$. Thus, if A is any given number (to be specified later) and $N > R_w^{(n)}(A)$ then there is an A-subset $Z \subset [1, N]$ all of whose n-subsets induce the same k-colouring of $\langle n, L, p \rangle$. In turn, this k-colouring induces a k-colouring of H_0 by $H_0 \xrightarrow{\bar{\phi}} \langle n, L, p \rangle \xrightarrow{\alpha_Y} \langle Y, L, p \rangle$ where $|Y| = n$. By the choice of H_0 there is a linearly coloured subgraph G'_0 in H_0 that is isomorphic to G_0. Let G'_1 be the subgraph of G'_0 corresponding to the subgraph G_1 of G_0. Note that $G'_1 = H_{j_0}$ for some $j_0 \in [1, s]$ and there is an index i_0 such that if $x \in H_{j_0}$ and $M = \phi(x)$ then $M(p - s + j_0) = i_0$.

Now consider all sets $T \in Z^{(p)}$ such that $T(p - s + j_0) = Z(\lfloor A/2 \rfloor) = a$. Each such set T gives rise to two sets, say T_1 and T_2, such that $T = T_1 \cup \{a\} \cup T_2$, $T_1 \subset [1, a-1]$, $T_2 \subset [a+1, N]$, $|T_1| = t_1 = p - s + j_0 - 1$ and $|T_2| = t_2 = s - j_0$. Let us colour each such pair (T_1, T_2) with the colour of the edge $x_{aj_0} T$. In this way we obtain a k-colouring of $Z_1^{(t_1)} \times Z_2^{(t_2)}$, where $Z_1 = Z \cap [1, a-1]$, $Z_2 = Z \cap [a+1, N]$. Lemma 6.4 implies that if A is sufficiently large (depending only on k, p and n) then there are subsets $U \subset Z_1$, $V \subset Z_2$, $|U| = i_0 - 1$ and $|V| = n - i_0$, such that all pairs (T_1, T_2) have the same colour, $T_i \subset Z_i$, $|T_i| = t_i$, $i = 1, 2$. Let G_i'' be the image of G_i' under the embedding

$$G_1' \subset G_0' \subset H_0 \xrightarrow{\Phi} \langle n, L, p \rangle \xrightarrow{\alpha_{U \cup \{a\} V \cup V}} \langle U \cup \{a\} \cup V, L, p \rangle$$

$$\subset \langle Z, L, p \rangle \subset H = H(N).$$

The G_0'' is linearly coloured by its choice and all the edges $x_{aj_0}M$, $M \in G_1''$, have the same colour. Consequently the subgraph of H induced by $V(G_0'') \cup \{x_{aj_0}\}$, which is isomorphic to G, has a linear colouring. ∎

The main theorem we have been seeking is an immediate consequence of the last two theorems.

THEOREM 6.10. *For every graph G and natural number k there is a graph H such that*

$$H \to (G)_k \quad and \quad cl(H) = cl(G).$$

Proof. By Theorems 6.8 and 6.9 there are graphs F and H such that $F \xrightarrow{v} (G)_k$, $H \xrightarrow{\text{lin}} (F)_k$ and $cl(G) = cl(F) = cl(H)$. Let c be a k-colouring of the edges of H. Then there is a subgraph F' isomorphic to F such that $V(F') = \{x_1, \ldots, x_m\}$ and if $i < j$, $x_i x_j \in E(F')$ then $c(x_i x_j) = \tilde{c}(x_j)$ for some k-colouring of the vertices of F'. In turn F' contains a subgraph G' isomorphic to G whose vertices get the same colour in the colouring $\tilde{c}$. This graph G' is clearly a monochromatic subgraph of H in the edge-colouring c. ∎

7. EXERCISES, PROBLEMS AND CONJECTURES

1^-. Show that G is l-degenerate (see p. 222) iff for some enumeration $x_1, x_2, \ldots, x_n$ of the vertices each x_i is joined to at most l of the vertices preceding it. (Cf. Theorem 1.1 and Corollary 1.2.)

2^-. Let G be a maximal l-degenerate graph of order $n, l > n$. Prove that

(i) $\delta(G) = l$;

(ii) if $x \in G$ and $d(x) = l$ then $G - x$ is maximal l-degenerate;

(iii) $e(G) = ln - \binom{l+1}{2}$;

(iv) there are at least $l + 1$ vertices of degree less than $2l$;

(v) $\text{cl}(G) = l + 1$;

(vi) $\kappa(G) = l$.

(Lick and White [LW1].)

3^-. Let k, l and r be integers such that $l \geqslant 0, r \geqslant 2$ and $k \geqslant \lceil r/(l + 1) \rceil$. Show that there exists a graph G satisfying $\chi_l(G) = k$ and $\text{cl}(G) = r$. (Simões-Pereira [S16].)

4^-. Deduce from Theorem 4.1 that for every $l \geqslant 0, k \geqslant 3$ and $g \geqslant 3$ there is a graph G with $\chi_l(G) = k$ and $g(G) = g$.

5^-. Let $n > \max\{k, l\}, k \geqslant 2, l \geqslant 0$. Show that a maximal (k, l)-partitionable graph is the join of k maximal l-degenerate graphs. Deduce that it is uniquely (k, l)-partitionable. (Simões-Pereira [S17].)

6. Let $J_{k,l}$ be a critically (k, l)-partitionable graph (that is $\chi_l(J_{k,l}) = k$ and $\chi_l(J_{k,l} - x) = k - 1$ for every vertex x). Show that

(i) $\kappa(J_{k,l}) \geqslant \begin{cases} 1 & \text{if } l \neq 1, \\ 2 & \text{if } l = 1, \end{cases}$

(ii) $\lambda(J_{k,l}) \geqslant \begin{cases} k - 1 & \text{if } l \neq 1, \\ 2k - 2 & \text{if } l = 1, \end{cases}$

(iii) $\delta(J_{k,l}) \geqslant (k - 1)(l + 1)$.

Let $L_{k,l}$ be a uniquely (k, l)-partitionable graph.
Show that

(iv) $\kappa(L_{k,l}) \geqslant k - 1$,

(v) $\lambda(L_{k,l}) \geqslant \min\{(k - 1)(l + 1), \binom{k}{2}\lambda_l\}$,

where $\lambda_l = \begin{cases} 1 & \text{if } l \neq 1 \\ 2 & \text{if } l = 1, \end{cases}$

(vi) $\delta(L_{k,l}) \geqslant (k - 1)(l + 1)$.

Let $M_{k,l}$ be a critically uniquely (k, l)-partitionable graph (cf. Corollary 4.9).

Show that

(vii) $\delta(M_{k,l}) \geqslant k(l + 1)$.

(Chartrand and Kronk [CK1], Kronk and Mitchem [KM2], Lick and White [LW1], Simões-Pereira [S17] and Thomason [T1c].)

7^-. Deduce from Ex. 6(iii) that

$$\chi_l(G) \leqslant 1 + \left\lfloor \frac{s(G)}{l+1} \right\rfloor,$$

where $s(G) = \max_{H \subset G} \delta(H) = \min\{s: G \text{ is } s\text{-degenerate}\}$.

8. Let $n = |G|$ and $\Delta = \Delta(G)$. Prove that there is a partition $V(G) = \bigcup_1^k V_i$, $k \leqslant \lfloor \Delta/2 \rfloor + 1$ such that $\Delta(G[V_i]) \leqslant 1$, $i = 1, 2, \ldots, k$, i.e. each $G[V_i]$ consists of independent edges and isolated vertices. (Gerencsér [G11].)

9. Prove the following generalization of Ex. 8.

Let G be an arbitrary graph. Let $d_1, d_2, \ldots, d_k$ be non-negative integers satisfying

$$\sum_1^k d_i = \Delta(G) - k + 1.$$

Then there is a partition $\bigcup_1^k V_i$ of $V(G)$ such that

$$\Delta(G[V_i]) \leqslant d_i, \qquad i = 1, 2, \ldots, k.$$

[Hint. Note first that it suffices to prove the result for $k = 2$. To do the case $k = 2$ consider a partition $V(G) = V_1 \cup V_2$ that minimizes

$$\phi(V_1, V_2) = d_1 e(G[V_2]) + d_2 e(G[V_1]).]$$

(Lovász [L13].)

10. Deduce from Lovász's theorem in Ex. 9 the following extension of the theorem of Brooks.

If $\Delta(G) = \Delta \geqslant 3$ and $\text{cl}(G) = r$ then

$$\chi(G) \leqslant \Delta + 1 - \left\lfloor \frac{\Delta + 1}{\max(4, r + 1)} \right\rfloor.$$

(Borodin and Kostochka [BK1], Catlin [C6].)

11. Prove that if $l \geqslant 1$,

$$r = \max\left\{2, \left\lceil \frac{\mathrm{cl}(G)}{l+1} \right\rceil \right\}, \qquad j = r(l+1)$$

and

$$\Delta + 1 \equiv d \pmod{j+1}, \qquad 0 \leqslant d \leqslant j,$$

then

$$\chi_l(G) \leqslant \left\lfloor \frac{\Delta+1}{j+1} \right\rfloor + \left\lfloor \frac{d-1}{l+1} \right\rfloor + 1.$$

(Bollobás and Thomason [BT1], Borodin and Kostochka [BK1]).

12. PROBLEM. Determine $\chi(\Delta, \mathrm{cl} \leqslant r)$, the *maximum* of the chromatic number of a graph G if $\Delta(G) = \Delta$ and $\mathrm{cl}(G) = \mathrm{r}$. Does $\chi(2r - 2, \mathrm{cl} \leqslant r) \leqslant r + 1$ hold?

In particular, give an upper bound on $\chi(\Delta, \mathrm{cl} \leqslant 2)$, the chromatic number of a triangle free graph with maximal degree Δ.

(Note that the theorem in Ex. 10 gives an upper bound on the chromatic number but it is unlikely to be near to the maximum.)

13^-. Suppose $\Delta(G) < k(l+1)$. Prove that $V(G) = \bigcup_1^k V_i$ such that $\Delta(G[V_i]) \leqslant l$ for every i.

14^+. Prove the following analogue of the theorem of Brooks for (k, l)-partitions.

Let $k \geqslant 2$, $l \geqslant 1$, $(k - 1)(l + 1) < \Delta \leqslant k(l + 1)$, and suppose $\Delta(G) = \Delta$ and $G \neq K^{\Delta+1}$. Then

$$\chi_l(G) \leqslant k = \left\lceil \frac{\Delta}{l+1} \right\rceil.$$

[Note that Ex. 13^- gives a considerably stronger result, unless $\Delta = k(l+1)$, and, as in the first proof of Theorem 1.6, the assertion follows immediately, unless G is $k(l + 1)$-regular.] (The case $l = 1$ was proved by Kronk and Mitchem [KM2], the general case by Mitchem [M31].)

15^+. Prove that for each $n \geqslant (k - 1)(l + 1) + 3$ there is a graph G of order n that is k-critical with respect to χ_l, that is $\chi_l(G) = k$ and $\chi_l(G - x) = k - 1$ for every vertex x. (The interesting part of this result is that the existence of such graphs is not trivial. The case $l = 1$ was proved by Bollobás and Harary [BH1], the general case by Thomason [T1a].)

16. CONJECTURE. Let $\Delta \geqslant 2$ be fixed. Let $\mathscr{G}_n = \{G : V(G) = \{1, 2, \ldots, n\}$

and $\Delta(G) \leqslant \Delta\}$, $\mathcal{H}_n = \{G \in \mathcal{G}_n : \chi'(G) = \Delta + 1\}$. Then $|\mathcal{H}_n|/|\mathcal{G}_n| \to 1$ as $n \to \infty$. As an easy exercise, prove this for $\Delta = 2$.

(Erdös and Wilson [W1] observed that, as $n \to \infty$, almost every graph $G(n, \lfloor n^2/4 \rfloor)$ has a unique vertex of maximum degree. Consequently, as $n \to \infty$, almost every graph G^n satisfies $\Delta(G^n) = \chi'(G^n)$, seemingly contradicting this conjecture.)

17. CONJECTURE. Let us say that a graph G is *critical* if $\chi'(G) = \Delta(G) + 1$ and $\chi'(G - e) = \Delta(G - e)$ for every edge $e \in E(G)$. If G is critical and has no isolated vertices then

$$e(G) \geqslant \tfrac{1}{2}(\Delta - 1)n + \tfrac{3}{2}$$

where $n = |G|$ and $\Delta = \Delta(G)$.

(The conjecture is due to Vizing [V6], [V8], who proved that if G is critical then every vertex that is adjacent to a vertex of degree $k < \Delta = \Delta(G)$ is also adjacent to at least $\Delta + 1 - k$ vertices of degree Δ. Using this result, Fiorini [F1a] proved $e(G) \geqslant \lfloor (\Delta + 1)/4 \rfloor n$.)

18. CONJECTURE. Every critical graph (in the sense of Conjecture 17) without isolated vertices has an *odd* number of vertices. (Beineke and Wilson [BW1], Jakobsen [J1], Wilson and Beineke [WB1].)

19[+]. Given a uniquely 4-edge-colourable graph G different from $K^{1,4}$, construct a uniquely 4-edge-colourable graph H with one of the following two properties:

  (i) H is 4-regular,
  (ii) H has two vertices of degree 2, incident with edge of the same two colours, and every other vertex of H has degree 4.

[Hint. Count the number of end vertices of two-coloured paths.] (Thomason [T24].)

20. CONJECTURE. Every uniquely 3-edge-colourable planar graph of order at least 5 contains a triangle. (Cantoni, see [T20].)

21. PROBLEM. For a fixed $k \geqslant 3$ determine the order of $u_k(n) = \min\{m : \exists G(n, m)$ which is uniquely k-vertex-colourable$\}$. In particular, improve on the trivial relations

$$\frac{k-1}{2} \leqslant \liminf_{n \to \infty} u_k(n)/n \leqslant \limsup_{n \to \infty} u_k(n)/n \leqslant k - 1.$$

22[-]. Let G be a critical k-chromatic graph. Prove by induction on k that any two vertices of G are joined by a path of length at least k. (Dirac [D20].)

23. PROBLEM. Determine the minimum size of a k-critical graph of order n. (Cf. Theorem 2.2.)

In the case $k = 4$ Gallai [G4] proved that this minimum is at least $(\frac{3}{2} + \frac{1}{26})n$ and Dirac [D6] showed that it is at most $\frac{5}{3}n - 4$. To see the latter assertion, form a "path" from l triangles by joining them with $l - 1$ edges, add a vertex and join it to every vertex of degree 2. The obtained graph is critical 4-chromatic, has $3l + 1$ vertices and $5l + 1$ edges.

24. PROBLEM. Determine or give a lower bound for $\min\{e(G^n): G^n$ is k-critical, $\mathrm{cl}(G^n) = r\}$. In particular, give bounds on the quantity above for $r = 2$. (For some results in this direction see Weinstein [W14a].)

25. CONJECTURE. As in §2, put $f_k(n) = \max\{e(G^n): G^n$ is k-edge-critical$\}$. Then

$$\lim_{n \to \infty} f_k(n)/n^2$$

exists.

26^-. Let G be a non-empty graph. Show that the minimal integer p for which there exist k-colourable ($k \geqslant 2$) graphs $G_1, \ldots, G_p$ such that $G = \bigcup_1^p G_i$, is given by the formula

$$p = \lceil \log_k (\chi(G)) \rceil,$$

where $\log_k$ denotes the logarithm to the base k.

27. CONJECTURE. Let k and g be natural numbers. There is a constant $C(k, g)$ such that if $\chi(G) \geqslant C(k, g)$ then G contains a subgraph H with $\chi(H) = k$ and $g(H) \geqslant g$.

Rödl has proved that $C(k, 4)$ exists for every k. (Erdös and Hajnal, see [E27].)

28. PROBLEM. Let G be a graph whose vertices are certain points of the plane and in which two vertices are joined iff they are at distance 1.

(i) What is the maximal chromatic number of such a graph G?

(ii) Is it true that if G does not contain a triangle then it is 3-colourable? (P. Erdös.)

29^-. Denote by $l(G, k)$ the minimum number of edges of G whose omission results in a k-colourable graph. Prove that if $t > k$ then $\min\{l(G, k): \chi(G) = t\}$ is attained for $G = K^t$. (P. Erdös.)

30^-. Denote by $p(G, k)$ the minimum number of edges whose addition to G results in a graph that is not $(k - 1)$-colourable, that is let

$$p(G, k) = \min\{e(H) - e(G): G \subset H, \chi(H) \geqslant k\}.$$

Put $p_k = \max\{p_k(G): \chi(G) < k\}$. Show that for $k \geqslant 4$ we have

$$p_k \leqslant \binom{k - 1}{2}.$$

31. CONJECTURE. (Ex. 30 ctd.)

$$p_k = \binom{k - 1}{2} \quad \text{if} \quad k \geqslant 4.$$

(Note that any graph G with $p_k(G) = \binom{k-1}{2}$ has girth at least $k + 1$, so a graph proving the conjecture has large girth and large chromatic number.)

32. Given a partially ordered finite set P, the graph of P, denoted by $G(P)$, has vertex set P and $x, y \in P$ are joined iff one of them is an immediate successor of the other, i.e. either $x < y$ and no z satisfies $x < z < y$ or else $x > y$ and no z satisfies $x > z > y$. Show that for every k there is a *lattice* L such that

$$\chi(G(L)) \geqslant k.$$

[Hint. Take a graph G satisfying $\chi(G) = k$ and $g(G) \geqslant 4k$. Define a partial order on $V(G)$ and adjoin a maximal element and a minimal element to this partially ordered set.] (Bollobás [B35].)

33. BERGE'S PERFECT GRAPH CONJECTURE. A graph is perfect iff neither the graph nor its complement contains an induced odd cycle of length at least 5. (Berge [B7].)

34^+. Let G be a connected graph such that $\mathrm{cl}(G) \leqslant r - 1$, $\Delta(G) \leqslant r$ and for $r = 4, 5$ $G \neq G_r$, where G_4 and G_5 are the graphs shown in Fig. 7.1.
(i) Prove that $\beta_0(G)/|G|$, the independence ratio of G, is greater than $1/r$.
(ii) Show that if in addition G is not r-regular then G has a non-uniform r-colouring, i.e. an r-colouring in which one of the colour classes has more than $|G|/r$ elements.
[Hint. To do (i) find an independent set X such that $\Gamma(X) \cup X \neq V(G)$ and $|X| > (1/r)|\Gamma(X) \cup X|$. Enlarge X step by step to a desired independent set; cf. Theorem 1.1. Part (ii) is easy.] (Albertson, Bollobás and Tucker [ABT1])

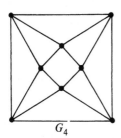 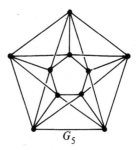

G_4 G_5

FIG. 7.1. The exceptional graphs G_4 and G_5.

35^+. (cf. Ex. 34). Let G be a connected graph such that $G \neq K^{r+1}$ and $\Delta(G) \leqslant r$ $(r \geqslant 3)$. Then G has an r-colouring in which one of the colour classes has $\beta_0(G)$ elements. (Catlin [C5], Mitchem [M31a].)

36^-. Show that if W is a maximal independent set of vertices of G then $\chi(G - W) = \chi(G) - 1$ need not hold.

37. PROBLEM. Prove or disprove the existence of a constant $\varepsilon > 0$ with the following property. If every subgraph H of a graph G can be made bipartite by the omission of at most $\varepsilon|H|$ edges, then $\chi(G) \leqslant 3$. (P. Erdös and A. Hajnal.)

38. PROBLEM. Let $0 < p < 1$ be fixed. Let G_n be random graph of order n in which the probability of an edge is p. Determine

$$\lim_{n \to \infty} \chi(G_n) \frac{\log n}{n}.$$

It was shown by Matula [M19], [M20], Grimmett and McDiarmid [GM1] and Bollobás and Erdös [BE10] that $cl(G_n)$ is almost determined by p and n. This implies that the limit above is between $\frac{1}{2} \log 1/(1 - p)$ and $\log 1/(1 - p)$. The first of these is thought to be the correct value but it has not even been proved that the limit is strictly less than $\log 1/(1 - p)$.)

39. Colour the edges of a K^n in such a way that no vertex is incident with more than d edges of the same colour. Show that in either of the following two cases K^n contains a Hamiltonian cycle with adjacent edges different colours.

(i) $d = 1$ and $n \geqslant 6$.

(ii) n is sufficiently large (depending on d).

[Hint to (ii). Show that, given d and k, if n is sufficiently large then K^n contains a K^k whose any two adjacent edges have different colours.] (Bollobás and Erdös [BE8], Chen and Daykin [CD1].)

40. Use Ramsey's theorem to prove the following assertion.

For every natural number r there are a finite set X and $\mathscr{A} \subset X^{(r)}$ such that

(i) $\bigcup \mathscr{A} = X$, (ii) if $A, B \in \mathscr{A}$ then $|A \cap B| \leqslant 1$ and (iii) if $Y \subset X$ and $Y \neq X$ then either Y or $X - Y$ contains an r-set belonging to $\mathscr{A}$.

[If (iii) is not satisfied then it is customary to say that $\mathscr{A}$ possesses property B.] (Abbott [A1].)

41^{-}. Prove that there exists a graph H such that $\mathrm{cl}(H) = 5$ and $H \rightarrow (K^3)_2$.

42. Prove that the minimal order of H in the previous exercise is 8.

[Hint. To prove the existence H with $|H| = 8$ let H be the graph obtained from K^8 by omitting the edges of a C^5.] (Graham [G16].)

43. Let $k \geqslant 3, m \geqslant 3, n \geqslant (k-1)(m-1) + 1$ and $n \geqslant k^2 - 2$. Show that if $\beta_0(G^n) < k$ then $G^n \supset C^m$. (Note that the exercise asserts that $r(C^m, K^k) \leqslant \max\{(k-1)(m-1) + 1, k^2 - 2\}$.) (Bondy and Erdös [BE11].)

44^{+}. Prove that there is a constant $c > 0$ such that

$$cs^2/(\log s)^2 \leqslant R(3, s).$$

[Hint. Use the counting ("probabilistic") method, as in the proof of Theorem 4.1.] (Erdös [E8])

45. (i) Prove that there is a constant $c > 0$ such that if $s = \lfloor c \log n/k \log k \rfloor$ then

$$R_k(s) \leqslant n.$$

(ii) Show that with $t = \lceil 2 \log n/\log 2 \rceil$ we have

$$n \leqslant R_2(t).$$

[Hint to (ii). Apply the counting method.] (Erdös and Szekeres [ES6], Greenwood and Gleason [GG2])

46. By substituting a graph into the vertices of another, prove that
$$(R_k(s) - 1)(R_l(s) - 1) \leqslant R_{k+l}(s) - 1.$$

Use this and Ex. 45(ii) to deduce the existence of a constant $c > 0$ such that if

$$s = \lfloor c \log n / \log k \rfloor$$

then

$$n \leqslant R_k(s)$$

(Abbott [A2].)

47. CONJECTURE. If $\chi(G) = s$ then

$$R(s, s) \leqslant r(G, G).$$

(P. Erdös.)

48. PROBLEM. Suppose $H \to (G_1, G_2)$ and $e(H) \leqslant \binom{n}{2}$. Does $K^n \to (G_1, G_2)$ hold? (P. Erdös.)

49. (For those who are familiar with the rudiments of algebraic topology.) Let r and k be natural numbers and let X be a finite set, $|X| \geqslant r$. Denote by $K = K(X, r, k)$ the simplicial complex with vertex set $X^{(r)}$, whose simplices are those sets $\{A_0, A_1, \ldots, A_m\}$ of (distinct) r-sets for which

$$\left| \bigcup_{i=0}^{m} A_i \right| \leqslant r + k.$$

Prove that K is $(k - 1)$-connected.

[Hint. Show that the imbedding of the $(k - 1)$-dimensional skeleton of K into $|K|$ is homotopic to a map into $|K(X, R, k - 1)| \subset |K|$.] (Lovász [L26].)

50^-. Use the result in the previous exercise and Theorem 4.5 to obtain an extension of Theorem 4.4.

51^-. Show that the proof of Theorem 4.4, the $(k + 2)$-colourability of the Kneser graph $K_{2r+k}^{(r)}$ and Borsuk's theorem imply that S^k does not have an r-fold covering with $2r + k - 1$ open hemispheres, that is Gale's theorem [G1a] is best possible.

52. An r-set colouring with s colours of a graph G is a function $\psi : V(G) \to \{1, 2, \ldots, s\}^{(r)}$ such that if $xy \in E(G)$ then $\psi(x) \cap \psi(y) = \varnothing$. The minimal number of colours required to r-set colour G is denoted by $\chi^{(r)}(G)$; thus $\chi^{(1)}(G) = \chi(G) = \chi_0(G)$.

Show that $\chi^{(r)}(G) \geqslant \chi(G) + 2r - 2$.

53. Let t, r be integers, $0 \leqslant t \leqslant r - 1$, and let X be a set with $|X| \geqslant 2r - t$ (or $|X| \geqslant 2r + 1$ if $t = 0$). Let A be a subset of $X^{(r)}$ with the property that if $\sigma \in A, \tau \in X^{(r)}$ and $|\sigma \cap \tau| = t$, then $\tau \in A$. Prove that $A = \varnothing$ or $A = X^{(r)}$.

54. Given graphs G and H, let $G \times H$ be the graph with vertex set $V(G) \times V(H)$ and edge set $E(G \times H) = \{(a, b)(a', b'): \; aa' \in E(G)$ and $bb' \in E(H)\}$. Put $G(k, r, s) = K_s^{(r)} \times K^k$, where $K_s^{(r)}$ is the Kneser graph. There is a canonical (1-set) colouring of $G(k, r, s)$ with k-colours achieved by colouring the ith copy of $V(K_s^{(r)})$ with colour i.

Show that if $\chi(G(k, r, s)) < k$ then there is no graph H with $\chi(H) = k$ and $\chi^{(r)}(H) = s$, and that if $G(k, r, s)$ is not uniquely colourable then there is no uniquely colourable graph H with $\chi(H) = k$ and $\chi^{(r)}(H) = s$.

[Hint. Given an r-set colouring ψ of H and a (1-set) colouring θ, construct the graph H^* by identifying vertices u and v of H if $\psi(u) = \psi(v)$ and $\theta(u) = \theta(v)$. By using Ex. 53, find a suitable imbedding of H^* in $G(k, r, s)$.] (Bollobás and Thomason [BT2].)

55. Show that if $G(k, r, s)$ has a non-canonical k'-colouring for some $k' \leqslant k$, then $\chi(K_s^{(r)}) \leqslant k$, and deduce that

$$\min\{\chi^{(r)}(G): \chi(G) = k\} = k + 2r - 2,$$

and that

$$\min\{\chi^{(r)}(G): G \text{ is uniquely } k\text{-colourable}\} = k + 2r - 1.$$

[Hint. By applying Ex. 53, show that for $\sigma \in \{1, 2, \dots, s\}^{(r)}$ there exist $x, y \in V(K^k)$ with $\psi((\sigma, x)) = \psi((\sigma, y))$. Define $\psi(\sigma) = \psi((\sigma, x))$.] (Bollobás and Thomason [BT2].)

56. CONJECTURE. Suppose G is such that $\chi(G - x - y) = \chi(G) - 2$ for every pair of adjacent vertices x and y. Then G is a complete graph. (L. Lovász)

57. PROBLEM. Given natural numbers m, k and n, define

$$f(m, k, n) = \max\{\chi(G): |G| = n \quad \text{and} \quad \chi(G[W]) \leqslant k$$

$$\text{for every } W \subset V, |W| = m\}.$$

Gallai [G4] proved that there exist 4-chromatic graphs in which every odd cycle has length at least $\lfloor \sqrt{n} \rfloor$, so $f(\lfloor \sqrt{n} \rfloor, 2, n) > 3$. Is there a constant c

such that

$$f(\lfloor c\sqrt{n} \rfloor, 2, n) = 3 \text{ for every } n?$$

(Erdös [E27].)

58. CONJECTURE. Given a natural number l, there exist constants $c_1 = c_1(l)$, $c_2 = c_2(l)$, such that, with the notation of Problem 57,

$$f(\lfloor c_1 n^{1/l} \rfloor, 2, n) = l + 1 < f(\lfloor c_2 n^{1/l} \rfloor, 2, n).$$

Perhaps the c_i can be chosen independently of l. (Erdös and Gallai, see [E27].)

59[+]. Let $g(H) = g$ and $V(H) = \{x_1, x_2, \ldots, x_r\}$. Then there exists a graph G with $g(G) = g$ such that whenever $y_1, y_2, \ldots, y_n$ is an enumeration of the vertices of G, we can find $i_1 < i_2 < \ldots < i_r$ such that $G[y_{i_1}, y_{i_2}, \ldots, y_{i_r}] \cong H$ and the isomorphism is given by $y_{i_j} \to x_j$.

[Hint. Use the counting method to select r-sets of $\{1, 2, \ldots, n\}$ so that the obtained r-graph does not contain a cycle of length less than g. Place copies of H into these r-sets so that the union G of the copies has the required property.] (Nešetřil and Rödl [NR6]).

60. PROBLEM. Show the existence of a function $f(n) \to \infty$ such that if G^n is critical 4-chromatic then it can not be made bipartite by the omission of fewer than $f(n)$ edges. (P. Erdös)

Chapter VI

Complete Subgraphs

Given a graph F_1, what is $ex(n; F_1)$, the maximum number of edges of a graph of order n not containing F_1 as a subgraph? For many graph theorists this is the most general extremal problem. It must be obvious by now that most extremal problems in graph theory are of a rather different kind. However, the question above is certainly a basic extremal problem, and in this chapter we shall discuss it. As the title of the chapter indicates, we shall not try to attack the problem head on but we shall build up a small theory concerning graphs without forbidden subgraphs and use this theory to start our attacks on particular cases.

§1 features the best known extremal result of graph theory, Turán's theorem [T5]. This result, proved in 1940 and always considered to be the first extremal theorem, answers the question above in the case $F_1 = K^r$. Most of the section is devoted to various non-trivial extensions of Turán's theorem and to other related results. In §2 we investigate analogous questions concerning bipartite and r-partite graphs ($r \geqslant 3$). The results are rather fragmentary; there are a good many unsolved problems in the area.

The basis of theory of graphs without forbidden subgraphs, to be presented in §4, is the theorem of Erdös and Stone [ES5], proved in 1946. The theorem, which will be proved in §3, states that for every $\varepsilon > 0$ and natural number t if the size of a graph $G = G^n$ is εn^2 greater than the maximal size of a graph G^n without a K^r then G not only contains a K^r but it even contains a $K_r(t)$, provided n is sufficiently large. The order of magnitude of the maximal t with this property was determined by Bollobás and Erdös [BE6] and the result was further refined in Bollobás, Erdös and Simonovits [BES1]. The main results of §5 were proved independently by Erdös [E22] and Simonovits [S19]; the proofs to be given are significantly simpler than the original ones.

The only aim of §5 is to present a theorem of Hajnal and Szemerédi [HS3].

This beautiful and deep theorem can be formulated as follows: if $G = G^n$ and $\Delta = \Delta(G)$ then G has a $(\Delta + 1)$-colouring in which each colour class has $\lfloor n/\Delta \rfloor$ or $\lfloor (n + 1)/\Delta \rfloor$ elements. The proof is a simplified version of the original one, but it is still rather hard. A special case of this result, proved earlier by Corrádi and Hajnal [CH7], implies that if $|G| \geqslant 3k$ and $\delta(G) \geqslant 2k$ then G contains k vertex disjoint cycles (Ex. 42).

Most results concerning the existence of a complete graph of given order are rather easy and can be proved neatly. One notable exception is the problem of the minimum size of a weakly r-saturated graph of order n (Conjecture 17). Perhaps it is even more curious that it is not known whether a somewhat unusual but simple condition is sufficient or not to guarantee a *triangle* (Conjecture 44).

As we have mentioned already the appearance of complete subgraphs of r-partite graphs is not understood too well. The best known unsolved problem in this area is due to Zarankiewicz (cf. Conjecture 15). Though the problem has been around for long, virtually no progress has been made with the general case. Another basic problem asks for the estimation of the minimal degree of an r-partite graph with n vertices in each class that forces the graph to contain a K^r (Conjecture 20).

We shall see that the order of $ex(n; F_1)$ is determined by the chromatic number $\chi(F_1)$, unless F_1 is bipartite. It would be interesting to decide, for which pairs of graphs F_1, F_2 $(\chi(F_1) = \chi(F_2) = 2)$ it is true that $ex(n; F_1) = ex(n; F_2)$ whenever n is sufficiently large (Problem 41).

1. THE NUMBER OF COMPLETE SUBGRAPHS

Given natural numbers n and q, denote by $T_q(n)$ the *complete q-partite* graph with

$$\left\lfloor \frac{n}{q} \right\rfloor, \left\lfloor \frac{n + 1}{q} \right\rfloor, \ldots, \left\lfloor \frac{n + q - 1}{q} \right\rfloor$$

vertices in the various classes. Note that $T_q(n)$ is the unique complete q-partite graph of order n whose colour classes are as equal as possible. Furthermore, as a q-partite graph of order n having $n_1, n_2, \ldots, n_q$ vertices in its colour classes has at most

$$\binom{n}{2} - \sum_1^q \binom{n_i}{2}$$

edges, $T_q(n)$ is the unique q-partite graph of order n with maximal size. Denote this size by $t_q(n)$:

$$t_q(n) = \binom{n}{2} - \sum_1^q \binom{n_i}{2} = \sum_{0 \leqslant i < j < q} \left\lfloor \frac{n+i}{q} \right\rfloor \left\lfloor \frac{n+j}{q} \right\rfloor.$$

An $(r-1)$-partite graph does not contain a K^r, a complete graph of order r, so $T_{r-1}(n)$ is a graph of order n and size $t_{r-1}(n)$ that does not contain a K^r. Turán [T5] (see also [T6]) proved in 1941 that every other graph of order n and size $t_{r-1}(n)$ contains a K^r. The case of the triangle ($r = 3$) was proved already in 1907 by Mantel [M16], but the general case was not considered at the time.

THEOREM 1.1. *Let r and n be natural numbers, $r \geqslant 2$. Then every graph of order n and size greater than $t_{r-1}(n)$ contains a K^r, a complete graph of order r. Furthermore, $T_{r-1}(n)$ is the only graph of order n and size $t_{r-1}(n)$ that does not contain a K^r.*

Proof. As $T_{r-1}(n)$ is a maximal graph without a K^r (i.e. every graph obtained from $T_{r-1}(n)$ by the addition of an edge contains a K^r), it suffices to prove the second assertion. We apply induction on r and then on n. For $r = 2$ and $n \leqslant r$ the assertion is trivial so suppose that $3 \leqslant r < n$, the assertion holds for smaller values of r and for r and smaller values of n.

Before starting the actual proof let us make an observation. The graph $T_{r-1}(n)$ contains a number of complete graphs of order $r-1$. Let $K = K^{r-1}$ be one of these. Then each vertex of $T_{r-1}(n)$ not in K is joined to exactly $r-2$ vertices of K and $T_{r-1}(n) - K$ is a $T_{r-1}(n-r+1)$.

Let $G = G(n, t_{r-1}(n))$ and suppose G does not contain a K^r. As $t_{r-2}(n) < t_{r-1}(n)$, the graph contains a K^{r-1}, say $L = K^{r-1} \subset G$, with vertex set $W = \{x_1, \ldots, x_{r-1}\}$. A vertex $x \in V(G) - W$ can be joined to at most $r-2$ of the vertices in L since otherwise x and L span a K^r. Therefore, by the observation above, $G - L$ has at least $t_{r-1}(n-r+1)$ edges. Since $G - L$ has order $n-r+1$ and does not contain a K^r, by the induction hypothesis it has at most $t_{r-1}(n-r+1)$ edges. Consequently $G - L$ is a $T_{r-1}(n-r+1)$ and each vertex of $G - L$ is joined to *exactly* $r-2$ vertices in W. Put

$$V_i = \{x \in V(G) : xx_j \in E(G) \text{ if } j \neq i, 1 \leqslant j \leqslant r-1\}, \qquad i = 1, 2, \ldots, r-1.$$

We have already established that $V_1, V_2, \ldots, V_{r-1}$ is a partitioning of

$V(G)$. If x, $y \in V_i$ then $xy \notin E(G)$ since otherwise $x_1, x_2, \ldots, x_{i-1}, x_{i+1}, \ldots, x_{r-1}$, x and y span a K^r. This shows that G is $(r'-1)$-partite with vertex classes $V_1, \ldots, V_{r-1}$. As $G - L$ is a $T_{r-1}(n - r + 1)$, the graph G is a $T_{r-1}(n)$ with vertex classes $V_1, \ldots, V_{r-1}$.   ∎

A variant of the proof above gives that a $G(n, t_{r-1}(n) + 1)$ not only contains a K^r but it almost contains a K^{r+1}. This was first shown by Dirac [D16].

THEOREM 1.2. *If $n \geq r + 1$ then every $G = G(n, t_{r-1}(n) + 1)$ contains a K^{r+1} from which an edge has been omitted.*

Proof. Apply induction on n. For $n = r + 1$ G is exactly a K^{r+1} minus an edge. Having proved it for graphs of order less than $n (\geq r + 2)$ pick a vertex $x \in G$ with $d(x) = \delta(G)$. It is easily checked that $\delta(G) \leq \delta(T_{r-1}(n))$ so $e(G - x) \geq t_{r-1}(n - 1) + 1$. By the induction hypothesis $G - x$ contains a K^{r+1} minus an edge.   ∎

As an immediate consequence of Theorem 1.1 we obtain the following result of Zarankiewicz [Z4].

COROLLARY 1.3. *If $K^r \not\subset G^n$ then $\delta(G^n) \leq (1 - 1/(r - 1)) n$.*   ∎

It is trivial that if G^n is $(r - 1)$-partite (i.e. $\chi(G^n) \leq r - 1$) then both $K^r \not\subset G^n$ and $\delta(G^n) \leq (1 - 1/(r - 1)) n$ hold. Andrásfai, Erdős and Sós [AES1] proved that if $\chi(G^n) \geq r$ and $K^r \not\subset G^n$ then the above estimate of $\delta(G^n)$ can be improved to

$$\delta(G^n) \leq \left(1 - \frac{1}{r - \frac{4}{3}}\right) n = \frac{3r - 7}{3r - 4} n.$$

For the rather long proof of this inequality we refer the reader to [AES1]. An extension of this inequality was proved by Erdős and Simonovits [ES3].

Erdős [E26] (see also [B30]) characterized the maximal degree sequences of a graph without a K^r. It is not very surprising that these are the degree sequences of complete $(r - 1)$-partite graphs. Though this result of Erdős extends Turán's theorem (Theorem 1.1 is an immediate consequence of Theorem 1.4), its proof is equally simple and pretty.

THEOREM 1.4. *Let G be a graph with vertex set $V = \{x_1, x_2, \ldots, x_n\}$. If G does not contain a K^r then there is an $(r - 1)$-partite graph G' with vertex set V such* *that*

$$d_G(x_i) \leqslant d_{G'}(x_i), \qquad i = 1, 2, \ldots, n.$$

Proof. We apply induction on r. For $r = 2$ the theorem is obvious. Suppose now that $r > 2$ and the result holds for smaller values of r. Pick a vertex of maximal degree and denote by W the set of vertices adjacent to this vertex. Put $H = G[W]$, i.e. let H be the subgraph of G spanned by W. Clearly H does not contain a K^{r-1}. By the induction hypothesis H can be replaced by an $(r - 1)$-partite graph H' in such a way that we do not decrease the degrees. Let G' be the graph obtained from H' by joining every vertex in $V - W$ to every vertex in W. It is immediate that the degree of a vertex is at least as large in G' as in G and, by construction, G' is $(r - 1)$-partite. ∎

Rather naturally, the case $r = 3$ of Theorem 1.1 has been investigated and extended most. In particular, Goodman [G15] (see also Sauvé [S11]) determined the minimal number of triangles (K^3s) contained in a graph and its complement provided the order of the graph is fixed. Moon and Moser [MM1] gave a lower bound on the number of triangles in a graph or order n and size m. Lovász [L20] showed that the sieve formula can be used to deduce both of these results. All these proofs differ only in their presentations.

Denote by $k_r(G)$ the number of K^rs in a graph G. Thus $k_2(G) = e(G)$ is the size of G and $k_3(G)$ is the number of triangles in G.

Denote by $(d_i)_1^n$ the degree sequence of a graph G of order n. Then the degree sequence of $\bar{G}$ is $(\bar{d}_i)_1^n = (n - 1 - d_i)_1^n$. There are $\Sigma_1^n \binom{d_i}{2}$ pairs of adjacent edges of G and there are $\Sigma_1^n \binom{n-1-d_i}{2}$ pairs of adjacent edges of $\bar{G}$. The sum of these two numbers can also be counted as follows. Each of the $k_3(G) + k_3(\bar{G})$ triangles in G and $\bar{G}$ contains three such pairs and each of the remaining $\binom{n}{3} - k_3(G) - k_3(\bar{G})$ triples of vertices contains exactly one such pair. Hence

$$\sum_1^n \binom{d_i}{2} + \sum_1^n \binom{n-1-d_i}{2} = 3k_3(G) + 3k_3(\bar{G}) + \binom{n}{3} - k_3(G) - k_3(\bar{G}).$$

Putting $e = e(G)$ and $\bar{e} = e(\bar{G})$, a rearrangement gives

$$k_3(G) + k_3(\bar{G}) = \binom{n}{3} - (n - 2)e + \sum_1^n \binom{d_i}{2} = \binom{n}{3} - (n - 2)\bar{e} + \sum_1^n \binom{\bar{d}_i}{2}. \quad (1)$$

COROLLARY 1.5. *A graph of order n and its complement contain at least $\frac{1}{24}n(n - 1)(n - 5)$ triangles.*

Proof. Note that the right-hand side of (1) is at least as large as

$$\binom{n}{3} - (n-2)\,\bar{e} + n\binom{2\bar{e}/n}{2} = \binom{n}{3} - \frac{2\bar{e}(n^2 - n - 2\bar{e})}{2n}$$

$$\geqslant \tfrac{1}{24}n(n-1)(n-5). \quad\blacksquare$$

Of course, this bound is attained if and only if $\bar{G}$ is δ-regular ($\delta = 2\bar{e}/n$) and so is G. In particular, one must have $n = 2\delta + 1$ and δ must be even since otherwise there is no δ-regular graph of order n. Thus if the bound is best possible then $n = 4k + 1$. It is easily seen that in this case Corollary 1.5 is indeed best possible. (For the other cases see Ex. 8.)

COROLLARY 1.6. *A graph G of order n and size e contains at least* $(e/3n)(4e - n^2)$ *triangles.*

Proof. With the notation of equation (1) we have

$$3k_3(\bar{G}) \leqslant \sum_i^n \binom{\bar{d}_i}{2}.$$

Taking into account that $\bar{e} = \binom{n}{2} - e$, (1) implies

$$k_3(G) \leqslant \binom{n}{3} - (n-2)\,\bar{e} + n\binom{2\bar{e}/n}{2} = \frac{e}{3n}(4e - n^2). \quad\blacksquare$$

Our next aim is to estimate the minimal number of complete subgraphs of order r that must be contained in every graph of order n and size m. Thus in a sense our aim is to extend and sharpen Theorem 1.1 stating that a graph of order n and size $t_{r-1}(n) + 1$ contains at least one K^r. The result we are going to present is in [B27]. The method we apply is actually suitable to attack the following more general problem. Recall that $k_r(G)$ denotes the number of complete graphs of order r in G. Let $2 \leqslant p < r \leqslant n$ and put

$$k_r(k_p^n \geqslant x) = \min\{k_r(G^n): k_p(G^n) \geqslant x\}.$$

Our aim is to estimate this function $k_r(k_p^n \geqslant x)$. Note that $k_2(G^n) = e(G^n)$ is the size of G^n, so $k_r(k_2^n \geqslant x)$ is the minimal number of K^r's in a graph of order n and size at least x. As suggested by the notation, the integers p, r and n are considered to be fixed and we investigate $k(x) = k_r(k_p^n \geqslant x)$ as a function of x. It will be convenient to consider graphs of order n with a fixed vertex set, say with $V = \{x_1, \ldots, x_n\}$. Denote by $\mathcal{G}^n$ the set of graphs with vertex set V. Let $\mathcal{C}^n$ consist of those graphs $G \in \mathcal{G}^n$ that are *complete q*-partite for some q. Thus a graph $G \in \mathcal{G}^n$ belongs to $\mathcal{C}^n$ if and only if every two non-adjacent vertices are joined to exactly the same set of vertices. Furthermore, put $\mathcal{T}^n = \{T_q(n): q = 1, 2, \ldots, n\} \subset \mathcal{C}^n$. Note that $G \in \mathcal{C}^n$ belongs to $\mathcal{T}^n$ if and only if the largest colour class has at most one more element than the smallest. To simplify the notation put $T_q = T_q(n)$.

Returning to the function $k(x) = k_r(k_p^n \geq x)$ observe that $G = T_{r-1}$ shows that

$$k(x) = k_r(k_p^n \geq x) = 0 \quad \text{for} \quad x \leq k_p(T_{r-1}),$$

so in the sequel we suppose that

$$k_p(T_{r-1}) \leq x \leq \binom{n}{p}.$$

Let $\psi(x) = \psi_r^p(x, n)$ be the *maximal convex function* defined on the interval $k_p(T_{r-1}) \leq x \leq \binom{n}{p}$ such that

$$\psi(k_p(T_q)) \leq k_r(T_q), \qquad q = r - 1, r, \ldots, n. \tag{2}$$

It is easily seen that for every q one has equality in (2).

THEOREM 1.7. $k(x) = k_r(k_p^n \geq x) \geq \psi(x) = \psi_r^p(x, n)$.

Proof. Let $c > 0$. Define a function f_c on $\mathcal{G}^n$ by setting

$$f_c(G) = k_p(G) - ck_r(G), \qquad G \in \mathcal{G}^n.$$

As the function $\psi(x)$ is convex, to prove the theorem it suffices to show that f_c *attains its maximum at a graph* $G \in \mathcal{T}^n$. To start with we replace f_c by a function $\tilde{f}_\gamma$ that can approximate f_c and we show this new function $\tilde{f}_\gamma$ attains its maximum at a graph $G \in \mathcal{T}^n$.

Let $\gamma, \alpha_1, \ldots, \alpha_n > 0$ be algebraically independent numbers. If K is a complete subgraph with vertex set $V(K)$, put

$$w(K) = \prod_{x_i \in K} (1 + \alpha_i),$$

$$w_q(G) = \sum_{K_q \subset G} w(K_q),$$

$$\tilde{f}_\gamma(G) = w_p(G) - \gamma w_r(G).$$

Furthermore, for $1 \leqslant i \leqslant n$ put

$$w_q^i(G) = \sum_{\{x_i\} \subset K_q \subset G} w(K_q),$$

$$\tilde{f}_\gamma^i(G) = w_p^i(G) - \gamma w_r^i(G).$$

Let $G_0 \in \mathscr{G}^n$ be a graph at which $\tilde{f}_\gamma$ attains its maximum. We may assume without loss of generality that every edge of G_0 is contained in at least one K^p. Suppose x_i and x_j are *non-adjacent* vertices of G_0. Then

$$\tilde{f}_\gamma(G_0) = \tilde{f}_\gamma^i(G_0) + \tilde{f}_\gamma^j(G_0) + A,$$

where A depends only on the graph $G_0 - x_i - x_j$. Suppose

$$\tilde{f}_\gamma^i(G_0)/(1 + \alpha_i) \neq \tilde{f}_\gamma^j(G_0)/(1 + \alpha_j);$$

say

$$\tilde{f}_\gamma^i(G_0)/(1 + \alpha_i) > \tilde{f}_\gamma^j(G_0)/(1 + \alpha_j).$$

Then omitting the edges incident with x_j and joining x_j to the vertices adjacent with x_i we obtain a graph G_0' for which

$$\tilde{f}_\gamma(G_0') > \tilde{f}_\gamma(G_0).$$

As this contradicts our assumption about G_0 we must have

$$(1 + \alpha_j) \tilde{f}_\gamma^i(G_0) = (1 + \alpha_i) f_\gamma^j(G_0). \tag{3}$$

It follows from (3) and the algebraic independence of $\gamma, \alpha_1, \ldots, \alpha_n$ that x_i and x_j are joined to *exactly the same vertices*. Consequently $G_0 \in \mathscr{C}^n$.

Note that $\tilde{f}_\gamma(G) \to f_c(G)$ as $\gamma \to c$ and $\alpha_1, \ldots, \alpha_n \to 0$, so $f_c(G)$ attains its maximum at *some* $G \in \mathscr{C}^n$. To complete the proof we have to show that G can be chosen from $\mathscr{T}^n$. We can suppose without loss of generality that c is *irrational*. We shall prove slightly more than necessary, namely that every graph $G \in \mathscr{C}^n$ at which f_c attains its maximum is also in $\mathscr{T}^n$. Let $G_1 \in \mathscr{C}^n$ be a graph at which $f_c(G)$, $G \in \mathscr{C}^n$, attains its maximum. Suppose G_1 is a complete q-partite graph and has $a_1, a_2, \ldots, a_q$ vertices in its classes, $0 < a_1 \leqslant a_2 \leqslant \ldots \leqslant a_q$. We can assume without loss of generality that $q \geqslant r$, otherwise the maximum is 0 and it is attained on $T_{r-1}(n)$. Then

$$f_c(G_1) = a_1 a_q B_1 - c a_1 a_q B_2 + C = a_1 a_q (B_1 - c B_2) + C, \qquad (4)$$

where B_1, B_2 and C are *non-zero* rational numbers, depending on $a_2, a_3, \ldots,$ a_{q-1} and $a_1 + a_q$. Consequently $B_1 - c B_2 \neq 0$. Note that now

$$B_1 - c B_2 > 0,$$

since otherwise we could replace a_1 and a_q by 0 and $a_1 + a_q$ (i.e. we could replace G_1 by a $(q-1)$-partite complete graph) and this would increase $f_c(G_1)$. Finally, $B_1 - c B_2 > 0$ implies that $a_q \leqslant a_1 + 1$, since otherwise we could increase $f_c(G_1)$ by replacing a_1 by $a_1 + 1$ and a_q by $a_q - 1$.

The inequality $a_q \leqslant a_1 + 1$ means exactly that $G_1 = T_q(n) \in \mathcal{T}^n$, as required. ∎

Noticing that $k_s(T_q(n)) \sim \binom{q}{s}(n/q)^s$, one can get a weaker but more tangible form of Theorem 1.7 by replacing $t_q(n; s)$ by $\binom{q}{s}(n/q)^s$. This can be made precise as follows.

Let G_1 be as above. Then

$$f_c(G_1) = \sum_{i_1 < \ldots < i_p} \prod_{j=1}^{p} a_{i_j} - c \sum_{i_1 < \ldots < i_r} \prod_{j=1}^{r} a_{i_j} = p(a_1, \ldots, a_q).$$

Consider the polynomial $p(a_1, \ldots, a_q)$ on the set

$$A = \left\{ a = (a_i)_1^q : a_i \geqslant 0, \sum_1^q a_i = n \right\}.$$

It is easily seen that p attains its maximum on A at a point $(a_i)_1^q$ such that $a_1 = \ldots = a_s$ and $a_{s+1} = a_{s+2} = \ldots = a_q = 0$. Thus

$$f_c(G_1) \leqslant \binom{s}{p}(n/s)^p - c \binom{s}{r}(n/s)^r$$

for some natural number s. This implies the following weaker form of Theorem 1.7.

THEOREM 1.7'. *Let* $\tilde{\psi}(x) = \tilde{\psi}_r^p(x; n)$ *be the maximal convex function defined in* $0 \leqslant x \leqslant \binom{n}{p}$ *such that*

$$\tilde{\psi}\left((n/q)^p \binom{q}{p}\right) \geqslant (n/q)^r \binom{q}{r}, \quad q = 1, 2, \ldots, n. \qquad (5)$$

Then

$$k_r(k_p^n \geqslant x) \geqslant \tilde{\psi}(x). \qquad \blacksquare$$

It is easily seen for each q one has equality in (5), so Theorem 1.7' has the following consequence.

COROLLARY 1.8. *Suppose* $r - 1 \leqslant q < n, 0 \leqslant y \leqslant 1$ *and*

$$x = y(n/q)^p \binom{q}{p} + (1 - y)(n/(q + 1))^p \binom{q + 1}{p}.$$

Then

$$k_r(k_p^n \geqslant x) \geqslant y(n/q)^r \binom{q}{r} + (1 - y)(n/(q + 1))^r \binom{q + 1}{r}. \qquad \blacksquare$$

The case $p = 2$, $r = 3$ and $q = 2$ of Corollary 1.8 was conjectured by Nordhaus and Stewart [NS1]. We state it here since it improves the bound given in Corollary 1.6.

COROLLARY 1.9. *A graph or order n and size e, $n^2/4 \leqslant e \leqslant n^2/3$, contains at least $(n/9)(4e - n^2)$ triangles.* $\qquad \blacksquare$

Let us mention another special case of Theorem 1.7, first proved by Erdös [E11], [E24] and rediscovered by Sauer [S10].

COROLLARY 1.10. *Let $2 \leqslant p \leqslant r \leqslant n$. If G^n contains more K^ps than $T_{r-1}(n)$ then it also contains a K^r.* $\qquad \blacksquare$

Considerable effort has been put into the determination of the function $k_r(k_p^n \geqslant x)$, estimated in Theorem 1.7, especially in the case when $p = 2, r = 3$ and x is not much larger than $t_2(n) = \lfloor n^2/4 \rfloor$. Rademacher (see [E6]) showed that every graph $G(n, t_2(n) + 1)$ contains not only one triangle, as asserted by Theorem 1.1, or $\lfloor 4n/9 \rfloor$ triangles, as asserted by Theorem 1.7, but in fact it contains at least $\lfloor n/2 \rfloor$ triangles. Extending this, Erdös [E6] showed that

$$k_3(G(n, t_2(n) + l)) \geqslant l\lfloor n/2 \rfloor, \qquad (6)$$

provided $l \leqslant 3$ and $l < n/2$. In [E9] Erdös showed that there is a constant $c > 0$ such that (6) holds whenever $l < cn$. In the same note Erdös conjectured that (6) holds whenever $l < n/2$. The graphs obtained from the complete bipartite graph $T_2(n)$ by putting l appropriate edges into the first class show that (6), if true, is best possible. Furthermore, if $n = 2k \geqslant 6$ and $l = k = n/2$, the graph $C^{k+1} + E^{k-1}$ shows that (6) need not hold.

A slightly weaker form of this conjecture of Erdös was proved recently by Lovász and Simonovits [LS2], showing that (6) holds whenever $n > n_0$ and $l < n/2$. (Lovász and Simonovits announced analogous best possible results for the number of K^rs in a graph.) Instead of giving the entire proof of this result, we reproduce only part of the proof from [LS2], showing that (6) holds if $l \leqslant n/4$. The proof is based on the inclusion–exclusion principle and is reminiscent of the proof of relation (1).

THEOREM 1.11. *Let* $G = G(n, m)$, *where* $m = t_2(n) + l$, $0 < l \leqslant n/4$. *Then* G *contains at least* $l\lfloor n/2 \rfloor$ *triangles.*

Proof. In order to avoid some unimportant numerical complications, we assume that n is *even*. Denote by $V = \{1, 2, \ldots, n\}$ the vertex set of G and let K be the complete graph with the same vertex set. Put $V_i = \Gamma_G(i)$, $W_i = V - V_i$, $d_i = d_G(i) = |V_i|$ and denote by $t^{(k)}$ the number of triangles in K that contain exactly k edges of $\bar{G}$. Thus $k_3(G) = t^{(0)}$. Let λ_i be the number of $V_i - W_i$ edges of $\bar{G}$. If $ij \in E(G)$, denote by $t_{ij}^{(k)}$ the number of triangles in K that contain ij and k edges of $\bar{G}$. Each of the following relations is an immediate consequence of the definitions:

$$\sum_1^n \lambda_i = 2t^{(2)}, \tag{7}$$

$$\sum_{ij \in E} t_{ij}^{(0)} = 3t^{(0)}, \tag{8}$$

$$\sum_{ij \in E} t_{ij}^{(2)} = t^{(2)}, \tag{9}$$

$$n = d_i + d_j - t_{ij}^{(0)} + t_{ij}^{(2)}, \quad \text{provided } ij \in E.$$

Summing the last equality over all edges ij, by (8) and (9) we obtain

$$nm = \sum_1^n d_i^2 - 3t^{(0)} + t^{(2)}, \tag{10}$$

since

$$\sum_{ij \in E} (d_i + d_j) = \sum_1^n d_i^2.$$

Putting $p_i = d_i - n/2$ we see that

$$\sum_1^n d_i^2 = \sum_1^n p_i^2 + n \sum_1^n p_i + \frac{n^3}{4} = \sum_1^n p_i^2 + 2n\left(m - \frac{n^2}{4}\right) + \frac{n^3}{4} = \sum_1^n p_i^2 + 2ln + \frac{n^3}{4}.$$

Taking into account (7) and (10), this gives

$$3t^{(0)} = \sum_1^n (p_i^2 + \lambda_i/2) + ln.$$

This relation gives the assertion of the theorem unless

$$\sum_1^n (p_i^2 + \lambda_i/2) < ln/2. \tag{11}$$

Therefore we may and will assume that (11) does hold. Then for some vertex i we have

$$p_i^2 + \lambda_i/2 < l/2. \tag{12}$$

There are

$$\frac{n^2}{2} + l - \left(\frac{n^2}{4} - p_i^2 - \lambda_i \right) = l + p_i^2 + \lambda_i$$

edges of G with both endvertices in V_i or both endvertices in W_i. Each such edge is in at least

$$\frac{n}{2} - |p_i| - \lambda_i \geqslant \frac{n}{2} - p_i^2 - \lambda_i$$

triangles of G the third vertex of which belongs to the other class. Hence, by (12),

$$k_3(G) > (l + p_i^2 + \lambda_i)\left(\frac{n}{2} - p_i^2 - \lambda_i \right) =$$

$$= l\frac{n}{2} + (p_i^2 + \lambda_i)\left(\frac{n}{2} - l - p_i^2 - \lambda_i \right)$$

$$\geqslant l\frac{n}{2} + (p_i^2 + \lambda_i)\left(\frac{n}{2} - 2l \right) > ln/2. \qquad \blacksquare$$

Let us estimate now the *maximal* number of complete subgraphs of order r contained in a graph of order n and size m. It turns out that the nature of this problem is quite different from the problems we have just discussed, in particular this maximum does not depend on n. It is slightly surprising that this problem has a complete solution. Applying a simple induction, Erdös and Hanani [E11] determined the maximum of the number of complete subgraphs contained in a graph of given size.

THEOREM 1.12. *Let* r *and* m *be natural numbers,* $3 \leqslant r$. *Put* $m = \binom{s}{2} + t$, $0 < t \leqslant s$. *Then*

$$\max\{k_r(G): e(G) = m\} = \binom{s}{r} + \binom{t}{r-1}.$$

Proof. Let G be obtained from a K^s by joining a new vertex to t of the vertices. Then $e(G) = m$ and $k_r(G) = \binom{s}{r} + \binom{t}{r-1}$. This shows that the maximum is at least as large as claimed, so it suffices to prove the converse inequality. To do that we apply induction on m. Notice that for $m \leqslant \binom{r}{2}$ the result is trivial. Assume now that $m > \binom{r}{2}$ (i.e. $s \geqslant r$) and the result holds for smaller values of m. Let G be a graph of size m. We may suppose without loss of generality that G has no isolated vertex. Suppose first that G has a vertex x such that $1 \leqslant d(x) = d < s$. Then there are at most $\binom{d}{r-1}$ K^rs containing x. If $d < t$ then by applying the induction hypothesis to $G - x$ we find that

$$k_r(G) \leqslant \binom{d}{r-1} + k_r(G - x) \leqslant \binom{d}{r-1} + \binom{s}{r} + \binom{t-d}{r-1} < \binom{s}{r} + \binom{t}{r-1}.$$

Similarly, if $t \leqslant d < s$ we have

$$k_r(G) \leqslant \binom{d}{r-1} + \binom{s-1}{r} + \binom{s+t-d}{r-1} < \binom{s}{r} + \binom{t}{r-1}.$$

Thus we may suppose that every vertex of G has degree at least s. Therefore the order n and size m of G satisfy $n \geqslant s + 1$ and $m \geqslant \frac{1}{2}s(s + 1)$. Then from $m \leqslant \binom{s+1}{2}$ we deduce that $G = K^{s+1}$, $t = s$, so

$$k_r(G) = \binom{s+1}{r} = \binom{s}{r} + \binom{s}{r-1},$$

as claimed.   ∎

Analogously to the problem discussed in Theorem 1.7 one has the following more general problem. Given p and r, $2 \leqslant p < r$, determine

$$k_r(k_p \leqslant x) = \max\{k_r(G): k_p(G) \leqslant x\}$$

for every x. (Though the determination of $k_r(k_p \leqslant x)$ would be equivalent to the determination of $k_p(k_r \geqslant x)$ which is the function in Theorem 1.7 with

p and r interchanged, it seems more suitable to choose this different notation to indicate that this is an essentially different problem.) This function has not been determined completely but a powerful result about hypergraphs goes most of the way towards it. A simple form of a theorem of Kruskal [K13] (rediscovered and successfully applied by Katona [K3]) states that if $0 \leqslant t < s, r \leqslant s, t = 0$ or $r - 1 \leqslant t$, then $\binom{s}{r} + \binom{t}{r-1} + 1$ different r-tuples contain at least $\binom{s}{p} + \binom{t}{p-1} + 1$ different p-tuples. Thus, as remarked in [B29], this theorem implies immediately the following extension of Theorem 1.12.

THEOREM 1.13. *If* $r \leqslant s$ *and* $t = 0$ *or* $r - 1 \leqslant t < s$ *then*

$$k_r\left[k_p \leqslant \binom{s}{p} + \binom{t}{p-1}\right] = \binom{s}{r} + \binom{t}{r-1}.$$

Proof. Kruskal's theorem implies that the left-hand side is at most as large as the right-hand side. The graph obtained by joining a new vertex to t vertices of a K^s shows that the converse holds as well. ∎

It is very likely that Theorem 1.12 can be strengthened to the following:

$$k_r\left[k_p \leqslant \binom{s}{p} + \binom{t+1}{p-1} - 1\right] = \binom{s}{r} + \binom{t}{r-1}.$$

Note that in view of Theorem 1.13 (or the example in its "proof") this would imply the value of $k_r(k_p \leqslant x)$ for every x.

Let us return to the investigation of the set of K^rs contained in every $G(n, m)$. The next question we are interested in is the following: at least how many edge disjoint K^rs are contained in every $G(n, m)$? The bound we give will not be a very good one but it will still extend Theorem 1.1. To formulate the result, which for $r = 3$ was proved by Erdős, Goodman and Pósa [EGP1] and for $r = 4$ was proved in [B27], we use slightly different terminology that will hide the crudeness of the result. We say that a graph G is *covered* by the subgraphs $G_1, \ldots, G_k$ if every *edge* of G is in at least one G_i.

THEOREM 1.14. *Let* $3 \leqslant r$. *Every graph of order n can be covered with at most* $t_{r-1}(n)$ *edge disjoint K^rs and edges. If $r > 3$ then $T_{r-1}(n)$ is the only graph that cannot be covered with fewer edge disjoint K^rs and edges. For $r = 3$ the extremal graphs are K^4, K^5 and $T_3(n), n = 1, 2, \ldots$.*

Proof. Clearly it suffices to prove the statement about the extremal graphs. Let us use induction on n. The result is trivial for $n \leqslant r$ so suppose $n > r$ and the assertion holds for graphs with less than n vertices. Note that the minimal degree in $T_{r-1}(n)$ is

$$\delta = \delta(T_{r-1}(n)) = t_{r-1}(n) - t_{r-1}(n-1) = \left\lfloor \frac{(r-2)n}{r-1} \right\rfloor$$

and Corollary 1.3 states that

$$\text{if} \quad \delta(G^n) > \delta(T_{r-1}(n)) \quad \text{then} \quad K^r \subset G^n. \tag{13}$$

Let $G = G^n$. Let x be a vertex of minimal degree, say of degree $\delta + k$. If $k < 0$ or $k = 0$ and x is a vertex of a K^r in G, the edges incident with x together with some other edges of G can be covered with at most $d - 1$ edge disjoint K^rs and edges, so the induction hypothesis implies the result. Similarly, if $k = 0$ and x is not contained in a K^r of G, the result follows from the induction hypothesis. For if we do need $t_{r-1}(n-1)$ edge disjoint K^rs and edges to cover $G - x$ then $G - x$ is a K^{n-1} ($n = 5, 6$) or $T_{r-1}(n-1)$. The first case is impossible since then G does have a K^r containing x. The second case leads to $G = T_{r-1}(n)$.

To complete the proof we shall show that if $k \geqslant 1$ and $r > 3$ then the edges incident with x and some other edges of G can be covered with *less than* δ edge disjoint K^rs and edges. The case $k \geqslant 1$ and $r = 3$ will be dealt with separately.

Let H be the subgraph spanning by the vertices adjacent to x. Then H has $\delta + k$ verticles and each vertex has degree at least $\delta + k - (n - \delta - k) = 2\delta + 2k - n = f$, since x was chosen to have minimal degree. Denote by h the *maximal* number of independent (i.e. vertex disjoint) K^{r-1}s of H, and let F be a subgraph of H obtained from H by omitting h independent K^{r-1}s. Then F does not contain a K^{r-1}, has $\delta + k - (r-1)h$ vertices and every vertex has degree at least $f - (r-1)h$. Consequently, by (13),

$$f - (r-1)h \leqslant \delta(T_{r-2}(\delta + k - (r-1)h)) = \left\lfloor \frac{r-3}{r-2}(\delta + k - (r-1)h \right\rfloor,$$

so

$$(r-2)(2\delta + 2k - n - (r-1)h) \leqslant (r-3)(\delta + k - (r-1)h),$$

that is

$$(r-1)\delta + (r-1)k - (r-2)n \leqslant (r-1)h.$$

As $(r - 1)\delta \geqslant (r - 2)n - (r - 2)$, this implies

$$k - \frac{r - 2}{r - 1} \leqslant h.$$

Thus $h \geqslant k$, i.e. H contains at least k independent K^{r-1}s. Consequently x is in at least k edge disjoint K^rs of G, so the $\delta + k$ edges at x can be covered with at most

$$k + (\delta + k - (r - 1)k) = \delta - (r - 3)k \tag{14}$$

edge disjoint K^rs and edges. Now if $r > 3$ this completes the proof since the expression in (14) is less than δ. If $r = 3$ and we do need $t_2(n)$ triangles and edges to cover G then (14) implies that we need $t_2(n - 1)$ triangles and edges to cover $G - x$. Thus $G - x$ is K^{n-1} ($n = 5, 6$) or $T_2(n)$. It is easily checked that this is possible only if G is K^4 or K^5. ∎

Marczewski [M16a] pointed out that every graph can be represented as an intersection graph. Theorem 1.14 has the following almost immediate consequence about the size of a set S such that G^n is the intersection graph of subsets of S. Once again the case $r = 3$ was proved by Erdös, Goodman and Pósa [EGP1] and the extension to $r \geqslant 3$ was shown in [B27].

COROLLARY 1.15. *Let G be a graph with vertices $x_1, \ldots, x_n, n \geqslant 4$. Then there is a set S, $|S| \leqslant t_{r-1}(n), r \geqslant 3$, containing distinct non-empty subsets $X_1, \ldots, X_n$, such that $x_i x_j$ is an edge of G if and only if $X_i \cap X_j \neq \emptyset$. Furthermore, $|X_i \cap X_j| \leqslant 1$ for all $i \neq j$ and every element of S is contained in 1, 2 or r of the subsets X_i.*

Proof. It is easily checked that the result holds for $n = 4, 5$. Suppose $n \geqslant 6$ and the result holds for smaller values of n. Then we can also assume that G has no isolated vertex. Let $\mathscr{A} = \{A_1, \ldots, A_s\}, s \leqslant t_{r-1}(n)$, be a set of edge disjoint K^rs and edges covering G. Then G can clearly be represented by subsets of $S = \mathscr{A}$ if we put

$$X_i = \{A_h : x_i \in V(A_h)\}, \qquad i = 1, , \ldots, n.$$

This representation does have all the required properties except two different vertices might be represented by the same subset, i.e. we might have $X_i = X_j$ for $i \neq j$. Suppose this is the case, i.e. $X_i = X_j$ for some i and j, $i \neq j$. Note that in this case both x_i and x_j are contained in only one of the sets in $\mathscr{A}$, say A_k. Then represent $G - x_i - x_j$ on a set S_1 with at most $t_{r-1}(n - 2)$ elements. Add three elements, say a, b and c to S_1 to obtain a set S. Add c to every set corresponding to a vertex of A_h and put $X_i = \{a, c\}, X_j = \{b, c\}$. The obtained representation of G has the desired properties and one can check that

$$|S| = |S_1| + 3 \leqslant t_{r-1}(n - 2) + 3 \leqslant t_r(n). \qquad \blacksquare$$

Let us conclude this section with some results that in a sense complement Theorem 1.1. Following Zykov [Z8] we say that a graph G is *r-saturated* (or K^r-*saturated*) if it does not contain a K^r but the addition of any edge results in a graph with a K^r. Furthermore, we say that G is *strongly r-saturated* if $k_r(G) < k_r(G^+)$ whenever G^+ is obtained from G by the addition of an edge. In this terminology Theorem 1.1 states that the *maximal* size of an r-saturated graph of order n is $t_{r-1}(n)$. Erdös, Hajnal and Moon [EHM1] determined the *minimal* size of an r-saturated graph of order n which also happens to be the minimal size of a strongly r-saturated graph of order n.

THEOREM 1.16. *The minimal size of a strongly r-saturated ($r \geqslant 3$) graph of order n is*

$$(r - 2)(n - r + 2) + \binom{r - 2}{2}.$$

The r-saturated graph $G_n = K^{r-2} + E^{n-r+2}$ is the only strongly r-saturated graph of order n and minimal size.

Proof. As G_n is a *minimal* r-saturated graph, it suffices to prove that if $G = G(n, m)$, $m = (r - 2)(n - r + 2) + \binom{r-2}{2}$, is strongly r-saturated then $G = G_n$. We prove this by induction first on r and then on n. If $r = 3$ then G is of size $n - 1$ and as it is strongly 3-saturated, its diameter is at most 2. Thus G is a tree of diameter at most 2 so $G = K^1 + E^{n-1}$, as claimed. Suppose now that $r \geqslant 4$. The result is trivial for $n \leqslant r$ so suppose $n > r$ and the assertion is true for smaller values of n.

Let x_1 and x_2 be non-adjacent vertices of G. Then, as G is strongly r-saturated, there are vertices $x_3, \ldots, x_r$ spanning a complete subgraph such that both x_1 and x_2 are joined to all these vertices. Let G^* be the graph obtained from $F = G - \{x_1, x_2\}$ by adding x_2 and joining it to every vertex of F that is joined to at least one of x_1 and x_2 in G. It is easily seen that G^* is a strongly r-saturated graph of order $n - 1$. Furthermore

$$e(G^*) \geqslant e(G) - (r - 2),$$

with equality if and only if no vertex $y \in V(G) - \{x_1, \ldots, x_r\}$ is joined to both x_1 and x_2 in G. Thus, by the induction hypothesis, $G^* = G_{n-1}$, i.e. G^* contains $r - 2$ vertices of degree $n - 2$ and $n - r + 1$ vertices joined to these. As $x_3, \ldots, x_r$ span a complete subgraph of G^*, at least one of these vertices, say x_r, has degree $n - 2$. Then the degree of x_r in G is $n - 1$. Consequently $G - x$ is strongly $(r - 1)$-saturated and has size $(r - 2)(n - r + 2) + \binom{r-2}{2}$ $-(n - 1) = (r - 3)(n - (r - 1) + 2) + \binom{r-3}{2}$. By the induction hypothesis $G - x = K^{r-3} + E^{n-r+2}$ and this implies $G = G_n$.   ∎

2. COMPLETE SUBGRAPHS OF r-PARTITE GRAPHS

Recall that $G_r(n_1, \ldots, n_r)$ denotes an r-partite graph with $n_1, \ldots, n_r$ vertices in its (colour) classes. In this section we discuss various conditions ensuring that a $G_r(n_1, \ldots, n_r)$ contains a $K_r(p_1, \ldots, p_r)$. Rather naturally we shall pay special attention to the cases $r = 2$ and $r = 3$.

Let us start with the following problem. Given m, n, s and t, what is the maximal size of a graph $G_2(m, n)$ if it does not contain a $K(s, t)$ with s vertices in the first class and t in the other? Denote this maximum by $z(m, n; s, t)$ and put $z(n; t) = z(n, n; t, t)$. To avoid the trivial cases we shall suppose that $2 \leqslant s \leqslant m$, $2 \leqslant t \leqslant n$. In 1951 Zarankiewicz [Z3] posed the problem of determining $z(n; 3)$ for $n = 4, 5, 6$ and the general problem has also become known as *the problem of Zarankiewicz*. The original problem was solved by Sierpinski [S14], additional numerical values (for $t = 3$) were supplied by Brzezinski (see [S15]), and Čulik [C12] (see Ex. 13). Some years later Guy [G21], [G22], [G23] and Guy and Znám [GZ2] determined a good number of further numerical values.

The first good bounds on $z(n; 2)$ were given by Hartman, Mycielski and Ryll-Nardzewski [HMR1]. Kővári, Sós and Turán [KST1] proved an upper bound for $z(n; t)$ that seems to give the correct order of magnitude. Hyltén-Cavallius [H29] observed that the same argument gives an upper bound for $z(m, n; s, t)$. In the case $m = n$, $s = t$ this upper bound was then improved slightly by Znám [Z6], [Z7]. It is somewhat discouraging that a trivial argument gives almost as good an upper bound as the best bound known at present, but we are still far from knowing the correct order of $z(n; t)$ for fixed values of $t > 3$.

LEMMA 2.1. *Let m, n, s, t, r, k be integers, $0 \leqslant s \leqslant m$, $0 \leqslant t \leqslant n$, $0 \leqslant k$, $0 \leqslant r \leqslant m$ and let $G = G_2(m, n)$ be a graph of size $z = my = km + r$ without* a $K(s, t)$. Then

$$m \binom{y}{t} \leqslant (m - r)\binom{k}{t} + r\binom{k + 1}{t} \leqslant (s - 1)\binom{n}{t}. \tag{1}$$

Proof. Denote by V_1, V_2 the colour classes of G. We shall say that a set T of V_2 with t elements (i.e. a t-set T of V_2) *belongs to* $x \in V_1$ if x is joined to every vertex in T. Clearly $\binom{d(x)}{t}$ t-sets belong to a vertex $x \in V_1$ and by assumption a t-set in V_2 belongs to at most $s - 1$ vertices of V_1. Consequently

$$\sum_{x \in V_1} \binom{d(x)}{t} \leqslant (s - 1)\binom{n}{t}. \tag{2}$$

As $\sum_{x \in V_1} d(x) = z = km + r$, $0 \leqslant r < m$, and $f(u) = \binom{u}{t}$ is a convex function of u, (2) implies (1). ∎

THEOREM 2.2. $z(m, n; s, t) < (s - 1)^{1/t}(n - t + 1) m^{1 - 1/t} + (t - 1) m$.

Proof. Apply Lemma 2.1 to an extremal graph of size $z(m, n; s, t) = z = my$. As $y \leqslant n$, (1) implies

$$(y - (t - 1))^t \leqslant (s - 1)(n - (t - 1))^t m^{-1}. \qquad ∎$$

Let us note the following consequence of Theorem 2.2 about the size of a graph of order n not containing a $K(s, t)$.

THEOREM 2.3. *Suppose G is a graph of order n not containing a $K(s, t)$, $2 \leqslant s$, $2 \leqslant t$. Then*

$$2e(G) \leqslant z(n, n; s, t)$$

so

$$e(G) \leqslant \tfrac{1}{2}(s - 1)^{1/t} (n - t + 1) n^{1 - 1/t} + \tfrac{1}{2}(t - 1) n.$$

Proof. Construct a graph $H = G_2(n, n)$ from G as follows. Take two copies, say V_1 and V_2, of the vertex class $V(G)$ of G. If $xy \in E(G)$, $x' \in V_1$ is the vertex corresponding to x and $y' \in V_2$ is the vertex corresponding to y then let $x'y'$ be an edge of H. Then $H = G_2(n, n)$ has $2e(G)$ edges and it does not contain a $K(s, t)$ (or a $K(t, s)$) so $2e(G) \leqslant z(n, n; s, t)$. ∎

The virtually trivial Lemma 2.1 allows us to recognize certain extremal graphs without actually examining any other graphs.

THEOREM 2.4. *Suppose $G = G_2(m, n)$ has colour classes V_1, V_2 and is such that if $x \in V_1$ then $d(x)$ is d or $d + 1$, and for every t-set T there nre exactly $s - 1$ vertices in V_1 joined to every vertex in T. Then $e(G) = z(m, n; s, t)$.* ∎

Of course the difficulty is that there are relatively few such clear cut extremal graphs and they seem to occur mostly in trivial situations, namely when m is very large compared to n so that we can have $d = t - 1$ (see Ex. 13).

The bound given in Theorem 2.2 for $z(n; t)$ is essentially the bound proved by Kővári, Sós and Turán [KST1]. Let us present now the improvement due to Znám [Z6].

THEOREM 2.5. (i) *If $2 \leqslant t < n$ then*

$$z(n; t) < (t - 1)^{1/t} n^{2 - 1/t} + \frac{t - 1}{2} n.$$

(ii) *If a graph of order n does not contain a $K(t, t)$ then its size is at most*

$$\tfrac{1}{2}\left((t - 1)^{1/t} n^{2 - 1/t} + \frac{t - 1}{2} n \right).$$

Proof. As in Theorem 2.3, part (i) implies part (ii). Furthermore, by Lemma 2.1, to prove (i) it suffices to show that if

$$U = \frac{t - 1}{2} + (t - 1)^{1/t} n^{1 - 1/t} \quad \text{then} \quad n \binom{U}{t} > (t - 1) \binom{n}{t}. \qquad (3)$$

Put $2h = t - 1$. Inequality (3) clearly follows if we prove that

$$n^{2/t}(U - k)(U - 2h + k) - (t - 1)^{2/t}(n - k)(n - 2h + k) > 0 \qquad (4)$$

for every integer k, $0 \leqslant k \leqslant t - 1$. The left-hand side of (4) is quadratic in k and as $t - 1 < n$ the coefficient of k^2 is negative. Thus it suffices to check (4) for $k = 0$ (and $k = t - 1$, which is the same), i.e.

$$n^{2/t} U(U - 2h) - (t - 1)^{2/t} n(n - 2h) > 0.$$

By the definition of U this reduces to

$$(2h)^{1 + 2/t} n > h^2 n^2 / t,$$

which holds since $h < n$. ∎

It is somewhat disheartening that for numerical applications the almost trivial Lemma 2.1 is the strongest, e.g. Theorem 2.5(i) gives $z(10; 3) \leqslant 68$ and Lemma 2.1 gives $z(10; 3) \leqslant 62$. It is considerably harder to prove a good *lower bound* for $z(n; t)$. Naturally Theorems 2.2 and 2.5 and Lemma 2.1 all give almost the same upper bound for $z(n; t)$ and it seems very likely that they actually give the correct order. In other words it seems very likely that there is a positive number C_t depending only on t ($\geqslant 2$) such that

$$C_t n^{2 - 1/t} < z(n, t)$$

for all $n \geqslant t$ (Conjecture 15). This has been proved, as we shall see presently, for $t = 2$ and 3. If $t > 3$ it is not known that $c_t > 0$ exists but we do not know either that we cannot take $c_t = \frac{1}{2}$, say, for all t if n is sufficiently large. For $t > 3$ the lower bound one can give for $z(n; t)$ (see Theorem 2.10) seems rather a poor one.

Let us turn now to the case $t = 2$. Thus we would like to determine the maximal size of a bipartite graph $G_2(m, n)$ if it does not contain a $K(2, 2)$, i.e. a quadrilateral. The next theorem is due to Reiman [R7].

THEOREM 2.6. (i) $z(m, n; s, 2) \leqslant \frac{1}{2}\{m + (m^2 + 4(s - 1) mn(n + 1))^{\frac{1}{2}}\}$. *In particular* $z(n; 2) \leqslant \frac{1}{2}(n + n(4n - 3)^{\frac{1}{2}})$.

(ii) *There are an infinite number of pairs* (m, n) *with* $m/n \to \infty$ *and* $n \to \infty$ *for which equality is attained in* (i).

Proof. (i) Put $z = \frac{1}{2}\{m + (m^2 + 4(s - 1) mn(n - 1))^{\frac{1}{2}}\}$. Note that

$$(z - m) z = (s - 1) mn(n - 1).$$

Suppose that there is a graph $G = G_2(m, n)$ of size greater than z that does not contain a quadrilateral. Denote the degrees of the vertices in the first colour class of G by $d_1, \ldots, d_m$. Then $\sum_1^m d_i = e > z$ and, as in the proof of Lemma 2.1, we have

$$(s - 1)\binom{n}{2} \geqslant \sum_1^m \binom{d_i}{2} = \frac{1}{2}\sum_1^m d_i^2 - \frac{1}{2}\sum_1^m d_i$$

$$\geqslant \frac{1}{2m}e^2 - \frac{e}{2} > \frac{z(z - m)}{2m} = (s - 1)\binom{n}{2}.$$

(ii) The above sequence of inequalities implies immediately when we can have equality in (i) (cf. Theorem 2.4). If $G = G_2(m, n)$ is a bipartite graph with colour classes V_1, V_2 showing that we can have equality in (i) then the vertices in V_1 have the same degree and for every pair of vertices in V_2 there are exactly $s - 1$ vertices in V_2 joined to both of them. If $x \in V_1$ let us call $\Gamma(x) \subset V_2$ a *line*. Then every line has the same number of elements and for every pair of vertices in V_2 there are *exactly* $s - 1$ *lines* containing both of them. In this way we are clearly led to the finite projective geometries to provide examples proving (ii).

Let q be a prime power and let $PG(r, q)$ be the r-dimensional projective geometry over the field of order q. Denote by $p(r, q)$ the number of points and by $l(r, q)$ the number of lines in $PG(r, q)$. Put $m = (s - 1) l(r, q)$ and $n = p(r, q)$. Let V_1 be the set of lines in $(s - 1)$ disjoint copies of $PG(r, q)$ and let V_2 be the set of points of $PG(r, q)$. Let $G = G_2(m, n)$ be the bipartite graph with vertex classes V_1 and V_2 obtained by joining a line $x \in V_1$ to a point $y \in V_2$ if and only if they are incident. By Theorem 2.4 this graph G shows that we have equality for $z(m, n; s, 2)$.

Finally, for the sake of completeness, let us substitute the explicit values of $p(r, q)$ and $l(r, q)$:

$$m = (s - 1) l(r, q) = (s - 1) \frac{(q^{r+1} - 1)(q^r - 1)}{(q^2 - 1)(q - 1)},$$

$$n = p(r, q) = \frac{q^{r+1} - 1}{q - 1}.$$

In particular, if p is a prime then

$$z(p^2 + p + 1; 2) = (p - 1)(p^2 + p + 1). \qquad \blacksquare$$

In view of (0.7) the construction in the proof of Theorem 2.6(ii) has the following consequence.

COROLLARY 2.7. *If n is sufficiently large then*

$$n^{\frac{3}{2}} - n^{\frac{4}{3}} < z(n; 2) \leqslant \tfrac{1}{2}(n + n\sqrt{4n - 3}).$$

In particular,

$$\lim_{n \to \infty} z(n; 2) n^{-\frac{3}{2}} = 1. \qquad \blacksquare$$

Theorem 2.3 implies that the size of a graph of order n that does not contain a $K(2, 2)$ (a quadrilateral) is at most $\tfrac{1}{4}(n + n\sqrt{4n - 3})$. In 1966 Erdös, Rényi and Sós [ERS1] noticed that certain graphs constructed in 1962 by Erdös and Rényi [ER4] show that this bound is essentially best possible, and so gave an asymptotic solution to a problem raised by Erdös [E4] in 1938. The same result was proved simultaneously and independently by Brown [B51].

THEOREM 2.8. *Denote by* $ex(n; K(2, 2))$ *the maximal size of a graph of order n not containing a quadrilateral. If q is any odd prime power then*

$$\tfrac{1}{2}q(q + 1)^2 \leqslant ex \ (q^2 + q + 1; K(2, 2)) \leqslant \tfrac{1}{2}q(q + 1)^2 + \frac{q + 1}{2}. \tag{5}$$

Furthermore, if n is sufficiently large then

$$\tfrac{1}{2}(n^{\frac{3}{2}} - n^{\frac{4}{3}}) \leqslant ex \ (n; K(2, 2)) \leqslant \tfrac{1}{4}(n + n\sqrt{4n - 3}).$$

Proof. By Theorem 2.3, Theorem 2.6(i) and (0.7) it suffices to prove the *first* inequality in (5). To prove this inequality we recall the graph G defined before Theorem IV.1.4. The vertices of G are the points of the finite projective plane $PG(2, q)$ over the field of order q .A point is joined to all the points on its polar with respect to the conic

$$x^2 + y^2 + z^2 = 0, \tag{6}$$

i.e. (a, b, c) and (α, β, γ) are joined iff (α, β, γ) is on the line $[a, b, c]$. A point not on the conic (6) is joined to exactly $q + 1$ points and the $q + 1$ points on this conic are joined to exactly q points each. Thus

$$2e(G) = q^2(q + 1) + (q + 1) q = q(q + 1)^2.$$

Finally, G does not contain a quadrilateral since there is exactly one line through two points so every vertex is uniquely determined by any two vertices adjacent to it. ∎

Brown [B51] made use of finite geometries to prove that the bound in Theorem 2.5(i) gives the correct order for $t = 3$ as well.

THEOREM 2.9. *Let p be an odd prime. Then*

$$p^5 - p^4 \leqslant z(p^3; 3) < 2^{\frac{1}{3}}p^5 + p^3. \tag{7}$$

In particular,

$$\lim_{n \to \infty} \inf z(n; 3) \, n^{-\frac{5}{3}} \geqslant 1.$$

Proof. By Theorem 2.5(i) and (0.7) it suffices to prove the first inequality in (7).

Consider the 3-dimensional affine space $AG(3, p)$ over the field F_p of order p. Choose a non-zero element α of F_p which is a quadratic residue if and only if -1 is not a quadratic residue. Let $S(x)$ be the sphere with centre $x = (x_i) \in AG(3, p)$ consisting of the points $y = (y_i)$ satisfying the equation

$$\sum_1^3 (x_i - y_i)^2 = \alpha. \tag{8}$$

Let us prove two properties of these spheres.

(a) *Every line l in $AG(3, p)$ meets $S(x)$ in at most two points.*

By translating the points of $AG(3, p)$, if necessary, we may suppose that l goes through the origin, i.e. l consists of the points

$$y = ta, \qquad t \in F_p,$$

$$a = (a_1, a_2, a_3) \neq (0, 0, 0). \tag{9}$$

If l meets $S(x)$ in more than two points then, substituting (9) into (8), we obtain the quadratic equation in t

$$\sum_1^3 (a_i t - x_i)^2 = (\Sigma a_i^2) t^2 - 2(\Sigma a_i x_i) t + \Sigma x_i^2 = \alpha$$

with more than two solutions in t. Consequently

$$\Sigma a_i^2 = \Sigma a_i x_i = 0, \qquad \Sigma x_i^2 = \alpha.$$

As $(a_1, a_2, a_3) \neq (0, 0, 0)$, we can suppose without loss of generality that $a_1 \neq 0$. Then

$$a_1^2 \alpha = a_1^2 \Sigma x_i^2 = (a_1 x_1)^2 + a_1^2(x_2^2 + x_3^2) = (a_2 x_2 + a_3 x_3)^2 + a_1^2(x_2^2 + x_3^2)$$

$$= x_2^2(a_1^2 + a_2^2) + x_3^2(a_1^2 + a_3^2) + 2a_2 a_3 x_2 x_3 = -(x_2 a_3 - x_3 a_2)^2,$$

contradicting the choice of α and proving (a).

(b) *If x, y and z are different points of $AG(3, p)$ then $S(x) \cap S(y) \cap S(z)$ contains at most 2 points.*

For suppose $a, b, c \in S(x) \cap S(y) \cap S(z)$. Then, by (a), x, y and z are not collinear. Consequently, as a, b and c must be on the intersection of the radical planes of the spheres $S(x)$, $S(y)$ and $S(z)$. the points a, b and c must be collinear, contradicting (a).

Let us construct a bipartite graph $G = G_2(p^3, p^3)$ with vertex sets V_1, V_2, which are two copies of the set of p^3 points in $AG(3, p)$. Connect a vertex $x \in V_1$ to $y \in V_2$ if and only if $x \in S(y)$ or, equivalently, $y \in S(x)$. Then (b) implies that G does not contain a $K(3, 3)$.

Note finally that $S(x)$ contains $p^2 - p$ points. Thus $e(G) = p^5 - p^4$. ∎

At the first sight one might hope that a variant of the construction in the proof above can be carried out in the case $t > 3$ as well. Such hopes are shattered by the simple observation that in the 4-dimensional Euclidean space there are circles C_1, C_2 of arbitrary radii such that the distance of any two points x, y, $x \in C_1$, $y \in C_2$, is the same.

We conclude the discussion of $z(m, n; s, t)$ with a lower bound providing a rather poor complement of Theorem 2.2.

THEOREM 2.10. *If* $2 \leqslant s \leqslant m$ *and* $2 \leqslant t \leqslant n$ *then*

$$\lfloor (1 - 1/s!t!)m^{1-\alpha}n^{1-\beta} \rfloor \leqslant z(m, n; s, t),$$

where

$$\alpha = \frac{s-1}{st-1}, \quad \beta = \frac{t-1}{st-1}.$$

In particular,

$$\lfloor (1 - (t!)^{-2}) n^{2-2/(t+1)} \rfloor \leqslant z(n; t).$$

Furthermore, there is a graph of order n *and size*

$$\lfloor \tfrac{1}{2}(1 - (1/s!t!))n^{2-(s+t-2)/(st-1)} \rfloor$$

that does not contain a $K(s, t)$.

Proof. Let V_1, V_2 be fixed disjoint sets $|V_1| = m$, $|V_2| = n$. Put $E = \lfloor m^{1-\alpha}n^{1-\beta} \rfloor$ and consider all bipartite graphs of size E with vertex classes V_1 and V_2. There are $\binom{mn}{E}$ such graphs. A fixed $K(s, t)$ with vertex classes in V_1 and V_2, respectively, is contained in $\binom{mn-st}{E-st}$ of these graphs and there are $\binom{m}{s}\binom{n}{t}$ different $K(s, t)$ subgraphs. Consequently there is a bipartite graph $H = G_2(m, n)$ of size E containing at most the following number of $K(s, t)$ subgraphs:

$$\binom{m}{s}\binom{n}{t}\binom{mn - st}{E - st}\Big/\binom{mn}{E} \leqslant \binom{m}{s}\binom{n}{t}\left(\frac{E}{mn}\right)^{st}$$

$$\leqslant \frac{1}{s!t!}\, m^{s(1 - \alpha t)}\, n^{t(1 - \beta s)} \leqslant \frac{1}{s!t!}\, m^{1 - \alpha}\, n^{1 - \beta}.$$

Delete one edge from each $K(s, t)$ in H. Then we obtain a graph $G = G(m, n)$ of size at least $\lfloor (1 - (1/s!t!))\, m^{1 - \alpha}\, n^{1 - \beta}\rfloor$ that does not contain a $K(s, t)$.

The second assertion can be proved analogously.   ∎

Let us investigate now the number of various complete subgraphs contained in r-partite graphs for $r \geqslant 3$. Unlike in the case $r = 2$, it is interesting to look for conditions on $G_r(n_1, \ldots, n_r)$ ensuring the existence of a K^r or a K^p, where $3 \leqslant p \leqslant r$ is fixed.

The next simple result, proved in a more general form by Bollobás, Erdös and Straus [BES2], is another extension of Theorem 1.1.

THEOREM 2.11. *Let* $3 \leqslant p \leqslant r$. *The maximal size of a graph* $G_r(n_1, \ldots, n_r)$ *not containing a* K^p *is the maximal size of a complete* $(p - 1)$-*chromatic graph* $G_r(n_1, \ldots, n_r)$, *i.e. it is*

$$\binom{n}{2} - \min \sum_{1}^{p-1} \binom{n(I_j)}{2},$$

where the minimum is taken over all partitions $\bigcup_{1}^{p-1} I_j$ *of* $\{1, \ldots, r\}$ *and* $n(I_j) = \sum_{i \in I_j} n_i$. *In particular,*

$$\max\{e(G): G = G_r(n, \ldots, n) \not\supset K^p\} = t_{p-1}(r)\, n^2.$$

Proof. The result can be proved by adapting the proof of Theorem 1.2 to this situation. We leave this to the reader.   ∎

The next results we present are due to Bollobás, Erdös and Szemerédi [BES3] and concern r-partite graphs with equal colour classes. We put $G_r(n) = G_r(n, \ldots, n)$ and denote the colour classes by V_i, $|V_i| = n$, where $i \in \mathbb{Z}_n$, the group of integers modulo n. As we have seen, it is very simple to determine the maximal size of a graph $G_r(n)$ without a K^r. The other natural way one makes sure that $G_r(n)$ contains a K^r is that one requires a large minimal degree. This leads to much more interesting and difficult problems.

Given $r \geqslant 3$ and $n \geqslant 1$ denote by $\delta_r(n)$ the *minimal integer* such that every $G_r(n)$ with $\delta(G_r(n)) > \delta_r(n)$ contains a K^r. Clearly $\delta_r(kl) \geqslant k\delta_r(l)$ for every k and l and so $\lim_{n \to \infty} \inf \delta_r(n)/n \geqslant \delta_r(k)/k$ for every k. Consequently $\lim_{n \to \infty} \delta_r(n)/n = c_r$ exists. The case $r = 3$ is reasonably well understood. It will be proved in particular that $c_3 = 1$. However for $r > 3$ we have only the following rather bad bounds on c_r (cf. Problem 20 and Conjecture 21).

THEOREM 2.12. (i) $\delta_r(n) \leqslant \left\{r - 2 + \dfrac{r - 2}{r}\right\} n$ so $c_r \leqslant r - 2 + \dfrac{r - 2}{r}$.

(ii) $c_4 \geqslant 2 + \frac{1}{9}$ and $c_r \geqslant r - 2 + \frac{1}{2} - \dfrac{1}{r - 2}$ *for* $r \geqslant 5$.

Proof. (i) If $\delta(G_r(n)) > (r - 2 + (r - 2)/r)n$ then $e(G_r(n)) > t_{r-1}(n)$ so by Theorem 2.11 $G_r(n)$ contains a K^r.

(ii) Let $n = 9k$ and construct a graph $F_4 = G_4(n)$ as follows.

Let $V_1 = X_1 \cup X_2 \cup X_3, |X_1| = k, |X_2| = |X_3| = 4k, V_i = A_i \cup B_i, |A_i| = 8k, |B_i| = k, i = 2, 3$, and $V_4 = A_4 \cup B_4, |A_4| = 2k, |B_4| = 7k$. Join every vertex of X_1 to every vertex of $A_2 \cup A_3 \cup V_4$; join every vertex of X_i to every vertex of $V_i \cup A_j \cup A_4, i, j = 2, 3, i \neq j$; join every vertex of A_4 to every vertex of $A_2 \cup A_3$; join every vertex of B_4 to every vertex of $V_2 \cup V_3$; and finally, join every vertex of A_i to every vertex of $B_j, i, j = 2, 3, i \neq j$. The obtained graph is F_4 (see Fig. 2.1).

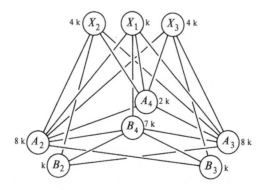

FIG. 2.1. The graph F_4 without a K^4.

It is easily checked that $\delta(F_4) = 19k = (2 + \tfrac{1}{9})n$ and $K^4 \not\subset F_4$ so $c_4 \geqslant 2 + \tfrac{1}{9}$.

To see the second inequality let $r \geqslant 5$, $k \geqslant 1$, $n = 2(r - 2)k$ and construct a graph $F_r = G_r(n)$ as follows. Let $V_i = A_i \cup B_i$, $|A_i| = |B_i| = (r - 2)k = n/2$, $i = 1, \ldots, r - 2$, and let

$$V_{r-1} = \bigcup_1^{r-2} A^j, \quad |A^j| = 2k,$$

$$V_r = \bigcup_1^{r-2} B^j, \quad |B^j| = 2k.$$

To construct F_r join two vertices of $\bigcup_1^r V_i$ that are in different classes unless one vertex is in A_i and the other is in $B_{i+1} \cup A^i$ or one vertex is in B_i and the other is in $A_{i+1} \cup B^i$ with the additional covention $A_{r-1} \equiv A_1$ and $B_{r-1} = B_1$ (see Fig. 2.2).

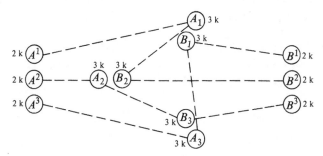

FIG. 2.2. The graph F_5 without a K^5. Vertices in classes joined by a dotted line are *not* joined.

It is easily checked that $\delta(F_r) = (r - 2 + \tfrac{1}{2} - 1/(r - 2))n$. Furthermore if $K^{r-2} \subset F_r - V_{r-1} \cup V_r$ then either every vertex of K^{r-2} belongs to $\bigcup_1^{r-2} A_i$ or every vertex of it belongs to $\bigcup_1^{r-2} B_i$. As no vertex in V_{r-1} is joined to a vertex in each A_i, $i = 1, \ldots, r - 2$, and no vertex in V_r is joined to a vertex in each B_i, $i = 1, \ldots, r - 2$, the graph F_r does not contain a K^r.   ∎

It is rather interesting that without knowing the value of c_r one can get a fairly good bound on the number of K^r's in a graph $G = G_r(n)$ provided $\delta(G) \geqslant (c_r + \varepsilon)n$.

THEOREM 2.13. *Let* $r \geqslant 4$ *and* $\varepsilon > 0$. *There exists a constant* $\delta = \delta(r, \varepsilon) > 0$ *such that if* $\delta(G_r(n)) > (c_r + \varepsilon)n$ *then* $G_r(n)$ *contains at least* δn^r K^r*s.*

Proof. Let $1 \leqslant m < n$, $G = C_r(n)$ and $\delta(G) > (c_r + \varepsilon)n$. We shall show that if $m = m(r, \varepsilon)$ and n is sufficiently large then for at least $\frac{1}{2}\binom{n}{m}^r$ choices of m-sets from the sets V_i the subgraph $G_r(m)$ spanned by the r m-sets contains a K^r. This will suffice to prove the theorem since then G contains at least

$$\frac{1}{2}\binom{n}{m}^r \bigg/ \binom{n-1}{m-1}^r = \frac{1}{2}m^{-r}n^r$$

K^rs, for one obtains at least $\frac{1}{2}\binom{n}{m}^r$ K^rs and each K^r occurs in at most $\binom{n-1}{m-1}^r$ choices.

We shall look for subgraphs $G_r(m)$ such that $\delta(G_r(m)) \geqslant (c_r + (\varepsilon/2))m$. Then, if m is large enough, the definition of c_r implies that $G_r(m)$ contains a K^r. In order to find such subgraphs we try to avoid putting a vertex $x \in V_i$ together with a set A_j of m vertices of V_j into the vertex set of $G_r(m)$ if x is joined to a much smaller proportion of the vertices in A_j than in V_j.

Let $x \in V_i$. Suppose x is joined to $u = d_j^{(x)}n$ vertices in V_j, $j \neq i$. As $c_r > r - 2$ we have $d_j^{(x)} > d > 0$ where d is an absolute constant. We say that an m-set $A_j \subset V_j$ is *bad with respect to* x or that (x, A_j) is a *bad pair* if x is joined to at most $(d_j^{(x)} - (\varepsilon/2r))m$ vertices in A_j. Clearly

$$\left(d_j^{(x)} - \frac{\varepsilon}{2r}\right)m \leqslant \left(1 - \frac{\varepsilon}{2r}\right)\frac{um}{n} = (1 - \varepsilon')\frac{um}{n}$$

and so the number of bad m-sets with respect to x (in V_j) is at most

$$\sum_{k \leqslant (1-\varepsilon')um/n} \binom{u}{k}\binom{n-u}{m-k}.$$

It is easily checked that this is at most

$$(1 - \eta)^m \binom{n}{m} < \frac{1}{2}(r^2m)^{-1}\binom{n}{m},$$

where $\eta = \eta(\varepsilon', d) > 0$. (To facilitate the proof of this inequality we may even suppose that n is sufficiently large compared to m.) Consequently the number of bad pairs is at most

$$rn(r - 1)\frac{1}{2}(r^2m)^{-1}\binom{n}{m} < \frac{n}{2m}\binom{n}{m}.$$

If (x, A_j), $x \in V$, is a bad pair, there are $\binom{n-1}{m-1}\binom{n}{m}^{r-2}$ choices of $\{A_h\}_1^r$, $A_h \subset V_h$, $|A_h| = m$ for which $x \in A_i$. (Of course, A_j is determined by the pair.) Thus the number of families $\{A_h\}_1^r$, $A_h \subset V_h$, $|A_h| = m$, $1 \leq h \leq r$, containing a bad pair is at most

$$\binom{n-1}{m-1}\binom{n}{m}^{r-2} \frac{n}{2m} \binom{n}{m} = \tfrac{1}{2}\binom{n}{m}^r.$$

Note now that if $\{A_h\}_1^r$ does not contain a bad pair then in the subgraph $G_r(m)$ spanned by $\bigcup_1^r A_h$ every vertex $x \in A_h$ has degree at least

$$\sum_{j \neq h} \left(d_j^{(x)} - \frac{\varepsilon}{2r} \right) m \geq \left(\frac{d(x)}{n} - \frac{\varepsilon}{2} \right) m \geq \left(c_r + \frac{\varepsilon}{2} \right) m.$$

Thus if m is sufficiently large then $G_r(m)$ contains a K^r.   ∎

As we have already mentioned, $c_3 = 1$. In fact Graver (see [BES2]) showed that if $\delta(G_3(n)) \geq n + 1$ then $G_3(n)$ contains a triangle (K^3). It is a somewhat unusual result that if $n \geq 4$ then $G_3(n)$ must contains at least 4 triangles and the result is best possible (see [BES2]).

THEOREM 2.14. *Suppose $G = G_3(n)$ has minimal degree $n + 1$. Then G contains at least $\min\{4, n\}$ triangles. This bound is best possible for every $n \geq 1$.*

Proof. (i) For $x \in V_i (i \in \mathbb{Z}_3)$ let $D^+(x)$ (resp. $D^-(x)$) be the set of vertices of V_{i+1} (resp. V_{i-1}) joined to x. Put $d^+(x) = |D^+(x)|$ and $d^-(x) = |D^-(x)|$. Since $d(x) = d^+(x) + d^-(x)$ is the degree of x, for every vertex x we have $d^+(x) + d^-(x) \geq n + 1$. We shall frequently rely on the following observations.

(*) Suppose $x \in V_i$, $y \in V_{i-1}$ and $xy \in E(G)$. Then there are at least

$$d^+(x) + d^-(y) - n$$

triangles containing the edge xy.   ∎

(**) Suppose $x \in V_i$ and $D' \subset D^-(x)$. Then there are at least

$$\sum_{y \in D'} (d^+(x) + d^-(y) - n)$$

triangles containing the vertex x.   ∎

Put $d_i^+ = \max\{d^+(x): x \in V_i\}$, $d_i^- = \max\{d^-(x): x \in V_i\}$. We may suppose without loss of generality that $d^+(x_1) = d_1^+ \geqslant \max(d_2^+, d_3^+)$ for some $x_1 \in V_1$.

Suppose $d_1^+ \leqslant n - 1$. Then $n \geqslant 3$ and $d^-(y) \geqslant 2$ for every y. Let $z \in D^-(x_1) \subset V_3$. If $d^+(z) = n - 1$ then by (**) (with $x = z, D' = D^-(z)$) there are at least 2 triangles with vertex z. If $d^+(z) \leqslant n - 2$ then by (*) at least 2 triangles of G contain the edge $x_1 z$. Thus at least 2 triangles contain each vertex of $D^-(x_1)$ and, as $|D^-(x_1)| \geqslant 1$, G has at least 4 triangles.

Suppose now that $d_1^+ = d^+(x_1) = n$ and the theorem holds for smaller values of n. Let us assume that G does not contain two triangles, say T_1 and T_2, such that $d^+(x_i) = n$ for a vertex x_i of T_i, $i = 1, 2$, where x_1 and x_2 are not necessarily distinct. By (*) there is a triangle T_1 containing x_1 and $D^-(x_1) = \{z_1\}$, $d^+(z_1) = n$, $d^-(z_1) = 1$, since otherwise (**) implies that there are two triangles containing x_1. Let $D^-(z_1) = \{y_1\}$. Then analogously $d^+(y_1) = n$ and $D^-(y_1) = \{x_1\}$. Let $G' = G_3(n - 1) = G - \{x_1, y_1, z_1\}$. Then $\delta(G') = n$ so G' contains at least $\min(4, n - 1)$ triangles implying that G contains at least $\min(4, n - 1) + 1 \geqslant \min\{4, n\}$ triangles.

Thus we may assume without loss of generality that G has two triangles each containing a vertex joined to every vertex of the next class. Similarly we may assume that G has two triangles each containing a vertex joined to every vertex of the previous class. If these four triangles are distinct we are home. Otherwise G contains a triangle $x_1 x_2 x_3$ such that

$$d^+(x_1) = n = \max\{d^-(x_1), d^-(x_2), d^-(x_3)\}.$$

We will show that in this case G contains at least n triangles. If $d^-(x_1) = n$ then for every edge yz, $y \in V_2$, $z \in V_3$, $x_i yz$ is a triangle. As there are at least n such edges G has at least n triangles. If $d^-(x_2) = n$ then G contains at least $d(x_3) - 1 \geqslant n$ triangles with vertex x_3. Finally if $d^-(x_3) = n$ then $x_1 x_3 y$ is a triangle for every $y \in V_2$. Thus G contains at least $\min\{4, n\}$ triangles, as claimed.

(ii) To show that G need not contain more triangles, put $G_3(1) = K^3$ and let $G_3(n)$ be obtained from $G_3(n - 1)$ with vertex classes V_1', V_2', V_3' by adding a vertex x_i to V_1' and joining it to x_{i+1}, x_{i+2} and to every vertex in V_{i+1}' (Fig. 2.3). This shows that G need not contain more than n triangles.

Let now $n \geqslant 5$. Instead of constructing a graph $G = G_3(n)$ with $\delta(G) = n + 1$ that contains only 4 triangles we shall construct $F(n, t) = G_3(n)$ with $n \geqslant 5t$ and $\delta(F) = n + t$ that contains only $4t^3$ triangles.

As always, the colour classes of $F(n, t)$ will be V_i, $|V_i| = n$, $i \in \mathbb{Z}_3$. Let $A_i \subset V_i$,

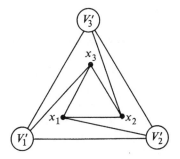

FIG. 2.3. The construction of $G_3(n)$ from $G_3(n-1)$.

$|A_i| = n - 2t$, $B_i = V_i - A_i$, $i \in \mathbb{Z}_3$ and $B_1 = \bar{B}_2 \cup \bar{B}_3$, $|\bar{B}_2| = |\bar{B}_3| = t$. Join every vertex of A_1 to every vertex of $A_2 \cup A_3$, join every vertex of $\bar{B}_j$ to every vertex of V_j, $j = 2, 3$ and join every vertex of B_i to every vertex of V_j for $i = 2$, $j = 3$ and $i = 3$, $j = 2$. Finally, join every vertex of $\bar{B}_i$ to t arbitrary vertices of A_j for $i = 2, j = 3$ and $i = 3$, $j = 2$. (In Fig. 2.4 a continuous line indicates that all the vertices of the corresponding classes are joined and a dotted line indicates that every vertex of $\bar{B}_i$ is joined to t arbitrary vertices of A_j.)

It is immediate that the triangles in $F(n, t)$ are of the form $x_i y_i z_j$, $x_i \in \bar{B}_i$, $y_i \in B_i$, $z_j \in A_j$, $i = 2, j = 3$ or $i = 3$, $j = 2$. Thus $k_3(F(n, t)) = 4t^3$, as claimed. ■

By making use of (**) in the proof of Theorem 2.14, one can prove a considerably more accurate version of Theorem 2.13 for the case $r = 3$.

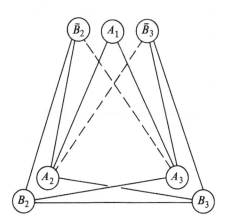

FIG. 2.4. The graph $F(n, t)$.

THEOREM 2.15. *If* $t \geqslant 1$ *then*

$$t^3 \leqslant \min\{k_3(G): G = G_3(n), \ \delta(G) \geqslant n + t\} \leqslant 4t^3.$$

Proof. The graph $F(n, t)$ constructed in the proof of Theorem 2.14 shows the second inequality. It is very likely that the second inequality is in fact equality (cf. Problem 21).

Let $G = G_3(n)$, $\delta(G) = n + t$. We can suppose without loss of generality that for some subset T_1 of V_1, $|T_1| = t$ we have

$$S = \sum_{x \in T_1} d^+(x) \geqslant \sum_{y \in T} d^+(y)$$

for every set $T \subset V_i$, $|T| = t$, $i \in \mathbb{Z}_3$.

Note that

$$d^-(y) - n \geqslant t - d^+(y) \quad \text{for every} \quad y \in V_i.$$

For $x \in T_1$ let $T_x \subset D^-(x)$, $|T_x| = t$. Then by (**) in the proof of Theorem 2.14, the number of triangles of G containing one vertex of T_1 is at least

$$\sum_{x \in T_1} \sum_{y \in T_x} (d^+(x) + d^+(y) - n) \geqslant \sum_{x \in T_1} \sum_{y \in T_x} (t + d^+(x) - d^+(y))$$

$$\geqslant \sum_{x \in T_1} \left(t^2 + td^+(x) - \sum_{y \in T_x} d^+(y)\right) \geqslant \sum_{x \in T_1} (t^2 + td^+(x) - S)$$

$$= t^3 + tS - tS = t^3. \qquad \blacksquare$$

It is easy to show that if $K^r \not\subset G_r(n)$ then $e(G_r(n)) \leqslant \left(\binom{r}{2} - 1\right)n^2$ and equality holds iff $G_r(n)$ is obtained from a $K_r(n)$ by omitting the edges joining two classes. The results are often similar if instead of a K^r we forbid another graph F. However, the problem becomes considerably more interesting if instead of asking for a bound on $e(G_r(n))$, we ask for a bound on the minimal number of edges joining two classes. As this kind of problem has not been investigated much, we only state the general problem. (See § 4 for the analogous general problem for graphs.)

Given an r-partite graph $G_r(n)$, denote by $e_{\min}(G_r(n))$ the minimum of the number of edges between two classes of $G_r(n)$. Let $\mathscr{F}$ be a finite family of graphs. Put

$$\text{ex}_r(n; \mathscr{F}) = \max\{e_{\min}(G_r(n)): F \not\subset G_r(n) \text{ if } F \in \mathscr{F}\}.$$

Determine or estimate $\text{ex}_r(n; \mathscr{F})$.

It is fairly surprising that even the case $\mathscr{F} = \{K_r(t)\}$, generalizing the problem of Zarankiewicz, is completely unresolved (see Problem 48).

To conclude this section we prove a result solving the natural complement of the problem of Zarankiewicz analogously to the way Theorem 1.13 complements Theorem 1.1 on graphs without complete subgraphs. Continuing the terminology of Theorem 1.13 we call a graph $G = G_2(m, n)$ $K(s, t)$-*saturated* if G does not contain a $K(s, t)$ but the addition of any edge joining vertices in different classes results in a graph containing a $K(s, t)$. Similarly $G = G_2(m, n)$ is *strongly $K(s, t)$-saturated* if the number of $K(s, t)$s increases whenever we add an edge joining vertices belonging to different classes. If $G = G_2(m, n)$ is such that $s - 1$ vertices in the first class have degree n and the others have degree $t - 1$, or is such that $t - 1$ vertices in the second class have degree m and the others have degree $s - 1$ then G is clearly $K(s, t)$-saturated and has size

$$(s - 1)n + (m - s + 1)(t - 1) = (t - 1)m + (n - t + 1)(s - 1).$$

The next result, due to Bollobás [B21] and Wessel [W15], shows that these graphs have minimal size among all the strongly $K(s, t)$-saturated graphs $G_2(m, n)$. We present this result mainly because of its proof, which might be applicable to several problems.

THEOREM 2.16. *A strongly $K(s, t)$-saturated graph $G = G_2(m, n)$ has size at most*

$$(s - 1)n + (m - s + 1)(t - 1) = (t - 1)m + (n - t + 1)(s - 1).$$

Proof. The result will be proved by assigning different weights of sets of vertices according to the existence of various edges. This method was first applied in [B19].

Let $G = G_2(m, n)$ and let $H = \overline{G_2(m, n)}$ be the complement of G in $K(m, n)$. Call an edge α of H $K(s, t)$-*essential* if $G + \alpha$ contains more $K(s, t)$ subgraphs than G. We shall prove the following assertion which seems slightly stronger than the theorem though is equivalent to it. The number of $K(s, t)$-essential edges of H is at most

$$(m - s + 1)(n - t + 1).$$

We apply induction on $s + t$. The assertion is trivial if $s = 1, t \geqslant 1$ or $s \geqslant 1$ and $t = 1$. Suppose $s > 1$, $t > 1$ and the assertion holds for all pairs (s', t')

with $s' + t' < s + t$. Let $H = G_2(m, n)$ and let $\alpha \in E(H)$ be a $K(s, t)$-essential edge. Then there is a set $S(\alpha)$ of s blue and t green vertices such that α is the only edge in the subgraph of H spanned by $S(\alpha)$. Let $b_1, b_2, \ldots, b_s$ be the blue vertices in $S(\alpha)$, where α is incident with b_1. Let the edge α (or more precisely the set $S(\alpha)$) assign weight $m - s + 1$ to b_1 and weight 1 to b_i, $i = 2, 3, \ldots, s$.

We claim that the $K(s, t)$-essential edges of H altogether assign at most

$$(m - s + 1)(n - t + 1)$$

weight to every blue vertex b. Suppose $d_H(b) = d$. If $d > n - t + 1$ then every $G_2(s, t)$ subgraph of H containing b has at least $d + t - n \geqslant 2$ edges, so no set $S(\alpha)$ contains b and hence no weight is assigned to b. Suppose now that $d \leqslant n - t + 1$. The edges of H incident with b assign at most $d(m - s + 1)$ weight to b. Denote by $\beta_1, \beta_2, \ldots, \beta_v$ the other edges of H assigning weight, and so weight 1, to the vertex b. No β_i has a common vertex with the edges (of H) incident with b so the β_is are in a bipartite subgraph $H^* = G(m - 1, n - d)$ of H. Furthermore, every $S(\beta_i) - b$ is contained in the vertex set of H^* so every β_i is a $K(s - 1, t)$-essential edge of H^*. Thus by the induction hypothesis

$$v \leqslant (m - s + 1)(n - d - t + 1).$$

Hence the weight assigned to a blue vertex is at most

$$d(m - s + 1) + (m - s + 1)(n - d - t + 1) = (m - s + 1)(n - t + 1).$$

On the other hand every $K(s, t)$-essential edge of H assigns exactly weight m to the set of blue vertices. Hence if e denotes the number of $K(s, t)$-essential edges of H then

$$m e \leqslant m(m - s + 1)(n - t + 1),$$

so

$$e \leqslant (m - s + 1)(n - t + 1). \qquad \blacksquare$$

The strongly $K(s, t)$-saturated graphs $G_2(m, n)$ of minimal size are exactly the graphs described before Theorem 2.16, as shown in [B22] and [W16]. The proof of this is left as an exercise (Ex. 16).

3. THE STRUCTURE OF GRAPHS

Recall that $t_q(n)$ denotes the maximal size of a q-partite graph of order n. There is a unique q-partite graph of order n and size $t_q(n)$, the complete q-partite graph $T_q(n)$, whose classes contain

$$n_1 = \left\lfloor \frac{n}{q} \right\rfloor, \; n_2 = \left\lfloor \frac{n+1}{q} \right\rfloor, \ldots, n_q = \left\lfloor \frac{n+q-1}{q} \right\rfloor$$

vertices, respectively. Thus

$$t_q(n) = \binom{n}{2} - \sum_1^q \binom{n_i}{2} = \tfrac{1}{2}\left(1 - \frac{1}{q}\right) n^2 + O(n). \tag{1}$$

Clearly $T_r(n)$ does not contain a K^{r+1} but according to Theorem 1.1 every graph $G(n, t_r(n) + 1)$ contains already a K^{r+1}. In fact, the proof of Theorem 1.1 or the proof of Theorem 1.2 also shows that every graph $G(n, t_r(n) + 1)$ contains a K^{r+2} from which one edge has been omitted. In other words every $G(n, t_r(n) + 1)$ contains a complete $(r + 1)$-partite graph one class of which contains two vertices and the other classes contain one vertex each. This simple observation is the first step towards a number of important results about the structure of extremal graphs not containing certain given subgraphs.

In 1946 Erdös and Stone [ES5] discovered the remarkable fact that for every natural number r and every $\varepsilon > 0$ if n is sufficiently large and

$$m \geqslant t_r(n) + \varepsilon n^2$$

then every $G(n, m)$ contains a complete $(r + 1)$-partite graph with *arbitrarily large vertex classes*.

To make this statement more precise, recall that $K_q(t)$ denotes the complete q-partite graph in which each class contains t vertices. Thus $K_q(1) = K^q$. Given n, r and $\varepsilon > 0$, put $g(n, r, \varepsilon) = \min\{t: \text{every } G(n, m) \text{ contains a } K_{r+1}(t) \text{ if } m \geqslant t_r(n) + \varepsilon n^2\}$. With this notation Erdös and Stone proved that

$$(l_r(n))^{\frac{1}{2}} \leqslant g(n, r, \varepsilon),$$

where l_s denotes the s times iterated logarithm. They also expected that the proper order of magnitude of $g(n, r, \varepsilon)$ is $l_r(n)$ if ε is small enough (see e.g.

Erdös [E18]). Later Erdös announced in [E2] that for given $\varepsilon > 0$ and $r \geqslant 2$ there exists a constant $c > 0$ such that

$$c \, (\log n)^{1/r} < g(n, r, \varepsilon) \tag{2}$$

and conjectured that the order of $g(n, r, \varepsilon)$ is $(\log n)^{1/r}$. However it turned out that for $r > 2$ the function $g(n, r, \varepsilon)$ is rather larger than expected. The proper order of $g(n, r, \varepsilon)$ was determined by Bollobás and Erdös [BE6], who proved that for any r and $0 < \varepsilon < 1/2r$ there are constants c_1 and $c_2 > 0$ such that

$$c_1 \log n \leqslant g(n, r, \varepsilon) \leqslant c_2 \log n \tag{3}$$

if n is sufficiently large. In (3) the upper bound for $g(n, r, \varepsilon)$ is more or less trivial, but it seems to be nearer to the truth (for small values of ε) than the lower bound, whose proof requires all the work. In particular, it was conjectured in [BE6] that there exists a constant c_r, depending only on r, such that

$$\frac{c_r \log n}{\log 1/\varepsilon} \leqslant g(n, r, \varepsilon),$$

i.e. for a fixed r the order of $g(n, r, \varepsilon)$ is $(\log n)/(\log 1/\varepsilon)$. This extension of the result in [BE6] was proved by Bollobás, Erdös and Simonovits [BES1].

THEOREM 3.1. (i) *There is an absolute constant $\alpha > 0$ such that if $0 < \varepsilon < 1/r$ and*

$$m > \left(1 - \frac{1}{r} + \varepsilon\right) \frac{n^2}{2} \tag{4}$$

then every $G(n, m)$ contains a $K_{r+1}(t)$, where

$$t = \left\lfloor \frac{\alpha \log n}{r \log 1/\varepsilon} \right\rfloor. \tag{5}$$

(ii) *Given a natural number r there exists a constant $\varepsilon_r > 0$ such that if $0 < \varepsilon < \varepsilon_r$ and $n \geqslant n(r, \varepsilon)$ is an integer then there exists a graph $G(n, m)$ satisfying (4) which does not contain a $K_{r+1}(t)$ with*

$$t = \left\lfloor 3 \frac{\log n}{\log 1/\varepsilon} \right\rfloor.$$

Proof. Let us start with the very easy proof of the second part which is in two steps. The first step is the proof of a result for $r = 1$ which is slightly stronger than the result we need.

Let $0 < \varepsilon < 1$ and $c_2 > 2/(\log 1/\varepsilon)$. Put $t = \lfloor c_2 \log n \rfloor$ and $m = \lfloor \varepsilon n^2/2 \rfloor$. Let us show that if n is sufficiently large then there exists a graph $G(n, m)$ that does not contain a $K_2(t)$.

The number of $K_2(t)$ graphs on n distinguishable vertices is

$$\frac{1}{2}\binom{n}{2t}\binom{2t}{t}$$

and there are

$$\binom{\binom{n}{2} - t^2}{m - t^2}$$

graphs with m edges containing a given set of t^2 edges. Thus the result follows if we show that for large enough n one has

$$\frac{1}{2}\binom{n}{2t}\binom{2t}{t}\binom{\binom{n}{2} - t^2}{m - t^2}\binom{\binom{n}{2}}{m}^{-1} < 1$$

As the left hand side is bounded by

$$(n(n-1)/2)^{-t^2} m^{t^2} n^{2t} \leqslant \left(\frac{\varepsilon n}{n-1}\right)^{t^2} n^{2t},$$

which tends to zero since $c_2 > 2/(\log 1/\varepsilon)$, the assertion follows.

To prove the result for $r > 1$ construct the following graph G if n is sufficiently large. Put $G_0 = T_r(n)$. Let $0 < \varepsilon < \varepsilon' < r^{-2}$. Add $\lfloor \varepsilon' n^2/2 \rfloor = \lfloor (r^2 \varepsilon')(n/r)^2/2 \rfloor$ edges to the largest class of G_0 in such a way that the class contains no $K_2(t)$ if

$$t \geqslant \frac{-5 \log n/r}{2 \log (r^2 \varepsilon')} = t_0$$

and n is sufficiently large. Note that if ε is sufficiently small, ε' is sufficiently near to ε and n is sufficiently large then the obtained graph $G = G(n, m)$ satisfies (4) and

$$t_0 < \frac{3 \log n}{\log 1/\varepsilon}.$$

This completes the proof of the second part.

Now let us turn to the essential part of the proof, i.e. to the proof of (i). To simplify the calculations we shall not choose $\alpha > 0$ immediately but we shall show that if $\alpha > 0$ is a *sufficiently small absolute constant* then the result holds.

First we deal with the case $r = 1$. Put $\alpha = 1$ and suppose that there are arbitrarily large values of n for which one can find a graph $G(n, m)$ satisfying (4) *without* a $K_2(t)$, where t is given by (5). Then, by Theorem 2.5 (ii)

$$2\varepsilon n^2 \leqslant (t - 1)^{1/t} n^{2 - 1/t} + tn,$$

and so

$$2 \leqslant 2\varepsilon n^{1/t} \leqslant (t - 1)^{1/t} + tn^{-1 + 1/t} < 2$$

if n is sufficiently large. Therefore $\alpha = 1$ will do in (5) if n is sufficiently large. This clearly implies the existence of the appropriate constant for $r = 1$. From now on we suppose that $r \geqslant 2$.

The following lemma will enable us to replace (4) by a more convenient condition.

LEMMA 3.2. *Let $\frac{1}{2} > c > \varepsilon > 0$. Suppose $n > 2c/\varepsilon$ and $e(G^n) > cn^2$. Then G^n contains a subgraph H of order p such that*

$$p > \varepsilon^{\frac{1}{2}} n,$$

$$\delta(H) > 2(c - \varepsilon)p$$

and

$$e(H) > e(G^n) - (c - \varepsilon)(n - p)(n + p + 1).$$

For sufficiently large n we may require that

$$p > (2\varepsilon)^{\frac{1}{2}} n.$$

Proof. Define a sequence of graphs $G_n = G^n$, $G_{n-1}, G_{n-2}, \ldots$ as follows. If $\delta(G_k) > 2(c - \varepsilon)k$ then stop the sequence with G_k. Otherwise pick a vertex $x_k \in G_k$ with $d_{G_k}(x_k) \leqslant 2(c - \varepsilon)k$ and put $G_{k-1} = G_k - x_k$. Let G_p be the last member of this sequence and put $H = G_p$. By construction

$$\binom{p}{2} \geqslant e(G_p) \geqslant e(G^n) - (c - \varepsilon) \sum_{k=p+1}^{n} k = e(G^n) - (c - \varepsilon)(n - p)(n + p + 1)$$

$$\geqslant cp^2 + \varepsilon(n^2 - p^2) - (c - \varepsilon)(n - p).$$

Hence

$$p^2(1 - 2c + 2\varepsilon) \geq 2\varepsilon n^2 - 2(c - \varepsilon)(n - p) + p > 2\varepsilon n^2 - 2cn,$$

and this implies both lower bounds on p. ∎

Lemma 3.2 shows that instead of Theorem 3.1 it suffices to prove the following assertion.

THEOREM 3.1'. *If $0 < \beta < 1$ is a sufficiently small absolute constant, $2 \leq r$, $0 < \varepsilon < 1/r$, and a graph G of order n satisfies*

$$\delta(G) \geq \left(1 - \frac{1}{r} + \varepsilon\right) n \tag{6}$$

then G contains a $K_{r+1}(M)$ where

$$M = \left\lfloor \beta \frac{\log n}{r \log(1/\varepsilon)} \right\rfloor.$$

Remark. It should be emphasized that though we have already spent some time on the proof of Theorem 3.1, all we have done was entirely routine. We have shown that the result (more precisely, part (i)), if true, is more or less best possible, we have proved part (i) in the almost trivial case $r = 1$ so that the slight peculiarities of this simple case should not interrupt the flow of the main argument, and we have justified the simple observation that instead of (4) we may use (6). Thus Theorem 3.1' is equivalent to Theorem 3.1 and we are about to begin the proof.

Proof of Theorem 3.1'. In our estimates we shall make use of the following part of inequality (1) of Ch. V. §4:

$$\binom{n}{k} \leq \frac{n^k}{k!} \approx \left(\frac{en}{k}\right)^k (2k)^{-\frac{1}{2}} < \left(\frac{en}{k}\right)^k.$$

The theorem is obvious if $M < 1$ so we can assume without loss of generality that $M \geq 1$, i.e.

$$n \geq (1/\varepsilon)^{r/\beta}. \tag{7}$$

In particular, we may assume that if $A > 3$ is an arbitrary constant then

$$n > A \quad \text{and} \quad n > A(\log n)^4 (1/\varepsilon)^4. \tag{7'}$$

We shall use induction on r. We have already proved the result for $r = 1$ so suppose that $r \geqslant 2$ and the result holds for smaller values of r.

Put

$$p_0 = \left\lfloor \frac{\beta \log n}{(r - 1) \log (1/\varepsilon')} \right\rfloor \geqslant \left\lfloor \frac{\beta \log n}{2(r - 1) \log r} \right\rfloor,$$

where $\varepsilon' = 1/(r - 1) - 1/r + \varepsilon$. As $\delta(G) \geqslant (1 - 1/(r + 1) + \varepsilon')n$, by the induction hypothesis G contains a $K_r(p_0)$.

In the sequel we shall make use of the intuitively obvious fact that if X is a small set of vertices of G then most vertices of G are joined to at least $(1 - 1/r + \varepsilon/2)|X|$ vertices in X. We state this as a lemma.

LEMMA 3.3. *Let $X \subset V(G)$ and denote by Y the set of those vertices of $G - X$ that are joined to at least $(1 - 1/r + \varepsilon/2)|X|$ vertices of X. Then*

$$\tfrac{1}{2}r\varepsilon n - |X| < |Y|.$$

Proof. Denote by S the number of edges joining vertices in X to the vertices in $V(G) - X$. Then, as $\delta(G) \geqslant (1 - 1/r + \varepsilon)n$,

$$|X|\left[\left(1 - \frac{1}{r} + \varepsilon\right)n - |X|\right] \leqslant S \leqslant |Y|\,|X| + (n - |X| - |Y|)\left(1 - \frac{1}{r} + \frac{\varepsilon}{2}\right)|X|.$$

Consequently

$$\frac{\varepsilon}{2}n|X| - \frac{|X|^2}{r} + \frac{\varepsilon}{2}|X|^2 \leqslant |Y|\left(\frac{1}{r} - \frac{\varepsilon}{2}\right)|X|,$$

implying the assertion of the lemma.   ∎

Let us continue the proof of Theorem 3.1'. Put $P = \lceil (2/\varepsilon)M \rceil < (3/\varepsilon)M$. Let us consider subgraphs $K_r(p_1, \ldots, p_r)$ of G such that

$$p_i \leqslant p + M, \qquad 1 \leqslant i \leqslant r, \tag{8}$$

where

$$p = \frac{1}{r}\sum_1^r p_i.$$

The rest of the proof is divided into two parts. In the first part we will show that if here we may have $p \leqslant P$ then G does contain a $K_{r+1}(M)$. In the second part we show that it cannot happen that $p < P$ for every $K_r(p_1, \ldots, p_r)$ satisfying (8) and G does contain a $K_{r+1}(M)$.

(a) Suppose that G contains a $K = K_r(p_1, \ldots, p_r)$ such that (8) holds and

$$P \leqslant p \leqslant P + 1.$$

(This is equivalent to supposing that there is a $K_r(p_1, \ldots, p_r)$ with $p \geqslant P$.)

If β is sufficiently small (in fact, if $\beta < \frac{1}{4}$) then by (7)

$$\varepsilon n > 4(P + 1). \tag{9}$$

Thus by Lemma 3.3 if Z denotes the set of vertices of $G - K$ joined to at least

$$\left(1 - \frac{1}{r} + \frac{\varepsilon}{2}\right) rp = p(r - 1) + \frac{\varepsilon pr}{2}$$

vertices of K then

$$|Z| \geqslant \frac{r}{2}(\varepsilon n - 2p) > \frac{\varepsilon rn}{4}.$$

A vertex of Z is joined to at least

$$p(r - 1) + \frac{\varepsilon rp}{2} - (r - 1)(p + M) \geqslant \frac{\varepsilon rP}{2} - (r - 1)M \geqslant M$$

vertices of each class of K, so it is joined to every vertex of a subgraph $K_r(M)$ of K. The number of $K_r(M)$ subgraphs of K is at most

$$\binom{P + 1 + M}{M}^r < \binom{2P}{M}^r < \left(\frac{2eP}{M}\right)^{rM} \leqslant \left(\frac{5e}{\varepsilon}\right)^{rM} \leqslant \left(\frac{1}{\varepsilon}\right)^{\beta \frac{\log n}{\log(1/\varepsilon)}} (5e)^{\beta \frac{\log n}{\log(1/\varepsilon)}}$$

$$< n^\beta e^{3\beta \frac{\log n}{\log(1/\varepsilon)}} < n^\beta n^{6\beta} = n^{7\beta}. \tag{10}$$

If β is sufficiently small then, by (7'),

$$n^{7\beta} < \frac{\varepsilon n}{\beta \log n} < \frac{\varepsilon n r^2 \log(1/\varepsilon)}{4\beta \log n} \leqslant \frac{r\varepsilon n}{4M} \leqslant \frac{|Z|}{M}.$$

Thus Z contains a set Z' of M vertices and K contains a $K_r(M)$ subgraph K' such that every vertex of Z' is joined to every vertex of K'. Consequently G contains a $K_{r+1}(M)$.

(b) Suppose that $p < P$ for every $K_r(p_1, \ldots, p_r)$ satisfying (8). Let $K = K_r(p_1, \ldots, p_r)$ be a subgraph for which p attains its maximum under the conditions (8). As G contains a $K_r(p_0)$, $M < p_0 \leqslant p < P$. Denote by U the set of vertices of G that are joined to at least M vertices of each class of K. Suppose first that this set U is large, say $|U| \geqslant n^{\frac{1}{2}}$. Then by (10) the number of $K_r(M)$ subgraphs of K is at most

$$\binom{p + M}{M}^r < n^{7\beta} < \frac{n^{\frac{1}{2}}}{M} < \frac{|U|}{M},$$

provided β is sufficiently small. Therefore, as in case (a), there are M vertices in U joined to every vertex of a $K_r(M)$ of K, showing that G does contain a $K_{r+1}(M)$.

Thus we may suppose that $|U| \leqslant n^{\frac{1}{2}}$. We will complete the proof of Theorem 3.1' by showing that this assumption leads to a contradiction. As in (a), denote by Z the set of vertices of $G - K$ joined to at least $(1 - (1/r) + (\varepsilon/2))\, rp$ vertices of K. (In fact we shall use only that each $z \in Z$ is joined to at least $(1 - (1/r))\, rp = (r - 1)\, p$ of the rp vertices of K. We have kept the definition of Z only to have a complete analogy with (a).) Then, as before, $|Z| > \varepsilon rn/4$. Put $W = Z - U$. If β is sufficiently small then

$$|W| > \frac{\varepsilon rn}{4} - n^{\frac{1}{2}} > \frac{\varepsilon rn}{5}.$$

Let us define an equivalence relation on W by putting $x \backsim y$ $(x, y \in W)$ if x and y are joined to exactly the same vertices of K. Our next aim is to show that there exists a large equivalence class, more precisely an equivalence class containing at least $\lfloor p + M \rfloor$ elements. We do this by showing that there are relatively few equivalence classes.

Let C_i denote the ith class of K. If $x \in W$ then there exists an index i_0, $1 \leqslant i_0 \leqslant r$, such that x is joined to less than M vertices of C_{i_0}. As x is joined to more than $(r - 1)p$ vertices of K, the number of vertices of $\bigcup_{j \neq i_0} C_j$ not joined to x is less than

$$(r - 1)(p + M) - ((r - 1)\, p - M) = rM.$$

Hence the number of equivalence classes in W is less than

$$\sum_{i=1}^{r} \left\{ \sum_{\lambda < M} \binom{P_i}{\lambda} \sum_{\mu < rM} \binom{rp - p_i}{\mu} \right\}$$

$$< r^2 M^2 \binom{2p}{M} \binom{rp}{rM} < r^2 M^2 \left(\frac{2P}{M}\right)^M \left(\frac{P}{M}\right)^{rM} e^{M(r+1)}$$

$$< r^2 M^3 e^{5rM} (1/\varepsilon)^{2rM} < 4 (\log n)^2 \, n^{5\beta/\log(1/\varepsilon)} \, n^{2\beta} < n^{\gamma},$$

where γ, obtained from (7'), depends only on β and tends to 0 as β does. In the above sequence of inequalities we have applied Stirling's formula and have used the fact that

$$P/M < \frac{3}{\varepsilon}, \quad rM \leqslant \frac{\beta \log n}{\log(/\varepsilon)} \quad \text{and} \quad 6.3^2 < (e^3)^2.$$

Therefore if β is sufficiently small then, again by (7'), the number of equivalence classes is less than

$$\frac{\varepsilon^3 n}{8 \log n} < \frac{\varepsilon r n}{10(3/\varepsilon)(\beta \log n/r\varepsilon)} < \frac{\varepsilon r n}{10 P} < \frac{\varepsilon r n}{5(p + M)}.$$

Consequently there exists a subset W_1 of W consisting of $\lfloor p + M \rfloor$ equivalent vertices.

We have arrived at the final step of the proof. We shall show that there is a $K' = K_r(q_1, \ldots, q_r)$ subgraph of G that contradicts the maximality of $K = K_r(p_1, \ldots, p_r)$. Denote by $\bar{C}_i$ the set of vertices of C_i (the ith class of K) joined to the vertices of W_1. We may suppose without loss of generality that $|\bar{C}_1| \leqslant |\bar{C}_2| \leqslant \ldots \leqslant |\bar{C}_r|$, in particular $|\bar{C}_1| < M$. We shall give different constructions for K' according to $|\bar{C}_2| \leqslant p$ or $|\bar{C}_2| > p$.

If $|\bar{C}_2| \leqslant p$, let the classes of $K' = K_r(q_1, \ldots, q_r)$ be defined as follows:

$$C_1^* = W_1, \quad C_2^* = C_1 \cup C_2 \quad \text{and} \quad C_j^* = \bar{C}_j, \, j = 3, \ldots, r.$$

Since

$$\left| \bigcup_1^r \bar{C}_j \right| > (r - 1)p,$$

we have

$$rp' = \left| \bigcup_1^r C_i^* \right| > rp.$$

Furthermore, $|C_i^*| \leqslant p + M < p' + M$. Thus K' satisfies (8) and contradicts the maximality of K.

If $|\overline{C}_2| > p$, select $q = \lfloor p + 1 \rfloor$ vertices from W_1 and also from each $\overline{C}_j, j = 2, \ldots, r$. These r vertex sets determine a $K_r(q)$ in G, contradicting the maximality of K.

This completes the proof of Theorem 3.1′ and so the proof of Theorem 3.1 as well. ∎

To conclude this section we present three lemmas about the existence of certain subgraphs. The first one is a variant of Lemma 2.1., the second is an essentially best possible result about the existence of an $(r + 1)$-partite graph $G_{r+1}(t, m_1, m_2, \ldots, m_r)$ such that for any $s \leqslant t$ vertices in the first class there are a large number of vertices in every other class adjacent to all of them. The third is a deep lemma due to Szemerédi [S29]. It will not be proved since we do not need it in the main body of the text. In fact, the sharp form of Lemma 3.5 is not needed either, but a weaker form of it will be applied a number of times, especially in §4 ,when investigating the structure of extremal graphs. It is slightly more convenient to state and prove these lemmas as assertions about systems of sets. In order to apply these lemmas to graphs one has to note only that a graph G with $V(G) = \{1, 2, \ldots, n\}$ is characterized by the set system $\{A_i : i = 1, 2, \ldots, n\}$, where $S_i = \Gamma(i), \quad i = 1, 2, \ldots, n$.

LEMMA 3.4. *Let* $X = \{1, \ldots, N\}, A_i \subset Y = \{1, \ldots, n\}$ *for each* $i \in X \sum_1^N |A_i|$
$\geqslant N\varepsilon n > 0, 0 < \alpha < 1$ *and* $(1 - \alpha)\varepsilon N \geqslant s$, *where* N, n *and* s *are natural numbers. Then there are indices* $i_1, i_2, \ldots, i_j \in X$ *such that*

$$\left| \bigcap_{k=1}^s A_{i_k} \right| \geqslant n(\alpha\varepsilon)^s.$$

Proof. For $i \in Y$ let $X_i = \{j : i \in A_j, j \in X\}, x_i = |X_i|$. We say that an s-set of X belongs to $i \in Y$ if $S \subset X_i$. Clearly $\binom{x_i}{s}$ s-sets belong to $i \in Y$. As $\sum_1^n x_i \geqslant N\varepsilon n$,

$$\sum_1^n \binom{x_i}{s} \geqslant n\binom{\varepsilon N}{s} \geqslant n\binom{N}{s}\binom{\varepsilon N}{s} \bigg/ \binom{N}{s} \geqslant n\left(\frac{\varepsilon N - 1}{N}\right)^s \geqslant \binom{N}{s}n(\alpha\varepsilon)^s.$$

Thus at least one s-set of X belongs to at least $n(\alpha\varepsilon)^s$ elements of Y, as asserted by the lemma. ∎

Suppose we are given subsets $A_1, \ldots, A_N$ of a set Y, where $|Y| = n$ and $|A_i| \geqslant \varepsilon n > 0$, and we would like to choose a certain number of these subsets in such a way that the various intersections of these subsets are as large as possible. If Y is a discrete probability space and the A_is are sets corresponding to *independent* events of probability ε then

$$\left| \bigcap_{j \in I} A_j \right| = \varepsilon^{|I|} n \quad \text{for every} \quad I \subset \{1, \ldots, N\}, \tag{11}$$

so (for a fixed N and arbitrary n) we cannot hope to do better than (11). The following consequence of Lemma 2.11 and the finite form of Ramsey's Theorem (Theorem V.6.2) shows that we can always do almost as well as (11), i.e. we *can select subsets behaving almost as well as if they were random.* We state this random selection lemma in a rather general form though for our purposes a considerably weaker version would also suffice.

LEMMA 3.5. *Let* $\varepsilon_1, \ldots, \varepsilon_k$ *be positive numbers, let* $0 < \alpha < 1$ *and let* t *be a natural number. Then there exists a natural number* $N = N(\varepsilon_1, \ldots, \varepsilon_k; \alpha; t)$ *with the following property.*

Let $V_1, \ldots, V_k$ *be sets,* $|V_i| = n_i$, *and let*

$$A_{ij} \subset V_i, \quad j \in \{1, \ldots, N\} = X,$$

$$\frac{1}{N} \sum_{j=1}^{N} |A_{ij}| \geqslant \varepsilon_i n_i, \quad i = 1, \ldots, k.$$

Then there is a subset $T \subset X, |T| = t$, *such that if* $S \subset T$ *then*

$$\left| \bigcap_{j \in S} A_{ij} \right| \geqslant (\alpha \varepsilon_i)^{|S|} n_i, \quad i = 1, \ldots, k.$$

Proof. Let $0 < \varepsilon' < \varepsilon$ and $0 < \alpha < \alpha' < 1$ be such that $\varepsilon \alpha = \varepsilon' \alpha'$. Suppose $A_1, A_2, \ldots, A_N$ are subsets of a set $V, |V| = n$ and

$$\frac{1}{N} \sum_{1}^{N} |A_j| \geqslant \varepsilon n.$$

If N' of the sets A_j have at least $\varepsilon' n$ elements then

$$N' + \varepsilon'(N - N') \geqslant N \varepsilon$$

so

$$N' \geqslant \frac{\varepsilon - \varepsilon'}{1 - \varepsilon'} N.$$

Furthermore, it clearly suffices to prove the lemma for $k = 1$ since $k > 1$ can be obtained by repeated application of the result for $k = 1$. Thus it suffices to prove the existence of an $N = N(\varepsilon, \alpha, s, t)$ such that if $A_1, \ldots, A_N$ are subsets of V, $|V| = n$ and

$$|A_j| \geqslant \varepsilon n \quad \text{for every} \quad j \in X = \{1, 2, \ldots, N\},$$

then there is a subset $T \subset X$, $|T| = t$, such that whenever S is an s-element subset of T we have

$$\left| \bigcap A_j \right| \geqslant (\alpha\varepsilon)^s n.$$

Let M be such that

$$(1 - \alpha)\varepsilon M > s. \tag{12}$$

Then by Lemma 3.4 if $Y \subset X$ and $|Y| = M$ then there is an s-element subset S of Y such that (12) holds. Thus, by Ramsey's theorem (V.6.2), if N is sufficiently large (depending only on M, t and s) there is a t-element subset T with the required property. ∎

The final lemma of the section is due to Szemerédi [S29]. This lemma was an important tool in Szemerédi's proof [S28] of the deep and important result that a sequence of integers with positive upper density contains arbitrarily long arithmetic progressions. Given subsets $U, W \subset V(G)$, denote by $e(U, W)$ the number of $U - W$ edges of G. The *density* of edges between U and W is

$$d(U, W) = \frac{e(U, W)}{|U||W|}.$$

The pair (U, W) is *ε-uniform* if

$$\left| d(U', W') - d(U, W) \right| < \varepsilon,$$

whenever $U' \subset U$, $W' \subset W$, $|U'| > \varepsilon|U|$ and $|W'| > \varepsilon|W|$.

LEMMA 3.6. (*Uniform Density Lemma*) *Given $\varepsilon > 0$ and an integer m, there is an $M = M(\varepsilon, m)$ such that the vertices of every graph G^n can be partitioned* *into classes* $V_1, V_2, \ldots, V_k$, *where* $m \leqslant k \leqslant M$ *and* $\lfloor n/k \rfloor \leqslant |V_i| \leqslant \lceil n/k \rceil$, *such that all but* $\lfloor \varepsilon k^2 \rfloor$ *of the pairs* $(V_i, V_j), 1 \leqslant i \leqslant j < k$, *are* ε-*uniform.*   ∎

4. THE STRUCTURE OF EXTREMAL GRAPHS WITHOUT FORBIDDEN SUBGRAPHS

Let $\mathscr{F}$ be a family of graphs, called the family of *forbidden* graphs. Denote by $EX(n; \mathscr{F})$ the set of graphs of order n not containing any forbidden graph and having *maximum* size. The elements of $EX(n; \mathscr{F})$ are the *extremal* graphs for the family $\mathscr{F}$. Write $ex(n; \mathscr{F})$ for the size of an extremal graph $H \in EX(n; \mathscr{F})$. If $\mathscr{F} = \{F_1, \ldots, F_k\}$ then we write $EX(n; F_1, \ldots, F_k)$ and $ex(n; F_1, \ldots, F_k)$. With this notation Theorem 1.1 states that

$$ex(n; K^{r+1}) = t_r(n) = \tfrac{1}{2}\left(1 - \frac{1}{r}\right)n^2 + O(n).$$

Put

$$r + 1 = \min_i \chi(F_i) = \chi(F_1).$$

We may suppose without loss of generality that $r \geqslant 1$. Erdös and Simonovits [ES1] observed that Theorem 3.1 has the following immediate consequence.

THEOREM 4.1.

$$\lim_{n \to \infty} ex(n; F_1, \ldots, F_k)n^{-2} = \tfrac{1}{2}\left(1 - \frac{1}{r}\right).$$

Proof. Note first that $F_i \not\subset T_r(n)$ since F_i is not r-colourable. Thus

$$ex(n; F_1, \ldots, F_k) \geqslant e(T_r(n)) = t_r(n) = \tfrac{1}{2}\left(1 - \frac{1}{r}\right)n^2 + O(n).$$

Conversely, if $\varepsilon > 0, m \geqslant \tfrac{1}{2}(1 - (1/r) + \varepsilon)n^2$ and n is sufficiently large then by Theorem 3.1 every $G(n, m)$ contains a $K_{r+1}(t)$ where t is the maximum of the order of the F_is. Then $K_{r+1}(t)$ contains F_1 since F_1 is $(t + 1)$-chromatic so

$$\limsup_{n \to \infty} ex(n, F_1, \ldots, F_k)n^{-2} \leqslant \tfrac{1}{2}\left(1 - \frac{1}{r}\right) + \varepsilon.$$ ∎

It is clear that the proof above is rather crude, all we needed was a very simple form of Theorem 3.1. However, it is equally clear that in order to prove finer results about $\mathrm{ex}(n; F_1, \ldots, F_k)$ and about the structure of the extremal graphs we should first concentrate on the case $k = 1, F_1 = K_r(t)$. This is exactly what we are going to do. The next results we shall present are due to Erdös [E21], [E22], Erdös and Simonovits [ES2] and Simonovits [S19]. The proofs we present are considerably simpler than the original ones. We start with a result asserting that if the size of a graph of order n is almost $t_r(n)$ and the graph does not contain a $K_{r+1}(t)$ (where t is small compared to n) then the graph is almost $T_r(n)$. This theorem is also an almost immediate consequence of Theorem 3.1 but, when dealing with extremal graphs, it is more powerful than Theorem 3.1.

THEOREM 4.2. *Let $r \geqslant 2$ and $t \geqslant 1$ be fixed. Suppose $G = G(n, m)$ does not contain a $K_{r+1}(t)$ and*

$$m = \tfrac{1}{2}\left(1 - \frac{1}{r} + o(1)\right)n^2.$$

Then the following assertions hold.

(i) *There is a $K_r(p_1, \ldots, p_r)$, $\displaystyle\sum_{1}^{n} p_i = n$, $p_i = (1 + o(1))\dfrac{n}{r}$, $i = 1, \ldots, r$.
   which can be obtained from G by adding and subtracting $o(n^2)$ edges.*
(ii) *G contains an r-partite graph of size $\tfrac{1}{2}(1 - 1/r + o(1))n^2$.*
(iii) *G contains an r-partite graph of minimal degree $(1 - 1/r + o(1))n$.*

Proof. Note first that (i) is equivalent to (ii) and (iii) implies (ii).

Throughout the argument we may discard any $o(n)$ of the vertices that do not please us. In particular, we may, and will, suppose that

$$\delta(G) = \left(1 - \frac{1}{r} + o(1)\right)n.$$

For if there are $n_1 = \lfloor \varepsilon n \rfloor$ vertices of degree at most $(1 - 1/r - \eta)n$ $(0 < 2\varepsilon < \eta)$ then after discarding these n_1 vertices we obtain a graph H of order $n_2 = n - n_1 = \lceil (1 - \varepsilon)n \rceil$ and size at least

$$\tfrac{1}{2}\left(1 - \frac{1}{r} + o(1)\right)n_2^2 + \tfrac{1}{2}\varepsilon^2 n_2^2.$$

By Theorem 3.1 this graph H contains a $K_{r+1}(t)$ if n is sufficiently large.

By applying the preceding argument to the r-partite subgraph guaranteed by (ii), we see also that (ii) implies (iii). Thus to prove the theorem it suffices to show that (ii) holds.

Note now that, again by Theorem 3.1, G contains a $K = K_r(s), s = \lfloor c \log n \rfloor$, where $c > 0$ depends on r. Denote by $C_1, \ldots, C_r$ the vertex classes of K. Let X be the set of vertices of $G - K$ that are joined to at least t vertices in each class of K. Each vertex of X is joined to every vertex of at least one of the $\binom{s}{t}^r K_r(t)$ subgraphs of K. Thus $|X| < t \binom{s}{t}^r = o(n)$ since otherwise X contains t vertices each of which is joined to every vertex of $K_r(t)$. Discard the vertices in X.

Let Y be the set of vertices in $G - K$ that are joined to at most

$$s_0 = (r - 1)s - \frac{s}{2t} + 2t$$

vertices of K. As there are $s(r - 1 + o(1))n$ edges joining K to $G - K$, we have $|Y| = o(n)$ so we may discard the vertices in Y as well.

Denote by V_0 the vertex set of $G - K$. Let V_i be the set of vertices of $G - K$ joined to less than t vertices of C_i, $i = 1, \ldots, r$. As each vertex of V_0 is joined to more than $s_0 > (r - 2)s + 2t$ vertices of K and less than t vertices of one of the C_is, this defines a partition of V_0 (Fig. 3.1).

If there are more than $o(n^2)$ edges (in fact if there are more than $O(n^{2 - 1/t})$ edges) joining vertices of some class V_i, say V_1, then $V_1[G]$ contains a $K_2(t)$,

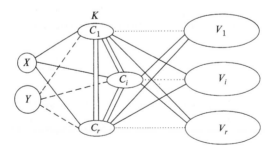

Fig. 3.1. The partition of $V(G)$. Each $x \in X$ is joined to at least t vertices in each C_i, each $y \in Y$ is joined to at most s_0 vertices in $\bigcup_1^r C_i$ and each $v_i \in V_i$ is joined to fewer than t vertices in C_i.

say with vertex set D_1. Each vertex of D_1 is joined to at least

$$(r - 1)s - \frac{s}{2t} + t - (r - 2)s$$

$$= s - \frac{s}{2t} + t$$

vertices in each $C_i (i \geqslant 2)$ so each C_i contains a set D_i of t vertices joined to every vertex of D_1. Then $\bigcup_{1}^{r} D_i$ spans a graph containing a $K_{r+1}(t)$. Thus there are only $o(n^2)$ edges in each set V_i so we may discard them. Finally, omit the vertices of K. The obtained subgraph of G is r-partite and has size $\frac{1}{2}(1 - 1/r + o(1))n^2$, proving (ii). $\blacksquare$

COROLLARY 4.3. *Let* $\mathscr{F} = \{F_1, \ldots, F_k\}, r + 1 = \min \chi(F_i)$ *and* $G \in EX(n; \mathscr{F})$. *Then*

$$\delta(G) = (1 - 1/r + o(1))n.$$

Proof. Suppose $\varepsilon > 0$ and there exists a vertex $x \in V(G)$ of degree at most $(1 - 1/r - \varepsilon)n$. Theorem 4.2 (iii) implies that G contains t vertices, say $x_1, \ldots, x_t$, where $t = \max |F_i|$, that are joined to $y_1, \ldots, y_q$, where $q = (1 - 1/r + o(1))n$. Let G^* be obtained from $G - x$ by joining x to each of the vertices $y_1, \ldots, y_q$. Clearly $e(G^*) > e(G)$. Suppose G^* contains F_i, say the subgraph G_i^* of G^* spanned by $\{x, z_1, \ldots, x_u\}, u < t$, contains F_i. Let x_j be a vertex not in $\{z_1, \ldots, z_u\}$. Then the subgraph G_i of G spanned by $\{x_j, z_1, z_2, \ldots, z_u\}$ contains G_i^* so it contains F_i as well. Thus G^* does not contain any of the forbidden subgraphs and has more edges than G, contradicting the extremality of G. $\blacksquare$

THEOREM 4.4. *Let* $\mathscr{F} = \{F_1, \ldots, F_k\}$ *and* $r + 1 = \chi(F_1) = \min \chi(F_i)$. *Then for every* $\varepsilon, 0 < \varepsilon < (1/2r)$, *there exists a constant* c_ε *with the following property. If* $G \in EX(n; \mathscr{F})$ *then the vertices of* G *can be partitioned into* r *classes, each with* $(1/r + o(1))n$ *vertices, such that with the exception of at most* c_ε *vertices each vertex of the graph is joined to all but* εn *of the vertices in the other classes. Furthermore, the* r-partite *subgraph of* G *spanned by these classes has size* $\frac{1}{2}(1 - 1/r + o(1))n^2$.

Proof. Let $W_1, W_2, \ldots, W_r$ be the vertex classes of an r-partite graph H of size $\frac{1}{2}(1 - 1/r + o(1))n^2$ and minimal degree $\delta(H) = (1 - 1/r + o(1))n$ contained in G, guaranteed by Theorem 4.2(iii). Let $\bigcup_{1}^{r} V_i$ be a partitioning of the vertices of G such that $W_i \subset V_i$ and if $x \in V(G) - \bigcup_1^r W_i$ belongs to V_i then x is not joined to more vertices of W_i than of $W_j, j \neq i$. In particular, by Corollary 4.3, each $x \in V_i$ is joined to at least $\frac{1}{2}((1/r) + o(1))n > (1/3r)$ vertices of each W_j, $j \neq i$, if n is sufficiently large.

Suppose $F_1 \subset K_{r+1}(t, u, u, \ldots, u)$. Denote by X the set of vertices in V_1 joined to at least εn vertices in V_1 and put $N = |X|$ and $\eta = (\varepsilon/2)^t$. Then by the selection lemma (Lemma 3.5) if $N \geq N(\varepsilon, r)$ one can find a set $X_0 \subset X$ of t vertices and subsets $Z_i \subset V_i, |Z_i| = \lfloor \eta n \rfloor$ such that every vertex of X_0 is joined to every vertex of Z_i, $i = 1, \ldots, r$.

As the r-partite subgraph of G spanned by the vertex classes $Z_1, \ldots, Z_r$ has size $\frac{1}{2}(1 - (1/r) + o(1))r^2\eta^2 n^2$, by Theorem 3.1 it contains a $K_r(u)$ if n is sufficiently large. This $K_r(u)$ together with X_0 spans a $K_{r+1}(t, u, u, \ldots, u)$, containing F_1. Consequently the assertion of the theorem holds with $c_\varepsilon = rN(r, \varepsilon)$. ∎

Remarks 4.5. (i) Note that the constant c_ε we obtained in Theorem 4.4 depends only on r, ε and t and it *does not depend on* u.

(ii) With a little more work one can prove the following slight extension of Theorems 4.2 and 4.4 (see [BESS1]).

Given $\varepsilon > 0$ and natural numbers r and t, there exist $c_\varepsilon = c(\varepsilon, r, t)$, $n_0 = n(\varepsilon, r, t)$ and $\eta = \eta(\varepsilon, r, t) > 0$ such that if $n > n_0$, $e(G^n) > \frac{1}{2}(1 - 1/r - \eta)n^2$ and $K_{r+1}(t) \not\subset G^n$ then the vertices of G^n can be partitioned into r classes, say $V_1, V_2, \ldots, V_r$, such that $\big| |V_i| - n/r \big| < \varepsilon n$, $e(G^n[V_i]) < \varepsilon n^2$, $e(V_i, V_j) > (r^{-2} + \varepsilon)n^2$ for $1 \leq i < j \leq r$, and, with the exception of at most c_ε vertices, each vertex of G^n is joined to at most εn vertices in its own class and to all but εn vertices in the other classes.

Armed with Theorem 4.4 we can improve Theorem 4.1 considerably.

THEOREM 4.6. *Let $F_1 = F_0 + K_{r-1}(u)$ where $\chi(F_0) = 2$ and let $F_2, \ldots, F_k$ be arbitrary graphs. Then*

$$ ex(n; F_1, \ldots, F_k) \leq \frac{1}{2}\left(1 - \frac{1}{r}\right)n^2 + (r + o(1))\, ex\left(\left\lfloor \frac{n}{r} \right\rfloor; F_0\right) + cn $$

where c depends only on F_0 and r.

Proof. We may suppose without loss of generality that $r + 1 = \min_i \chi(G_i)$. Let G be an extremal graph of order n. Put

$$ p_0 = |F_0| \quad \text{and} \quad \varepsilon = \frac{1}{3rp_0}. $$

This choice of ε makes sure that if each of p_0 vertices is joined to all but εn of a set V_i of $((1/r) + o(1))n$ vertices then V_i contains a subset W_i, $|W_i| = \lceil \eta n \rceil$, $\eta = (1/2r)$, such that each of those p_0 vertices is joined to every vertex in W_i.

By Theorem 4.4 we may delete $c = c_\varepsilon$ vertices of G (c depends only on r and p_0) in such a way that the remaining graph G^* has the following properties. The vertices of G^* can be partitioned into r classes, say $V_1, \ldots, V_r$, $|V_i| = ((1/r) + o(1))n$, such that each vertex $x \in V_i$ is joined to all but εn vertices of V_j for every $j \neq i$. Furthermore, the r-partite subgraph H^* of G^* spanned by the vertex classes $V_1, \ldots, V_r$ has size $\frac{1}{2}(1 - (1/r) + o(1))n^2$.

Denote by G_i the subgraph of G^* spanned by V_i^*. Suppose one of the G_is, say G_1, contains an F_0. Then by the choice of ε we can find

$$W_j \subset V_j, \quad |W_j| = \lceil \eta n \rceil, \quad j = 2, \ldots, r,$$

such that each vertex of F_0 is joined to each vertex of W_j, $j = 2, \ldots, r$. Clearly the $(r - 1)$-partite subgraph of H^* spanned by $\bigcup_2^r W_j$ has

$$\frac{1}{2}\left(1 - \frac{1}{r - 1} + o(1)\right)(r - 1)^2 \eta^2 n^2$$

edges so it contains a $K_{r-1}(u)$ if n is sufficiently large. The vertices of this $K_{r-1}(u)$ together with the vertices of the F_0 in G_1 span a subgraph of G containing F. Consequently no G_i contains an F_0 and so

$$e(G) \leqslant \frac{1}{2}\left(1 - \frac{1}{r}\right)n^2 + r\ ex\left((1 + o(1))\frac{n}{r}; F_0\right) + cn.$$

To complete the proof it suffices to show that

$$ex(m'; F_0) \leqslant \frac{m}{m - 1}(1 + \varepsilon)^2 ex(m; F_0) \tag{1}$$

if $m' = (1 + \varepsilon)m$. Let H be a graph of order $m' = (1 + \varepsilon)m$ without an F_0. Then each of the $\binom{m'}{m}$ subgraphs of H spanned by m of the vertices has at most $ex(m; F_0)$ edges. As an edge is contained in $\binom{m'-2}{m-2}$ of these subgraphs

$$e(H) \leqslant ex(m, F_0)\binom{m'}{m} \Big/ \binom{m' - 2}{m - 2} \leqslant \frac{m}{m - 1}(1 + \varepsilon)^2 ex(m; F_0),$$

proving (1). ∎

COROLLARY 4.7. *Let* $r + 1 = \min\limits_{1 \leqslant i \leqslant k} \chi(F_i) \geqslant 2$ *and suppose* F_1 *has an* $(r + 1)$-*colouring in which one of the colour classes contains* t_1 *vertices and another contains* t_2 *vertices* $(t_1 \leqslant t_2)$. *Then*

$$ex(n; F_1, \ldots, F_k) < \tfrac{1}{2}\left(2 - \frac{1}{r}\right)n^2 + cn^{2 - 1/t_1},$$

where the constant c *depends only on* t_1 *and* t_2.

Proof. Clearly $F_1 \subset K(t_1, t_2) + K_{r-1}(u)$ for some u. Put

$$F_0 = K(t_1, t_2), \qquad F_1' = F_0 + K_{r-1}(u).$$

By Theorem 2.3 we have

$$ex\left(\left\lfloor \frac{n}{r} \right\rfloor; F_1\right) < c \left(\frac{n}{r}\right)^{2 - 1/t_1},$$

where c depends only on t_1 and t_2. Apply Theorem 4.6 to $\{F_1', F_2, \ldots, F_k\}$. ∎

To conclude the discussion of the structure of extremal graphs not containing a set of graphs $\{F_1, \ldots, F_k\}$ let us summarize our results.

THEOREM 4.8. *Let* $r + 1 = \min\limits_{1 \leqslant i \leqslant k} \chi(F_i)$ *and suppose* F_1 *has an* $(r + 1)$-*colouring in which one of the colour classes contains* t *vertices. Let* $G = G^n$ *be an extremal graph with respect to the forbidden graphs* $\{F_i\}_1^k$. *Then*

$$\delta(G) = \left(1 - \frac{1}{r} + o(1)\right)n,$$

the vertices of G *can be partitioned into* r *classes such that each vertex is joined to at most as many vertices of its own class as of another class, for every* $\varepsilon > 0$ *there are at most* $c_\varepsilon = c(\varepsilon, F_1, \ldots, F_k)$ *vertices joined to at least* εn *vertices of the same class, there are* $O(n^{2 - 1/t})$ *edges joining vertices belonging to the same class and each class has* $(n/r) + O(n^{1 - 1/2t})$ *vertices.*

Proof. We have already proved everything except the last statement. Note that if an r-partite graph of order n has size at least $\tfrac{1}{2}(1 - (1/r))n^2 + O(n^{2 - 1/t})$ then each of the vertex classes has $(n/r) + O(n^{1 - 1/2t})$ vertices. ∎

If $F_1, \ldots, F_k$ are given graphs then starting with Theorem 4.8 we may well

be able to determine the extremal graphs *exactly* provided they have sufficiently large order. As an illustration we shall deduce the following beautiful
result first proved by Simonovits [S19] by a different method. The case $r = 2$
had been proved earlier by Erdös [E9a] and Moon [M35] had obtained
a partial solution of the problem.

THEOREM 4.9. *Let r and s be fixed integers. If n is sufficiently large then*

$$G^* = K^{s-1} + T_r(n - d + 1)$$

is the unique extremal graph of order n for sK^{r+1}, i.e. for s disjoint copies of K^{r+1}.

Proof. Let us apply induction on s. For $s = 1$ this is just Theorem 1.1. Suppose $s \geqslant 2$ and the result holds for smaller values of s.

Let $G = G^n$ be an extremal graph and consider the partition of $V(G)$ given
in Theorem 4.8. Let $0 < \varepsilon < (1/4sr)$ and suppose there is a vertex x joined
to εn vertices of its own class. Then $G - x$ contains a $K^r(u)$ where u is an
arbitrarily large constant, say $u = s(r + 1)$, such that x is joined to every
vertex of this $K_r(u)$. This implies that $G - x$ does not contain an $(s - 1)K^{r+1}$
so the theorem follows from the induction hypothesis.

Therefore we may suppose that if $0 < \varepsilon < (1/4sr)$ is fixed and n is sufficiently
large then no vertex is joined to εn vertices of its own class. Consequently
every vertex is joined to all but, say, $2\varepsilon n$ vertices of the other classes. In
particular, if ε is sufficiently small and there is an edge xy joining two vertices
of the same class then $G - \{x, y\}$ contains a $K_{r-1}(u)$ (u arbitrarily large
constant) such that x and y are joined to every vertex of this $K_r(u)$. This
implies that if we delete the edges joining vertices belonging to different
classes then we obtain a graph H that does not contain s independent edges
and has maximal degree at most εn. We could invoke Theorem II. 4.6.(iii) to
put an upper bound on the number of edges but the following trivial estimate
will suffice to complete the proof. Take a maximal system of $s' < s$ independent edges. Then each edge of H is incident with at least one of the $2s'$
endvertices of these edges, so

$$e(H) \leqslant 2s'\varepsilon n < 2s\varepsilon n < \frac{n}{2r} < e(G^*) - t_r(n),$$

contradicting the extremality of G.   ∎

We conclude this section with a result of somewhat different nature but closely connected to the theorems of this section. We shall determine the maximal size of a graph of order n that does not contain a graph F_n, where F_n depends on n and its order tends to infinity with n. More precisely we shall take $F_n = K(2, 2, \lfloor \alpha n \rfloor)$, where $\alpha > 0$ is a fixed constant. Similar results could be obtained with $F_n = K_r(t, t, \ldots, t, \lfloor \alpha n \rfloor)$, where r, t and $\alpha > 0$ are fixed but they would be in much less explicit form.

Take a complete bipartite graph of order n. Add edges to both colour classes in such a way that one does not obtain a quadrilateral in either class. Denote by $Q(n)$ the set of graphs one can obtain in this way and let $\bar{q}(n)$ be the maximal size of a graph belonging to $Q(n)$. It follows from Theorem 2.8 that if n is sufficiently large then

$$\tfrac{1}{4}n^2 + 2^{-\frac{3}{2}}n^{\frac{3}{2}} - n^{4/3} < \bar{q}(n) < \tfrac{1}{4}n^2 + 2^{-\frac{3}{2}}n^{\frac{3}{2}} + n.$$

Furthermore, if n is sufficiently large, there is a graph $G \in Q(n)$ with $\bar{q}(n)$ edges in which every vertex is joined to less than $\sqrt{n}$ vertices in the same class. Consequently the maximal t for which this G contains a $K(2, 2, t)$ is less than $\sqrt{n}$. Bollobás, Erdős, Simonovits and Szemerédi [BESS1] proved the rather surprising fact that *one more* edge ensures the existence of a $K(2, 2, \alpha n)$, where α is an absolute constant.

THEOREM 4.10. *There is an absolute constant n_0 such that every graph of order $n \geqslant n_0$ and size $\bar{q}(n) + 1$ contains a $K(2, 2, t)$ with $t \geqslant 10^{-3} n$.*

Proof. Once again we shall not choose n_0 immediately but we suppose that n_0 is chosen in such a way that all the conditions imposed on n throughout the proof are satisfied provided $n \geqslant n_0$. It will be clear that this can be done.

Let $G = G(n, \bar{q}(n) + 1)$. The definition of $\bar{q}(n)$ implies that $\bar{q}(n) - n/2 \geqslant \bar{q}(n - 1)$. Hence, by applying the proof of Lemma 3.2, one can show easily that G contains a subgraph of order $p \geqslant n/8$ and size at least $\bar{q}(p) + 1$ such that $\delta(H) \geqslant (\tfrac{1}{2} - 10^{-2})p$. (Lemma 3.2 would give only $p \geqslant n/10$.)

Let us apply Theorem 4.2(iii) to the graph H of size at least $p^2/4$ and to the subgraph $K_3(3) = K(3, 3, 3)$. In particular, Theorem 4.2(iii) asserts the existence of a constant n_1 such that if $p \geqslant n_1$ (so, e.g. if $n \geqslant 8n_1$) then H contains either a $K(3, 3, 3)$ or a bipartite subgraph with minimal degree at least $(\tfrac{1}{2} - 10^{-4})p$. We shall discuss the two possibilities separately.

1. *The graph H contains $K_0 = K(3, 3, 3)$.* Note that K_0 contains 108 quadrilaterals. We say that a vertex of $R = H - K_0$ *covers* a quadrilateral Q in K_0 if it is joined to every vertex of Q. It is easily checked that

a vertex joined to $4 + r$ vertices of K_0 covers at least $3r$ quadrilaterals. *Suppose $p \geqslant 10^4$*. Then every vertex of K_0 is joined to at least $(\frac{1}{2} - 10^{-4})\, p - 8$ $> (\frac{1}{2} - 10^{-3})\, p$ vertices of R so the vertices of R altogether cover at least

$$3\{9(\tfrac{1}{2} - 10^{-3})\, p - 4p\} > p$$

quadrilaterals. Consequently there is a quadrilateral Q_0 covered by at least

$$t \geqslant \frac{p}{108} > 10^{-3}\, n$$

vertices of R. These t vertices and Q_0 clearly span a subgraph containing a $K(2, 2, t)$.

2. *The graph H of order p contains a bipartite graph B such that $\delta(B) > (\frac{1}{2} - 10^{-4})\, p$.*

Denote by W_1 and W_2 the colour classes of B. Let S be the set of vertices in H that are joined to at least $10^{-4}\, p$ vertices in each class of B. We shall use the random selection lemma (Lemma 3.5) to show that if $|S|$ is larger than a certain absolute constant then B contains a $K(2, 2, t)$, $t \geqslant n/80$.

We may suppose that at least half of the vertices in S are joined to at least $10^{-4}\, p$ vertices in W_1 and to at least $\frac{1}{2}(\frac{1}{2} - 3 . 10^{-4})\, p > \frac{1}{4}(1 - 10^{-3})|W_2|$ vertices in W_2. Thus by Lemma 3.5 if $|S| \geqslant s_1$ (s_1 is an absolute constant) then one can find two vertices in S, say x_1 and x_2, such that there are vertices x_3, $x_4 \in W_1$ and $y_1, \ldots, y_l \in W_2$, $l > (\frac{1}{4}(1 - 10^{-2}))^3 |W_2| > p/9$, adjacent to both x_1 and x_2 (Fig. 4.1). As $\delta(B) \geqslant (\frac{1}{2} - 10^{-4})\, p$, at most $4 . 10^{-4}\, p$ of the y_is may not be adjacent to both x_3 and x_4. Consequently at least $p/10$ of the y_is are adjacent to all four vertices of the $K(2, 2)$ spanned by x_1, x_2, x_3, x_4, so B contains a $K(2, 2, t)$ where $t \geqslant p/10 \geqslant n/80$.

Therefore we may suppose without loss of generality that $|S| < s_1$, that is less than s_1 (absolute constant) vertices of H are joined to at least $10^{-4}\, p$ vertices in both W_1 and W_2. Denote by S' the set of vertices of H joined to at least $p/10$ vertices in both W_1 and W_2. Clearly $S' \subset S$. We shall complete the proof of the theorem by discussing the cases "S' is non-empty" and "S' is empty" separately.

(a) *S' is non-empty, say $a \in S'$.* Denote by Z_i the set of vertices of W_i adjacent to a, $i = 1, 2$. Then we may suppose that

$$z_1 = |Z_1| \geqslant \frac{p}{10}, \qquad z_2 = |Z_2| \geqslant \tfrac{1}{2}(\tfrac{1}{2} - 3 . 10^{-4})\, p \geqslant \frac{p}{6}.$$

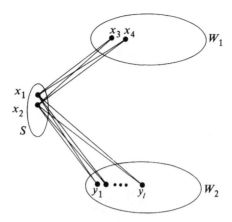

FIG. 4.1. The structure of G if $|S| \geqslant s_1$.

Put $\tilde{H} = H - S$. We shall partition the vertices of $\tilde{H}$ into two classes $\tilde{W}_1$, $\tilde{W}_2$ according to their relationship to W_1 and W_2. Let $\tilde{W}_i$ be the union of $W_i - S$ and the set of vertices of $\tilde{H}$ joined to less than $10^{-3}\,p$ vertices of W_i, $i = 1, 2$.

Note that the assertion of the theorem holds if a vertex $c \in \tilde{W}_i$ is joined to two vertices of Z_i, say b and d (Fig. 4.2). For then $abcd$ is a quadrilateral in H, a is joined to at least $p/10$ vertices of $W_i (j \neq 1)$, each of b and d is not joined to at most $2.10^{-4}p$ vertices of W_j and c is not joined to at most $3.10^{-4}p$ vertices of W_j. Thus at least $t \geqslant p(10^{-1} - 4.10^{-4} - 3.10^{-4}) > p/11$ vertices of W_j are joined to each vertex of the quadrilateral $abcd$. So H contains a $K(2, 2, t), t > p/11 > 10^{-2}n$.

Consequently we may assume without loss of generality that no vertex of $\tilde{W}_i$ is joined to two vertices of Z_i, $i = 1, 2$. This implies that there are at most p edges joining a vertex in Z_i to a vertex in $\tilde{W}_i$. Put $w_i = |\tilde{W}_i|$, $i = 1, 2$. Then $w_1 + w_2 = p - |S| \geqslant p - s_1$. Put $q(x) = ex(x; K(2, 2)) = ex(x; C^4)$, that is denote by $q(x)$ the maximal size of a graph of order x without a quadrilateral. By Theorem 2.8 $q(x) < x^{\frac{3}{2}} + x$. It is easily seen that

$$\frac{p^2}{4} + q(w_1 - z_1) + q(w_2 - z_2) + p < q(p) - s_1 p \leqslant e(\tilde{H}).$$

Consequently $\tilde{H}$ contains a quadrilateral with all four vertices in the same class W_i. Then at least $(\frac{1}{2} - 10^{-4} - 4 \cdot 3.10^{-4})\,p > p/3$ vertices of the other class are joined to every vertex of this quadrilateral and so $\tilde{H}$ contains a $K(2, 2, t)$ with $t > p/3 \geqslant n/24$.

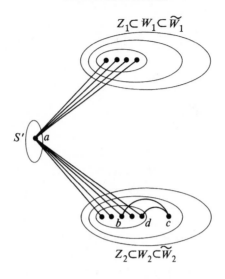

FIG. 4.2. The structure of G if $|S| < s_1$ and there is a vertex of a H joined to at least $10^{-4}p$ vertices in each of W_1 and W_2.

(b) *The set S' is empty*, i.e. no vertex of H is joined to $p/10$ vertices in each class W_i of B. Denote by V_i the set of vertices of H not joined to $p/10$ vertices of W_i, $i = 1, 2$. Then clearly $V_1 \cup V_2$ is a partitioning of the vertex set of H and $W_1 \subset V_1$, $W_2 \subset V_2$. Consequently

$$(\tfrac{1}{2} - 10^{-4})\,p \leqslant |V_i| \leqslant (\tfrac{1}{2} + 10^{-4})\,p$$

and as $\delta(H) \geqslant (\tfrac{1}{2} - 10^{-4})\,p$, every vertex of one class V_i is not joined to at most

$$(2 \cdot 10^{-4} + 10^{-1})\,p$$

vertices of the other class. The definition of the function $q(x)$ and $e(H) \geqslant q(p) + 1$ imply that at least one of the classes V_1, V_2, say V_1, spans a subgraph containing a quadrilateral. Then at least

$$(\tfrac{1}{2} - 10^{-4} - 8 \cdot 10^{-4} - 4 \cdot 10^{-1})\,p > \frac{p}{11}$$

vertices of V_2 are joined to every vertex of this quadrilateral. Thus H contains a $K(2, 2, t)$ with $t > p/11 > n/100$.   ■

The constant 10^{-3} in Theorem 4.10 is clearly not best possible. We have made no attempt to determine the best constant following from the proof above since it is rather far from being best possible (cf. Ex. 38 and Conjecture 39).

5. INDEPENDENT COMPLETE SUBGRAPHS

We have proved (Theorem 4.9) that if r and s are given and n is sufficiently large then $K^{s-1} + T_r(n - s + 1)$ is the unique graph of maximal size among all graphs of order n without s independent K^{r+1}s. If n is not too large, i.e. it is not much larger than $s(r + 1)$, the first value of n for which G^n might contain sK^{r+1}, then the assertion clearly fails. The aim of this section is to present a theorem of Hajnal and Szemerédi [HS3] which is rather similar to Theorem 4.9 but holds for every $n \geqslant s(r + 1)$. The theorem was conjectured by Erdös [E19] and states that the graph $K^{s-1} + T_r(n - s + 1)$ is extremal among all graphs of order n without sK^{r+1} in the following sense: if $\delta(G^n) > \delta(K^{s-1} + T_r(n - s + 1))$ then $G^n \supset sK^{r+1}$. It turns out that the case $n = s(r + 1)$ implies the general case so we shall prove the theorem for $n = s(r + 1)$. Furthermore it will be slightly more convenient to work with the complement of G^n since for large s it has much smaller size than G^n. For the complement of G^n the theorem of Hajnal and Szemerédi states that if $n = s(r + 1)$ and $\Delta(G^n) \leqslant r$ then $G^n \subset T_{r+1}(n)$. Thus we can produce a colouring of G with equal colour classes if we are allowed to use one more colour than the minimum guaranteed by Brooks' theorem (Theorem V.1.6). Various special cases of this assertion had been proved earlier by Dirac [D7] ($s = 2$), Corrádi and Hajnal [CH7] ($s = 3$), Zelinka [Z5] ($r = 1, 2$), Grünbaum [G19] ($r = 3$) and Sumner [S25] ($r = 4, 5$). Furthermore, Corrádi [C11] showed that G^n is contained in a $T_{r+1}(n)$ to which very few edges have been added. However, the difficulties encountered in these results (with the exception of [CH7] and [C11]) are not really comparable with the difficulties arising in the proof of the general case. Our proof of the theorem of Hajnal and Szemerédi [HS3] is somewhat simpler than the original proof though the essence is all in [HS3].

THEOREM 5.1. *If $n = s(r + 1)$ and $\Delta(G^n) \leqslant r$ then G^n has an $(r + 1)$-colouring with equal colour classes, i.e. $G^n \subset T_{r+1}(n)$.*

Proof. We fix s and apply induction on r. For $r = 0$ there is nothing to prove. Suppose now that $r > 0$, the theorem holds for smaller values of r but it does not hold for r. Let $G = G^n$ be a *minimal counterexample* to the assertion. Let $x_1 x_2 \in E(G)$. Then $G - x_1 x_2 \subset T_{r+1}(n)$ and x_1 and x_2 belong to the same colour class of $T_{r+1}(n)$. As $d(x_1) \leqslant \Delta(G^n) \leqslant r$, there is a colour class no vertex of which is adjacent to x_1. Adding x_1 to that colour class we obtain an $(r + 1)$-colouring of G in which one of the colour classes has $s - 1$ vertices, $r - 1$ colour classes have s vertices and the last colour class has $s + 1$ vertices.

It will be convenient to use special terminology in the proof. We say that $(V_j)_1^{r+1}$ is a *chain* if $V(G) = V = \bigcup_1^{r+1} V_j$ is a partition of V into independent sets V_j, $|V_1| = s - 1$, $|V_j| = s$ if $2 \leqslant j \leqslant r$ and $|V_{r+1}| = s + 1$. By the remark above there exist chains in G.

For $X \subset V$ and $x \in V$ denote by $\Gamma(x, X)$ the set of vertices in X adjacent to x and let $d(x, X) = |\Gamma(x, X)|$. A set of $X \subset V$ is said to have a *uniform partition* if X is the disjoint union of independent sets each containing s independent vertices. As G is a counterexample to the assertion of the theorem, $V = V(G)$ does not have a uniform partition. We say that a vertex $x_k \in V_k$ is *accessible* in the chain $(V_j)_1^{r+1}$ if there exist $x_{i_l} \in V_{i_l}$, $1 \leqslant l \leqslant m$, $i_1 = 1$, $i_m = k$, such that $d(x_{i_l}, V_{i_{l-1}}) = 0$ for every l, $2 \leqslant l \leqslant m$. Note that in this definition we may always assume $i_j \neq i_l$ if $j \neq l$. A set V_j is *accessible* if it contains an accessible vertex. Suppose that V_{r+1} is accessible, i.e. we can have $k = r + 1$ in the previous definition. Then $(V_j')_1^{r+1}$ is a uniform partition of V, where

$$V_1' = V_1 \cup \{x_{i_2}\}, \quad V_{i_l}' = V_{i_l} \cup \{x_{i_{l+1}}\} - \{x_{i_l}\}, \, 2 \leqslant l \leqslant m - 1, \quad V_k' = V_k - \{x_{i_m}\},$$

and $$V_t' = V_t, \qquad 2 \leqslant t \leqslant r, \qquad t \neq i_l, \qquad 1 \leqslant l \leqslant m,$$

As this contradicts our original assumption, V_{r+1} *is not accessible*. The union of all accessible sets in a chain $(V_j)_1^{r+1}$ will be denoted by A and we put $B = V - A$. Note that if $x \in B$ then $d(x, V_j) \geqslant 1$ for every accessible set V_j and so the induction hypothesis implies that

$$B - \{x\} \text{ has a uniform partition.} \tag{1}$$

Let us say that V_t is a *terminal set* of the chain $(V_j)_1^{r+1}$ if V_t is accessible and for every other accessible set V_k there exist x_{i_l}, $1 \leqslant l \leqslant m$, $i_1 = 1$, $i_m = k$, where $i_j \neq i_l$ if $j \neq l$, such that $d(x_{i_l}, V_{i_{l-1}}) = 0$ *and* $i_l \neq t$, $2 \leqslant l \leqslant m$. Informally, an accessible set is a terminal set if it is not needed in the justification of the accessibility of any other set of the chain. Note that if x_t is an accessible vertex of a terminal set V_t, $V_t \neq V_1$, then

$$(A - V_t) \cup \{x_t\} \text{ has a uniform partition,} \tag{2}$$

and so

$$B \cup V_t - \{x_t\} \text{ does not have a uniform partition.} \tag{3}$$

Call a chain $(V_j)_1^{r+1}$ a *u-chain* if it has exactly $u + 1$ accessible sets. If $(V_j)_1^{r+1}$ is a u-chain and $x \in V_{r+1}$ then, as x is not accessible, there are at most

$$r - (u + 1) = r - u - 1 \quad \text{sets} \quad V_j \quad \text{such that} \quad d(x, V_j) = 0. \tag{4}$$

Denote by v the *minimal* integer for which *there is a v-chain*. Suppose $v = 0$, that is $A = V_1$ in some chain $(V_j)_1^{r+1}$. Then the number of edges joining vertices in A to vertices in B is

$$\sum_{x \in B} d(x, V_1) \geqslant |B| = rs + 1.$$

On the other hand

$$\sum_{x \in B} d(x, V_1) = \sum_{a \in A} d(a) \leqslant (s - 1) r,$$

so this is impossible. Consequently $v \geqslant 1$ and so V_1 is not a terminal set.

The following two lemmas go a long way towards proving Theorem 5.1.

LEMMA 5.2. *Let V_t be a terminal set of a v-chain $(V_j)_1^{r+1}$. Denote by A the union of the accessible sets, put $B = V - A$ and $q = r - v$. Then there is a vertex $x \in V_t$ such that $d(x, A) < v - 2q$. In particular, $v \geqslant 2q + 1$. Furthermore, $(V_j)_1^{r+1}$ has at least $2q + 1$ terminal sets.*

Proof. Suppose $d(x, A) \geqslant v - 2q$ for every $x \in V_t$. Put

$$Y_0 = \{y \in V_t : v \leqslant d(y, A)\},$$

$$Y_1 = \{y \in V_t : v - q \leqslant d(y, A) < v\},$$

$$Y_2 = \{y \in V_t : v - 2q \leqslant d(y, A) < v - q\}.$$

Then by the assumption $\bigcup_0^2 Y_i$ is a partition of V_t. Note that the vertices in $Y_1 \cup Y_2$ are all accessible since if $y \in Y_1 \cup Y_2$ then $d(y, V_j) \geqslant 1$ cannot hold for each of the v accessible sets V_j different from V_t. Put furthermore

$$X_0 = \{x \in B: \ d(x, Y_0) \geqslant 1\},$$

$$X_1 = \{x \in B - X_0: \ d(x, Y_1) \geqslant 2\}.$$

$$X_2 = \{x \in B - X_0 - X_1: \ d(x, Y_1 \cup Y_2) \geqslant 3\}.$$

Suppose first that $B \neq \bigcup_0^2 X_i$, say $x \in B - \bigcup_0^2 X_i$. Then $d(x, Y_0) = 0$, $d(x, Y_1) \leqslant 1$ and $1 \leqslant d(x, Y_1 \cup Y_2) \leqslant 2$. Let $y \in \Gamma(x, Y_1 \cup Y_2)$ be such that $\Gamma(x, Y_1) - \{y\} = \varnothing$. If $\Gamma(x, Y_1 \cup Y_2) = \{y\}$ then the uniform partitions of $A \cup \{y\} - V_t$ and $B - \{x\}$, which exist by (2) and (1), together with the set $V_t' = V_t \cup \{x\} - \{y\}$ form a uniform partition of $V = V(G)$, contradicting the assumption. Thus $\Gamma(x, Y_1 \cup Y_2) = \{y, z\}$ where $x \in Y_2$ (see Fig. 5.1).

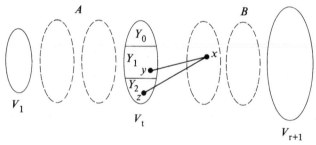

FIG. 5.1. The only vertices of $Y_1 \cup Y_2$ joined to $x \in B - \bigcup_0^2 X_i$ are y and z, $y \in Y_1$, $z \in Y_2$.

We construct a new chain $(V_j')^{r+1}$ as follows. Let $V_1' = V_t \cup \{x\} - \{y, z\}$. Take a uniform partition of $(A - V_t) \cup \{y\}$ whose existence is guaranteed by (2) since y is accessible and V_t is terminal. This partition has v classes so at least one of them is such that z is not adjacent to any of its vertices. Add z to one of those classes to obtain V_{r+1}' ($|V_{r+1}'| = s + 1$). By (1) we can add to these sets a uniform partition of $B - \{x\}$ to obtain $(V_i')_1^{r+1}$. By the definition of v the chain $(V_i')_1^{r+1}$ is a u-chain where $u \geqslant v \geqslant 1$. As $d(z, A) \leqslant v - q - 1$, by the construction of $(V_i')_1^{r+1}$ there are at least $v - 1 - (v - q - 1) = q > r - v - 1$ sets V_j' such that $d(z, V_j') = 0$, contradicting (4). Therefore $B = \bigcup_0^2 X_i$, as claimed.

As $\Delta(G) \leqslant r$,

$$d(y, B) \leqslant r - v + iq \quad \text{if} \quad y \in Y_i.$$

Put $X_2' = \{x \in X_2 : d(x, Y_1) = 1\}$, $X_2'' = X_2 - X_2'$. Then, counting the number of $B - Y_i$ edges first from B and then from Y_i, we find that

$$|X_0| \le (r - v)|Y_0| = q|Y_0|,$$

$$2|X_1| + |X_2'| \le (r - v + q)|Y_1| = 2q|Y_1|,$$

$$2|X_2'| + 3|X_2''| \le (r - v + 2q)|Y_2| = 3q|Y_2|.$$

Consequently

$$3(qs + 1) = 3|B| \le 3|X_0| + 3|X_1| + \tfrac{7}{2}|X_2'| + 3|X_2''| \le 3q|V_t| = 3qs,$$

and this contradiction completes the proof of the first part.

To see the second part note that there exists a terminal set V_t for which there is a *unique* accessible set V_j such that $d(y, V_j) = 0$ for some $y \in V_t$ and V_j is not terminal. ∎

Let m be the maximal integer for which there is an m-chain with at least $2q + 1$ terminal sets. By Lemma 5.2 this number m is well defined.

LEMMA 5.3. *Let V_t be a terminal set of an m-chain $(V_j)_1^{r+1}$ with at least $2q + 1$ terminal sets. Put*

$$C_t = \{x \in B : d(x, V_t) = 1\}.$$

Let $y \in V_t$, $C_y = \Gamma(y, C_t)$. Then C_y spans a complete subgraph of G.

Proof. Suppose first that $C_y \ne \varnothing$, say $x \in C_y$, and y is accessible. Then, by (1) and (2), $B - \{x\}$ has a uniform partition and $(A - V_t) \cup \{y\}$ also has a uniform partition. Furthermore, $V_t \cup \{x\} - \{y\}$ is an independent set, so $V = V(G)$ has a uniform partition which is not supposed to exist. Consequently,

$$\text{if } C_y \ne \varnothing \text{ then } y \text{ is inaccessible.} \tag{5}$$

Let us turn now to the proof of the lemma. Suppose the assertion is false, i.e. there exist $x_1, x_2 \in C_t$ such that $x_1 x_2 \notin E(G)$. Then by (5) the vertex y is inaccessible so there exists an accessible vertex $z \in V_t - \{y\}$. We shall distinguish two cases according to the value of $\max\{d(x_i, B): i = 1, 2\}$. Put $p = r - m$.

(a) Suppose first that the maximum above is at least $p - 1$, say $d(x_1, B) \geqslant p - 1$. Then $d(x_1, V_j) = 1$ for each of the $d + 1$ accessible sets V_j. We construct a new chain $(V_j')_1^{r+1}$ that will contradict the maximality of m. Put $V_t' = V_t \cup \{x_1\} - \{y\}$ and let $V_j' = V_j$ if V_j is accessible. By (1) the set $B - \{x_1\}$ has a uniform partition. The vertex y can be added to a class V_l' of this partition since $d(y, V_j') \geqslant 1$ for each of the $m + 1$ indices j for which V_j is accessible and so

$$d(y, B - \{x_1\}) \leqslant r - (m + 1) < r - m = \frac{|B - \{x_1\}|}{s}.$$

In this new chain V_j' is accessible if V_j was and the set V_h' containing x_2 *is also accessible* since $d(x_2, V_t') = 0$. Thus $(V_j')_1^{r+1}$ is $(m + 1)$-accessible. Furthermore, if V_j was a terminal set of $(V_i)_1^{r+1}$ then V_j' is a terminal set of the chain $(V_i')_2^{r+1}$, with the possible exception of V_t'. However V_h' is certainly a terminal set of $(V_i')_1^{r+1}$ so this new chain has at least as many terminal sets as $(V_i)_1^{r+1}$, contradicting the maximality of m.

(b) Suppose now that $\max\{d(x_i, B): i = 1, 2\} \leqslant p - 2$. Put $B' = B \cup \{y\} - \{x_1, x_2\}$, $A' = V - B'$. If $x \in B'$ and $\Gamma(x, V_t) \neq \{y\}$ then $d(x, A') \geqslant m + 1$ so $d(x, B') \leqslant p - 1$. If $x \in B'$ and $\Gamma(x, V_t) = \{y\}$ then $x \in C_y$ so by part (a) either $d(x, B) \leqslant p - 2$ or x is adjacent to x_1 and x_2 (Fig. 5.2). In both cases it follows that $d(x, B') \leqslant p - 1$. Consequently $|B'| = ps$ and $d(x, B') \leqslant p - 1$ for every $x \in B'$.

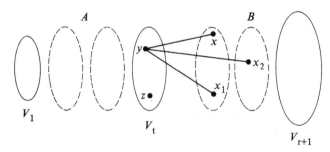

FIG. 5.2. The vertex $z \in V_t$ is accessible and $y \in V_t$ is inaccessible.

Therefore, by the induction hypothesis, B' has a uniform partition. We can add to this a uniform partition of $(A - V_t) \cup \{z\}$ together with the independent set $V_t \cup \{x_1, x_2\} - \{y, z\}$ to obtain a uniform partition of $V = V(G)$. This is impossible since G is a counterexample to the theorem. ∎

Let us return to the proof of Theorem 5.1. Let $(V_j)_1^{r+1}$ be an m-chain with at least $2q + 1$ terminal sets. Denote by T the set of indices of terminal sets and for $t \in T$ put

$$C_t = \{x \in B: \quad d(x, V_t) = 1\},$$
$$D_t = B - C_t.$$

If $x \in D_t$ then $d(x, V_t) \geqslant 2$ and $d(x, V_j) \geqslant 1$ for every accessible set V_j. Thus if x is contained in t_x sets D_t then $d(x, B) \leqslant r - (m + 1 + t_x) = p - 1 - t_x$. Consequently

$$\sum_{x \in B} (p - 1 - d(x, B)) \geqslant \sum_{x \in B} t_x = \sum_{t \in T} |D_t| \geqslant (2q + 1)|D_{t_0}|, \tag{6}$$

where $t_0 \in T$ is such that $|D_{t_0}| = \min_{t \in T} |D_t|$.

Our final aim is to prove an inequality contradicting (6). Let $t \in T$ be fixed. For $y \in V_t$ put

$$d_y = d(y, C_t).$$

If $d_y > 0$ then by (5) y is inaccessible. Thus $d(y, A) \geqslant m$ and so $d_y \leqslant p = r - m$. Put

$$Y_r = \{y \in V_t : d_y = r\},$$

$$|Y_r| = s_r, \qquad r = 0, \ldots, p.$$

Then $V_t = \bigcup_0^p Y_r$ is a partition of V_t so

$$s = \sum_0^p s_r.$$

Furthermore put

$$X_r = \bigcup_{y \in Y_r} \Gamma(y, C_t), \qquad r = 0, \ldots, p. \tag{7}$$

As (7) is a partition of X_r and $C_t = \bigcup_0^p X_r$ is a partition of C_t, we have

$$|C_t| = \sum_0^p rs_r,$$

$$|D_t| = ps + 1 - \sum_0^p rs_r = p\sum_0^p s_r + 1 - \sum_0^p rs_r = \sum_0^p (p - r)s_r + 1.$$

By Lemma 5.3

$$d(x, B) \geqslant r - 1 \quad \text{if} \quad x \in X_r,$$

so

$$\sum_{x \in B} (p - 1 - d(x, B)) \leqslant \sum_0^p (p - r) |X_r| + (p - 1) |D_t| = \sum_0^p (p - r) r s_r$$

$$+ (p - 1) |D_t|. \tag{8}$$

Note now that

$$\sum_0^p (p - r) r s_r < \sum_0^p (p - r) p s_r + p = p\left(\sum_0^p (p - r) s_r + 1\right) \leqslant (2q - p) |D_t|.$$

Thus by (8)

$$\sum_{x \in B} (p - 1 - d(x, B)) < (2q - 1) |D_t|. \tag{9}$$

Taking $t = t_0 \in T$ in inequality (9) we obtain an inequality contradicting (6). Thus the proof of Theorem 5.1 is complete. ∎

Let us formulate two other versions of Theorem 5.1.

THEOREM 5.1′. *Suppose* $G = G^n$ *and* $\Delta(G) = \Delta$. *Then* $G \subset T_{\Delta+1}(n)$.

Proof. Let s be the integer such that $n \leqslant (\Delta + 1)s < n + \Delta + 1, (\Delta + 1)s - n = p$. Put $H = G \cup K^p$. Then $\Delta(H) = \Delta$, H has order $(\Delta + 1)s$ so by Theorem 5.1

$$H \subset T_{\Delta+1}((\Delta + 1)s) = T.$$

Thus

$$G \subset T - K^p = T_{\Delta+1}(n),$$

since a class of T can contain at most one vertex of $K^p \subset T$. ∎

THEOREM 5.1″. *Let* $K = K(s - 1, s_1, s_2, \ldots, s_r)$, *where*

$$s \leqslant s_1 \leqslant s_2 \leqslant \ldots \leqslant s_r \leqslant s_1 + 1, \quad n = s - 1 + \sum_1^r s_i \geqslant s(r + 1).$$

Then $K \not\supset sK^{r+1}$ *but if* $G = G^n$ *and*

$$\delta(G) > \delta(K) = n - s_r$$

then

$$G \supset sK^{r+1}.$$

Proof. The assertion $K \not\supset sK^{r+1}$ is obvious. Suppose $G = G^n$ and $\delta(G) = \delta(K) + 1 = n - s_r + 1$. Note that $s_r - 1 \geqslant s$. If $n \geqslant (s_r - 1)(r + 1)$ then, as $\Delta(\bar{G}) = s_r - 2$,

$$\bar{G} \subset T_{s_r-1}(n),$$

so

$$G \supset (s_r - 1) K^{\lfloor n/(s_r - 1) \rfloor} \supset sK^{r+1}.$$

If $n < (s_r - 1)(r + 1) = n'$, put

$$H = G + K^p, \quad \text{where} \quad p = n' - n.$$

Then

$$\delta(H) = n' - s_r + 1,$$

so

$$H \supset (s_r - 1) K^{\lfloor n'/(s_r - 1) \rfloor} \supset (s_r - 1) K^{r+1}.$$

Consequently

$$G \supset (s_r - 1 - p) K^{r+1}.$$

Note finally that

$$(s_r - 1)r + s \leqslant n$$

so

$$p \leqslant s_r - 1 - s,$$

i.e.

$$G \supset sK^{r+1}. \qquad \blacksquare$$

6. EXERCISES, PROBLEMS AND CONJECTURES

1^-. Prove the following extension of Theorem 1.1. If $2 \leqslant r < n' \leqslant n$, G has order n and size $t_r(n) + p$, where $p \leqslant 1$, then G contains a subgraph of order n' and size at least $t_r(n') + p$. (Dirac [D16].)

2^+. Suppose $G = G(n, t_r(n)) \neq T_r(n)$ and $1 \leqslant p \leqslant r - 1$. Then G contains a subgraph H_p of order $r + p$ and size at least

$$\binom{r + p}{2} - p + 1$$

(i.e. less than p edges are *missing* from H_p). (Dirac [D16].)

3. Prove that if $l \geqslant 0$ and a graph $G(n, t_2(n) - l)$ contains a triangle, then it contains at least $\lfloor n/2 \rfloor + l - 1$ triangles. State and prove a generalization of this assertion to a result about the number of complete graphs of order p. (Erdős [E15], Moon [M34].)

4. Let S be the simplex in $\mathbb{R}^n$ given by $S = \{\mathbf{x} = (x_i)_1^n : x_i \geqslant 0, \sum_1^n x_i = 1\}$. Given a graph $G = G^n$ with

$$V(G) = \{1, 2, \ldots, n\} \quad \text{and} \quad E(G) = E,$$

put

$$F_G(\mathbf{x}) = \sum_{ij \in E(G)} x_i x_j, \quad \mathbf{x} \in S.$$

(Note that $x_i x_j$ is the *product* of the real numbers x_i and x_j.) Prove that

$$\max_{\mathbf{x} \in S} F_G(\mathbf{x}) = \tfrac{1}{2}\left(1 - \frac{1}{r}\right)$$

where $r = cl(G)$. (Motzkin and Straus [MS1].)

5. (Ex. 4 ctd.) Prove that $F_G(\mathbf{x})$ has a maximum in the interior of S iff G is a complete r-partite graph. (Motzkin and Straus [MS1].)

6. Deduce Theorem 1.1 from the results of the previous two exercises.

7. Apply Theorem 1.1 and Theorem I.5.13 to deduce that a graph $G(n, m)$ with $n \geqslant 2k - 2$ and $m > (2k - 4)(n - k + 1)$ contains a k-connected subgraph. (Cf. Corollary I.5.14.)

8. Refine the proof of Corollary 1.3 to show that the minimal number of triangles contained in a graph of order n and its complement is

$$2\binom{l}{3} \quad \text{if} \quad n = 2l$$

$$\tfrac{2}{3}k(k - 1)(4k + 1) \quad \text{if} \quad n = 4k + 1$$

$$\tfrac{2}{3}k(k + 1)(4k - 1) \quad \text{if} \quad n = 4k + 3.$$

(Goodman [G15]), see also Sauvé [S11].)

9. Let $G = G^n$ and suppose $\bar{G}$ does not contain a triangle. Prove that

$$k_3(G) \geqslant \binom{\left\lfloor \dfrac{n}{2} \right\rfloor}{3} + \binom{\left\lfloor \dfrac{n+1}{2} \right\rfloor}{3}$$

and the result is best possible if n is even or $n > 9$. (Lorden [L11].)

10. Suppose $G = G^n$ is self-complementary, i.e. it is isomorphic to its complement. Deduce from the result of the previous exercise that

$$k_3(G) \geqslant \begin{cases} \dfrac{2k}{3}(k-1)(2k-1) & \text{if } n = 4k, \\[2mm] \dfrac{k}{3}(k-1)(4k+1) & \text{if } n = 4k+1, \end{cases}$$

and this inequality is best possible. (Clapham [C10].)

11. PROBLEM. Given n, p and q determine

$$\min\{k_p(G): \quad G = G^n, \overline{G} \not\supset K^q\}.$$

In particular, prove that if G has $3k$ vertices, no four of which are independent, then G contains at least $3\binom{k}{3}$ triangles. (Erdös [E11].)

12. Given $1 \leqslant s \leqslant t$, let V_1, V_2 be disjoint sets, $W_i \subset V_i, |V_i| = t, |W_i| = s$, $i = 1, 2$. Let $H_{s,t}$ be the bipartite graph with colour classes V_1, V_2, containing every edge xy, $x \in W_i$, $u \in V_j$, $i \neq j$. By refining the proof of Theorem 2.3, show that there exists a constant $c_{s,t}$ such that

$$\text{if } e(G^n) \geqslant c_{s,t} n^{2-1/s} \quad \text{then} \quad G^n \supset H_{s,t}.$$

(P. Erdös.)

13^-. Make use of Theorem 2.4 to prove

$$z(m,n;s,t) = (t-1)m + (s-1)\binom{n}{t}$$

if $m \geqslant (s-1)\binom{n}{t}$. (Čulik [C13].)

14. Given a natural number n, consider sequences $1 < a_1 < a_2 < \ldots < a_k < n$ such that the products $a_i a_j$, $1 \leqslant i \leqslant j \leqslant k$, are all distinct. Show that there exist positive constants c_1, c_2 such that

$$\pi(n) + c_1 n^{\frac{3}{4}}/(\log n)^{\frac{3}{2}} < \max k < \pi(n) + c_2 n^{\frac{3}{4}}/(\log n)^{\frac{3}{2}}$$

if n is sufficiently large, where $\pi(n)$ denotes the number of primes not greater than n.

[Hint. Every integer $m < n$ is of the form $m = m_1 m_2$ where $m_1 < n^{\frac{1}{3}}$ and

m_2 is a prime or $m_2 < n^{\frac{3}{4}}$. Construct a graph with vertex set $\{2, 3, \ldots, \lfloor n^{\frac{3}{4}} \rfloor\} \cup \{p : p \text{ is a prime}, n^{\frac{3}{4}} < p < n\}$ by joining two vertices if their product appears in the sequence $(a_i)_1^k$. Recall that $\dfrac{\pi(n) \log n}{n} \to 1$.] (Erdös [E4], [E25].)

15. CONJECTURE. Let t be a fixed natural number. Then

$$\lim_{n \to \infty} z(n; t) \, n^{-2 + 1/t}$$

exists and is positive. At present this conjecture is known to be true only for $t = 2$ (Corollary 2.7).

The following weak form of this conjecture is known only for $t \leq 3$ (cf. Theorem 2.9).

There is a constant $C_t > 0$, depending only on t, such that

$$z(n; t) > C_t n^{2 - 1/t}.$$

16. Make use of the method of the proof of Theorem 2.16 to show that if $G = G_2(m, n)$ is a strongly $K(s, t)$-saturated graph of size

$$(s - 1)n + (m - s + 1)(t - 1)$$

then either $m - s + 1$ vertices in the first class of G have degree $t - 1$ (and so the others have degree n) or $n - t + 1$ vertices in the second class of G have degree $s - 1$ (and so the others have degree m). (Bollobás [B22] and Wessel [W16].)

17. CONJECTURE. A graph $G = G^n$ is *weakly r-saturated* if there is a nested sequence of graphs $G_0 = G \subset G_1 \subset \ldots \subset G_l = K^n$ such that $e(G_{i+1}) = e(G_i) + 1$ and $k_r(G_{i+1}) > k_r(G_i)$ for each i, $0 \leq i < l$. (Note that r-saturated and strongly r-saturated graphs are also weakly r-saturated.)

A weakly r-saturated graph of order n $(3 \leq r \leq n)$ has at least

$$(r - 2)(n - r + 2) + \binom{r - 2}{2}$$

edges.

[The conjecture is known to be true for $r \leq 7$. If true, the conjecture is a substantial extension of Theorem 1.13 and there are many extremal graphs.] (Bollobás [B23].)

18. Given natural numbers $n_1, n_2, \ldots, n_r (r \geq 3)$ let

$$\delta_1 = \max\left\{ \sum_{i \in I} n_i : I \subset \{1, \ldots, r\}, \sum_{i \in I} n_i \leq \tfrac{1}{2} \sum_1^r n_i \right\}.$$

Prove that if $\delta(G_r(n_1, \ldots, n_r)) \geqslant \delta_1 + 1$ then $G_r(n_1, \ldots, n_r)$ contains a triangle. (Bollobás, Erdös, Straus [BES2].)

19^+. Let $c > 0$ and $\alpha \geqslant 0$. Prove that there exists a constant $C = C(c, \alpha) > 0$ with the following property. If n is sufficiently large and

$$\delta(G_3(n)) \geqslant n + cn(\log n)^{-\alpha}$$

then $G_3(n)$ contains a $K_3(s)$ with $s \geqslant C(\log n)^{1 - 3\alpha}(\log \log n)^{-1}$. (Bollobás, Erdös and Szemerédi [BES3].)

20. CONJECTURE. Let c_r be the constant defined in the paragraph preceding Theorem 2.12. Then

$$\lim_{r \to \infty} (r - c_r) = \tfrac{3}{2}.$$

It has not even been disproved that the limit above is 1. (Bollobás, Erdös and Szemerédi [BES3].)

21. PROBLEM. Let t be a natural number. Prove or disprove that if n is sufficiently large, $G = G_3(n)$ and $\delta(G) = n + t$ then G contains at least $4t^3$ triangles. (Cf. Theorem 2.14.)

22. Let G be a fixed non-empty graph. Let $a = a(G)$ be the smallest non-negative number with the following property. If $\varepsilon > 0$ then there is a sequence of graphs $G_1, G_2, \ldots$ with $e(G_i) \to \infty$ such that *every* subgraph of G_t with size at least $(a + \varepsilon) e(G_t)$ contains G.

Prove that $a = 1 - [1/(k - 1)]$, where $k = \chi(G)$. (Erdös [E30].)

23. CONJECTURE. Let G be a bipartite graph of size at least 2. Then there exists a rational number $\alpha = \alpha(G)$, $1 \leqslant \alpha < 2$, for which

$$\lim_{n \to \infty} f(n; G)/n^\alpha = c_\alpha(G) > 0.$$

Conversely, for every rational α, $1 \leqslant \alpha < 2$, there is a bipartite graph G such that the equality above holds. (Erdös and Simonovits, see [E28], [E29].)

24. CONJECTURE. Suppose $G = G^{5n}$ does not contain a triangle. Then G can be made bipartite by the omission of at most n^2 edges. (Erdös [E27]).

25. CONJECTURE. Let $c > 0$. Suppose $G = G(n, \lfloor cn^2 \rfloor)$ is such that $e(G[W]) \geqslant c(n/2)^2 (1 + o(1))$ for every set $W \subset V$, $|W| = \lfloor n/2 \rfloor$. Then for every fixed k and sufficiently large n the graph G contains a K^k. (P. Erdös.)

26. Show that if n is sufficiently large then there is a graph G^n such that

$$\beta_0(G^n) \leqslant \mathrm{cl}(G^n) \leqslant \frac{2\log n}{\log 2}.$$

[Hint. Apply the counting method.] (Erdös [E5].)

27^+. Show that there exist positive constants c_1 and c_1 such that

$$c_1 \frac{n}{(\log n)^2} < \max_{|G|=n} \frac{\chi(G)}{\mathrm{cl}(G)} < c_2 \frac{n}{(\log n)^2}.$$

[Hint. Deduce the first inequality from Ex. 26; to prove the second make use of an estimate of $R(k, l)$ and the fact that $\binom{a+b}{b} \geqslant n$ implies $ab > c_3(\log n)^2$.] (Erdös [E20].)

28^-. Let $f(k, l)$ be the minimum of the size of a graph with clique number k and chromatic number l. Show that

$$f(k, k) = \binom{k}{2} \quad \text{and} \quad f(k-1, k) = \binom{k+2}{2} - 5.$$

29. (Ex. 28 ctd.) Show that there exist positive constants c_3 and c_4 such that

$$c_3 k^3 < f(2, k) < c_4 k^3 (\log k)^3.$$

(Erdös [E20].)

30. PROBLEM. (Ex. 28 ctd.) Does $f(2, k)/k^3$ tend to a finite limit as k tends to ∞? (Erdös [E20].)

31. Put

$$f(n, k, l) = \max\{m : G = G(n, m), \mathrm{cl}(G) < k, \beta_0(G) = \mathrm{cl}(\bar{G}) < l\}.$$

Show that

$$f(n, 3, l) \leqslant \frac{ln}{2}.$$

(Erdös and Sós [ES4].)

32^+. (Ex. 31 ctd.) Let $r \geqslant 2$. Show that if $l = o(n)$ then

$$f(n, 2r+1, l) \leqslant \frac{1}{2}\left(1 - \frac{1}{r}\right) n^2(1 + o(1)).$$

[Hint. Work with a subgraph H for which $\delta(H)/|H|$ is large.] (Erdös and Sós [ES4].)

33^{+}. (Ex. 31 ctd.) Prove that if $l = o(n)$ then

$$f(n, 4, l) \leqslant (1 + o(1))\frac{n^2}{8}.$$

(Szemerédi [S27].)

34^{+}. (Ex. 31 ctd.) Prove that equality holds in the inequality of Ex. 33.

[Hint. With an appropriate choice of $\varepsilon > 0$ and k define a graph whose vertex set is a set of uniformly distributed points of the $(k + 1)$-dimensional unit sphere by joining two points if their distance is at least $2 - \varepsilon k^{-\frac{1}{2}}$.] (Bollobás and Erdös [BE7].)

35. PROBLEM. (Cf. Ex. 34). Are there graphs $G = G(n, \lfloor n^2/8 \rfloor)$ such that $\mathrm{cl}(G) = 3$ and $\beta_0(G) = \mathrm{cl}(\bar{G}) = o(n)$? (Bollobás and Erdös [BE7].)

36. Let $r \geqslant 3$ be a natural number. Show that if n is sufficiently large then every $G(n, t_{r-1}(n) + 2)$ contains a $K^1 + 2K^{r-1}$, i.e. two complete graphs of order r with exactly one vertex in common.

[Hint. Apply Theorem 4.8 with the single forbidden subgraph $K^1 + 2K^{r-1}$. The assertion of the exercise was conjectured by Busolini [B59].)

37. (Ex. 36 ctd.) Let $r \geqslant 3$ and $k \geqslant 1$ be natural numbers. Show that if n is sufficiently large then every $G(n, t_{r-1}(n) + k)$ contains k complete graphs of order r, say $K_1, K_2, \ldots, K_k$ such that $\bigcup_1^k K_i$ is connected and any two K_is have at most one vertex in common.

38. Let $c > \frac{1}{16}$. By considering random graphs show that there are $\varepsilon > 0$ and n_0 such that for every $n \geqslant n_0$ and $t \geqslant cn$ there is a graph G of order n and size at least $(1 + \varepsilon) n^2/4$ that does not contain a $K(4, t)$. Note that $G \not\supset K(2, 2, t)$.

39. CONJECTURE. Every $c < \frac{1}{16}$ will do in Theorem 4.19 in place of 10^{-3}, that is there is an $n(c)$ such that if $n \geqslant n(c)$ then $G(n, \bar{q}(n) + 1) \supset K(2, 2, t)$. (Note that by Ex. 38 this cannot hold for $c > \frac{1}{16}$.)

40. Denote by $f(n)$ the maximal number of cliques in a graph of order n. Show that for every $n \geqslant 5$, $T_r(n)$ is an extremal graph for some r. Deduce that,

for $n \geqslant 2$,

$$f(n) = \begin{cases} 3^{n/3} & \text{if} \quad n \equiv 0 \ (\text{mod } 3), \\ 4 \cdot 3^{\lfloor n/3 \rfloor - 1} & \text{if} \quad n \equiv 1 \ (\text{mod } 3), \\ 2 \cdot 3^{\lfloor n/3 \rfloor} & \text{if} \quad n \equiv 2 \ (\text{mod } 3), \end{cases}$$

(Moon and Moser [MM3].)

41. PROBLEM. Give conditions on a pair of graphs (F_1, F_2) ensuring that $\text{ex}(n; F_1) = \text{ex}(n; F_2)$, provided n is sufficiently large. Conversely, what can you say about F_1 and F_2 if $\text{ex}(n; F_1) = \text{ex}(n; F_2)$ for sufficiently large values of n? (Erdös and Simonovits, see [E27].)

42. Show that if $|G| \geqslant 3k$ and $\delta(G) \geqslant 2k$ then G contains k vertex disjoint cycles.

[Hint. Deduce it from the case $|G| = 3k$ which is exactly the special case of Theorem 5.1 proved by Corrádi and Hajnal [CH7].]

43. Prove that if the number of vertices of degree at least $2k$ in a graph G is at least $k^3 + 2k - 4$ greater than the number of vertices of degree at most $2k$ then G contains k vertex disjoint cycles. (Dirac and Erdös [DE1].)

44. CONJECTURE. Let G be a graph such that $|\Gamma(X)| \geqslant \min\{\frac{3}{2}|X|, |G|\}$ holds for every set $X \subset V(G)$. Then G contains a triangle. (Woodall [W24].)

45. PROBLEM. Let $t \geqslant 4$ and put $\psi_t(n) = \max\{\delta(G^n): K^3 \not\subset G^n, \chi(G^n) \geqslant t\}$. Then $\frac{1}{3} \leqslant \overline{\lim\limits_{n \to \infty}} \ \psi_t(n)/n \leqslant \frac{2}{5}$.

Prove that $\overline{\lim\limits_{n \to \infty}} \ \psi_t(n)/n = \frac{1}{3}$.

(Erdös and Simonovits (ES3].)

46.$^+$ Prove that if $e(G^n) > cn^2$ and G^n does not contain a triangle then G^n contains a $K_2(r, m)$, where $m = 2^{2r-1} c^r n + o(n)$. [Hint. Choose a small $\varepsilon > 0$ and apply the Uniform Density Lemma [Lemma 3.6]. Let R^k be the graph with vertex set $\{V_1, V_2, \ldots, V_k\}$ and edge set $\{V_i V_j: (V_i, V_j) \text{ is } \varepsilon\text{-uniform and } d(V_i, V_j) > \varepsilon\}$. Show that one may assume that each edge of G^n is a $V_i - V_j$ edge, where $V_i V_j \in E(R^k)$. Show that R^k does not contain a triangle. Let V_1 be the class incident with most edges in G^n and let $V_2, \ldots, V_s$ be the classes joined to it in R^k. Prove that the bipartite subgraph of G^n spanned by the classes V_1 and $\bigcup\limits_{2}^{s} V_i$ contains a $K_2(r, m)$.] (Bollobás, Erdös, Simonovits and Szemerédi [BESS1].)

47.[+] Prove that if the shortest odd cycle in G^n has length at least $2k + 3$ then G^n can be made bipartite by the omission of not more than $n^2/(2k)$ edges, provided n is sufficiently large. [Hint. Show that we may assume that for every partition $V(G^n) = U \cup W$ we have $d(U, W) \geqslant 2/(2k + 1)$. As in Ex. 46[+], apply the Uniform Density Lemma and construct a graph R^k. Prove that R^k contains an odd cycle of length at most $2k + 1$.] Bollobás, Erdös, Simonovits and Szemerédi [BESS1].)

48. *Problem.* Estimate $ex_r(n; K_r(t))$ for $r \geqslant 3$. (The notation was introduced after Theorem 2.15).

49. *Problem.* Prove the following extension of Turán's theorem (Theorem 1.1). For every $r \geqslant 3$ there is a constant $c_r > 0$ such that every $G(n, t_r(n) + 1)$ contains a vertex x of degree $d \geqslant c_r n$ for which $e(G[\Gamma(x)]) \geqslant t_{r-1}(d) + 1$. (Erdös [E28].)

Chapter VII

Topological Subgraphs

Let $G_1, G_2, \ldots, G_l$ be fixed multigraphs. Give conditions ensuring that a graph G contains a TG_i, i.e. a subgraph determining the same topological space as G_i. Questions of this kind have already been discussed in the book. In particular, a set of l vertex disjoint cycles is exactly a $T(lK^3)$, a set of l vertex disjoint paths joining two vertices is a topological multiple edge with multiplicity l. In the chapter we present results naturally formulated in terms of topological subgraphs and related concepts. It is somewhat embarrassing to use the term "topological" since it might suggest some connection with genuine topological problems. In fact, there is no area of graph theory where topological or analytical difficulties arise and the subject of this chapter is no exception.

In §1 we investigate contractions of graphs. We are especially interested in conditions ensuring that a graph is subcontractible to a complete graph of given order. This problem is clearly rather important in its own right but its connection with Hadwiger's conjecture further enhances its importance.

Mader was the first to show that for every integer k there is an integer $c_k > 0$ such that every $G(n, c_k n)$ contains a TK^k. This result implies that if $\{G_1, G_2, \ldots, G_l\}$ is a finite set of graphs then for some integer $c = c(G_1, \ldots, G_l) > 0$ every $G(n, cn)$ contains a TG_i, $i = 1, 2, \ldots, l$. Our main aim is to estimate the constants c_k and c. The deepest result of the chapter, a theorem of Mader [M9], will be presented in §2; it gives an upper bound on the constant c_k.

The last section (§3) is devoted to the search for subgraphs H half way between ordinary graphs and topological graphs. Such a graph H is obtained from a graph H_0 by subdividing a *given* set of edges of H_0. The problems to be discussed include a number of well-known and natural questions.

Though this chapter contains some deep results, the entire theory seems to be only at the beginning of its development. One hopes that new tools will be invented to tackle the problems in the area and soon the present results will be only fragments of a rich theory.

1. CONTRACTIONS

The interest in contractions and subcontractions of graphs is mostly due to Hadwiger's conjecture: $\chi(G) = k$ implies $G > K^k$. We have mentioned that the conjecture is trivially true for $k = 3$, and for $k = 4$ it was proved by Dirac [D13]. Furthermore, by Wagner's equivalence theorem (Theorem V.3.11) the truth of the 4CC implies it for $k = 5$. Here we shall deduce the case $k = 4$ of Hadwiger's conjecture from a stronger result of Mader [M3].

Let $\mathscr{F}_5$ be the following set of finite graphs:

$$\mathscr{F}_5 = \{G : |G| \geqslant 6 \text{ and } a, b \in G, d(a) < 5 \text{ and } d(b) < 5 \text{ imply } ab \in E(G)\}.$$

In other words if $G \in \mathscr{F}_5$ then $|G| \geqslant 6$, G has at most 5 vertices of degree at most 4 and these vertices span a complete subgraph. If $G > H$ is considered to mean that G is greater than H, then the relation $>$ defines a partial order on the set of all finite graphs, in particular on $\mathscr{F}_5$. Recall that if H is a subgraph of G then $G > H$; furthermore we write G/xy for the graph resulting from G if we contract the edge xy.

THEOREM 1.1. *The set $\mathscr{F}_5$ has exactly four minimal graphs with respect to $>$, namely K^6, $B_1 = (K^2 \cup K^1) + C^4$, $B_2 = E^3 + C^5$ and I_{12}, the graph of the icosahedron (see Fig. 1.1).*

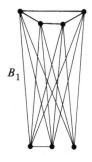
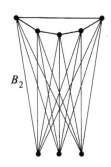
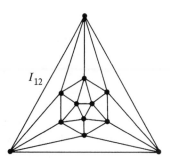

FIG. 1.1. The minimal graphs $B_1 = (K^2 \cup K^1) + C^4$, $B_2 = E^3 + C^5$ and I_{12}, the graph of the icosahedron.

Proof. Let G be a minimal graph of $\mathscr{F}_5$. We have to show that G is one of the four graphs enumerated in the theorem. $|G| = 6$ implies $G = K^5$ so we assume that $|G| \geq 7$.

The proof will be divided into three lemmas. Each lemma is proved by discussing a number of cases based on the fact that *if* $xy \in E(G)$ *then* $G/xy \notin \mathscr{F}_5$. Though the proof of the second lemma is not too short either, we leave it to the reader since its nature is rather similar to that of the first lemma.

LEMMA 1.2. *Let* $X = \{x \in G : d(x) < 5\}$. *Then* $|X| \leq 1$.

Proof. Suppose $|X| > 1$. Put $H = G - X$. Every vertex $x \in X$ is joined to H since otherwise $G - x \in \mathscr{F}_5$. Let $x \in X$ and $\{y \in H : xy \in E(G)\} = \{y_0, y_1, \ldots, y_l\}$. Then $l \geq 1$ and $y_0 y_m \in E(G)$ for some $1 \leq m \leq l$ since otherwise $G/xy_0 \in \mathscr{F}_5$. This shows, in particular, that $|X| \leq 3$.

(a) Suppose first that $X = \{x_1, x_2, x_3\}$. Then there exist vertices y_i^j $i = 1, 2, 3$, $j = 1, 2$, such that $x_i y_i^1$, $x_i y_i^2$ and $y_i^1 y_i^2 \in E(G)$, $i = 1, 2, 3$. Since $G/x_1 x_2 \notin \mathscr{F}_5$, there exists a vertex $y_{12} \in H$ that is joined to x_1 and x_2 and is not joined to x_3. We can find vertices y_{23} and y_{31} with analogous properties. Hence we may assume that $y_1^2 = y_2^1 = y_{12}, y_2^2 = y_3^1 = y_{23}$ and $y_3^2 = y_1^1 = y_{31}$. Note now that $y_{12} y_{23} \in E(G)$ since otherwise $G/y_{12} x_2 \in \mathscr{F}_5$. Similarly $y_{23} y_{31}$ and $y_{31} y_{12} \in E(G)$, that is y_{12}, y_{23} and y_{31} form a triangle. Put $G' = G - \{x_1, x_2, x_3\}$. Then G' contains at least one vertex z distinct from y_{12}, y_{23} and y_{31}. Since $d_{G'}(z) = d_G(z) \geq 5, |G'| \geq 6$. Hence $G' \in \mathscr{F}_5$, contradicting the minimality of G.

(b) Suppose now that $X = \{x_1, x_2\}$. Since $G/x_1 x_2 \notin \mathscr{F}_5$ there are at least two vertices, say y_1 and y_2, that are joined to both x_1 and x_2. If there were a third vertex, y_3 say, joined to both x_1 and x_2, then, as at the end of (a), $y_1 y_2 y_3$ would have to form a triangle and we would find $G - \{x_1, x_2\} \in \mathscr{F}_5$, contradicting the minimality of G. Hence y_3 does not exist. If $d(x_1) = 3$ then $G/x_1 y_1 \in \mathscr{F}_5$. Consequently $d(x_1) = 4$ and, similarly, $d(x_2) = 4$, that is there exist vertices $z_1, z_2 \in G - \{x_1, x_2, y_1, y_2\}$ such that $x_1 z_1, x_2 z_2 \in E(G)$. Note now that $y_1 z_1 \in E(G)$ since otherwise $G/x_1 y_1 \in \mathscr{F}_5$. Similarly $y_1 z_2, y_2 z_1$ and $y_2 z_2$ are also edges. Our next observation is that if $z_1 z_2 \in E(G)$ then contracting the triangle $x_1 x_2 y_1$ to a vertex the resulting graph belongs to $\mathscr{F}_5$. Thus z_1 is not joined to z_2. Each of the vertices y_1, y_2, z_1, z_2 has degree 5 since otherwise omitting an edge joining it to $\{x_1, x_2\}$ the resulting graph belongs to $\mathscr{F}_5$. Denote by u_1 and u_2 the additional vertices adjacent to y_1 and y_2. If $u_1 = u_2$ then contracting the edge $y_1 u_1$ to a vertex and contracting

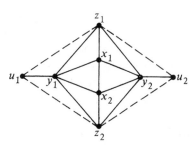

FIG. 1.2. The subgraph $G_0 = G[x_1, x_2, y_1, y_2, z_1, z_2, u_1, u_2]$. The dotted edges may or may not belong to G_0.

$G[x_1, x_2, z_1, y_2]$ to a vertex the resulting graph belongs to $\mathscr{F}_5$. Hence $u_1 \neq u_2$. Thus the subgraph G_0 spanned by the set $\{x_1, x_2, y_1, y_2, z_1, z_2, u_1, u_2\}$ has the form shown in Fig. 1.2.

Since $G/y_1z_1 \notin \mathscr{F}_5$, there must be a vertex, distinct from x_1, joined to both y_1 and z_1. This vertex must be u_1 so $u_1z_1 \in E(G)$. Similarly u_1z_2, u_2z_1, u_2z_2 are also edges. Hence the dotted edges in Fig. 1.2 do belong to G_0. Contracting $G_0 - u_2$ to a vertex we obtain a graph belonging to $\mathscr{F}_5$ since the only vertices of G_0 that may be joined to vertices not in G_0 are u_1 and u_2. We have found once again that our assumptions contradict the minimality of G. ∎

Lemma 1.3. *Suppose $\delta(G) < 5$. Then $G \cong B_1 = (K^2 \cup K^1) + C^4$.* ∎

Note that by Lemma 1.2 the graph G of Lemma 1.3 has exactly one vertex x_0 of degree at most 4. The proof of Lemma 1.3 is based on a not too short investigation of the structure of G near to x_0. As we have mentioned already, for the details we refer the reader to the original paper of Mader [M3].

Lemma 1.4. *Suppose $\delta(G) \geqslant 5$. Then G is either $B_2 = E^3 + C^5$ or I_{12}.*

Proof. If α is an edge incident with a vertex of degree at least 6 then $G - \alpha \in \mathscr{F}_5$. Hence $\Delta(G) = 5$, i.e. G is 5-*regular*.

Since $G \neq K^6$, $\mathrm{cl}(G) \leqslant 5$. If $\mathrm{cl}(G) = 5$ let $x, y \in K^5 \subset G$ be such that x and y are not joined to the same vertex $z \in G - K^5$. Then $G/xy \in \mathscr{F}_5$. Hence $\mathrm{cl}(G) \leqslant 4$.

Suppose now that $\mathrm{cl}(G) = 4$ and let $K = K^4 \subset G$, $V(K) = \{x_1, x_2, x_3, x_4\}$. For every edge $x_ix_j \in E(K)$ there is a vertex y_{ij} of $G - K$ joined to both since otherwise $G/x_iy_j \in \mathscr{F}_5$. If a vertex x of K is joined to two vertices not belonging to K, say to y and z, then y is joined to z since otherwise $G/xy \in \mathscr{F}_5$.

Since G does not contain a K^5, $y_{12} \neq y_{34}$, $y_{13} \neq y_{24}$ and $y_{14} \neq y_{23}$. However, y_{12}, y_{13} and y_{14} cannot all be distinct for otherwise $d(x_1) \geqslant 6$; assume then that $y_{12} = y_{13} = y_1$, say. Since y_1 is adjacent to each of y_{14}, y_{24} and y_{34}, and $d(y_1) = 5$, we must have $y_{14} = y_{24} = y_2$, say. Then, since $K^5 \subset G$, the vertices y_1, y_2 and y_{34} are all distinct and adjacent; but then $G[x_1, x_2, x_3, x_4, y_1, y_2, y_{34}] = B_1 \in \mathscr{F}_5$. Thus $\mathrm{cl}\,(G) \leqslant 3$.

For every edge $xy \in E(G)$ there exist at least two vertices joined to both x and y. Hence $\delta(G[\Gamma(x)]) \geqslant 2$ for every vertex $x \in G$. In particular $G[\Gamma(x)]$ contains a cycle. Since $\mathrm{cl}\,(G) = 3$, $G[\Gamma(x)]$ does not contain a triangle. This implies that either $G[\Gamma(x)]$ contains a quadrilateral or $G[\Gamma(x)]$ is exactly a C^5.

Suppose now that for some vertex $x_0 \in G$ the graph $H = G[\Gamma(x_0)]$ contains a quadrilateral. Then $\delta(H) \geqslant 2$ implies that $H = K^{2,3}$. Let $V_1 = \{x_1, x_2\}$ and $V_2 = \{y_1, y_2, y_3\}$ be the two vertex classes of $H = K^{2,3}$. There is a unique vertex z_1 in $G - \{x_0\} - H$ that is joined to x_1. Then $x_1 y_1 \in E(G)$ since there are at least two vertices joined to both x_1 and y_1. Similarly z_1 is joined to y_2 and y_3 as well. If $z_1 x_2 \in E(G)$ then by identifying x_0, y_1 and z_1 with x_1 and y_2 with x_2 the resulting contraction of G belongs to $\mathscr{F}_5$. Thus z_1 is not joined to x_2. Hence there is a unique vertex $z_2 \neq z_1$ in $G - \{x_0\} - H$ that is joined to x_2. As before, z_2 is joined to each of y_1, y_2 and y_3. Then $V(G) = \{x_0, x_1, x_2, y_1, y_2, y_3, z_1, z_2\}$ must hold since otherwise the identification of the vertices in $H \cup \{x_0, z_1\}$ results in a contraction of G that belongs to $\mathscr{F}_5$. Then z_1 must be joined to z_2 and G is indeed $B_2 = E^3 + C^5$.

Suppose finally that $G[\Gamma(x)]$ is a C^5 for *every* vertex $x \in G$. To complete the proof of the lemma we have to show that G is exactly I_{12}, the graph of the icosahedron. Let $x_0 \in G$ and let the pentagon $x_1 x_2 \ldots x_5$ be $G[\Gamma(x_0)]$. Since $G[\Gamma(x_1)]$ is also a pentagon, G contains two more vertices, say y_1 and y_2, such that $G[\Gamma(x_1)]$ is the pentagon $x_2 x_0 x_5 y_1 y_2$. Using the fact that $G[\Gamma(x_1)]$ is a pentagon we find another vertex, say y_3, such that $G[\Gamma(x_2)]$ is $x_3 x_0 x_1 y_2 y_3$. Continuing in this way we find $y_1, y_2, \ldots, y_5$ such that $y_1 y_2 \ldots y_5$ is a pentagon and y_i is also joined to x_i and x_{i-1} ($x_0 \equiv x_5$). Taking $G[\Gamma(y_2)]$ we see that G contains another vertex, y_0, say, and y_0 is joined to y_1, y_2 and y_3. Then $y_0 \in \Gamma(y_3)$ so y_0 is joined to y_4 and $y_0 \in \Gamma(y_1)$ so y_0 is joined to y_5. The subgraph of G spanned by the vertices enumerated so far is exactly I_{12}. Hence G cannot have any more vertices and so it is I_{12}. ∎

The proof of Theorem 1.1 is nearly complete. For Lemmas 1.3 and 1.4 imply that if G is a minimal graph of $\mathscr{F}_5$ with respect to $>$ then G is one of the graphs K^6, B_1, B_2 and I_{12}. Therefore we have to prove only that there is no subcontraction of one of these graphs to another. Now $B_1 \nsucc K^6$ since

$B_1/xy \not\cong K^6$ (see Fig. 1.1). We see also that $B_2 \not> K^6$ since $B_2 - y_1$ (see Fig. 1.1) is planar so it has no contraction to K^5. Furthermore, I_{12} is planar so it has no subcontraction to any of the graphs K^6, B_1 or B_2. Finally $B_2 \not> B_1$ since every edge xy of B_2 is contained in at least 2 triangles so B_2/xy contains at least 2 vertices of degree 4. ∎

Let us list now some immediate consequences of Theorem 1.1. Most of these results have a more natural form than Theorem 1.1 itself.

COROLLARY 1.5. *If* $\delta(G) \geqslant 5$ *then* $G > K^{6^-}$ *or* $G > I_{12}$.

Proof. Each of the graphs K^6, B_1 and B_2 is subcontractible to K^{6^-}, the complete hexagon from which an edge has been omitted. ∎

COROLLARY 1.6. *If G is planar and* $\delta(G) = 5$ *then* $G > I_{12}$. ∎

The next consequence of Theorem 1.1 was first proved by Dirac [D13] (see also Wagner [W3]).

COROLLARY 1.7. *If* $\delta(G) \geqslant 3$ *then* $G > K^4$. *In particular, if* $\chi(G) \geqslant 4$ *then* $G > K^4$.

Proof. Suppose $\delta(G) \geqslant 3$. Then $E^2 + G \in \mathscr{F}_5$ so $E^2 + G$ has a subcontraction to a minimal graph H of $\mathscr{F}_5$. Hence there are vertices $x, y \in H$ such that $G > H - \{x, y\}$. It is easily checked that each of these graphs $H - \{x, y\}$ has a subcontraction to K^4. ∎

Analogously to Corollary 1.7 we find the following result due to Halin and Jung [HJ1].

COROLLARY 1.8. *If* $\delta(G) \geqslant 4$ *then* $G > K^5$ *or* $G > K_3(2) = K(2, 2, 2)$. ∎

Corollaries 1.7 and 1.8 imply, in particular, that if $\delta(G) = \delta$, $\delta = 3$ or 4, then G has a subcontraction to a δ-connected graph. Mader [M3] observed that this is not true for $\delta \geqslant 5$. Take two disjoint copies of B_1 and identify the two vertices of degree 4, one in each copy. The resulting graph, B_3 say, is such that $\delta(B_3) = 5$, $\kappa(B) = 1$ and B_3 has no proper subcontraction to a graph with minimal degree at least 5.

Hadwiger's conjecture is that K^k is the only minimal graph in $\{G : \chi(G) \geqslant k\}$

with respect to $>$. Though, as we mentioned already, this is not known for $k \geqslant 6$, the following result of Dirac [D20] sheds some light on the minimal graphs.

THEOREM 1.9. *Let G_0 be a minimal k-chromatic graph with respect to $>$. Then either $G_0 = K^k$ or else $\delta(G_0) \geqslant k$.*

Proof. Suppose $G_0 \neq K^k$. Then $|G_0| \geqslant k + 1$ and $\text{cl}(G_0) \leqslant k - 1$. Furthermore, since G_0 is a minimal graph of the class $\{G : \chi(G) \geqslant k\}$, every proper subgraph of G_0 is $(k - 1)$-colourable, i.e. G_0 is critical k-chromatic. In particular $\delta(G_0) \geqslant k - 1$. Assume now that $\delta(G) \geqslant k$ does not hold and let $x \in G_0$ be a vertex of degree $k - 1$. Put $\Gamma(x) = \{x_1, x_2, \ldots, x_{k-1}\}$. Since $\text{cl}(G_0) \leqslant k - 1$, we may suppose without loss of generality that $x_1 x_2 \notin E(G_0)$. Let G^* be the graph obtained from G_0 by contracting the path $x_1 x x_2$ to a vertex. Then $\chi(G^*) \leqslant k - 1$. Since x_1 is not joined to x_2, a $(k - 1)$-colouring of G^* induces a proper $(k - 1)$-colouring of $G_0 - x$ in which x_1 and x_2 are assigned the same colour. This implies that at most $k - 2$ of the colours are used to colour the vertices in $\Gamma(x)$. Hence this $(k - 1)$-colouring of $G_0 - x$ can be extended to a $(k - 1)$-colouring of G_0, contradicting $\chi(G_0) = k$. ∎

Putting together Corollaries 1.5, 1.7 and Theorem 1.9, we get the following information about the case $k = 5$ of Hadwiger's conjecture, extending some results of Dirac [D17], [D18] and Wagner [W3].

THEOREM 1.10. *If $\chi(G) = 5$ then either $G > K^5$ or $G > I_{12}$.* ∎

Wagner [W4] showed that there is a connection between the chromatic number of a graph and its subcontractibility to K^p. Namely Wagner proved the following weak form of Hadwiger's conjecture: there is a function $\psi(p)$ such that $\chi(G) \geqslant \psi(p)$ implies $G > K^p$ (cf. Ex. 2). At present when trying to deduce that a graph G has a subcontraction to K^p (p large) we cannot go considerably further with the condition $\chi(G) \geqslant k$ than with the much less restrictive conditions $\delta(G) \geqslant k - 1$ or $e(G) \geqslant [(k - 1)/2]|G|$. Thus it is natural that in the rest of the section we concentrate on the following problem: what is the maximal size of a graph of order n if it has no subcontraction to K^p? More precisely, we shall investigate a slightly less sensitive function, the supremum of the density of these graphs. Mader [M2] was the first to notice that if the density $e(G)/|G|$ of a graph G is sufficiently large then $G > K^p$. We deduce this from the following lemma that will be useful in the sequel as well.

Recall that for a natural number d we defined

$$\mathscr{D}_d = \left\{ G : |G| \geqslant d, e(G) \geqslant d|G| - \binom{d+1}{2} + 1 \right\}.$$

LEMMA 1.11. *Let G be a minimal element of $\mathscr{D}_d$ $(d \geqslant 2)$ with respect to $>$. Then $d + 1 \leqslant \delta(G) < 2d$ and if $x \in G$ and $G_x = G[\Gamma(x)]$ then $\delta(G_x) \geqslant d$.*

Proof. It follows from (0.5) that $\delta(G) \geqslant d + 1$ and so $|G| \geqslant d + 2$. Since $\binom{d+1}{2} > 1$ we have $\delta(G) \leqslant 2e(G)/|G| < 2d$.

Let $x \in G$, $xy \in E(G)$. Then $G/xy \notin \mathscr{D}_d$ so $e(G) - e(G/xy) \geqslant d + 1$. This means that at least d vertices of G are joined to both x and y, i.e. $d_{G_x}(y) \geqslant d$. ∎

COROLLARY 1.12. *Let $d \geqslant 2$ be a natural number such that if $|G| < 2d$ and $\delta(G) \geqslant d$ then $G > K^p$. Then every $G \in \mathscr{D}_d$ has a subcontraction to K^{p+1}.*

Proof. Given $G \in \mathscr{D}_d$ let G_0 be a minimal graph in $\mathscr{D}_d$ such that $G > G_0$. Pick a vertex $x \in G_0$ with $d(x) = \delta(G_0)$ and put $G_1 = G_0[\Gamma(x)]$. Then $0 < |G_1| < 2d$ and $\delta(G_1) \geqslant d$ so $G_1 > K^p$. Since $G_0 \supset G_1 + K^1$ (where x is the vertex of K^1), this implies that $G_0 > K^{p+1}$. ∎

COROLLARY 1.13. *If $n \geqslant 4$ then*

$$\max\{e(G^n) : G^n \not> K^5\} = 3n - 6.$$

Proof. Corollaries 1.7 and 1.12 imply that every $G \in \mathscr{D}_3$ has a subcontraction to K^5 so the maximum is at most $3n - 6$. Every maximal planar graph proves the converse inequality. ∎

THEOREM 1.14. *Every $G = G(n, 2^{p-3}n)$, $p \geqslant 3$, has a subcontraction to K^p.*

Proof. Apply induction on k. For $p = 3$ the result is trivial and Corollary 1.12 proves the induction step. ∎

For $p \geqslant 2$ let

$$c(p) = \inf\{c : \text{ if } e(G) \geqslant c|G| \text{ then } G > K^p\}.$$

Theorem 1.14 implies that in this definition of $c(p)$ we take the infimum of a non-empty set of real numbers and, in fact, $c(p) \leqslant 2^{p-3}$. Let $G =$

$P^{n-p+3} + K^{p-3}$. Then $G = G(n, m)$, where $m = (p - 2)(n - p + 3) - 1$ and $G \not> K^p$. Hence $c(p) \geqslant p - 2$. One can say slightly more, namely that

$$c(p + 1) \geqslant c(p) + 1$$

for every p. For let $\varepsilon > 0$ and let G be a graph such that $e(G) \geqslant (c(p) - \varepsilon)|G|$ and $G \not> K^p$. Put $H = K^1 + mG$. Then $H \not> K^{p+1}$ and, as $m \to \infty$,

$$e(H)|H|^{-1} \to e(G)|G|^{-1} + 1 \geqslant c(p) + 1 - \varepsilon.$$

In the proof of Theorem 1.14 we used a weak form of Corollary 1.12. In [M3] Mader made use of the full force of Corollary 1.12 to prove a considerable strengthening of Theorem 1.14. In the proof of this theorem, which is the main result of the section, we shall need the following lemma as well, giving a sufficient condition on a graph to have a contraction to K^p.

LEMMA 1.15. Let $\mathscr{C}_m = \{G : 2\delta(G) - |G| \geqslant 2m\}$. If $m \geqslant t\lfloor \log_2 t \rfloor \geqslant 2$ and $G \in \mathscr{C}_m$ then $G > K^t$.

Proof. Suppose $m \geqslant t\lfloor \log_2 t \rfloor, t \geqslant 2$ and $G \in \mathscr{C}_m$. Note first that if $W \subset U \subset V$, $|U| + \log_2|W| \leqslant 2m$ then there are at least $|W|(\delta(G) - 2m + 1) > \frac{1}{2}|W||V - U|$ edges between W and $V - U$. Consequently a vertex $u_1 \in V - U$ is joined to more than half of the vertices in W, say $W_1' = \Gamma(u_1) \cap W$, $|W_1'| < \frac{1}{2}|W|$. Then $W_1 = W - W_1'$ and $U_1 = U \cup \{u_1\}$ satisfy exactly the same conditions as W and U, so we can find a vertex $u_2 \in V - U_1$ joined to more than half of the vertices in W_1. Repeating this process we see that there are vertices $u_1, \ldots, u_s \in V - U$, $s < \log_2|W|$ (so $s \leqslant \lfloor \log_2|W| \rfloor$) such that each vertex of W is joined to at least one u_i. Put $H = G - U$. Then

$$\delta(H) \geqslant \delta(G) - |U| \geqslant \tfrac{1}{2}|G| + m - |U| \geqslant \tfrac{1}{2}(|G| - |U|).$$

Hence diam $H \leqslant 2$; in particular the vertex u_1 can be joined by a path of length at most 2 to each u_i, $1 < i \leqslant s$. This implies that there is a set

$$S \subset V - U, \quad \{u_1, \ldots, u_s\} \subset S, \quad |S| \leqslant 2\lfloor \log_2|W| \rfloor - 1$$

such that $G[S]$ is connected.

To prove the lemma we will use this process to select disjoint subsets

$V_1, V_2, \ldots, V_t$ of V with the following properties:

$$|V_1| = |V_2| = 1, \quad |V_j| \leqslant 2\lfloor \log_2(j - 1)\rfloor - 1, \quad j \geqslant 3, \tag{1}$$

$$\text{if } 1 \leqslant i < j \leqslant t \text{ then there exists a } V_i - V_j \text{ edge}, \tag{2}$$

$$G[V_i] \text{ is connected for each } i, \quad 1 \leqslant i \leqslant t. \tag{3}$$

Pick an edge $xy \in E(G)$ and put $V_1 = \{x\}$ and $V_2 = \{y\}$. Suppose we have already selected $V_1, V_2, \ldots, V_l$ ($l < t$) such that (1), (2) and (3) are satisfied. Put $U = \bigcup_1^l V_i$ and pick a vertex from each V_i to form a set W, $|W| = l$. Then

$$|U| + \log_2 l \leqslant 2 + \sum_{j=3}^{l} (2\lfloor \log_2(j - 1)\rfloor - 1) + \log_2 l \leqslant 2t\lfloor \log_2 t\rfloor,$$

so there is a set $V_{l+1} \subset V - U$, $|V_{l+1}| \geqslant 2\log_2 l - 1$ such that $G[V_{l+1}]$ is connected and each vertex of W is joined to at least one vertex of V_{l+1}. Thus $V_1, V_2, \ldots, V_{l+1}$ also satisfy (1), (2) and (3). ∎

THEOREM 1.16. *If $p \geqslant 4$ then $c(p) \leqslant 8(p - 2)\lfloor \log_2(p - 2)\rfloor$.*

Proof. Put $m = (p - 2)\lfloor \log_2(p - 2)\rfloor$. Because of Corollary 1.12 it suffices to prove that if $|G| \leqslant 16m$ and $\delta(G) \geqslant 8m$ then $G > K^{p-1}$. Let

$$\mathscr{E}_m = \{H: \ 16m \geqslant |H| \geqslant 8m, e(H) \geqslant 5m|H| - 16m^2\}.$$

Then $G \in \mathscr{E}_m$. Let $G_0 < G$ be a minimal graph in $\mathscr{E}_m$ with respect to $<$. It suffices to prove that $G_0 > K^{p-1}$.

1. Suppose first that $|G_0| > 8m$. Then $e(G_0) = 5m|G_0| - 16m^2$ so $\delta(G_0) \leqslant 8m$. Let $x \in G_0$, $d(x) = \delta(G_0)$ and let $xy \in E(G_0)$. Since $G_0/xy \notin \mathscr{E}_m$, there are at least $5m$ vertices of G_0 joined to both x and y. Therefore if $G_1 = G_0[\Gamma(x)]$ then $|G_1| \leqslant 8m$ and $\delta(G_1) \geqslant 5$. Hence $G_1 \in \mathscr{E}_m$ and so $G_1 > K^{p-2}$ by Lemma 1.15. This implies $G_0 > K^{p-1}$.

2. Let us consider now the case $|G_0| = 8m$. We have

$$e(G_0) = 24m^2 > 4m|G_0| - \binom{4m + 1}{2}$$

so $G_0 \in \mathscr{D}_{4m}$. Let $H_1 < G_0$ be a minimal graph in $\mathscr{D}_{4m}$ and let $H_2 = H_1[\Gamma(x)]$, where $x \in H_1$, $d(x) = \delta(H_1)$. Since $e(H_1) = 4m|H_1| - \binom{4m+1}{2}$ and $|H_1| \leqslant |G_0| \leqslant 8m$, we see that $|H_2| = \delta(H_1) < 6m$.

By Lemma 1.11 we have $\delta(H_2) \geqslant 4m$ so $H_2 \in \mathscr{C}_m$. Hence $H_2 > K^{p-2}$, $G_0 > H_1 > K^{p-1}$.   ∎

For small values of p the graphs occurring in Corollary 1.12 and Lemma 1.11 are rather small and by straightforward discussion of the cases one can obtain a much better estimate of $c(p)$ than the one provided by Theorem 1.16. Notably Mader [M3] showed that $c(p) = p - 2$ holds for $p = 6$ and 7 as well. It is rather hard not to conjecture that $c(p)$ is always $p - 2$ or, at least, that $c(p) = O(p)$. In the proof of Theorem 1.16 we constructed graphs such that the edges formed higher and higher proportion of all pairs. By continuing this process one arrives at the following result that might be useful in the proof of $c(p) = O(p)$.

THEOREM 1.17. *Suppose there exist* $\varepsilon > 0$ *and* $a > 0$ *such that*

$$\delta(G) \geqslant (1 - \varepsilon)|G| > a p \quad \text{implies} \quad G > K^p.$$

Then $c(p) = O(p)$.   ∎

2. TOPOLOGICAL COMPLETE SUBGRAPHS

We shall give necessary conditions on a graph to contain a TK^p. In particular we are interested in the function $ex(n; TK^p)$. As one would expect, exact results exist only for small values of p. Notably $ex(n; TK^3) = n - 1 (n \geqslant 3)$ trivially, and Dirac [D13] proved that $ex(n; TK^4) = 2n - 3$ $(n \geqslant 3)$. As we shall see, TK^4 has the rather interesting property, shared by TK^3, the cycle, that every maximal graph of order n *without* a TK^4 has the same size.

THEOREM 2.1. *If* $\kappa(G) \geqslant 3$ *then* $G \supset TK^4$.

Proof. Let a, $b \in G$ and let P_1, P_2 and P_3 be independent $a - b$ paths. Since $G - a - b$ is connected, there exists a $c - d$ path, say Q, such that, possibly with a change of notation, $V(Q) \cap V(P_1) = \{c\}$, $V(Q) \cap V(P_2) = \{d\}$ and $V(Q) \cap V(P_3) = \varnothing$. Then the vertices a, b, c and d, and appropriate segments of P_1, P_2, P_3 and Q form a topological K^4.   ∎

THEOREM 2.2. *Let G be a maximal graph of order* $n \geqslant 3$ *that does not contain a* TK^4. *Then*

$$e(G) = 2n - 3.$$

Proof. We apply induction on n. The assertion being trivial for $n = 3$ we proceed to the induction step. Clearly $\kappa(G) \geqslant 1$. Recall that $W(H)$ is the set of branchvertices of H (i.e. vertices of degree at least 3).

Suppose first that $\kappa(G) = 1$. Then there exist vertices $x \in G$ and $y, z \in \Gamma(x)$ such that y and z belong to different components of $G - x$. It is easily checked that $G^* = G + yz$ does not contain a TK^4 either. For the only role yz could play in a TK^4 in G^* is to appear in an xy (or xz) path. Since xy and $xz \in E(G)$ this would imply that already G contained a topological K^4. Hence $\kappa(G) \geqslant 2$ and so, by Theorem 2.1, $\kappa(G) = 2$.

Note that if $H \supset TK^4$ and (V_1, V_2) is a 2-separator of H then either $W(TK^4) \subset V_1$ or $W(TK^4) \subset V_2$. Hence if $V_1 \cap V_2 = \{x, y\}$ and $xy \in E(H)$ then either $H[V_1]$ or $H[V_2]$ contains a TK^3.

Let $G = G_1 \cup G_2$, $V(G_1) \cap V(G_2) = \{x, y\}$. Suppose $xy \notin E(G)$. Then $G^* = G + xy$ contains a topological K^4, say $T = TK^4$. By the remark above we may assume that $T \subset G_1 + xy$. Then in T the edge xy can be replaced by an $x - y$ path in G_2, which exists because of $\kappa(G) \geqslant 1$, so G also contains a topological K^4, contrary to the assumption. Hence we may assume that $xy \in E(G_1) \cap E(G_2)$. Again by the remark above both G_1 and G_2 are *maximal* graphs without a TK^4. Thus by the induction hypothesis

$$e(G) = e(G_1) + e(G_2) - 1 = 2|G_1| - 3 + 2|G_2| - 3 - 1$$

$$= 2(|G_1| + |G_2| - 2) - 3 = 2|G| - 3. \qquad \blacksquare$$

COROLLARY 2.3. *If* $n \geqslant 3$ *then* $ex(n; TK^4) = 2n - 3$. $\qquad \blacksquare$

Denote by K^5_{-2} a K^5 from which two *adjacent* edges have been deleted. Thomassen [T2] proved that one more edge than the number of edges ensuring a TK^4 already ensures a TK^5_{-2} as well. Since $\bar{\kappa}(TK^5_{-2}) = 4$, this result extends Theorem I.5.1(ii). Not surprisingly, the proof is also very similar and the extremal graphs turn out to be the same. As in the proof of Theorem I.5.1(ii), we need some properties of a connected graph whose blocks are wheels. The fairly short case by case examination proof of the following lemma is left to the reader (Ex. 7).

LEMMA 2.4. *Let G be a connected graph whose blocks are wheels. Then G has the following properties.*

(i) $e(G) = 2|G| - 2$ *and G is a maximal graph without a TK^5_{-2}.*

(ii) *If $xy \in E(G)$ and G^* is obtained from G by adding a new vertex z and joining it to x and y then G^* contains a TK^5_{-2}.*

(iii) *If $x_1, x_2, x_3 \in G$, $x_1 x_2 \in E(G)$ and G^* is obtained from $G - x_1 x_2$ by adding to it a new vertex y and joining y to x_1, x_2 and x_3 then G^* contains a TK^5_{-2}.* ∎

THEOREM 2.5. *If $n \geqslant 4$ then*

$$ex\ (n;\ TK^5_{-2}) = 2n - 2$$

and the extremal graphs are the connected graphs of order n whose blocks are wheels.

Proof. We apply induction on n. For $n = 4$ and 5 the assertion is easily checked so suppose $n \geqslant 6$ and the result holds for smaller values of n.

Let $G = G^n$ be such that $G \not\supset TK^5_{-2}$ and $e(G) = 2n - 2$. By Lemma 2.4 (i) it suffices to prove that G is connected and its blocks are wheels.

As in the proof of Theorem 1.2 one can show that $\kappa(G) \geqslant 2$. Suppose that $d(x) = 2$ for some $x \in G$. Then $G - x$ is an extremal graph of order $n - 1$ so by the induction hypothesis it is a connected graph whose blocks are wheels. Lemma 2.4(i) and (ii) imply that G contains a TK^5_{-2}. Since this contradicts our assumption, $\delta(G) \geqslant 3$.

Since $e(G) = 2|G| - 2$ there exists a vertex $a_0 \in G$ of degree three. Say $\Gamma(a_0) = \{a_1, a_2, a_3\}$. Let $b \in G - \{a_0, a_1, a_2, a_3\}$. Then $\kappa(G) \geqslant 2$ implies that there exist two independent paths from y to $\{a_1, a_2, a_3\}$, say P_1 and P_2, $V(P_1) \cap V(P_2) = \{b\}$. This implies that a_1, a_2 and a_3 do not span a triangle since then G contains the TK^5_{-2} formed by $G[a_0, a_1, a_3, a_3]$ and P_1, P_2. Thus we may assume that $a_1 a_2 \notin E(G)$. As in the proof of Theorem I.5.1(ii), put $H = G - a_0 + a_1 a_2$. Then $e(H) = 2(n - 1) - 2$, $|H| = n - 1$, and $H \not\supset TK^5_{-2}$. Hence H is an extremal graph of order $n - 1$ so by the induction hypothesis it is connected and its blocks are wheels. Then Lemma 2.4(iii) implies that G itself is connected and its blocks are wheels. ∎

For $x \in G$ let $\varepsilon(x) = \max\{0, 3 - d(x)\}$. Extending a theorem of Pelikán [P3], Thomassen [T2] proved the following result.

THEOREM 2.6. *If* $|G| > 3$ *and* $\sum_{x \in G} \varepsilon(x) \leqslant 3$ *then* $G \supset TK^{5-}$.

Proof. Let us assume that the theorem fails and let G_0 be a counter-example of minimal order, say order n, and minimal size. It is easily checked that $n \geqslant 6$, $\delta(G_0) = 3$ and if x is a vertex of degree 3 then $G[\Gamma(x)]$ is a triangle. Let now $G_1 \supset G_0$ be a maximal graph of order n without a TK^{5-}. Then $\mathrm{cl}\,(G_1) \geqslant \mathrm{cl}(G_0) \geqslant 4$, say $A = \{a_0, a_1, a_2, a_3\}$ is the vertex set of a K^4 in G_0. As in the proof of Theorem 2.5 one can check that G_1 is 2-connected and if a set $\{x, y\}$ separates G_1 then $xy \in E(G_1)$. Hence every component of $G_2 = G_1 - A$ is joined to at least 2 of the vertices in A. On the other hand, $G_1 \not\supset TK^{5-}$ implies that a component of G_2 cannot be joined to more than 2 of the vertices in A. Using the fact that G_1 is a counterexample of minimal order it follows that G_2 is, in fact, connected.

We may assume that a_2 and a_3 are the vertices of A joined to G_2 and so $d(a_0) = d(a_1) = 3$. Now since $G - \{a_0, a_1\} \not\supset TK^{5-}$ and it is *not* a counter-example to the theorem, $\varepsilon(a_2) + \varepsilon(a_3) > \varepsilon(a_0) + \varepsilon(a_1) = 2$. Hence, because of $\kappa(G_1) \geqslant 2$, we may assume that $\Gamma(a_2) \cap V(G_2) = \{a_4\}$ and $|\Gamma(a_3) \cap V(G_2)| \geqslant 2$. Since $\{a_3, a_4\}$ separates G_1, a_3 is joined to a_4. Thus, again because of $\kappa(G_1) \geqslant 2$ we have $\Gamma(a_3) \cap V(G_2) = \{a_4, a_5\}$ (see Fig. 2.1). By assumption $G_1 + a_2 a_5$ contains a TK^{5-}. Since $\{a_4, a_5\}$ separates G_1, this implies that either $G_3 = G[a_0, a_1, \ldots, a_5]$ or G_2 contains a TK^{5-}. The graph G_3 is shown in Fig. 2.1 and it does not contain a TK^{5-}. Hence $G_2 \subset G_1$ does, contrary to the assumption. ∎

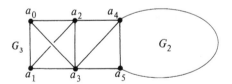

$$a_0 \qquad a_2 \qquad a_4$$
$$G_3 \qquad\qquad G_2$$
$$a_1 \qquad a_3 \qquad a_5$$

FIG. 2.1. The structure of G_1.

A planar graph does not contain a TK^5 so $ex(n; TK^5) \geqslant 3n - 6(n \geqslant 4)$. It seems likely that equality holds here (cf. Conjecture 9) but the best upper bound of $ex(n; TK^5)$, obtained by Thomassen [T2], is rather far from this lower bound. Because of the inductional proof it turns out to be more convenient to prove slightly more than a pure upper bound of $ex(n; TK^5)$.

THEOREM 2.7. *Suppose* $|G| = n \geqslant 6$ *and* $e(G) \geqslant 4n - 10$. *Then for every vertex* $x \in G$ *there is a* TK^5 *in* G *such that* $x \notin W(TK^5)$.

Proof. We apply induction on n. For $n = 5$ the assertion is trivial since $G(5, 9) = K^{6-}$. Suppose $n > 5$ and the assertion holds for smaller values of n.

Let $|G| = n$, $e(G) = 4n - 10$ and $x \in G$. Put $G_0 = G[\Gamma(x)]$. If $\delta(G_0) \geqslant 4$ then Theorem 2.6. implies that $G_0 \supset TK^{5-}$. If y and z are the branchvertices of degree 3 in this TK^{5-} then we can add the path yxz to it to obtain a desired TK^5. Thus we may assume that $d_{G_0}(y) \leqslant 3$ for some vertex $y \in G_0$. Let $G^* = G/xy$ and denote by $x^* \in G^*$ the contraction of xy. Then

$$e(G^*) \geqslant e(G) - d_{G_0}(y) - 1 \geqslant e(G) - 4 = 4|G^*| - 10.$$

Hence by the induction hypothesis G^* contains a topological K^5 in which x^* is not a branchvertex. This implies that G contains a TK^5 such that $x \notin W(TK^5)$.   ∎

COROLLARY 2.8. *If $n \geqslant 5$ then*

$$3n - 6 \leqslant ex(n; TK^5) \leqslant 4n - 11.$$ ∎

The main aim of this section is to give bounds on $ex(n; TK^p)$. Of course there is no $p \geqslant 5$ for which its value is known for every n. In fact it is not entirely trivial that for every p there is a constant c_p such that

$$ex(n; TK^p) \leqslant c_p n, \qquad n \geqslant p. \tag{1}$$

Jung [J2] pointed out that if

$$m < \frac{p}{2} + \binom{p/2}{2} = \frac{p(p + 2)}{8}$$

then $K^{m, m}$ does not contain a topological K^p. Hence if there is such a c_p, then

$$c_p \geqslant \frac{p(p + 2)}{16}.$$

Dirac [D18], [D21] and Jung [J1] proved that there is a function $\phi(p)$ such that $\chi(G) \geqslant \phi(p)$ implies $G \supset TK^p$. Mader [M2] was the first to prove the existence of c_p for each p; notably he showed that

$$f(n; TK^p) < p2^{\binom{p-1}{2}} n.$$

After a sharpening due to Halin [H8] the inequality was improved substantially by Mader [M9], who proved that $c_p = O(2^p)$. In the rest of the section we present this theorem of Mader.

In the proof of this result we shall make use of the concept of an admissible triple (G, K, E_0), defined in Chapter I, §5 (p. 41). Let $K = K^{p+1} \subset H$, $V(K) = \{x_0, x_1, \ldots, x_p\}$. Define

$$U = \Gamma(x_0) - V(K) \cup \bigcap_1^p \Gamma(x_i)$$

and for $x \in U$ put

$$g(x) = \min\{i: 1 \leqslant i \leqslant p + 1, xx_i \notin E(H)\},$$

$$F_0 = \{xx_{g(x)}: x \in U\}.$$

Denote by $H(K^{\leqslant})$ the graph obtained from $H - x_0$ by adding F_0 to the edge set. The notation does not express that $H(K^{\leqslant})$ depends on the linear order of the vertices of K, but will do for our purpose. Note that if $K_0 = K - x_0$ then by the construction

$$(H(K^{\leqslant}), K_0, F_0)$$

is an admissible triple. We shall need the following lemma, which is a consequence of Lemma I.5.9.

LEMMA 2.9. Let $K = K^{p+1} \subset H$, $V(K) = \{x_0, x_1, \ldots, x_p\}$, $K_0 = K - x_0$ and let $H(K^{\leqslant})$ be the graph and F_0 the edge set constructed above. Let $T_1 = TK^r \subset H(K^{\leqslant}) - K_0 = H - K$ and suppose this T_1 is contained in a TK^{r+1}, say in $T_2 = TK^{r+1} \subset H(K^{\leqslant})$ such that $W(T_2) \subset V - V(K_0)$. Then there exists another TK^{r+1}, say $T_0 = TK^{r+1}$, such that $T_1 \subset T_0 \subset H$, $W(T_0) = W(T_2)$.

Proof. Let $W(T_2) - W(T_1) = \{a\}$. Let H^* be the graph obtained from $H - (V(T_1) - W(T_1))$ by adding to it a vertex b and joining b to each of the r vertices in $W(T_1)$. Let $G = H(K^{\leqslant})^*$ be the graph obtained from $H(K^{\leqslant})$ in a similar way. Denote by E_0 the set of edges in F_0 that belong to G as well: $E_0 = F_0 \cap E(G)$. Then by the construction (G, K_0, E_0) is an admissible triple. The existence of T_2 means exactly that

$$\kappa(a, b; G) = r.$$

Analogously T_0 exists iff

$$\kappa(a, b; H^*) = r.$$

Note that H^* is obtained from $G - E_0$ by adding the vertex x_0 to it and joining it to some vertices, *including all the endvertices of the edges in* E_0. Thus if $\kappa(a, b; G - E_0) = r$, the proof is finished. Therefore by Lemma I.5.9 we may assume that $G - E_0$ contains $r - 1$ independent $a - b$ paths, say P_1, $P_2, \ldots, P_{r-1}$, and two more paths, say Q_1 and Q_2, such that Q_1 joins a to an endvertex of an edge in E_0 and Q_2 joins b to an endvertex of an edge in E_0 and only a and b may be common to any two paths. Then in H^* we can join x_0 to the endvertices of Q_1 and Q_2 and so we can string them together to obtain another $a - b$ path, independent of each P_i, $i = 1, 2, \ldots, r - 1$. Hence $\kappa(a, b; H^*) = r$, as required. ∎

Before stating the main theorem of the section, we prove another lemma, which is essentially the induction step in the proof of the theorem.

LEMMA 2.10. *Let r and s be natural numbers, $t = r + s - 1$. Suppose $\delta(H) \geqslant s$ implies that there exists a vertex $a \in H$ such that $H - a$ contains a TK^r. Then if $G \in \mathcal{D}_t$ and $K = K^{p+1} \subset G$, where $0 \leqslant p < r$, there exist $T_1 = TK^r \subset T_0 = TK^{r+1} \subset G$ such that $W(T_0) \cap V(K) = \varnothing$ and $V(T_1) \cap V(K) = \varnothing$.*

Proof. We apply induction on $|G|$, $G \in \mathcal{D}_t$. By the definition of $\mathcal{D}_t$, $G \in \mathcal{D}_t$ implies $|G| \geqslant t$ and, in fact, $|G| = t$ or $t + 1$ cannot hold either. Thus suppose $G \in \mathcal{D}_t$, $|G| \geqslant t + 2$ and the assertion holds for graphs of order less than $|G|$ in $\mathcal{D}_t$. We distinguish two cases.

1. *There is a clique of order at most r containing K.* In this case we may assume without loss of generality that K itself is this clique. Put $V(K) = \{x_0, x_1, \ldots, x_p\}$ and consider the graph $H = G(K^{\leqslant})$. Since K is a clique,

$$\bigcap_{i=1}^{p} \Gamma_G(x_i) = \varnothing.$$

Hence

$$|H| = |G| - 1 \quad \text{and} \quad e(H) = e(G) - p \geqslant e(G) - r - 1,$$

so $H \in \mathcal{D}_t$. By the induction hypothesis there exist $T_1 = TK^r \subset T_2 = TK^{r+1} \subset H$ such that $W(T_2) \cap V(K) = \varnothing$ and $V(T_1) \cap V(K) = \varnothing$. Then Lemma 2.9 implies the existence of the required $T_0 = TK^{r+1}$.

2. *Every clique containing K has order at least $r + 1$.* We may assume without loss of generality that $|K| = p + 1 = r$, $V(K) = \{x_1, x_2, \ldots, x_r\}$. Since

K is not a clique,

$$U = \bigcap_1^r \Gamma(x_i) \neq \varnothing.$$

Put $H = G[U]$ and suppose first that $\delta(H) < s$, say $x \in H$, $d_H(x) < s$. Let $x_{r+1} = x$ and note that the vertices $x_1, x_2, \ldots, x_{r+1}$ span a complete subgraph K_0 of G. Using this linear order of the vertices of K_0 construct the graph $F = G(K_0^{\leqslant})$. Since

$$\left| \bigcap_1^{r+1} \Gamma(x_i) \right| = d_H(x_{r+1}) < s,$$

$$e(F) - e(G) = \left| \bigcap_1^{r+1} \Gamma(x_i) \right| + r \leqslant t.$$

Thus $F \in \mathscr{D}_t$. The induction hypothesis applied to F and Lemma 2.9 imply again the existence of the required T_1 and T_0.

Therefore we may assume that $\delta(H) \geqslant s$. Then by the assumption of the lemma there exists a vertex $a \in H$ such that $H - a$ contains a TK^r, say T_1. Put $W(T_1) = \{y_1, y_2, \ldots, y_r\}$. Let T_0 be the TK^{r+1} obtained from T_1 by adding to it the vertex a together with the paths $ax_i y_i$, $i = 1, 2, \ldots, r$. Then T_1 and T_0 have the required properties. ∎

Armed with Lemma 2.10 we are ready to present the main theorem of this section, proved by Mader [M9].

THEOREM 2.11. *Let* $p \geqslant 4$ *and put* $t(p) = 3 . 2^{p-3} - p$, $\quad d(p) = 2t(p)$.

(i) *Every* $G \in \mathscr{D}_{t(p)}$ *contains a* TK^p.
(ii) *If* $\delta(G) \geqslant d(p)$ *then there is a vertex* $a \in G$ *such that* $G - a$ *contains a* TK^p.

Proof. When proving (ii) we may assume that $\delta(G) = d(p)$. Let $a \in G$, $d(a) = d(p)$. Then $|G| \geqslant d(p) + 1 > t(p)$ and

$$e(G - a) \geqslant t(p) |G| - 2t(p)$$

$$= t(p) |G - a| - t(p) \geqslant t(p) |G - a| - \binom{t(p) + 1}{2}.$$

Hence $G - a \in \mathscr{D}_{t(p)}$. This shows that if (i) holds for a certain value of p then so does (ii).

Let us prove now (i) by induction on p. For $p = 4$ the assertion follows from Theorem 2.2. Suppose now that $p > 4$ and (i) holds for smaller values of p. By the remark above so does (ii). In particular, the conditions of Lemma 2.10 are satisfied for $r = p - 1$ and $s = d(p - 1)$.

Then $t = r + s - 1 = p - 1 + 3.2^{p-3} - 2(p - 1) - 1 = t(p)$. Thus if $G \in \mathcal{D}_{t(p)}$ then Lemma 2.10 implies that G contains a TK^p. ∎

Let us restate the first part of Theorem 2.11 in terms of $ex(n; TK^p)$.

THEOREM 2.11'. *If $p \geqslant 4$ and $n \geqslant t(p) = 3.2^{p-3} - p$ then*

$$ex(n; TK^p) \leqslant t(p) n - \binom{t(p) + 1}{2}. \qquad \blacksquare$$

3. SEMI-TOPOLOGICAL SUBGRAPHS

The following simple result was proposed by Pósa [P8] as an exercise in a Hungarian journal. If $n \geqslant 4$ and $e(G^n) \geqslant 2n - 3$ then G^n contains a cycle and one of its diagonals. The graph $K(2, n - 2)$ shows that the result is best possible; in fact if $n \geqslant 6$ then $K(2, n - 2)$ is the unique extremal graph (see [C2]). In his solution to this problem Czipszer [C14] observed that if $\delta(G) \geqslant k + 2$ then G contains a cycle and k diagonals incident with a vertex. For if $\delta(G) \geqslant k + 2$ and $P = x_0 x_1 \ldots x_p$ is a longest path in G then let $1 = i_0 < i_1 < \ldots < i_{k+1}$ be indices such that $x_0 x_{i_j} \in E(G)$. Then $x_0 x_1 \ldots x_{i_{k+1}}$ is a cycle and $x_0 x_{i_1}, x_0 x_{i_2}, \ldots, x_0 x_{i_k}$ are its diagonals. This implies the following results.

THEOREM 3.1. *Let $k \geqslant 1$ and denote by $g_k(n)$ the maximal size of a graph of order n that does not contain a cycle and k diagonals incident with a vertex. Then*

$$g_k(n + 1) \leqslant g_k(n) + k + 1, \qquad n \geqslant 1. \qquad (1)$$

If $n \geqslant k + 1$ then

$$(k + 1) n - (k + 1)^2 \leqslant g_k(n) \leqslant (k + 1) n - \binom{k + 2}{2}. \qquad (2)$$

Furthermore, if $n \geqslant 2k + 2$ and $k = 1, 2$ or 3 then

$$g_k(n) = (k + 1) n - (k + 1)^2. \qquad (3)$$

Proof. Inequality (1) follows from the observation preceding the theorem. Since $g_k(k + 1) = \binom{k+1}{2} = (k + 1)^2 - \binom{k+2}{2}$, (1) implies (2). Similarly, to prove (3) one has to check only that $g_1(4) = 4$, $g_2(6) = 9$ and $g_3(8) = 16$. ∎

Erdös conjectured that if $n \geqslant 2(k + 1)$ then $g_k(n)$ is in fact always equal to $(k + 1)n - (k + 1)^2$, the lower bound in (2). This was disproved by M. Lewin (Ex. 12). However, it is rather likely that there is a function $n(k)$ for which $n \geqslant n(k)$ implies that $g_k(n) = (k + 1)n - (k + 1)^2$ does hold (Conjecture 13).

The graph formed by a cycle and k diagonals incident with a vertex is an example of a semi-topological graph. Given a *multigraph* F and a spanning subgraph $F_0 \subset F$ denote by $\mathrm{ST}(F, F_0)$ a *graph* obtained from F by subdividing some of the edges *not* belonging to F_0. A graph $\mathrm{ST}(F, F_0)$ is a *semi-topological graph F with kernel F_0*. Thus if $F = K^1 + P^{k+1}$ and the edges of F_0 are exactly the edges incident with the vertex of K^1 then a graph formed by a cycle and k diagonals incident with a vertex is exactly an $\mathrm{ST}(F, F_0)$.

Given a graph F_0 which is not a forest there is no constant c such that $e(G) \geqslant c|G|$ implies $G \supset F_0$. In particular, given $F \supset F_0$, there is no constant c such that whenever $e(G) \geqslant c|G|$ the graph G contains an $\mathrm{ST}(F, F_0)$. We shall see shortly that, as shown in [B37], this cannot happen if F_0 is a forest.

Let F_0 be a forest and let d be such that if $G \in \mathscr{D}_d$ and $x_0 \in G$ then $G - x_0 \supset F_0$. Since every $G \in \mathscr{D}_d$, $d \geqslant 1$, contains a subgraph H with $\delta(H) \geqslant d + 1$, such a d always exists; in fact if $|F_0| \geqslant 2$ the minimal such d is at most $|F_0| - 1$.

THEOREM 3.2. *Let $F_0 \subset F_1 \subset F_2 \subset \ldots$, where each F_i is a multigraph with vertex set $V(F_0)$ and $e(F_i) - e(F_0) = i$. Put $d_i = 2^i d$.*

If $G \in \mathscr{D}_{d_i}$ and $x_0 \in G$ then G contains an $\mathrm{ST}(F_i, F_0)$ such that $x_0 \notin F_0$.

Proof. We apply induction on i. For $i = 0$ there is nothing to prove so suppose $i > 0$ and the results hold for smaller values of i. Now we apply induction on $n = |G|$. As $G \in \mathscr{D}_{d_i}$ implies $|G| \geqslant d_i + 2$, we may assume that $|G| \geqslant d_i + 2$ and the assertion holds for graphs of smaller order.

Put $G_0 = G[\Gamma(x_0)]$. Suppose first that $\delta(G_0) \geqslant d_i$. Then $G_0 \in \mathscr{D}_{d_{i-1}}$ so G_0 contains an $\mathrm{ST}(F_{i-1}, F_0)$, say S_1. Denote by a and b the vertices of S_1 such that joining them by a path independent from S_1 the resulting graph is an $\mathrm{ST}(F_i, F_0)$. Since $e(F_i) - e(F_{i-1}) = 1$, these vertices a and b exist. Then in G we can join a to b by the path $a\,x_0b$ to obtain a required $\mathrm{ST}(F_i, F_0)$ subgraph of G.

Assume now that $\delta(G_0) \leqslant d_i - 1$, say $y \in G_0$ is such that $d_{G_0}(y) \leqslant d_i - 1$.

Put $G^* = G/x_0 y$ and denote by $x^* \in G^*$ the contraction of $x_0 y$. Then $|G^*| = |G| - 1 \geqslant d_i + 2$ and

$$e(G^*) - e(G) = d_{G_0}(y) + 1 \leqslant d_i.$$

Hence $G^* \in \mathscr{D}_{d_i}$. Therefore by the induction hypothesis G^* contains an $ST(F_i, F_0)$, say S_1, such that x^* is not a vertex of the kernel F_0. Consequently x^* appears on at most one path joining two vertices of F_0. The inverse image of this path in G contains a path joining the same two vertices of the kernel F_0 so the inverse image of S_1 in G is again an $ST(F_i, F_0)$ such that x is not a vertex of F_0. ∎

COROLLARY 3.3. *Let F_0 be a tree of order $p \geqslant 3$. If*

$$e(G) \geqslant 2^{\binom{p-1}{2}}(p-1)|G|$$

then G contains a TK^p, say T_1, such that T_1 contains a tree with vertex set $W(T_1)$ isomorphic to F_0. ∎

A slightly weaker form of the next consequence of Theorem 3.2 was proved in [B34] by a different method.

COROLLARY 3.4. *Let d and s be natural numbers, $t = ds2^{d-1}$ and let $G \in \mathscr{D}_t$. Then G contains a vertex x_0, a path P not containing x_0 and d paths of length s joining x_0 to P, say $P_1, P_2, \ldots, P_d$, such that*

$$V(P_i) \cap V(P_j) = \{x_0\}, \quad 1 \leqslant i < j \leqslant d$$

and

$$|V(P_i) \cap V(P)| = 1, \qquad 1 \leqslant i \leqslant d.$$

Proof. Let F_0 be the tree obtained from the star $K^1 + E^d$ by dividing each edge into s edges. Add $d - 1$ edges to F_0 so that in the resulting graph F the vertices of E^d span a path of length $d - 1$. Then $e(F) - e(F_0) = d - 1$ and if $G \in \mathscr{D}_{t_0}$, where $t_0 = ds$ and $x_0 \in G$ then $G - x_0$ contains an F_0. Hence Theorem 3.2 implies the assertion. ∎

Corollary 3.4 can be used to prove the following result, first proved in [B34], that answers a question raised by Burr and Erdös [E28].

THEOREM 3.5. *Let k and l be natural numbers, $l \leqslant k$. Put $c_{kl} = (k+1)l\,2^k$. If $G \in \mathscr{D}_{c_{kl}}$ then G contains a cycle of length $2l$ modulo k.*

Proof. Let $G \in \mathcal{D}_{c_{kl}}$. Then the conditions of Corollary 3.4 are satisfied with $d = k + 1$ and $s = 1$. Let $\{x_0, P, P_1, P_2, \ldots, P_{k+1}\}$ be the system guaranteed by Corollary 3.4. If P_i is an $x_0 - x_i$ path $(1 \leqslant i \leqslant k + 1)$ then there exist x_i, $x_j, 1 \leqslant i \leqslant j \leqslant k + 1$, whose distance on P is 0 modulo k. Then the cycle formed by P_i, P_j and the $x_i - x_j$ segment of P has length $2l$ modulo k. ∎

If k is odd, $\{2, 4, \ldots, 2k\}$ is a complete set of residues modulo k. Thus Theorem 3.5 has the following consequence.

THEOREM 3.6. *Let k be an odd natural number and let G be a graph satisfying*

$$e(G) \geqslant k(k + 1)\, 2^k |G|.$$

Then for every natural number l the graph G contains a cycle of length l modulo k. ∎

It is very likely that Theorems 3.5 and 3.6 are very far from being best possible. However, the restriction to odd k in Theorem 3.6 is essential. The graph

$$T_2(n) = K\!\left(\left\lfloor \frac{n + 1}{2} \right\rfloor, \left\lfloor \frac{n}{2} \right\rfloor\right)$$

shows that a graph of order n and size $\lfloor n^2/4 \rfloor$ need not contain an odd cycle. Consequently if k is even then there is no constant c_k such that if $e(G) \geqslant c_k |G|$ and l is a natural number then G contains a cycle of length l modulo k.

Let us investigate now the existence of another semi-topological subgraph. In the rest of the section we write S_r for the semi-topological graph consisting of a cycle C, a vertex x_0 not on C and r edges joining x_0 to C. Trivially, $ex(n; S_1) = n$ and graphs of maximal degree $r - 1$ show that $ex(n; S_r) \geqslant \lfloor \frac{1}{2}(r - 1)n \rfloor$ if $n \geqslant r$.

$$ex(n; S_r) \leqslant sn - \binom{s + 1}{2}, \text{ provided } n \geqslant r + 1 \geqslant 3,$$

where $s = (r + 1)\, 2^r$. Though it is easy to improve this upper bound, the only values of $r > 1$ for which $ex(n; S_r)$ has been determined for every (or for every sufficiently large) value of n are 2 and 3. The case $r = 2$ is due to Bollobás and Erdös [BE5] and the case $r = 3$ was settled by Thomassen [T2a]. Note first that $ex(n; S_2) \geqslant e(n; \overline{\kappa} \leqslant 2)$ since if G contains an S_2 then it has two vertices joined by three independent paths so then $\bar{\kappa}(G) \geqslant 3$. In [BE5] it is proved that in fact $ex(n; S_2) = e(n; \bar{\kappa} \leqslant 2)$. This result extends Theorem I.5.1 (i) since it shows that if the size of a graph G is sufficiently large to ensure that there are two vertices joined by 3 independent paths then there are also two vertices which are joined by 3 independent paths, one of which has length exactly 2.

THEOREM 3.7. *If* $n \geqslant 1$ *then* $ex(n; S_2) = \lfloor \frac{3}{2}(n - 1) \rfloor$. *If* $n > 1$ *is odd then each extremal graph is connected and its blocks are triangles.*

Proof. We shall make use of the fact that if E is a connected graph whose blocks are triangles and E^* is obtained from E by joining two vertices of E by a path independent of E, then

$$E^* \text{ contains an } S_2. \tag{4}$$

We apply induction on n to prove the assertions of the theorem simultaneously. For $n \leqslant 4$ the assertion is easily checked so suppose $n > 4$ and the theorem has been proved for smaller values of n.

Let us show first that $ex(n; S_2) = \lfloor \frac{3}{2}(n - 1) \rfloor$. As we remarked earlier, the extremal graphs of Theorem I.5.1(i) show that $ex(n; S_2) \geqslant \lfloor \frac{3}{2}(n - 1) \rfloor$. Let us assume that, contrary to the assertion, we have $ex(n; S_2) > \lfloor \frac{3}{2}(n - 1) \rfloor$. Let $G = G(n, m)$, $m = \lfloor \frac{3}{2}(n - 1) \rfloor + 1$ be a graph not containing an S_2. The induction hypothesis implies that $\delta(G) \geqslant 2$, and since $e(G) = m < \frac{3}{2}n$, in fact $\delta(G) = 2$. Let x_0 be a vertex of degree 2. Then

$$\lfloor \tfrac{3}{2}(n - 1) \rfloor - 1 = e(G - x_0) \leqslant ex(n - 1; S_2) = \lfloor \tfrac{3}{2}(n - 2) \rfloor.$$

This implies that n is even and $G - x_0$ is an extremal graph. By the induction hypothesis $G - x_0$ is connected and its blocks are triangles. Since G is obtained from $G - x_0$ by the addition of the path $x_1 x_0 x_2$, (4) implies that G contains an S_2.

The proof of the second assertion will take somewhat longer. (Note that the proof of the first assertion will not be complete until we accomplish this, since the induction hypothesis includes both assertions.) Thus let $n = 2u + 1$ and let $G = (2u + 1, 3u)$ be a graph not containing an S_2. The induction hypothesis implies that G is connected so our aim is to show that its blocks are triangles or, equivalently, that *each edge is contained in a triangle*.

As before, $\delta(G) = 2$. Let $x_0 \in G$ have degree 2 and let $\Gamma(x_0) = \{x_1, x_2\}$.

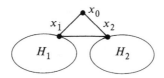

FIG. 3.1. The structure of G when $x_1 x_2 \in E(G)$.

Assume first that $x_1 x_2 \in E(G)$. Then in $H = G - x_0 - x_1 x_2$ the vertices x_1 and x_2 belong to different components, since otherwise G contains an S_2. Suppose H is the disjoint union of H_1 and H_2 (Fig. 3.1).

Putting $n_1 = |H_1|$, $n_2 = |H_2|$, we have $n = n_1 + n_2 + 1$ and

$$\lfloor \tfrac{3}{2}(n-1) \rfloor = e(G) = e(H_1) + e(H_2) + 3 \leqslant \lfloor \tfrac{3}{2}(n_1 - 1) \rfloor + \lfloor \tfrac{3}{2}(n_2 - 1) \rfloor +$$

This inequality implies that n_1 and n_2 are both odd and

$$e(H_1) = \lfloor \tfrac{3}{2}(n_1 - 1) \rfloor, \quad e(H_2) = \lfloor \tfrac{3}{2}(n_2 - 1) \rfloor,$$

i.e. H_1, H_2 are extremal graphs of odd order. Hence by the induction hypothesis each edge of H_1 and H_2 is contained in a triangle. Since G is obtained from $H_1 \cup H_2$ by the addition of a triangle, each edge of G is contained in a triangle, as required.

Assume now that $x_1 x_2 \notin E(G)$. In this case our aim is to arrive at a contradiction. If $H = G - x_0$ is disconnected, say it is the disjoint union of G_1 and G_2, $n_1 = |G_1| \geqslant 1$, $n_2 = |G_2| \geqslant 1$, where $n_1 + n_2 = n - 1 = 2u$, then

$$3 = e(G) = e(G_1) + e(G_2) + 2 \leqslant \lfloor \tfrac{3}{2}(n_1 - 1) \rfloor + \lfloor \tfrac{3}{2}(n_2 - 1) \rfloor + 2 \leqslant 3u - 1.$$

This contradiction shows that H is connected. There is no cycle C in H containing both x_1 and x_2 since C together with x_0 and the edges $x_0 x_1$, $x_0 x_2$ of G would form a semi-topological graph S_2. Hence there is a vertex $x_3 \in H$ separating x_1 from x_2, say $H = H_1 \cup H_2$ where $x_1 \in H_1 - x_3$, $x_2 \in H_2 - x_3$ and $V(H_1) \cap V(H_2) = \{x_3\}$ (Fig. 3.2). Since $|H| = |H_1| + |H_2| - 1$ is even, we may assume that $|H_1| = 2p \geqslant 2$ and $|H_2| = n - 2p + 1 = 2u - 2p + 1$ $\geqslant 2$. Then by the induction hypothesis

$$3u = e(G) = e(H_1) + e(H_2) + 2 \leqslant 3p - 2 + 3u - 3p + 2 = 3u.$$

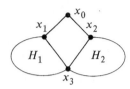

FIG. 3.2. The structure of G when $x_1x_2 \notin E(G)$.

This implies that $e(H_1) = 3p - 2$ and $e(H_2) = 3u - 3p$, i.e. H_1 and H_2 are extremal graphs. Furthermore, since $|H_2|$ is odd, H_2 is connected and its blocks are triangles. The graph G contains an $x_2 - x_3$ path independent of H_2, going through x_0 and x_1, so (4) implies that G contains an S_2, contradicting the hypothesis. ∎

A semi-topological graph S_3 is a topological K^4 so, by Corollary 2.3, we have $ex(n; S_3) \geqslant ex(n; TK^4) = 2n - 3$. It was conjectured in [BE5] and proved by Thomassen [T2a] that, in fact, $ex(n; S_3) = 2n - 3$. This theorem of Thomassen (Theorem 3.9) is a considerable extension of Corollary 2.3, since it says that if the size of a graph is sufficiently large to ensure the existence of a TK^4 then the graph contains an S_3, which is a special subdivision of K^4. The proof of Theorem 3.9 is rather involved. Following the original presentation in [T2a], we start with a lemma. Both the lemma and the proof of Theorem 3.9 are divided into several parts. In some cases the straightforward proofs are left to the reader.

A *cockade* is defined recursively as follows:

(i) K^3 and $K(3, 3)$ are cockades,

(ii) if G_1 and G_2 are disjoint cockades and $x_iy_i \in E(G_i)$, $i = 1, 2$, then the graph obtained from $G_1 \cup G_2$ by identifying x_1 with x_2 and y_1 with y_2 is a cockade.

Note that in (ii) the edge x_1y_1 is identified with x_2y_2. It is easily seen that a cockade of order n is 2-connected, does not contain an S_3 and has $2n - 3$ edges.

LEMMA 3.8. (i) *If a 3-connected graph contains a triangle then it contains an S_3 as well.*

(ii) *Let a and b be non-adjacent vertices of a cockade G and let G' be obtained from G by joining a to b by a path independent of G. Then G' contains an S_3.*

(iii) *Let a, b and c be vertices of a cockade G. Suppose $ab \in E(G)$, $bc \notin E(G)$ and $ac \notin E(G)$. Write G' for the graph obtained from $G - ab$ by adding to it a vertex d and joining d to each of a, b and c. Then G' contains an S_3.* ∎

THEOREM 3.9. *If $n \geqslant 3$ then $ex(n, S_3) = 2n - 3$. The extremal graphs are the cockades.*

Proof. Let $G = G(n, 2n - 3)$, $n \geqslant 3$, be a graph without an S_3. Since every cockade is a maximal graph without an S_3, to prove the theorem it suffices to show that G is a cockade. We do this by induction on n. For $n = 3, 4$ the assertion is trivial so we proceed to the induction step.

Let W be a cutset of G and write t for the number of edges of $G[W]$. Then $G = G_1 \cup G_2$, where G_1 and G_2 have $|W|$ vertices and t edges in common. Hence

$$e(G) = 2n - 3 = e(G_1) + e(G_2) - t \leqslant 2|G_1| - 3 + 2|G_2| - 3 - t$$

$$= 2n - (6 + t - 2|W|).$$

This implies that $|W| \geqslant 2$, that is G is 2-connected. Furthermore, if $|W| = 2$ and $t = 1$ then both G_1 and G_2 are extremal graphs, that is cockades. Then, by the definition of a cockade, G is also a cockade. If $|W| = 2$ and $t = 0$ then W consists of two non-adjacent vertices. Since these vertices are joined by a path in G_2, by Lemma 3.8(ii) the subgraph G_1 is not a cockade, so $e(G_1) \leqslant 2|G_1| - 4$. Similarly $e(G_2) \leqslant 2|G_2| - 4$, implying the following contradiction:

$$e(G) = 2n - 3 \leqslant 2|G_1| - 4 + 2|G_2| - 4 = 2n - 4.$$

Thus we may assume that G is 3-connected. Since, according to Lemma 3.8(i), G does not contain a triangle, by Theorem VI.1.1 we have $2n - 3 \leqslant \lfloor n^2/4 \rfloor$ so $n \geqslant 6$ and if $n = 6$ then $G = K(3, 3)$. Hence we may assume that $n \geqslant 7$.

To complete the proof we shall show that the assumptions that G is 3-connected and $n \geqslant 7$ lead to a contradiction. This is the longest and most important part of the proof. We divide it into four steps, the first of which can be proved by a modification of the argument applied above.

(a) *If ab and bc are edges of G then $G - \{a, b, c\}$ is connected.*
(b) *Write V_3 for the set of vertices of degree 3. Then $V_3 \neq \varnothing$, $V_3 \neq V$ and $G[V_3]$ is a forest.*

The first two relations follow from $\delta(G) \geqslant 3$, $n \geqslant 7$ and $e(G) = 2n - 3$. Suppose that $G[V_3]$ is not a forest. Let C be a cycle of $G[V_3]$ of minimal length m. Put $G' = G - C$ and denote by W the set of vertices of G' joined to C. Every vertex of C is joined to exactly one vertex of W and every vertex of W is joined to at most 2 vertices of C, otherwise G would contain an S_3. Thus $|W| \geqslant m/2$. Hence $|W| \geqslant 3$, since if $|W| \leqslant 2$ then $m = 4$ and so W disconnects G. There is no triangle in G so W contains two non-adjacent vertices a and b. The graph G contains an $a - b$ path independent of G' so G' is not a cockade. Hence, by the induction hypothesis, $e(G') \leqslant 2|G'| - 4$ $= 2(n - m) - 4$ and so $e(G) \leqslant e(G') + 2m \leqslant 2n - 4$. This contradiction completes the proof of (b).

By (b) there is a vertex x_0 of degree 3 adjacent to at most one vertex of degree 3. Let $\Gamma(x_0) = \{x_1, x_2, x_3\}$, where $d(x_1) \geqslant 4$ and $d(x_2) \geqslant 4$. By Lemma 3.8(iii) the graph $H = G - x_0 + x_1 x_2$ contains an S_3. Since G does not contain an S_3, we may assume without loss of generality that H contains a cycle C and that $x_2 \in C$, $x_1 \notin C$ and x_1 is adjacent to three vertices of C: x_2, y_1 and y_2. Our next aim is to prove that $x_3 \in C$ can also be assumed.

(c) $G - \{x_0, x_1\}$ contains a cycle C' such that $x_2, x_3, y_1, y_3 \in C'$.

We may assume that $x_3 \notin C$, otherwise $C' = C$ will do. The vertices x_2, y_1 and y_2 divide C into three arcs. Denote by P the $y_1 - y_2$ arc of C. Since $G - x_0$ is 2-connected, it contains two independent paths from x_3 to $V(C) \cup \{x_1\}$, say P_1 and P_2, where P_1 is an $x_3 - z_1$ path and P_2 is an $x_3 - z_2$ path. It is easily checked that $z_1, z_2 \in P$ must hold, for otherwise G contains an S_3. Then C contains a $z_1 - z_2$ arc P' not contained in P and the cycle $C' = P_1 \cup P_2 \cup P'$ has the required properties (see Fig. 3.3).

(d) G contains an S_3.

Since $d(x_1) \geqslant 4$, there is a vertex $x_4 \in G - \{x_0, x_2, x_3, y_2, y_3\}$ adjacent to x_1. By (a) the graph $G - \{x_1, y_1, y_2\}$ is connected and contains a path from x_4 to C'. This path, the cycle C' and the edges $x_1 y_1$, $x_1 y_2$ and $x_1 x_4$ form a topological K^4, say T_4; x_1 is a branchvertex of T_4. In a TK^4 a branchvertex and any two other vertices are contained in a cycle. Consequently there is a cycle in T_4 which includes x_1, x_2 and x_3. The vertex x_0 is not on this cycle so G indeed contains an S_3.

Since assertion (d) contradicts our initial assumption, the proof of the theorem is complete. ∎

To conclude this section we present a theorem from [B39] about the

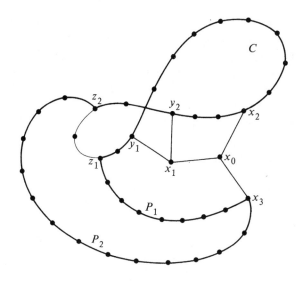

FIG. 3.3. Finding a cycle C'.

existence of two nested cycles, i.e. edge disjoint cycles C_1 and C_2 such that $V(C_2) \subset V(C_1)$. The result answers a problem of Erdös [E27]. Most of the work in the proof is put into the following lemma.

LEMMA 3.10. *If $\delta(H) \geqslant 7$ then H contains a cycle C and a path P such that*

$$E(C) \cap E(P) = \varnothing \quad and \quad V(C) \subset V(P).$$

Proof. Let P be a *longest* path, say $P = x_0 x_1 \ldots x_p$. Note that x_0 is joined only to vertices of P and if $x_0 x_i \in E(G)$ then $x_{i-1} x_{i-2} \cdots x_0 x_i x_{i+1} \ldots x_p$ is also a longest path. Put

$$X(P, x_0) = \{x_i : x_0 x_i \in E(G), i > 1\} = \Gamma(x_0) - \{x_1\}.$$

Note that if $x \in X(P, x_0)$ then x is adjacent to at least two vertices that are endvertices of longest paths. Let

$$X = \bigcup X(P, x_0)$$

where the union is taken over all longest paths P and both endvertices of these paths. Let Y be the set of endvertices of the longest paths in G. Since $\delta(H) \geqslant 7$, $|X(P, x_0)| \geqslant 6$. Consequently if $y \in Y$ then y is joined to at least 6 vertices in X. By the definition of X each vertex $x \in X$ is joined to at least 2 vertices in Y.

Let $L = H[X \cup Y]$ and let M be obtained from L by omitting the edges joining vertices in $Z = X - Y$. Then M has $|Y| + |Z|$ vertices and by the observations above

$$2e(M) \geqslant 6|Y| + 2|Z|. \tag{5}$$

Since each edge of M is incident with a vertex in Y, we have

$$|E(M) \cap E(P)| \leqslant 2|Y|. \tag{6}$$

Inequalities (5) and (6) imply

$$e(M - E(P)) \geqslant |Y| + |Z|.$$

Consequently $M - E(P)$ contains a cycle C. This cycle C and the path P have the required properties.   ∎

THEOREM 3.11. *If $|G| = n \geqslant 7$, $e(G) \geqslant 7n - 30$ and x is any vertex of G then there is a cycle C_1 in G and a cycle C_2 in $G - x$ such that C_1 and C_2 are edge disjoint and $V(C_2) \subset V(C_1)$.*

Proof. We apply induction on n. It is easily checked that the theorem holds for $n = 7$. Suppose now that $n \geqslant 8$ and it holds for graphs of order less than n. Let H be the subgraph of G spanned by the vertices adjacent to x. We distinguish two cases according to the magnitude of $\delta(H)$.

(a) *Suppose first that $\delta(H) \leqslant 6$*, say a vertex y in H is joined to $d = d_H(y) \leqslant 6$ vertices in H. Let $G^* = G/xy$ and denote by x^* the vertex of G^* that is the contraction of xy.

Clearly $e(G^*) = e(G) - d - 1 \geqslant e(G) - 7$ and since $|G^*| = |G| - 1 \geqslant 7$, we see that G^* satisfies the conditions of the theorem. Therefore by the induction hypothesis there are edge disjoint cycles C_1^* in G^* and C_2^* in $G^* - x^*$ such that $V(C_2^*) \subset V(C_1^*)$. Since $G^* - x^* = G - \{x, y\}$, the cycle C_2^* is also in $G - x$. It is easily seen that the cycle C_1^* can be changed slightly (replacing at most 2 edges incident with x^* by a path of length 2 or 3) in such a way that the resulting cycle C_1 satisfies $E(C_1) \cap E(C_2^*) = \varnothing$ and $V(C_2^*) \subset V(C_1)$.

(b) *Suppose now that $\delta(H) \geqslant 7$.* Let C_2 and P be the cycle and the path guaranteed by Lemma 3.10. Close P in G by joining x to the two endvertices of

P. If the resulting cycle is denoted by C_1 then we have $E(C_1) \cap E(C_2) = \varnothing$ and $V(C_2) \subset V(P) \subset V(C_1)$, as required.   ∎

It is not known whether Theorem 3.11 can be extended to a set of k nested cycles (cf. Conjecture 17).

4. EXERCISES, PROBLEMS AND CONJECTURES

1. Prove Lemma 1.3 on the lines of the proof of Lemma 1.2. (Mader [M3].)

2^-. Note that if $\chi(G) = k$ and $S_r = \{x \in G : d(x_0, x) = r\}$ then for some r we have

$$\chi(S_r) + \chi(S_{r+1}) \geqslant k.$$

Deduce from this the following weaker form of Theorem 1.14. There is a function $\psi(p)$ such that

$$\chi(G) \geqslant \psi(p) \quad \text{implies} \quad G > K^p.$$

(Wagner [W4] see also [C3].)

3^-. Deduce from Theorem 1.10 that if $\chi(G) = 5$, $G \not> K^5$ and G_m is obtained from G by deleting m of its vertices then

  (i) $G_2 > K^{5-}$,
  (ii) G_3 has a subcontraction to the graph of the triangular prism.
(Dirac [D17], [D18].)

4^-. Let $c(p)$ be the function defined after Theorem 1.14. Show that if $G = G(n, c(p)n)$ then $G > K^p$.
(Mader [M2].)

5. CONJECTURE. If $p \geqslant 3$ and $G = G(n, (p-2)n)$ then $G > K^p$. (Note that, in view of the example preceding Lemma 1.15, this implies $c(p) = p - 2$.)

6. Let $c_1 > 1/(2r+2)$, $r \geqslant 2$. Then there is a constant $c_2 > 0$ such that if $|G| = n$, $p \leqslant c_2 n^{1/r}$ and $e(G) \geqslant c_1 n^2$ then $G \supset TK^p$.
[Hint. Imitate the proof of Lemma 1.15, taking into account Ramsey's theorem (Theorem V.6.2).] (Erdös and Hajnal [EH1].)

7. Prove Lemma 2.4.

8. PROBLEM. Let G be the graph obtained from $E^{2m} + K^3$ by putting a

1-factor into E^{2m}. Then $|G| = n = 2m + 3$, $e(G) = 7m + 3 = \frac{7}{2}n - \frac{15}{2}$ and G does not have the property stated in the conclusion of Theorem 2.7. Is this example worst possible? In other words: does Theorem 2.7 remain valid if $4n - 10$ is replaced by $\frac{7}{2}n - \frac{13}{2}$? (Thomassen [T2].)

9. CONJECTURE. $ex(n, TK^5) = 3n - 6$ for $n \geqslant 6$. (G. A. Dirac.)

10. Following Chartrand, Geller and Hedetniemi [CGH1] we say that a graph G has property P_k if $G \not\supset TK(\lfloor k/2 \rfloor + 1, \lceil k/2 \rceil + 1)$. Show that for $1 \leqslant k \leqslant 4$ a graph G has property P_k iff $G \not\supset K^{k+1}$ and $G \not\supset K(\lfloor k/2 \rfloor + 1, \lceil k/2 \rceil + 1)$. (Halin [H5] and Wagner [W2].)

11. Call a graph *outerplanar* if it can be embedded in the plane in such a way that all the vertices lie on the same face.

Prove that a graph is outerplanar iff it has property P_3 defined in Ex. 10. (Chartrand and Harary [CH2].)

12. Let $g_k(n)$ be the function defined in Theorem 3.1. Prove that if $k \geqslant 6$ and $2k \leqslant n < \frac{5}{2}(k - 1)$ then

$$g_k(n) > (k + 1)n - (k + 1)^2.$$

[Hint. To construct a graph G^n showing the inequality start with a complete bipartite graph of order n and add to it some edges joining vertices in the larger class.] (M. Lewin)

13. CONJECTURE. There is a function $n(k)$ such that if $n \geqslant n(k)$ then

$$g_k(n) = (k + 1)n - (k + 1)^2.$$

14^-. Show that to prove Conjecture 13 it suffices to show that for every k there is an integer $n = n(k)$ such that

$$g_k(n) = (k + 1)n - (k + 1)^2.$$

15. PROBLEM. Estimate the function $ex(n; S_r)$ (see Theorems 3.7 and 3.9).

16. Show that the bound $7n - 30$ in Theorem 3.11 cannot be replaced by $4n - 13$. (Bollobás [B39].)

17. CONJECTURE. For every natural number k there is a constant c_k such that if $e(G) \geqslant c_k|G|$ then G contains k nested cycles, i.e. edge disjoint cycles $C_1, C_2, \ldots, C_k$ such that $V(C_1) \supset V(C_2) \supset \ldots \supset V(C_k)$. (Bollobás [B39].)

18. PROBLEM. Given a natural number k is there a natural number $u(k)$ such that $G \in \mathcal{D}_{u(k)}$ contains a Hamiltonian subgraph H that belongs to $\mathcal{D}_k$? [Note that a positive answer to this question would imply the truth of Conjecture 17]. (Bollobás [B39].)

19[+]. Use Mader's theorem (Theorem 2.11) and Menger's theorem (or the result of Ex. I.12) to deduce the following extension of Ex. I.12.
For every natural number k there is a natural number $f(k)$ such that if $\kappa(G) \geqslant f(k)$, $A = \{a_1, \ldots, a_k\}$, $B = \{b_1, \ldots, b_k\}$, $A \cup B \subset V(G)$ and $A \cap B = \varnothing$ then G contains disjoint paths $P_1, \ldots, P_k$, where P_i joins a_i to b_i, $1 \leqslant i \leqslant k$. (Larman and Mani [LM1].)

20. PROBLEM. Let $n > k$ be natural numbers. Determine $ex(n; k\text{ — reg}) = \max\{m : \exists G(n, m) \text{ without a } k\text{-regular subgraph}\}$. Is it true that if $\varepsilon > 0$ and $n > n_0(\varepsilon, k)$ then

$$ex(n; k\text{-reg}) < n^{1 + \varepsilon}?$$

Trivially $ex(n; 2\text{-reg}) = n - 1$ but even the inequality above is not known if $k \geqslant 3$.

Erdős and Simonovits proved that if Q is the graph determined by the edges of a cube then $ex(n; Q) < cn^{\frac{8}{5}}$ for some constant c. This result clearly implies $ex(n; 3\text{-reg}) < cn^{\frac{8}{5}}$.

By joining three vertices suitably to C^{2n}, prove that $ex(2n + 3 : 3\text{-reg}) \geqslant 6n$, as observed by Chvátal. (Erdős and Sauer, [E28], [E29], see also Conjecture 36 in Ch. II.§6).

21. PROBLEM. Szemerédi suggested the following variation of the preceding problem. Let $n > k$ be natural numbers. Estimate

$$ex(n; k\text{-reg}!) = \max\{m : \exists G(n, m) \text{ without a } k\text{-regular } \textit{induced} \text{ subgraph}\}.$$

22. *The thrackle conjecture of Conway.* A *thrackle* is a finite set of nodes in the plane together with a set of closed Jordan curves joining some pairs of nodes, satisfying the following condition: each pair of curves has *exactly* one point in common which is either a common *end point* or a *crossing* point of the curves.

J. H. Conway conjectures that a thrackle contains at most as many curves as nodes.

(Assuming the thrackle conjecture Woodall [W22] classified the thrackle-able graphs.)

23. PROBLEM. Let $h_1(n)$ be the maximal integer for which there is a graph

$G(n, h_1(n))$ without two edge disjoint cycles having the same vertex set. Define $h_2(n)$ analogously for cycles $x_1 x_2 \ldots x_s$ and $x_{i_1} x_{i_2} \ldots x_{i_t}$, $1 \leqslant i_1 < i_2 < \ldots < i_t \leqslant s$. Estimate $h_1(n)$ and $h_2(n)$ for every n. Estimate the analogous functions for several cycles. (Erdös [E29].)

Chapter VIII

Complexity and Packing

The results to be presented in this chapter are of a rather different nature from those we have presented so far. These results, which are very recent, fall into two loosely connected groups.

We have discussed numerous conditions implying that a graph has certain properties. The set of graphs with a fixed vertex set having a given property is easily identifiable with the property itself. Using this set we shall define the so-called computational complexity of the property. In §§1 and 2 we present some general theorems concerning the computational complexity, together with some results about the complexity of specific properties. The main result of these sections is a startling theorem of Rivest and Vuillemin [RV2], giving a strong lower bound on the computational complexity of rather general properties.

In the next two sections we shall discuss results and problems concerning the packing of graphs. Given a set of graphs $G, G_1, G_2, \ldots, G_l$, we say that $G_1, G_2, \ldots, G_l$ can be packed into G if we can find inclusions $V(G_i) \subset V(G)$, $i = 1, 2, \ldots, l$, such that $e \in E(G_i)$ implies $e \in E(G) - \bigcup_{j \neq i} E(G_j)$. The inclusions above are said to form a *packing* of $G_1, G_2, \ldots, G_{l-1}$ and G_l into G. We may always suppose that $|G| = |G_1| = \ldots = |G_l| = n$ since if $|G_i| < |G|$ then we may add isolated vertices to G_i to increase $|G_i|$ to $|G|$. If $G = K^n$ then we say simply that there is a packing of $G_1, G_2, \ldots, G_l$. Most of the results and problems of the chapter concern the packing of two graphs of order n, say G_1 and G_2. In this case the inclusions $V(G_1) \subset V(K^n)$ and $V(G_2) \subset V(K^n)$ can be replaced by the single inclusion $V(G_1) \subset V(G_2)$ induced by them. This inclusion has the property that $e \in E(G_1)$ implies $e \notin E(G_2)$, i.e. it effectively makes G_1 a *subgraph* of $\bar{G}_2$ so it might be considered more natural to say that G_1 is a subgraph of $\bar{G}_2$. However, we have several reasons to retain the phrase: there is a packing of G_1 and G_2. Firstly, our graphs have many fewer edges

401

than their complements and we always work with the graphs themselves. Another reason is that the results are connected to packings of several graphs. The third and most important reason is that we wish to distinguish the problems about packings from the more usual problems about the existence of certain subgraphs. The results of Chapter III, §4 about Hamiltonian cycles and Hamiltonian paths, the results of Chapter II, §§1, 4 about sets of independent edges and the results of Chapter VI, §4 about forbidden subgraphs could all be considered packing results. However, the nature of these problems and the method of proofs are essentially different from those to be discussed in this chapter. In most packing results we claim that *each* member of a *large* family of graphs contains *each* member of another *large* family. The least we mean by large is that the number of non-isomorphic graphs of order n in the family grows faster than any polynomial of n. Recall that, e.g. in the results about forbidden subgroups it was essential that we had only *finitely* many forbidden subgraphs.

The strongest packing theorem we present is in §3. It is due to Bollobás and Eldridge [BE2] and it gives a bound on $e(G_1) + e(G_2)$ ensuring that there is a packing of G_1 and G_2, provided $\max\{\Delta(G_1), \Delta(G_2)\} < n - 1$. In §4 we present results about the possibility of packing G_1 and G_2, provided $e(G_1)$ is small. In the final section we apply some of the packing results, including the main packing theorem, to certain questions concerning computational complexity. In particular we find a non-trivial property of oriented graphs with minimal computational complexity.

As the results we are going to present have been proved only recently, it is hardly surprising that there are a great number of fundamental unsolved problems in the area. It is very likely that the theory of packing will develop considerably in the near future. Among the problems of special importance are a possible generalization of the theorem of Hajnal and Szemerédi (Conjecture 18), an extension of the main packing theorem to graphs with maximal degree $n - k$ (Conjecture 22) and a conjecture about packing n trees into K^n (Conjecture 23).

1. THE COMPLEXITY OF GRAPH PROPERTIES

Denote by $\mathscr{G}^n$ the set of all graphs G^n with a fixed vertex set V, say $V = \{1, 2, \ldots, n\}$. A *property* $\mathscr{P}$ of G^n is a subset of $\mathscr{G}^n$ such that $G \in \mathscr{P}$ whenever an isomorphic copy of G belongs to $\mathscr{P}$. We say that $G \in \mathscr{G}^n$ has property $\mathscr{P}$ if $G \in \mathscr{P}$. With a slight abuse of notation we define a *game* $\mathscr{P}$ as follows. The game is between two players, called the Constructor and Algy or the Hider and the Seeker. Algy asks questions from the Constructor about a hypothetical graph. Each question (or probe) is whether or not a certain pair of vertices is an edge or not and each question is answered by the Constructor. When posing a question Algy takes into account all the information he has received up to that point. The game is over when Algy can decide whether or not the graph the Constructor has been constructing has property $\mathscr{P}$. The aim of the Constructor is to keep Algy guessing as long as possible and, naturally, Algy tries to decide as soon as possible whether the graph of the Constructor has property $\mathscr{P}$ or not. The number of moves (i.e. the number of questions) in this game, assuming that both players play optimally, is the *computational* (or *argument*) *complexity of* $\mathscr{P}$, briefly the *complexity* of $\mathscr{P}$, and is denoted by $c(\mathscr{P})$.

In a rather natural version of the game the Constructor wins if he can force Algy to ask *all* the pairs. Assuming that both of them play optimally this means that the Constructor wins if $c(\mathscr{P}) = \binom{n}{2}$. In this case the property $\mathscr{P}$ is called *elusive*. If $c(\mathscr{P}) \geq k$ then $\mathscr{P}$ is said to be *k-elusive*.

The complexity of a property $\mathscr{P}$ of directed graphs, directed graphs without loops or oriented graphs can be defined analogously. If $\mathscr{P}$ is a property of directed graphs without loops then a question is of the following form: is there a directed edge *from a to b*? Of course, we call $\mathscr{P}$ elusive if $c(\mathscr{P}) = n(n-1)$. We *do not consider* directed graphs with loops since if $\mathscr{P}$ is a property depending only on the loops of the graph then $c(\mathscr{P}) = n$. If $\mathscr{P}$ is a property of oriented graphs then the natural form of a question is the following: what is the relation between a and b? (And so there can be 3 answers: 1. they are not joined, 2. there is an $\overrightarrow{ab}$ edge, 3. there is a $\overrightarrow{ba}$ edge.) Then $\mathscr{P}$ is elusive if $c(\mathscr{P}) = \binom{n}{2}$. As in the rest of the book, unless otherwise mentioned, we deal with graph properties. However, it will be appropriate to make a few remarks about directed graphs as well.

A property $\mathscr{P}$ is *trivial* if either no graph has $\mathscr{P}$ or every graph has $\mathscr{P}$. We shall *always* assume (though sometimes we reiterate it) that the properties at hand are non-trivial since for a trivial property $\mathscr{P}$ we have $c(\mathscr{P}) = 0$.

Holt and Reingold [HR1] were the first to give bounds on the complexity of certain properties of directed graphs. Hopcroft and Tarjan [HT2] and Kirkpatrick [K6] obtained other lower bounds on the complexity of various properties. A certain ambiguity present in the subject was cleared up by Milner and Welsh [MW1], [MW2] and Best, van Emde Boas and Lenstra [BBL1], who also proved several new results.

In this section we investigate the efficiency of a certain straightforward strategy of the Constructor, and deduce that a number of natural graph properties are elusive. We show that a non-elusive property $\mathscr{P}$ contains at least three non-isomorphic graphs. We prove by an example that the complexity of a non-trivial graph property may be $O(n)$.

Before presenting any of the results let us formulate the definition of the complexity in a more general setting. Let T be a t element set. We shall call an ordered pair $S = (E, N)$ of disjoint subsets of T a *preset*. We think of a preset $S = (E, N)$ as our information about a subset X of T in the following sense: E is the set of elements known to belong to X and N is the set of elements known not to belong to X. We call $P = E \cup N$ the set of *probed* elements of the preset S. Accordingly a set $U \subset T$ is *identified* with the preset $(U, T - U)$. We say that $S_2 = (E_2, N_2)$ *contains* or *extends* $S_1 = (E_1, N_1)$ if $E_1 \subset E_2$ and $N_1 \subset N_2$. In this case we write $S_1 < S_2$. An *algorithm* is a function ϕ assigning to each preset $S = (E, N), P = E \cup N \neq T$, an element $x = \phi(S) \in T - P$. This element x is called the *probe* prescribed by the algorithm for S.

Let $\mathscr{F}$ be a collection of subsets of T, i.e. a subset of the set $\{0, 1\}^T$ of all subsets of T. Given an algorithm ϕ and a set $X \subset T$, define a sequence $X_0 = (E_0, N_0), (E_1, N_1), \ldots$ of presets by putting $X_0 = (E_0, N_0) = (\varnothing, \varnothing)$ and

$$X_{i+1} = (E_{i+1}, N_{i+1}) = \begin{cases} (E_i \cup \phi(X_i), N_i) & \text{if} \quad \phi(X_i) \in X, \\ (E_i, N_i \cup \phi(X_i)) & \text{if} \quad \phi(X_i) \notin X, \end{cases}$$

$i = 0, 1, \ldots, t$. Then clearly $X_0 < X_1 < X_2 < \ldots < X_t = X$. Put

$$c(\mathscr{F}, \phi, X) = \min\{m: \text{either } Y \in \mathscr{F} \text{ for every } X_m < Y \subset T$$

$$\text{or } Y \notin \mathscr{F} \text{ for every } X_m < Y \subset T\},$$

i.e. define $c(\mathscr{F}, \phi, X)$ as the minimal number of steps the algorithm ϕ needs to decide whether X belongs to $\mathscr{F}$ or not. Thus

$$c(\mathscr{F}, \phi) = \max_{X \subset T} c(\mathscr{F}, \phi, X)$$

is the maximal number of probes ϕ can be forced to make before it can find out whether a subset X of T belongs to $\mathscr{F}$ or not. We call $c(\mathscr{F}, \phi)$ the *length of the $\mathscr{F}$-deciding algorithm* ϕ. Finally

$$c(\mathscr{F}) = \min_{\phi} c(\mathscr{F}, \phi),$$

where the minimum is taken over all algorithms, is the *computational complexity* of $\mathscr{F}$. It is the minimal number of probes in which an algorithm can decide about every subset of T whether it belongs to $\mathscr{F}$ or not. If $\mathscr{P}$ is a property of graphs with vertex set $V = \{1, 2, \ldots, n\}$ then choosing T to be the set of all unordered pairs of vertices, i.e. $T = V^{(2)}$, $\mathscr{G}^n$ is naturally identified with $\{0, 1\}^T$ and $\mathscr{P}$ with a subset $\mathscr{F}$ of $\{0, 1\}^T$. From now on we usually call $\mathscr{F}$ a *property* of the subsets of T. With these identifications the two definitions of the computational complexity of $\mathscr{P}$ clearly agree. In this situation we speak of *pregraphs* instead of presets; a graph G with edge set E is identified with the pregraph $(E, V^{(2)} - E)$. Properties of directed graphs can be interpreted analogously.

It is very easy to combine this algorithmic approach with the game approach. A *strategy* or *construction* is a function ψ assigning to a pair (S, x), where $S = (E, N)$ is a preset and $x \in T - P = T - E \cup N$, a preset $\psi(S, x) = (E', N')$ such that $\psi(S, x) > S$ and $E' \cup N' = E \cup N \cup \{x\}$. If $x \in E'$ then we say that ψ chooses x an element and if $x \in N'$ then ψ chooses x a non-element. (When discussing graph properties, naturally instead of an element and a non-element we speak of an edge and a non-edge.)

Given an algorithm ϕ and a construction ψ, define a sequence $X_0 = (E_0, N_0)$, $X_1 = (E_1, N_1), \ldots$ of presets by putting $X_0 = (\varnothing, \varnothing)$ and $X_{i+1} = \psi(X_i, \phi(X_i))$, $i = 0, 1, \ldots, t - 1$. If $\mathscr{F}$ is a collection of subsets of T, define

$$c(\mathscr{F}, \phi, \psi) = \min\{m : \text{either } Y \in \mathscr{F} \text{ for every } X_m < Y \subset T$$

$$\text{or } Y \notin \mathscr{F} \text{ for every } X_m < Y \subset T\},$$

and

$$c(\mathscr{F}) = \min_{\phi} \max_{\psi} (\mathscr{F}, \phi, \psi).$$

Once again $c(\mathscr{F})$ is the computational complexity of $\mathscr{F}$. If $T = V^{(2)}$ and $\mathscr{F}$ corresponds to a graph property $\mathscr{P}$ then an algorithm ϕ represents the activity of Algy (i.e. the Seeker) and a strategy ψ that of the Constructor (i.e. the Hider). We call a strategy ψ a *winning strategy* for the Constructor if $c(\mathscr{F}, \phi, \psi) = t = |T|$ for every algorithm ϕ, i.e. if ψ is a strategy showing that $\mathscr{F}$ is elusive.

Denote by $\mathscr{F}^c$ the *complement* of $\mathscr{F}$ in $\{0, 1\}^T$. Then $c(\mathscr{F}^c) = c(\mathscr{F})$ by definition. A property $\mathscr{F}$ is called *monotone* if $X \in \mathscr{F}$ and $Y \subset X$ imply $Y \in \mathscr{F}$. Sometimes the complement of a monotone property is also called monotone. Since $c(\mathscr{F}^c) = c(\mathscr{F})$, we may adopt this more restrictive definition without loss of generality. For each property $\mathscr{F}$ on T define the *dual* property $\mathscr{F}^*$ by

$$\mathscr{F}^* = \{F : F \subset T \text{ and } T - F \notin \mathscr{F}\}.$$

Clearly $\mathscr{F}^*$ is monotone iff $\mathscr{F}$ is. Furthermore, if $\mathscr{F}$ is a property of graphs then so is $\mathscr{F}^*$. The second dual is the original property. Finally if $\mathscr{F}$ is any non-trivial property with $T \notin \mathscr{F}$ put $\mathscr{F}^{\text{mon}} = \{X : X \subset Y \in \mathscr{F} \text{ for some } Y\}$. Clearly $\mathscr{F}^{\text{mon}}$ is a non-trivial monotone property.

Given a property $\mathscr{F}$ with $T \notin \mathscr{F}$, define a strategy ψ_0, called the *simple strategy*, as follows. If $S = (E, N)$ is a preset and $x \in T - E \cup N$, let $\psi(S, x) = (E \cup \{x\}, N)$ iff some extension of $(E \cup \{x\}, N)$ has property $\mathscr{F}$, i.e. iff $(E \cup \{x\}, N) < X \subset T$ holds for some $X \in \mathscr{F}$. The following extension of a theorem of Milner and Welsh [MW1] gives a necessary and sufficient condition ensuring that ψ_0 is a winning strategy.

THEOREM 1.1. *Let $\mathscr{F}$ be a non-trivial property of the subsets of T such that $T \notin \mathscr{F}$. Then ψ_0 is a winning strategy for the Constructor in the game $\mathscr{F}$ iff whenever $x \in X \in \mathscr{F}$ there are an element $y \in T - X$ and a set $Y \in \mathscr{F}$ such that $(X - \{x\}) \cup \{y\} \subset Y$.*

Remark. Equivalently we could have required that the condition is satisfied for every *maximal* set X in $\mathscr{F}$.

Proof. (i) Suppose first that whenever $x \in X \in \mathscr{F}$ there are an element $y \in T - X$ and a set $Y \in \mathscr{F}$ such that $(X - \{x\}) \cup \{y\} \subset Y$. Assume that ψ_0 is not winning, so $c(\mathscr{F}, \phi, \psi_0) = m < t$ for some algorithm ϕ. Let

$$X_0 < X_1 < \ldots < X_m, \quad X_k = (E_k, N_k),$$

be the sequence of presets constructed by ϕ and ψ_0. By the definition of ψ_0 there is a set U such that $X_m < U \in \mathscr{F}$. Hence $V \in \mathscr{F}$ must hold for *every* set V, $X_m < V \subset T$, in particular for the set $X = T - N_m$. Let

$$x \in T - (E_m \cup N_m) \subset T - N_m = X.$$

Then by the assumption there are an element $y \in T - X = N_m$ and a set $Y \in \mathscr{F}$ such that $(X - \{x\}) \cup \{y\} \subset Y$. Now if $k + 1 = \min\{l : y \in N_l\}$ then

$$(E_k \cup \{y\}, N_k) < (E_m \cup \{y\}, N_m) < (X - \{x\}) \cup \{y\} \subset Y \in \mathscr{F},$$

contradicting the fact that ψ_0 chose y a non-element.

(ii) Suppose now that ψ_0 is a winning strategy for the Constructor in the game $\mathscr{F}$. Note that ψ_0 is a winning strategy for the Constructor in the game

$\mathscr{F}$ iff it is winning in the game $\mathscr{F}^{\mathrm{mon}}$. Thus we assume without loss of generality that $\mathscr{F}$ is monotone. Suppose now that there exist $X \in \mathscr{F}$ and $x \in X$ such that if $y \in T - X$ then $(X - \{x\}) \cup \{y\}$ is not contained in a set belonging to $\mathscr{F}$. Put $s = |X|$ and let ϕ be the algorithm that in the first $s - 1$ moves asks the elements of $X - \{x\}$ and whose last (i.e. tth) move asks whether x is an element or not. Then in $t - 1$ moves the Constructor, using the simple strategy ψ_0, constructs the preset $(X - \{x\}, T - X)$. There are only 2 sets extending this preset, namely $X - \{x\}$ and X. Since both belong to $\mathscr{F}$, $c(\mathscr{F}, \phi, \psi_0) \leqslant t - 1$, contradicting the assumption that ψ_0 is a winning strategy for the Constructor. ∎

Milner and Welsh [MW2] pointed out that already Theorem 1.1 implies that a number of graph properties are elusive. (Some of these results were proved independently by Best, van Emde Boas and Lenstra [BBL1].)

THEOREM 1.2. *The following properties of graphs G^n are elusive.*

   (i) *G^n has at most k edges $(0 \leqslant k < \binom{n}{2})$.*
   (ii) *G^n is a spanning tree.*
  (iii) *G^n is a forest with k edges $(k < n)$.*
  (iv) *G^n is acyclic $(n \geqslant 3)$.*
   (v) *G^n is connected.*
  (vi) *G^n is 2-connected.*

Proof. Each of these properties or its complement satisfies the conditions of Theorem 1.1, so the simple strategy shows that the properties are elusive. ∎

Theorem 1.2 might give the impression that Theorem 1.1 can be applied to most natural graph properties, i.e. to most of the graph invariants commonly used. This is certainly not the case; the simple strategy cannot be applied to any of the following properties:

$$\chi(G) \geqslant k \ (k \geqslant 3), \qquad cl(G) \leqslant k \ (k \geqslant 3), \qquad \kappa(G) \leqslant k \ (k \geqslant 3),$$

$$\lambda(G) \leqslant k \ (k \geqslant 2),$$

$$\mathrm{diam}\, G \geqslant d, \qquad F(G) = 0, \qquad \delta(G) \leqslant \delta \ (0 < \delta < n - 1)|$$

$$\Delta(G) \leqslant \Delta \ (0 < \Delta < n - 1).$$

In the next section we shall determine the complexity of some of these properties.

Best, van Emde Boas and Lenstra found the following non-trivial application of Theorem 1.1.

THEOREM 1.3. *For $n \geq 5$ planarity is an elusive property of G^n.*

Proof. Let $\mathscr{P}$ be the property of being planar. The total set T corresponds to K^n. As $n \geq 5$, $K^n \notin \mathscr{P}$. Let us show that the simple strategy ψ_0 is winning for the game $\mathscr{P}$. Let us assume that this is not so. Then, by Theorem 1.1, there exist a maximal planar graph G and $ab \in E(G)$ such that G is the *only* maximal planar graph (of order n) containing $G - ab$. Fix an imbedding of G in the plane. Since G is maximal planar, all the faces of this imbedding are triangles. In particular, ab is the edge of two neighbouring triangular faces, say that of abc and abd. Then $cd \in E(G)$ since otherwise $G - ab + cd$ would be another maximal planar graph containing $G - ab$.

Let cde and cdf be the two faces containing cd. If $\{c, f\} = \{a, b\}$ then abc, abd, cde, cdf are all the faces so $n = 4$, contradicting our assumption.

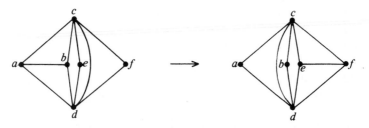

FIG. 1.1. Replacing G by $G - ab + ef$.

Hence we may assume that $a \notin \{e, f\}$. Then $ef \notin E(G)$ since it would intersect either cd or the path cad. Change the drawing of $G - ab$ as follows: join c to d inside the quadrilateral $acbd$ instead of inside the quadrilateral $ecfd$. Then we can join e to f inside the quadrilateral $ecfd$ (Fig. 1.1), contradicting the assumption that G is the only maximal planar graph containing $G - ab$. ∎

Given a property $\mathscr{P}$ of G^n (i.e. $\mathscr{P} \subset \mathscr{G}^n$) define the *order of* $\mathscr{P}$ as the number of non-isomorphic graphs in $\mathscr{P}$. Milner and Welsh [MW1] noticed every property of order 1 is elusive but there is a property $\mathscr{P}_3$ of graphs G^7 which is not elusive and its order is 3. The non-isomorphic graphs belonging to $\mathscr{P}_3$ are shown in Fig. 1.2. We leave it to the reader to show that $\mathscr{P}_3$ is not elusive

FIG. 1.2. The three non-isomorphic graphs in $\mathscr{P}_3$.

(Ex. 2). In fact for every $n \geqslant 7$ there is a non-elusive property of G^n with order 3 (see Ex. 3). Milner and Welsh conjectured that every property of order 2 is elusive. This was proved by Bollobás and Eldridge [BE3].

THEOREM 1.4. *Let $\mathscr{P}$ be a property of G^n with order at most 2. Then $\mathscr{P}$ is elusive.*

Proof. Suppose $G \in \mathscr{P}$ iff $G \cong H_1$ or $G \cong H_2$, where H_1 and H_2 may be isomorphic. We shall show that in the role of the Constructor we can force Algy to make $\binom{n}{2}$ probes. If the graphs H_1, H_2 do not differ by exactly one edge then we choose a fixed copy of H_1 and answer edge iff Algy probes an edge of this copy of H_1. We thus force Algy to make $\binom{n}{2}$ probes.

Suppose that for a fixed copy of H_1 there is a unique edge e with $H_2 \cong H_1 + e$. We say non-edge to the first probe and choose a fixed copy of H_2 with this edge playing the role of e. Then we say edge if and only if Algy probes an edge of this copy of H_2. We thus forced Algy to make $\binom{n}{2}$ probes.

Finally suppose that for a fixed copy of H_1 there are two distinct edges e and f such that $H_1 + e \cong H_2 \cong H_1 + f$. We say *edge* to the kth probe iff some copy of H_2 will be an extension of the pregraph arising at that point. Algy will be forced to make $\binom{n}{2}$ probes unless the edge set we construct after $\binom{n}{2} - 1$ probes is exactly the edge set of a graph H_1 and the very last probe of Algy is a pair which can play the role of f. But this is impossible since in that case we would have had to choose the pair playing the role of e an edge. ∎

Most graph properties we have mentioned until now are elusive and not a single one of them has complexity $o(n^2)$. One may be tempted to conjecture that $c(\mathscr{P}) = o(n^2)$ can never hold but the following example, due to Best, van Emde Boas and Lenstra [BBL1], shows that this is very far from being true. Call a graph G of order n a *scorpion graph* if it contains a vertex b (the *body*) of degree $n - 2$, the only vertex of G not adjacent to b, say t (the *tail*), has degree 1 and is adjacent to a vertex u of degree 2 (Fig. 1.3). The other $n - 3$ vertices of G are joined or not joined to each other arbitrarily.

FIG. 1.3. A scorpion graph.

THEOREM 1.5. *If $\mathscr{P}$ is the property of being a scorpion graph then $c(\mathscr{P}) \leqslant 6n$.*

Proof. As before, we use the game terminology. However, as our aim is to show that $c(\mathscr{P})$ is small, now we play the role of Algy. Since $c(\mathscr{P}) \leqslant \binom{n}{2}$ we may assume that $n \geqslant 14$. Define the *weight* of a candidate body x (resp. tail) as two minus the number of edges incident with x which have been refused (resp. given). Thus each candidate has weight 2 or 1.

First we ask for the edges $12, 23, \ldots, (n-1)n, n1$. By these n probes we can partition $V = \{1, 2, \ldots, n\}$ into three parts: B_2, the set of candidate bodies of weight 2 (these are *not* candidate tails), T_2, the set of candidate tails of weight 2 (these are *not* candidate bodies) and $R = V - B_2 - T_2$. For each vertex $x \in R$ one edge incident with x has been given and one has been refused. By making at most $r = |R|$ more probes each of which is incident with at least one vertex in R, we can partition R into 2 parts, say B_1 and T_1 such that B_1 consists of candidate bodies of weight 1 and T_1 consists of candidate tails of weight 1. The vertices in B_1 (resp. T_1) are not candidate tails (resp. bodies).

At this stage of the game, after at most $n + r$ probes, the set of candidate bodies, $B = B_1 \cup B_2$, is disjoint from the set of candidate tails, $T = T_1 \cup T_2$, and the sum of the weights is $2n - r$. Algy now asks for edges between candidate bodies and candidate tails. This part of the game, which takes at most $2n - r - 2$ questions, since each question reduces the total weight by at least one, terminates when all edges between the remaining candidate bodies and candidate tails have been asked for. Denote by b (resp. t) the number of remaining candidate bodies (resp. tails). Since at most b of the edges joining them have been refused and at most t have been given, $b + t \geqslant bt$. Hence either $bt = 0$ or $\min\{b, t\} = 1$, in which case we are home, or else $b = t = 2$. In this last case let x_1, x_2 be the candidate bodies and y_1, y_2 the candidate tails. We may suppose that x_1y_1, x_2y_2 have been refused and x_1y_2, x_2y_1 have been given. Let us ask for the edges x_1x_2, y_1y_2. We may assume that x_1x_2 is given and y_1y_2 is refused since otherwise the Constructor cannot construct a scorpion graph. Now we ask for an edge x_1z. If it is given, we know that y_2 cannot be the tail of a scorpion graph and x_2 cannot be the body. If it is refused then x_1 cannot be the body and y_1 cannot be the tail. This shows that after at most $n + r + 2n - r - 2 + 3 = 3n + 1$ probes we can find the *unique vertex* that might be the body or the tail of a scorpion graph. It is clear that Algy needs at most $3(n - 1) - 3$ more probes to decide whether the Constructor has a scorpion graph or not.   ■

2. MONOTONE PROPERTIES

Recall that a property $\mathscr{P}$ is said to be *monotone* if $G \in \mathscr{P}$ and $H \subset G$ implies $H \in \mathscr{P}$. After some initial difficulties Aanderaa, Rosenberg, Lipton and Snyder (see [R13] and [LS1]) advanced the conjecture that there is an $\varepsilon > 0$ such that $c(\mathscr{P}) > \varepsilon n^2$ for every monotone property $\mathscr{P}$ of graphs or directed graphs (or order n) *without loops*. If we allowed loops then the conjecture would fail since, as we have already remarked, if $\mathscr{P}$ is a property depending only on the loops of the graph then $c(\mathscr{P}) \leqslant n$. Discarding monotonicity the conjecture fails again, as shown by Theorem 1.5 (cf. also Ex. 4).

Best, van Emde Boas and Lenstra [BBL1] proved a number of results about the complexity of monotone properties and conjectured that every (non-trivial) monotone property of graphs or directed graphs is *elusive*. They also showed that there is a non-trivial property $\mathscr{P}$ with $c(\mathscr{P}) = O(n)$. The conjecture of Aanderaa, Rosenberg, Lipton and Snyder was proved by Rivest and Vuillemin [RV1], [RV2]. Even more, they proposed an even stronger conjecture and they proved that conjecture for infinitely many values of the parameter. The main aim of this section is to present the results of Rivest and Vuillemin. In addition to that we give concrete best possible strategies of the Constructor for some natural graph properties not covered by general results. In particular we show by a definite strategy that if k is any natural number then the properties $cl(G) \leqslant k$ and $\chi(G) \leqslant k$ are elusive.

We shall use the notation of the previous section: T is a set of t elements and $\mathscr{F}$ is a property of the subsets of T, i.e. $\mathscr{F} \subset \{0, 1\}^T$. Define the *enumerating polynomial* $p(z) = p_{\mathscr{F}}(z)$ by

$$p(z) = \sum_{X \in \mathscr{F}} z^{|X|} = \sum_{0 \leqslant i \leqslant t} N(\mathscr{F}, i) z^i,$$

where $N(\mathscr{F}, i)$ is the number of sets in $\mathscr{F}$ with i elements. The following result was proved independently by Best, van Emde Boas and Lenstra [BBL1] and Rivest and Vuillemin [RV1], [RV2].

THEOREM 2.1. *If* $c(\mathscr{F}) = k$ *then* $(1 + z)^{t-k}$ *divides* $p(z)$.

Proof. Let ϕ be an algorithm with $c(\mathscr{F}, \phi) = c(\mathscr{F}) = k$. Using ϕ, from each set $X \in \mathscr{F}$ one produces a sequence $X_0 < X_1 < \ldots < X_k$ of presets such that $X_k < Y \subset T$ implies $Y \in \mathscr{F}$. Let this sequence be ξ and denote by $\mathscr{S}_\xi$ the set of all subsets of T that give rise to this sequence ξ. Then $\mathscr{S}_\xi \subset \mathscr{F}$. Put $l_\xi = |E_k| \leqslant k$, where $X_k = (E_k, N_k)$. In $\mathscr{S}_\xi$ there are exactly $\binom{t-k}{m}$ sets X with $l_\xi + m$ elements so

$$\sum_{X \in \mathscr{S}_\xi} z^{|X|} = z^{l_\xi}(1 + z)^{t-k}.$$

Hence

$$p(z) = \sum_{X \in \mathscr{F}} z^{|X|} = \sum_\xi z^{l_\xi}(1 + z)^{t-k},$$

so $p(X)$ is divisible by $(1 + z)^{t-k}$.   ∎

COROLLARY 2.2. *If $p_{\mathscr{F}}(-1) \neq 0$, i.e. if the number of sets in $\mathscr{F}$ with an even number of elements is not equal to the number of sets in $\mathscr{F}$ with an odd number of elements, then $\mathscr{F}$ is elusive.*   ∎

This corollary allows one to prove that another set of graph properties are elusive (cf. Exx. 7–9). In order to find more applications of Corollary 2.2, from now on we shall not consider completely general properties $\mathscr{F}$ but only those that resemble slightly the graph properties. Denote by $\Gamma(\mathscr{F})$ the group of permutations on T that leave $\mathscr{F}$ invariant, i.e. under which every set belonging to $\mathscr{F}$ is mapped into a set belonging to $\mathscr{F}$. It seems that the weakest and most natural condition one should impose on $\mathscr{F}$ is that $\Gamma(\mathscr{F})$ be *transitive* on T. Note that when $\mathscr{F}$ is a property of graphs with vertex set V then $T = V^{(2)}$, the set of all unordered pairs of elements in V, and $\Gamma(\mathscr{F})$ contains the group of permutations on T induced by the full group of permutations on V. Hence in this case $\Gamma(\mathscr{F})$ is trivially transitive. If $\mathscr{F}$ is a nontrivial monotone graph property then, in particularly, $\varnothing \in \mathscr{F}$ and $T \notin \mathscr{F}$. Thus the following conjecture of Rivest and Vuillemin [RV1] is a very powerful extension of the conjecture of Best, van Emde Boas and Lenstra.

CONJECTURE 2.3. *If $\Gamma(\mathscr{F})$ is transitive on T, $\varnothing \in \mathscr{F}$ and $T \notin \mathscr{F}$ then $\mathscr{F}$ is elusive.*   ∎

It was a great step forward when Rivest and Vuillemin proved this conjecture in the case when $t = |T|$ is a prime power. Though $|V^{(2)}| = \binom{n}{2}$ is not a prime power if $n \geqslant 4$, Rivest and Vuillemin showed also that this result implies easily the original conjecture of Aanderaa, Rosenberg, Lipton and Snyder. By the same method, Kleitman and Kwiatkowski [KK2] improved slightly the bounds given by Rivest and Vuillemin. We present some of these results, starting with the key theorem of Rivest and Vuillemin.

THEOREM 2.4. *Let* $t = |T| = p^r$, *where* p *is a prime. If* $\Gamma(\mathcal{F})$ *is transitive on* T, $\varnothing \in \mathcal{F}$ *and* $T \notin \mathcal{F}$ *then* $\mathcal{F}$ *is elusive.*

Proof. Consider an orbit θ of subsets of T under the action of $\Gamma = \Gamma(\mathcal{F})$. The subsets in this orbit θ have the same size (i.e. number of elements), say m. Denote by c the number of sets in θ containing a given element of T. As Γ is transitive, every element of T is contained in exactly c of the sets in θ. Hence

$$\sum_{X \in \theta} |X| = |\theta| m = tc = p^r c.$$

Thus, either m is p^r or 0, or else p divides $|\theta|$. In other words, there are exactly two orbits with size not divisible by p: the orbit consisting of the single set T and the orbit consisting of the empty set. If $\mathcal{F}_i = \{X \in \mathcal{F} : |X| \equiv i \,(\mathrm{mod}\,2)\}$, $i = 0, 1$, then each of $\mathcal{F}_0$ and $\mathcal{F}_1$ consists of entire orbits only. As exactly one of them contains an orbit of size 1 and all other orbits have size divisible by p, $|\mathcal{F}_0| \neq |\mathcal{F}_1|$. Hence, by Corollary 2.2, $\mathcal{F}$ is elusive. ∎

THEOREM 2.5. *If* $\mathcal{P}$ *is a non-trivial monotone property of graphs of order* $n = 2^m$ *then* $c(\mathcal{P}) \geq n^2/4$.

Proof. Put $H_i = 2^{m-i} K^{2^i}$, i.e. let H_i be the disjoint union of 2^{m-i} copies of K^{2^i}. Then $H_0 = E^n \subset H_1 = (n/2) K^2 \subset \ldots \subset H_m = K^n$. Since $\mathcal{P}$ is monotone, there is an index j such that $H = H_j \in \mathcal{P}$ and $H_{j+1} \notin \mathcal{P}$. Put $J = 2^{m-j-1} K^{2^j}$. Then $H = 2J$ and $H_{j+1} \subset K = H + H$ so $K \notin \mathcal{P}$.

Playing the role of the Constructor, we will be rather generous to Algy. We give away that the graph G we are constructing satisfies $H \subset G \subset K$. However, when answering questions about the edges in $T = E(K) - E(H)$ we are trying to play as well as possible. Define a property $\mathcal{F}$ of the subsets of T as follows:

$$\mathcal{F} = \{E(G) - E(H): \ H \subset G \subset K \quad \text{and} \quad G \in \mathcal{P}\}.$$

Then $\varnothing \in \mathcal{F}$, $T \notin \mathcal{F}$ and $c(\mathcal{F}) \leq c(\mathcal{P})$. In fact, if Algy knows without any probes that $H \subset G \subset K$, then he needs exactly $c(\mathscr{F})$ probes to decide whether $G \in \mathscr{P}$ or $G \notin \mathscr{P}$.

Let us check that T and $\mathscr{F}$ satisfy the conditions of Theorem 2.4. We have seen already that $\varnothing \in \mathscr{F}$ and $T \notin \mathscr{F}$. If $\alpha, \beta \in E(K) - E(H)$ then there is a permutation of V mapping H and K into themselves, that maps α into β. Therefore $\Gamma(\mathscr{F})$ is transitive on $T = E(K) - E(H)$. Finally T is exactly the edge set of a bipartite graph with $n/2$ vertices in each of the classes so $|T| = n^2/4 = 2^{2m-2}$. Thus, by Theorem 2.4, $c(\mathscr{P}) \geqslant c(\mathscr{F}) = n^2/4$, as claimed.   ∎

THEOREM 2.6. *If $\mathscr{P}$ is a nontrivial monotone property of graphs of order n then* $c(\mathscr{P}) \geqslant n^2/16$.

Proof. Denote by $c(n)$ the minimum of $c(\mathscr{P})$ as $\mathscr{P}$ ranges over all nontrivial monotone properties of graphs of order n. The assertion is an immediate consequence of Theorem 2.5 if we prove the following inequality: if $2^m < n < 2^{m+1}$ then

$$c(n) \geqslant \min\{c(n-1), 2^{2m-2}\}. \tag{1}$$

To see (1), consider a monotone property $\mathscr{P}$ of graphs of order n. If $K^1 \cup K^{n-1} \notin \mathscr{P}$ or $K^1 + E^{n-1} \in \mathscr{P}$ then, as in the proof of Theorem 2.5, if the Constructor gives away that a certain vertex has degree 0 or $n-1$, Algy still needs at least $c(n-1)$ probes. Hence we may assume that $K^1 \cup K^{n-1} \in \mathscr{P}$ and $K^1 + E^{n-1} \notin \mathscr{P}$.

Put $r = 2^{m-1}$ and $s = n - 2r$. Then the monotonicity of $\mathscr{P}$ implies that $(K^r + K^s) \cup E^r$ has property $\mathscr{P}$ since it is a subgraph of $K^1 \cup K^{n-1}$. Similarly $K^r + (K^s \cup E^r) \notin \mathscr{P}$ since it contains $K^1 + E^{n-1}$. As in the proof of Theorem 2.5, $\mathscr{P}$ can be used to define a transitive property $\mathscr{F}$ on the set T of edges joining K^r to E^r with $c(\mathscr{F}) \leqslant c(\mathscr{P})$. The property $\mathscr{F}$ satisfies the conditions of Theorem 2.4 so $c(\mathscr{F}) \geqslant r^2$, completing the proof of (1).   ∎

As we have mentioned already, the proofs of Theorem 2.5 and 2.6 can be refined to give better lower bounds on the complexity of monotone graph properties (cf. Ex. 13). At the end of §5, using an entirely different technique, we give a lower bound on the complexity of all graph properties.

Kirkpatrick [K6] showed that the property of containing a triangle is elusive. It is somewhat surprising that it is not too easy to prove that $\mathrm{cl}(G) \geqslant r$ or $\chi(G) \geqslant r$ are also elusive. In the rest of the section we present a result of Bollobás [B28] asserting that the Constructor can win both of the games $\mathrm{cl}(G) \geqslant r$ and $\chi(G) \geqslant r$ by playing the same strategy.

Our graphs have vertex set V, $|V| = n$. The presets of the total set $T = V^{(2)}$ are called *pregraphs*; in order to conform to our graph notations the pregraph corresponding to the present (E, N) of T is denoted by $\tilde{G} = (V, E, N)$. We denote by $P = E \cup N$ the set of probed edges. A pair $(a, b) \in V^{(2)}$ is denoted simply by ab.

THEOREM 2.7. *Let* $2 \leqslant r \leqslant n$. *Then the property of containing a complete graph of order* r *is elusive.*

Proof. We shall define a strategy ψ and we prove that it is a winning strategy for the Constructor in the game $\mathrm{cl}(G) \geqslant r$.

Put $U = \{1, \ldots, r - 2\}$, $W = V - U$. Let $\tilde{G} = (V, E, N)$ be a pregraph and let $e \in V^{(2)} - P = V^{(2)} - E \cup N$. Put $G = (V, E)$ and let $H = G[W]$. For $Z \subset V$ put

$$D_Z = \{w \in V: \ wz \in P \quad \text{for all} \quad z \in Z - \{w\}\}.$$

If $e = wz$, $z \in Z$, and $\{wz: z \in Z - \{w\}\} \subset P \cup \{e\}$ then e is said to be the *last probe from* w *to* Z. If $Z = V$ then e is the *last probe from* w and we put $D = D_V$.

We say that a vertex $w \in W$ is a $\tilde{G}$-*critical* vertex if $w \notin D$, the component T of $H - D$ containing w is a *tree with centre* w *and radius at most* 1 and every vertex of $T - \{w\}$ belongs to D_w.

In order to decide whether or not ψ should make $e = uv$ an edge, we consider five cases. Recall that u and v are interchangeable since e is an unordered pair.

C_0. $P \cup \{e\} \supset V^{(2)} - U^{(2)}$.

C_1. $u, v \in U$.

C_2. $u, v \in W$, e is not the last probe from u but it is the last probe from u to W, and u is an isolated vertex of $H - D$.

C_3. $u \in U$, $v \in W$ and e is not the last probe from v to U.

C_4. $u \in U$, $v \in W$, e is the last probe from v to U, v is $\tilde{G}$-critical and in G no vertex of W is joined to every vertex of $\{v\} \cup U$.

Let ψ *be the strategy that chooses* e *to be an edge iff* $\bigcup_0^4 C_i$ *holds.*

Let ϕ be an algorithm and let $\tilde{G}_0 < \tilde{G}_1 < \ldots < \tilde{G}_t$, $t = \binom{n}{2}$ be the sequence of pregraphs arising in the game if Algy applies ϕ and the Constructor ψ. Put $\tilde{G}_i = (V, E_i, N_i)$ and let $G_i = (V, E_i)$, $H_i = G_i[W]$, $P_i = E_i \cup N_i$, D_i, D_Z^i be the graphs and sets we define from $\tilde{G}_i$ as at the beginning of the proof.

To complete the proof of the theorem we have to prove that G_{t-1} does not contain a K^r and G_t does. We divide the proof of this assertion into a number of lemmas.

Denote by $T_m(x)$ the component of H_m that contains the vertex $x \in W$.

LEMMA 2.8. $T_t(x)$ is a tree for every $x \in W$.

Proof. Let $x_1, x_2, \ldots, x_m \in W (m \geqslant 3)$. We can suppose without loss of generality that for some j we have $x_1 x_m \in P_j$ and $x_i x_{i+1} \notin P_j$ for $1 \leqslant i < m$. Then $x_1 x_m$ is not the last probe from either x_1 or x_m in W, so C_2 does not hold when probing $x_1 x_m$. Thus $x_1 x_m$ is not an edge of H_j. Consequently H_t does not contain a circuit. ∎

LEMMA 2.9. *Suppose* $x_0 \in W - D^m$ *and* $x_0 x_1, \ldots, x_k$ *is a path in* H_m. *Then* $x_3, x_4, \ldots, x_k \in D^m$.

If we also have $x_0 \in W - D_W^m$ *then* $x_1 \in D_W^m$ *and* $x_2 \in D^m$.

In particular, if $x \in W - D_W^m$ *then* x *is the only* $\tilde{G}_m$-*critical vertex of* $T_m(x)$ *and every vertex of* $T_m(x) - D^m - \{x\}$ *is joined to* x.

Proof. Let $\alpha_i = x_{i-1} x_i, 1 \leqslant i \leqslant k$. If $k < 3$ there is nothing to prove in the first assertion. Otherwise let $3 \leqslant p \leqslant k$. When α_2 is probed x_1 cannot play the role of u in C_2 since $x_0 \notin D^m$. Therefore, x_2 plays the role of u when α_2 is probed. Thus α_3 is probed before α_2 and x_3 plays the role of u in C_2 for this probe. Continuing this argument we see that x_{p-1} plays the role of u in C_2 when α_{p-1} is probed. Therefore α_p is probed before α_{p-1} and by C_2 we must have $x_p \in D^m$.

Let now $x_0 \in W - D_W^m$. To prove the second part of the assertion note that when α_1 is probed, x_1 must play the role of u in C_2 and hence $x_1 \in D_W^m$, α_2 is probed before α_1, and $x_2 \in D^m$.

The third assertion of the lemma follows immediately from the second. ∎

LEMMA 2.10. *Suppose no vertex of* $T_m(x)$ *is joined to every vertex of* U *in* G_m. *Then* $T_m(x)$ *contains a* $\tilde{G}_m$-*critical vertex.*

Proof. The assertion holds for $m = 0$. Suppose $m \geqslant 1$ and it holds for smaller values of m. By the induction hypothesis we can assume that x is $\tilde{G}_{m-1}$-critical. If $\{\alpha\} = P_m - P_{m-1}$ and α is not adjacent to x, then x is also $\tilde{G}_m$-critical. So we can assume that $\alpha = xy$.

(a) Suppose $y \in W$. Then by Lemma 2.9 we can suppose that α is the last probe from x to W. Furthermore, x and y are $\tilde{G}_l$-critical for every $l < m$, so $x \notin D^m$ since otherwise the last probe from x to U would have given an edge. Thus if $\alpha \in N_m$, we have finished, x itself is $\tilde{G}_m$-critical. Therefore we can suppose without loss of generality that $\alpha \in E_m$. Consequently when α was probed, C_2 had to hold. By symmetry we may suppose that y played the role of u in C_2. Then y is an isolated vertex of $T_{m-1}(y) - D^{m-1}$ and so x is not only $\tilde{G}_{m-1}$-critical but also $\tilde{G}_m$-critical.

(b) Suppose now that $y \in U$. Then α cannot be the last probe from x to U since then by C_4 the vertex x would be joined to every vertex of U in G_m. Consequently $x \notin D^m$ and so x is indeed $\tilde{G}_m$-critical.   ■

For the fairly straightforward proofs of the next three lemmas we refer the reader to [B28] (cf. Ex. 15).

LEMMA 2.11. *Suppose $x \in W$, $\{\alpha\} = E_{j+1} - E_j$ and α is the last probe from x to U. Then for every $m \geq j$ every vertex of $T_m(x) - D^m - \{x\}$ is joined to x and every vertex of $T_m(x) - x$ belongs to D_W^m.*   ■

Denote by k the minimal index for which $P_{k+1} \supset V^{(2)} - U^{(2)}$.

LEMMA 2.12. *At most one vertex of $T_k(x)$ is joined to every vertex of U in G_k.*   ■

LEMMA 2.13. *$\chi(G_{t-1}) \leq r - 1$. In particular, $\mathrm{cl}(G_{t-1}) \leq r - 1$.*   ■

LEMMA 2.14. *G_t contains a K_r.*

Proof. Let $\{\alpha\} = P_{k+1} - P_k$. In other words α is the first pair that ψ makes an edge because case C_0 holds. (In fact α is the *only* pair that becomes an edge *only* because of C_0.)

(a) Suppose $\alpha = xy$, x, $y \in W$. We claim that x and y are joined to every vertex of U in G_k. For if, say, x is not joined to every vertex of U in G_k then there is a vertex $z \in W$, $xz \in E_k$ such that in G_k, z is joined to every vertex of U. Furthermore, z had to be joined to every vertex of U before xz was probed. However, then C_2 does not hold when xz is probed so $xz \notin E_k$. This contradiction shows that x and y are joined to every vertex of U in G_k, so the set $U \cup \{x, y\}$ spans a K_r in G_t.

(b) Suppose now that $\alpha = xy$, $x \in U$, $y \in W$. By Lemma 2.11 if a vertex z of $T_k(y)$ is joined to every vertex of U in G_k then $zy \in E_k$, so the vertex set $U \cup \{y, z\}$ spans a K_r in G_t. Thus we can assume without loss of generality that in G_k no vertex of $T_k(y)$ *is joined to every vertex of U*.

By considering the last probe from y to W one can see immediately that y is *not an isolated vertex of H_k* otherwise ψ would make that last probe an edge.

Put $W_0 = \{z \in W : (z, y) \in E_k\}$. Then $W_0 \neq \emptyset$. Let $z_i \in W_0$ and let $\gamma_i = z_i t_i$ be the last probe from z_i to U. If $\{\gamma_i\} = P_{l_i+1} - P_{l_i}$ then z_i cannot be $\tilde{G}_{l_i}$-critical for otherwise $z_i \in T_k(y)$ is joined to every vertex of U in G_k. Thus by Lemma 2.9 we have $z_i \in D_W^{l_i}$. Let $z_0 t_0$ be the last probe of the form zt, $z \in W_0$, $t \in U$ and put $l = l_0$. Then $yz \in P_l$ (implying $yz \in E_l$), $z \in D_W^l$ and $z \in D^{l+1}$ *for every $z \in W_0$.*

We also have $y \in D_W^l$. For suppose yw is the last probe from y to W and $yw \notin P_l$. When probing yw, y can play the role of u in C_2, so yw becomes an edge, giving $w \in W_0$ and $yw \notin E_l$. This contradiction shows that $y \in D_W^l$.

Now by Lemma 2.10 at least one of y and z_0 is $\tilde{G}_l$-critical. We have established that y is an endvertex of $T_l(y) - D$ and $y \in D_W^l$. Consequently z_0 is $\tilde{G}_l$-critical. This implies that the strategy ψ makes γ_0 an edge, so in G_k the vertex z_0 of $T_k(y)$ is joined to every vertex of U.   ∎

Theorem 2.7 is clearly an immediate consequence of Lemmas 2.13 and 2.14.

THEOREM 2.15. *The property $\chi(G) \geqslant r$ is elusive for $2 \leqslant r \leqslant n$.*

Proof. Lemmas 2.13 and 2.14 also imply that ψ is a winning strategy for the Constructor.   ∎

3. THE MAIN PACKING THEOREM

Let $\mathscr{P}$ be a property of graphs of order n. We have seen that if $\mathscr{P}$ is monotone then $c(\mathscr{P}) \geqslant n^2/16$ but in general $c(\mathscr{P})$ need not exceed $6n$. It seems that one needs essentially different techniques to give a good lower bound on $c(\mathscr{P})$ if $\mathscr{P}$ is an arbitrary property. Kirkpatrick [K6] proved that $c(\mathscr{P}) \geqslant \sqrt{2n} - 2$ for every nontrivial property $\mathscr{P}$. Milner and Welsh [MW1] noticed that results of the following kind might be useful to give lower bounds on $c(\mathscr{P})$: if G_1 and G_2 are graphs of order n satisfying certain weak conditions then there are edge disjoint copies of G_1 and G_2 with the same vertex set. In particular, they conjectured that the condition $e(G_1) + e(G_2) \leqslant \lfloor \frac{3}{2}(n - 1) \rfloor$ is sufficient for this. Milner and Welsh also pointed out that the conjecture improves the bound given by Kirkpatrick to $c(\mathscr{P}) \geqslant \lfloor \frac{3}{2}(n - 1) \rfloor$. This conjecture was proved by Sauer and Spencer [SS2]. Though we shall also give a separate simple proof of this result (Corollary 3.3(iv)), the main aim of this section is to present a strong extension of this, proved by Bollobás and Eldridge [BE2].

Let G_1 and G_2 be subgraphs of G. In accordance with the general use of the term "cover" (cf. introduction to Chapter II.), the set of edges of G_2 incident with at least one vertex of G_1 is said to be *covered* by G_1. The *density* of a graph G is $e(G)/|G|$. Recall that a *star* is a tree all of whose edges are incident with one vertex, which we call the *centre* of the star.

Consider graphs with n vertices. If G_1 is a star and $\delta(G_2) = 1$ then there is no packing of G_1 and G_2. We quickly come to the conclusion that the bad feature of this example is that one graph has a vertex of degree $n - 1$. So we consider graphs with no vertices of degree $n - 1$. If G_1 is the disjoint union of a star and an isolated vertex, G_2 is a disjoint union of cycles and $n > 3$ then neither G_1 nor G_2 has a vertex of degree $n - 1$ and there is no packing of G_1 and G_2. The main result of this section, proved by Bollobás and Eldridge [BE2], shows that this situation is essentially worst possible, i.e. if we impose the extra condition $e(G_1) + e(G_2) \leqslant 2n - 3$ then, with only finitely many exceptions, there is a packing of G_1 and G_2.

THEOREM 3.1. *Suppose that* $\{G_1, G_2\}$ *satisfies* $|G_1| = |G_2| = n$,

$$\Delta(G_1), \Delta(G_2) < n - 1, \qquad e(G_1) + e(G_2) \leqslant 2n - 3$$

and $\{G_1, G_2\}$ *is not one of the following pairs:* $\{2K^2, E^1 \cup K^3\}$, $\{E^2 \cup K^3, K^2 \cup K^3\}$, $\{3K^2, E^2 \cup K^4\}$, $\{E^3 \cup K^3, 2K^3\}$, $\{2K^2 \cup K^3, E^3 \cup K^4\}$, $\{E^4 \cup K^4, K^2 \cup 2K^3\}$, $\{E^5 \cup K^4, 3K^3\}$ *(see Fig. 3.1). Then there is a packing of* G_1 *and* G_2.

Proof. We shall prove the result by induction on n. Clearly it is true for $n \leqslant 4$ so we shall assume that $n > 4$ and that the result is true for smaller values. The bulk of the work in the proof of the induction step is contained in the following lemma.

LEMMA 3.2. *Let* p *be a natural number such that* $2 \leqslant 2p \leqslant n$ $(n \geqslant 5)$. *Let* T *be a tree and let* G *be a graph such that* $|T| = p$, $|G| = n$, $\Delta(G) < n - 1$ *and* $n - 1 \leqslant e(G) \leqslant n + n/p - 3$. *Then there is a packing of* T *and* G *such that* T *covers at least* $p + 1$ *edges of* G *and* $\Delta(G - T) < n - p - 1$.

Proof. We first prove this result for $p \leqslant 2$. If $p = 1$, i.e. T is a single vertex, then $e(G) < 2n - 3$. Put T on a vertex u of degree $\Delta(G)$. Then T covers $\Delta(G) \geqslant 2$ edges. If $G - T = G - u$ has a vertex v of degree $n - 2$ then put

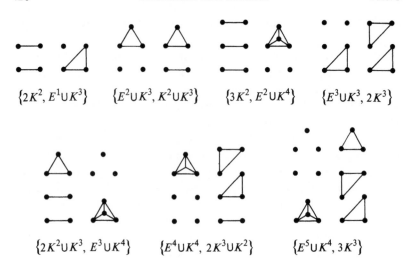

$$\{2K^2, E^1 \cup K^3\} \qquad \{E^2 \cup K^3, K^2 \cup K^3\} \qquad \{3K^2, E^2 \cup K^4\} \qquad \{E^3 \cup K^3, 2K^3\}$$

$$\{2K^2 \cup K^3, E^3 \cup K^4\} \qquad \{E^4 \cup K^4, 2K^3 \cup K^2\} \qquad \{E^5 \cup K^4, 3K^3\}$$

FIG. 3.1. The exceptional pairs.

T on a vertex w adjacent to both u and v and of maximum possible degree. Then $d(w) \geqslant 2$ so T again covers at least 2 edges and $\Delta(G - T) < n - 2$.

Now suppose $p = 2$ and T is the edge xy. Then $e(G) \leqslant 2n - 3 - n/2$. Put x on a vertex u of degree $\Delta(G)$ and put y on a vertex v not adjacent to u and of maximum possible degree. Then T covers at least 3 edges. If $G - T$ has a vertex w of degree $n - 3$ then u is adjacent to w for otherwise both u and v have degree at least $n - 3$ and so $3(n - 3) \leqslant \frac{3}{2}(n - 2)$, i.e. $n \leqslant 4$. Consequently u and w have degree $n - 2$ so $2(n - 2) - 1 \leqslant \frac{3}{2}(n - 2)$, i.e. $n \leqslant 4$. So $\Delta(G - T) < n - 3$.

We may suppose now that $p \geqslant 3$ and so $n \geqslant 6$. We complete the proof of the lemma by induction on n. Suppose $n \geqslant 6$ ($p \geqslant 3$) and the result holds for smaller values of n.

Before presenting the essential part of the proof let us eliminate a rather easy case. Suppose that G has an isolated vertex, say u. If $n = 6$ then T is a path xyz. Place y on u, place x on a vertex of degree $\Delta(G)$ and place z on a vertex of maximal degree in $G - \{x, y\}$. As $e(G) = 5$, T does cover at least 4 edges since if $\Delta(G) \leqslant 2$ then $\Delta(G - \{x, y\}) \geqslant 2$. Then $G - T$ has at most one edge so $\Delta(G - T) < 3 - 1$. Suppose now that $n \geqslant 7$. Let v be a vertex of maximal degree in G, say $d(v) = \Delta = \Delta(G) \geqslant 2$. Place an endvertex x of T on v and the vertex y adjacent to x on u. Put $T' = T - \{x, y\}$, $G' = G -$

$\{u, v\}$. Then

$$\Delta(G') < n - 3$$

since otherwise $n - 3 \leqslant \Delta(G') \leqslant \Delta$ so

$$2(n - 3) \leqslant e(G) \leqslant n + n/p + 3 \leqslant 2n - 3 - \tfrac{2}{3}n < 2n - 7.$$

Add $s \leqslant \Delta - 2$ edges to G' to obtain a graph G^* with at least $n - 3$ edges such that

$$\Delta(G^*) < n - 3.$$

We also have

$$n - 3 \leqslant e(G^*) \leqslant e(G) - 2 \leqslant n - 2$$
$$+ n/p - 3 < n - 2 + (n - 2)/(p - 2) - 3.$$

Thus if T^* if a tree of order $p = 2$ containing T' then the pair (T^*, G^*) satisfies the conditions of the lemma since $n - 2 \geqslant 5$. Therefore by the induction hypothesis there is a packing of T^* and G^* such that $\Delta(G^* - T^*) < n - 2 - (p - 2) - 1$ and T^* covers at least $p - 2 + 1 = p - 1$ edges of G^*. Note that at least $p - 1 - s \geqslant p + 1 - \Delta$ of these edges are also edges of G. The packing of T^* into G together with x into v and y into u gives a packing of T and G such that

$$\Delta(G - T) = \Delta(G^* - T^*) < n - p - 1$$

and T covers at least

$$\Delta + p + 1 - \Delta = p + 1$$

edges of G.

From now on we may suppose that $3 \leqslant p$ and G has no isolated vertex. We now concentrate on showing that we can pack T into $\bar{G}$ in such a way that T covers at least $p + 1$ edges of G.

Omit a vertex u of degree $\Delta(G)$ and a vertex not adjacent to u of maximum possible degree. The resulting graph G' has no vertex of degree $n - 3$ for otherwise G has at least $2(n - 3) + 2$ edges so

$$2(n - 3) + 2 \leqslant n + n/p - 3 \leqslant \tfrac{4}{3}n - 3,$$

so $n \leqslant 1$. Furthermore $e(G') + p - 1 \leqslant 2(n - 3)$ so we can apply the induction hypothesis of the lemma to show that there is a packing of T and G'. Thus there is a packing of T and G. Choose a packing for which T covers the maximum possible number of edges. Suppose that this is at most p edges.

Denote by W the vertices covered by T and put $Z = V(G) - W$. The graphs T and G together form a new graph with n vertices and we shall distinguish the edges contributed by T by colouring them *red* and colouring the other edges *green*. Suppose $x \in W$ is not joined by a green edge to Z and $y \in Z$ is not joined by a green edge to W. We can get another packing by putting the vertex originally placed on x on y instead. Since G has no isolated vertex, in the new packing T covers at least one more edge. Thus in our packing either each vertex of W is joined by a green edge to Z or each vertex of Z is joined by a green edge to W. In each case T must cover at least p edges and *there can be no green edge with both vertices in W*. Since G has no isolated vertices we can see that in fact each vertex of W is joined by exactly one green edge to Z. Suppose all the vertices of W are adjacent to the same vertex of Z. Then, since $e(G) \geqslant n - 1$ and $\Delta(G) < n - 1$, Z has another vertex incident with at least 2 green edges. We put any vertex of T on this vertex instead. Then T covers at least $p + 1$ edges. Suppose each vertex of W is adjacent to a different vertex of Z. Since T covers the maximum possible number of edges it follows that $\Delta(G) \leqslant 1$. But then G has fewer than $n - 1$ edges. So we may suppose that there are vertices w, w', $w'' \in W$ such that w, w' are both joined to the same vertex $x \in Z$ and w'' is not joined to x. Now consider an alternative packing of T in which we place an endvertex on x and its neighbour on w'' and the other vertices arbitrarily on $W - \{w', w''\}$. Redefine W and Z to agree with this packing. T cannot cover fewer than p edges so T does cover the maximum number of edges but *there is a green edge xw with both vertices in W*. This contradicts the assertion we proved earlier. Consequently there is a packing of T and G such that T *covers at least $p + 1$ edges of G*.

Suppose that no such packing satisfies $\Delta(G - T) < n - p - 1$. Then G must have at least $(n - p - 1) + (p + 1) = n$ edges and so

$$n \leqslant n + n/p - 3, \quad \text{i.e.} \quad 3p \leqslant n.$$

Again consider a packing of T and G in which T covers the *maximum* possible number of edges of G. Define the sets W and Z and red and green edges as before. By our assumption there is a vertex $v \in Z$ joined to all the other vertices of Z. We have already shown that either each vertex of W is joined by a green edge to a vertex of Z or each vertex of Z is joined by a green edge to a vertex of W. Consequently not all the vertices of Z are joined only to v. Furthermore v is the only vertex of Z which is joined to all the other vertices of Z for otherwise G has at least $p + 2(n - p - 1)$ edges and so

$$p + 2(n - p - 1) \leqslant n + n/p - 3.$$

But this is false. The $n - p - 1$ vertices of $Z - \{v\}$ span a subgraph with at most

$$n/p - 3 \leqslant 2(n - p - 1) - 3 - (p - 1)$$

edges so, by our induction hypothesis for the theorem, T can be packed into $Z - \{v\}$. Since not all these vertices have degree 1 we can choose such a packing in which T covers at least $p + 1$ edges of G. By assumption $G - T$ has a vertex w of degree $n - p - 1$. Since $\Delta(G) < n - 1$, $w \neq v$ so G has at least $2(n - p - 1)$ edges. Hence

$$2(n - p - 1) \leqslant n + n/p - 3.$$

Therefore

$$3(p - 1) \leqslant n(p - 1)/p \leqslant 2p - 1,$$

so

$$p \leqslant 2.$$

Thus we can always pack T into $\bar{G}$ in such a way that T *covers at least $p + 1$ edges of G and $\Delta(G - T) < n - p - 1$.* ∎

Proof of Theorem 3.1 concluded. Without loss of generality we may assume that $e(G_1) + e(G_2) = 2n - 3$ and $e(G_1) \leqslant e(G_2)$. So $e(G_1) \leqslant n - 2$ and a least dense component T of G_1 is a tree with say p vertices where $((p - 1)/p) n + e(G_2) \leqslant 2n - 3$. Thus the pair $\{T, G_2\}$ satisfies the conditions of the lemma. Choose a packing of T into $\bar{G}_2$ such that T covers at least $p + 1$ edges of G_2 and $\Delta(G_2 - T) < n - p - 1$. Provided $G_1 - T$ is not a star and $\{G_1 - T, G_2 - T\}$ is not one of the pairs of Figure 3.1 then we may complete a packing of G_1 and G_2 by applying the induction hypothesis to $\{G_1 - T, G_2 - T\}$. It is straightforward to check that in the latter case there is a packing of G_1 and G_2. As an example we shall check this for the case when $\{G_1 - T, G_2 - T\} = \{E^5 \cup K^4, 3K_3\}$. Clearly $G_1 - T \neq 3K^3$ since then G_1 would have $n - 1$ edges. If $G_1 - T = E^5 \cup K^4$ then T must be an isolated vertex and $G_1 = E^6 \cup K^4$. Then G_2 is obtained from $3K^3$ by adding an extra vertex adjacent to two edges. But then there is a packing of G_1 and G_2.

To complete the proof of the theorem it is sufficient to prove the assertion when G_1 is the disjoint union of a tree T and a star S with $|S| \geqslant |T| = p$. We leave the rather straightforward proof to the reader (cf. Ex. 16). ∎

COROLLARY 3.3. *Suppose G_1 and G_2 are graphs of order n. Let us say that (*) holds if $\Delta(G_1) = n - 1$ and $\delta(G_2) \geqslant 1$ or else $\Delta(G_2) = n - 1$ and $\delta(G_1) \geqslant 1$.*

 (i) *If* (∗) *holds then there is no packing of* G_1 *and* G_2.
 (ii) *Suppose* (∗) *does not hold,* $\{G_1, G_2\}$ *is not one of the pairs of Fig. 3.1 and* $e(G_1) + e(G_2) \leqslant 2n - 3$. *Then there is a packing of* G_1 *and* G_2.
 (iii) *If* (∗) *does not hold and* $e(G_1) + e(G_2) \leqslant 2n - 4$ *then there is a packing of* G_1 *and* G_2.
 (iv) *If* $e(G_1) + e(G_2) \leqslant \lfloor \frac{3}{2}(n - 1) \rfloor$ *then there is a packing of* G_1 *and* G_2.

Proof. It is sufficient to prove (ii). Suppose $\Delta(G_1) \leqslant \Delta(G_2)$. If $\Delta(G_2) < n - 1$ the result follows by Theorem 3.1. If $\Delta(G_2) = n - 1$ then G_1 has an isolated vertex x. We put this on a vertex y of degree $\Delta(G_2)$. Then $e(G_1 - x) + e(G_2 - y) \leqslant n - 2$ so (e.g. again by Theorem 3.1) one can easily find a packing of $G_1 - x$ into $G_2 - y$, completing the proof. ∎

The last part of this corollary, due to Sauer and Spencer [SS2], can be proved directly, and fairly simply, without making use of Theorem 3.1 which seems to be a more powerful result. The main idea of the proof is similar to that of Theorem 3.1. The result is obvious if $n \leqslant 4$ so we suppose that $n \geqslant 5$ and the result is true for smaller values. Without loss of generality we may assume that $e(G_1) + e(G_2) = \lfloor 3(n - 1)/2 \rfloor$ and $e(G_1) \leqslant e(G_2)$. If one graph has an isolated vertex x and the other has a vertex y of degree at least 2 we can put x on y and complete the packing by induction. So we may assume that this does not happen. $\Delta(G_2) \geqslant 2$ since otherwise $e(G_1) + e(G_2) \leqslant \frac{1}{2}n + \frac{1}{2}n = n < \lfloor \frac{3}{2}(n - 1) \rfloor$. Therefore G_1 has no isolated vertex. Let T be a least dense component of G_1. Then T has density less than $3/4$ so T is a path with 2 or 3 vertices. If $T = xy$ we may put x on a vertex u of maximal degree in G_2 and y on any vertex not adjacent to u. We can do this since $e(G_2) \leqslant \lfloor \frac{3}{2}(n - 1) \rfloor - \frac{1}{2}n < n - 1$. Then T covers at least 2 edges and the packing can be completed by induction. If $T = xyz$ then $\Delta(G_1) \geqslant 2$ so G_2 has no isolated vertex. On the other hand $e(G_2) \leqslant \lfloor \frac{3}{2}(n - 1) \rfloor - \frac{2}{3}n < n - 1$ so G_2 has at least 2 components. Put y on a vertex u of maximal degree in G_2 and x and z on any vertices in another component. Then T covers at least 3 edges and the packing can be completed by induction. ∎

Note that the example given before Theorem 3.1, in which G_1 is a star and $\delta(G_2) = 1$, shows that this result is best possible.

It is of some interest to determine conditions under which a packing of G_1 and G_2 may be partly specified in advance. As the last result of this section we state a theorem of this kind. It is not best possible and its unnatural form is justified only by its applicability to computational complexity. For its rather long and tedious proof we refer the reader to [BE2].

THEOREM 3.4. *Let G_1 and G_2 be graphs such that $|G_1| = |G_2| = n$ and $e(G_1) + e(G_2) \leqslant \frac{3}{2}(n - 2)$. Furthermore let $x_i \in G_i$, $d_{G_i}(x_i) \leqslant \frac{1}{2}(n - 2)$, $i = 1, 2$. Then G_1 can be packed into $\bar{G}_2$ in such a way that x_1 is put on x_2.* ∎

One cannot replace the condition $d_{G_1}(x_2) \leqslant \frac{1}{2}(n - 2)$ of Theorem 3.4 by $d_{G_1}(x_1) + d_{G_2}(x_2) \leqslant n - 2$. For let $r \geqslant 2$, $n \geqslant 2(r^2 + 1)$, $G_1 = K^{r+1} \cup E^{n-r-1}$ with x_1 as a vertex of the K^{r+1} and $G_2 = S \cup K^{r+1}$ where S is a star of order $n - r - 1$ with centre x_2. Then $|G_1| = |G_2| = n$, $d_{G_1}(x_1) + d_{G_2}(x_2) = n - 2$ and $e(G_1) + e(G_2) \leqslant \frac{3}{2}(n - 2)$ but one cannot pack G_1 into $\bar{G}_2$ in such a way that x_1 is put on x_2.

4. PACKING GRAPHS OF SMALL SIZE

In this section we present other sufficient conditions for packing two graphs. We begin with the following theorem of Sauer and Spencer [SS2], first announced by Catlin [C4].

THEOREM 4.1. *If $2\Delta(G_1)\,\Delta(G_2) < n$ then there is a packing of G_1 and G_2.*

Proof. The assertion is trivial for $n \leqslant 2$ so assume that $n \geqslant 3$. Identify $V(G_1)$ with $V(G_2)$ in such a way that the graphs G_1 and G_2 have minimal number of edges in common. Suppose $V(G_1) = \{x_1, x_2, \ldots, x_n\}$, $V(G_2) = \{y_1, \ldots, y_n\}$ and we identify x_i with y_i. We may assume that G_1 and G_2 are *not* edge disjoint, say $x_1 x_2 \in E(G_1)$ and $y_1 y_2 \in E(G_2)$. We shall show that this is impossible since there exists an index $i > 2$ such that by flipping x_2 and x_i, i.e. mapping x_2 into y_i and x_i into y_2, the number of common edges decreases.

Let L be the set of indices $l > 2$ for which there is an index j such that either $x_2 x_j \in E(G_1)$ and $y_j y_l \in E(G_2)$ or $y_2 y_j \in E(G_2)$ and $x_j x_l \in E(G_1)$. Since $x_1 x_2 \in E(G_1)$ and $y_1 y_2 \in E(G_1)$,

$$|L| \leqslant 2\Delta(G_1)\,\Delta(G_2) - 2 < n - 2.$$

Hence there is a natural number i, $3 \leqslant i \leqslant n$, such that $i \notin L$. This index i has the required properties. For if we change the identification of $V(G_1)$ and $V(G_2)$ by interchanging the images of x_2 and x_i then $x_1 x_2$ is no longer on an edge of

G_2. Even more, the condition on i is equivalent to the condition that no edge of G_1 incident with either x_2 or x_i is on an edge of G_2. Furthermore, by this interchange we do not place any other edge of G_1 on an edge of G_2, so the number of common edges of G_1 and G_2 indeed decreases.   ∎

Bollobás and Eldridge [BE2] pointed out that this result is rather near to being best possible. Let $d_1 \leqslant d_2 < n$ be natural numbers and suppose $n \leqslant (d_1 + 1)(d_2 + 1) - 2$. Let $n = p_i(d_i + 1) + r_i$, $1 \leqslant r_i \leqslant d_i + 1$, and put $G_i = p_i K^{d_i + 1} \cup K^{r_i}$, $i = 1, 2$. Then $\Delta(G_1) \leqslant d_1$, $\Delta(G_2) \leqslant d_2$ and there is no packing of G_1 and G_2. Thus if $n \leqslant (\Delta(G_1) + 1)(\Delta(G_2) + 1) - 2$ then there need not be a packing of G_1 and G_2. Bollobás and Eldridge [BE3a] conjectured that this is in fact the worst possible case (see also [BE3] for a slightly incorrect version of the conjecture).

CONJECTURE 4.2. *If* $|G_1| = |G_2| = n$ *and* $(\Delta(G_1) + 1)(\Delta(G_2) + 1) \leqslant n + 1$ *then there is a packing of* G_1 *and* G_2.   ∎

The conjecture holds for $\Delta(G_1) = 1$ (Ex. 19). For $\Delta(G_2) = 2$ it would imply the theorem of Corrádi and Hajnal [CH7] and its general case would be a considerable extension of Theorem V.5.1, due to Hajnal and Szemerédi [HS3].

The rest of the section is contained in [BE2]. The problem of packing a graph of constant size into graphs of increasing size resembles slightly Turán's problem about complete subgraphs. One has the following essentially best possible result. Recall that $t_r(n)$ is the size of the Turán graph $T_r(n)$, so $t_r(n) \sim \frac{1}{2}(1 - 1/r)\, n^2$.

THEOREM 4.3. *Let* m *be a fixed natural number and let* s *be given by*

$$\binom{s}{2} \leqslant m < \binom{s+1}{2}.$$

If n *is sufficiently large,* $|G_1| = |G_2| = n$, $e(G_1) = m$ *and*

$$e(G_2) \leqslant \binom{n}{2} - t_{s-1}(n) - 1$$

then there is a packing of G_1 *and* G_2. *The example of* $G_1 = K^s \cup E^{n-s}$ *and* $G_2 = T_{s-1}(n)$ *shows that the result is best possible.*

Proof. Put $\chi = \chi(G_1)$. If $\chi < s$ then Theorem VI.3.1 implies that $\bar{G}_2 \supset G_1$ if n is sufficiently large.

Suppose now that $\chi = s$. It is easily seen that then either $G_1 \supset K^s$ or else $s = 3$ and $G_1 = C^5$. One can check that in the second case $\bar{G}_2 \supset G_1$. Thus one may assume that G_1 is a K^s and $s - 1$ more edges. Furthermore, as every $G(n, t_{s-1}(n) + 1)$ contains a K^{s+1} from which an edge has been omitted (Theorem 1.2) one may suppose also that G_1 is not a K^{s+1} minus an edge.

Given any $\varepsilon > 0$ we can suppose without loss of generality that every vertex of $\bar{G}_2$ has degree at least $\{(s - 2)/(s - \varepsilon)\} n$. As $\bar{G}_2$ has at least $t_{s-1}(n) + 1$ edges, it does contain a K^s. To complete the proof it suffices to show that if $U \subset W \subset V(\bar{G}_2), |U| \leqslant s - 2$ and $|W| < 3s$ then there is a vertex $x \in V(\bar{G}_2) - W$ that is joined to every vertex in U in the graph $\bar{G}_2$. By the assumption on the degree in $\bar{G}_2$ at least $\{1 - |U|/(s - 1) - |U| \varepsilon\} n$ vertices are joined to each vertex in U. Consequently there is a suitable x provided $\varepsilon < \{(s - 1)(s - 2)\}^{-1}$ and n is sufficiently large. ∎

The theorem almost certainly holds for every n, but the proof is likely to involve the examination of some cases.

Let us turn now to the main result of this section.

THEOREM 4.4. *Let* $0 < \alpha < \frac{1}{2}$. *If n is sufficiently large,* $|G_1| = |G_2| = n$,

$$e(G_1) \leqslant \alpha n \qquad and \qquad e(G_2) \leqslant \tfrac{1}{5}(1 - 2\alpha) n^{3/2},$$

then there is a packing of G_1 and G_2.

Proof. Let us suppose that $\frac{1}{4} \leqslant \alpha < \frac{1}{2}$ and let us prove the result under the condition $e(G_2) \leqslant \frac{1}{2}c(1 - 2\alpha)n^{\frac{3}{2}}$, where $0 < c < \frac{1}{2}$.

Let P be a set of $\lfloor(1 - 2\alpha)n\rfloor = p$ isolated vertices of G_1. Choose a set of p vertices of G_2 incident with the maximum possible number of edges and put them into P. Denote by G_1^*, G_2^* the remaining graphs and put $\Delta_2 = \Delta(G_2^*)$. Then

$$|G_1^*| = |G_2^*| = m = n - p \geqslant 2\alpha n, \qquad e(G_1^*) \leqslant m/2$$

and

$$(1 - 2\alpha)n\Delta_2/2 \leqslant e(G_2), \qquad so \qquad \Delta_2 \leqslant cn^{\frac{1}{2}}.$$

If $\Delta_2 = 0$, we have nothing to prove. Suppose $\Delta_2 \geqslant 1$. Theorem VI.1.1 implies that G_2^* contains $\lceil m/(\Delta_2 + 1)\rceil$ independent vertices. Let Q be a set of $q = \lceil m/(cn^{\frac{1}{2}} + 1)\rceil$ independent vertices. Choose q vertices of G_1^* having maximum degree in G_1 and put them on Q. Let $W = V(G_1) - P - Q$.

Then

$$\Delta_1 = \max_{x \in W} d_{G_1}(x) \leqslant \left\lfloor \frac{m}{q+1} \right\rfloor.$$

Complete the identification of the vertices of G_1 and G_2 in such a way that these graphs have minimal number of common edges. Let us show that this is a packing, i.e. G_1 and G_2 have no edge in common.

Suppose otherwise. If $v \in G_1$, denote by v' the vertex of G_2 identified with v and use a similar convention for sets of vertices. Then there are vertices $x \in W = V(G_1) - P - Q$ and $y \in V(G_1) - P$ such that $xy \in E(G_1)$ and $x'y' \in E(G_2)$.

Suppose $z \in W - \{x\}$. Interchanging x' and z' the graphs G_1^* and G_2^* must have a common edge ending at x or z since otherwise we would certainly decrease the number of common edges of G_1 and G_2. Consequently with

$$Z_1 = \{z \in W: \ \exists v \in G_1^*, zv \in E(G_1), v'x' \in E(G_2)\},$$

$$Z_2 = \{z' \in W': \ \exists v \in G_1^*, z'v' \in E(G_2), vx \in E(G_1)\},$$

we must have $Z_1 \cup Z_2 \supset W$, $Z_1' \cup Z_2' \supset W'$.

As $d_{G_2^*}(x') \leqslant \Delta_2$ and $e(G_1^*) \leqslant m/2$,

$$|Z_1| \leqslant \Delta_2 + m/2.$$

Also $d_{G_1}(x) \leqslant \Delta_1$ and $d_{G_2}(v') \leqslant \Delta_2 (v \in G_1^*)$ imply

$$|Z_2| \leqslant \Delta_1 \Delta_2.$$

Thus

$$\Delta_2 + m/2 + \Delta_1 \Delta_2 \geqslant m - q,$$

$$\Delta_2 (\Delta_1 + 1) \geqslant \frac{m}{2} - q.$$

It is easily seen that this cannot hold for large values of n. ∎

If $G_1 = K^{t+1} \cup E^{n-t-1}$ and $G_2 = tK^{n/t}$ then clearly there is no packing of G_1 and G_2. This example shows that the theorem is near to being best possible (cf. Conjecture 20).

We have seen that we can pack G_1 into $\bar{G}_2$ provided that $e(G_1) + e(G_2)$ $\leqslant \lfloor \frac{3}{2}(n - 1) \rfloor$. But if we ensure that G_1 and G_2 have no vertices of degree $n - 1$ then we can pack G_1 into $\bar{G}_2$ if we only have $e(G_1) + e(G_2) \leqslant 2n - 4$. It is rather unexpected that by restricting the maximum degrees even more, say to $n - k$ for some constant k, the bound $2n - 4$, ensuring a packing, cannot be increased substantially, in fact, it cannot even be increased to $(2 + \varepsilon)n - c$. However, Theorem 3.1 may have an extension to the packing of several graphs, as conjectured by Bollobás and Eldridge (Conjecture 22).

Gyárfás and Lehel [GL1], having proved some results about packing trees into K^n, stated the interesting conjecture that given n trees of order $1, 2, \ldots, n$, we can pack them into K^n (Conjecture 23). If there is such a packing then clearly every edge of K^n is needed to accommodate the edges of the trees.

5. APPLICATIONS OF PACKING RESULTS TO COMPLEXITY

The results of this section are due to Bollobás and Eldridge [BE3] and they concern mostly oriented graphs. As in §§ 1 and 2, the oriented graphs and graphs we consider will have the fixed vertex set $V = \{1, 2, \ldots, n\}$. We use $\vec{E}$ to denote a set of ordered pairs of distinct elements of V such that at most one of $\vec{ab}$ and $\vec{ba}$ is in $\vec{E}$. Thus $\vec{G} = (V, \vec{E})$ is exactly an oriented graph with vertex set V. A set $\vec{E}$ defines in a natural way a subset $E = \{ab : \vec{ab} \in \vec{E}\}$ of $V^{(2)}$. If $N \subset V^{(2)}$ is disjoint from E then we call $\tilde{G} = (V, \vec{E}, N)$ an *oriented pregraph*; $\vec{E}$ is the set of edges and N is the set of non-edges of $\tilde{G}$. Naturally we call $\tilde{G} = (V, \vec{E}, N)$ an *extension* of $\tilde{G}' = (V, \vec{E}', N')$ iff $\vec{E}' \subset E$ and $N' \subset N$. In notation $\tilde{G}' < \tilde{G}$. If $E \cup N = V^{(2)}$ then we identify $\tilde{G} = (V, \vec{E}, N)$ with the oriented graph $\vec{G} = (V, \vec{E})$.

When defining the computational complexity $c(\mathscr{P})$ of a property $\mathscr{P}$ of oriented graphs then, as we mentioned already, the Constructor can give three answers to a probe ab, namely $\vec{ab} \in E$, $\vec{ba} \in E$ or neither. Thus if the Constructor plays with a certain strategy ψ and Algy with an algorithm ϕ, then a sequence $\tilde{G}_0 < \tilde{G}_1 < \ldots$ of oriented pregraphs will arise. Algy can decide after p probes whether the graph under construction has $\mathscr{P}$ or not iff either every oriented graph $\vec{G}$ with $\tilde{G}_p < \vec{G}$ has $\mathscr{P}$ or none of these has $\mathscr{P}$.

Before stating the first result we need some more definitions. The degree of a vertex v in $\tilde{G} = (V, \vec{E}, N)$ is defined by $d_{\tilde{G}}(v) = d_G(v)$ where G is the graph (V, E). The *total degree* of v in $\tilde{G}$ is defined by $\text{tot deg}_{\tilde{G}} v = d_{G^+}(v)$, where $G^+ = (V, E \cup N)$. A vertex v is a *sink* in $\tilde{G}$ if there is no vertex u with $\vec{vu} \in \vec{E}$

or $uv \in N$. A *source* is defined similarly. Clearly if v is a sink of an oriented graph $\vec{G}$ and $\tilde{G} < \vec{G}$ then v is a sink in $\tilde{G}$ as well. Note also that an oriented graph contains a sink iff it has a certain subgraph. In particular, the property of not containing a sink is monotone. With the usual definition of a sink this is false for directed graphs.

THEOREM 5.1. *Let $\mathscr{P}$ be the property "the oriented graph contains a sink". Then $c(\mathscr{P}) = 2n - \lfloor \log_2 n \rfloor - 2$.*

Proof. As usual, we denote by $\tilde{G}_0 < \tilde{G}_1 < \dots$ the sequence of oriented pre-graphs arising in the game. We first show that Algy can always find a sink in $2n - \lfloor \log_2 n \rfloor - 2$ probes. For the first stage he successively probes $\lfloor n/2 \rfloor$ independent pairs. After this there are at most $\lceil n/2 \rceil$ sinks in $\tilde{G}_{\lfloor n/2 \rfloor}$. Suppose at the ith stage he is left with $m_i (> 1)$ sinks. He successively probes $\lfloor m_i/2 \rfloor$ independent pairs of these vertices making sure that if there is a vertex z at which he makes no probe then he does not probe a pair xy with tot deg x > tot deg z. After $n - 1$ probes there is clearly at most one sink left, of total degree at least $\lfloor \log_2 n \rfloor$. Hence $c(\mathscr{P}) \leqslant 2n - \lfloor \log_2 n \rfloor - 2$.

Next in the role of the Constructor we describe a strategy which shows that $c(\mathscr{P}) \geqslant 2n - \lfloor \log n \rfloor - 2$. In fact Algy's most effective algorithm against this strategy will prove to be the one he used above. Let xy be the kth probe. If neither x nor y is a sink in $\tilde{G}_{k-1}$ we say yx is a non-edge. If x is a sink and y is not a sink we say yx is an edge. If both x and y are sinks and

$$\text{tot deg}_{\tilde{G}_{k-1}} x = \text{tot deg}_{\tilde{G}_{k-1}} y$$

we say $\vec{xy}$ or $\vec{yx}$ is an edge arbitrarily. If both x and y are sinks and

$$\text{tot deg}_{\tilde{G}_{k-1}} x > \text{tot deg}_{\tilde{G}_{k-1}} y$$

we say $\vec{xy}$ is an edge (i.e. we keep the sink with lower degree). Algy will waste a probe if he probes xy for x and y non-sinks.

A probe xy for x a sink and y not a sink can be made later if necessary and in fact may not be needed if at some later stage x is not a sink. Therefore we may assume that all the first probes have the form xy for x and y sinks, until there is only one sink left.

Until this stage, if the (total) degree of a sink in $\tilde{G}_k$ is l then *there are at least 2^l vertices in its component in $\tilde{G}_k$.* We prove this by induction on k. It is certainly true if $k = 0$. Suppose $k > 0$ and the sink is x and xy was the last probe at x. When this last probe was made our strategy shows that x and y were sinks of degree at least $l - 1$ and hence had at least 2^{l-1} vertices in their components. Hence x does have at least 2^l vertices in its component in $\vec{G}_k$.

When we have only one sink left, after $n - 1$ probes, and its degree is l then this shows that $2^l \leqslant n$, that is $l \leqslant \lfloor \log_2 n \rfloor$. Clearly Algy needs at least $n - l - 1$ more probes to show that the eventual graph does contain a sink. So $c(\mathscr{P}) \geqslant 2n - \lfloor \log_2 n \rfloor - 2$.   ∎

THEOREM 5.2. *The computational complexity of a non-trivial monotone property of oriented graphs with n $(\geqslant 8)$ vertices is at least*

$$2n - \lfloor \log_2 n \rfloor - 2.$$

Remark. With a little more work one may show that this bound is attained only by the properties of containing a sink (or a source) and their complements. Indeed this is probably true for all non-trivial properties.

Proof. Replacing the property by its complement we have a property $\mathscr{P}$ that is preserved under *addition* of edges. We have to show that $c(\mathscr{P}) \geqslant 2n - \lfloor \log_2 n \rfloor - 2$. We shall prove that in most cases $c(\mathscr{P}) \geqslant 2n - 4$. Let us assume that $c(\mathscr{P}) < 2n - 4$. As our aim is to show that this is hardly possible, we play the role of the Constructor, i.e. we are trying to apply an effective strategy.

The sequence of pregraphs arising is again $\tilde{G}_0 < \tilde{G}_1 < \ldots; \tilde{G}_i = (V, \vec{E}_i, N_i)$, $G_i = (V, E_i)$ and $G_i^+ = (V, E_i \cup N_i)$.

If $\mathscr{P}$ contains a graph with maximal degree less than $n - 1$, choose $\vec{G} \in \mathscr{P}$ with $\Delta(\vec{G}) < n - 1$ and minimal number of edges. If every graph with property $\mathscr{P}$ has a vertex of degree $n - 1$ choose $\vec{G} \in \mathscr{P}$ with minimal number of edges. Since $\mathscr{P}$ is non-trivial, $e = e(\vec{G}) > 0$. We divide the proof into a number of cases depending on the form of $\vec{G}$. The strategies we adopt usually take the following form. We use a rule to produce pregraphs up to some $\tilde{G}_k$ and then use a packing theorem to pack G and G_k^+ (or $G_k^+ - e$ where $e \in E_k$). We can then choose a copy of $\vec{G}$ which shares no edges with $E_k \cup N_k$ (or at most one). The strategy then chooses $\vec{ab}$ to be an edge if and only if $\vec{ab}$ is an edge of this copy of $\vec{G}$. We note that if $e \geqslant 2n - 4$ we can immediately choose a fixed copy of $\vec{G}$ and thus force Algy to make at least $2n - 4$ probes. So we may assume that $e < 2n - 4$. In particular if $\Delta(\vec{G}) = n - 1$ then $\vec{G}$ has only one vertex of degree $n - 1$.

Case 1. $\vec{G}$ has no vertex of degree $n - 1$ and $e \geqslant n - 1$. We answer non-edge to the first $2n - 4 - e$ probes. By Corollary 3.3(iii) we can pack G and G^+_{2n-4-e}. We can thus force Algy to make at least $2n - 4$ probes.

Case 2. $\vec{G}$ has no vertex of degree $n - 1$ and $e < n - 1$. If $\vec{G}$ has an isolated vertex v we can proceed as in Case 1. (If G^+_{2n-4-e} has a vertex of degree $n - 1$ we can put v on it.) So we may assume that $\vec{G}$ has no isolated vertex and hence that $e(\vec{G}) \geqslant n/2$. As before the least dense component of G is a tree with at most $(n/2) - 1$ vertices. So since G has no isolated vertex it has, in this component, a vertex v of degree 1 whose neighbour u has degree at most $(n/2) - 2$. If Algy never probes all the pairs at one vertex during the first $2n - 4 - e$ probes we can proceed as in Case 1. But if Algy does probe the last pair wz at w during the first $2n - 4 - e$ probes then we answer edge, placing v on w and u on z. We then return to the strategy of answering non-edge until Algy has made $2n - 3 - e$ probes. Notice that he cannot have probed all the pairs at any other vertex. We now wish to pack $G - v$ into $G^+_{2n-3-e} - v$ with u on a preassigned vertex. The degree of u in $G - V$ is at most $(n/2) - 3 \leqslant \frac{1}{2}\{(n - 1) - 2\}$ and the degree of u in $G^+_{2n-3-e} - v$ is

$$2n - 3 - e - (n - 1) \leqslant \frac{n}{2} - 2 \leqslant \frac{1}{2}\{(n - 1) - 2\}.$$

Since also

$$e(G - v) + e(G^+_{2n-3-e} - v) \leqslant e - 1 + 2n - 3 - e - (n - 1)$$
$$\leqslant n - 3 \leqslant \tfrac{3}{2}\{(n - 1) - 2\},$$

Theorem 3.4 shows that we can complete the packing. We can thus force Algy to make at least $2n - 4$ probes.

Case 3. $\vec{G}$ has a vertex v of degree $n - 1$ which is neither a sink nor a source and $e' = e - (n - 1) > (n/2) - 3$. We reply non edge to the first $n - e' - 3$ ($< n/2$) probes. We put v on an isolated vertex of $G^+_{n-e'-3}$. We now want to pack two graphs with $n - 1$ vertices and altogether $n - 3$ edges, so this can be done. We can thus force Algy to make at least $n - e' - 3 + e = 2n - 4$ probes.

Case 4. $\vec{G}$ has a vertex v of degree $n - 1$ which is neither a sink nor a source and $e' = 3 - (n - 1) \leqslant (n/2) - 3$. For the first $n - 2$ probes we proceed as follows. If xy is probed we say non edge unless $\max\{\text{tot deg } x, \text{tot deg } y\} < (n/2) - 2$ when we say edge arbitrarily. By considering the least dense component of G^+_{n-2} we see that G^+_{n-2} has either a vertex y of degree 0 and a vertex

z of degree at most $\frac{1}{2}(n - 1) - 1$ or a vertex y of degree 1 joined by an edge (not a non-edge) to a vertex z of degree at most $\frac{1}{2}(n - 1)$. We put v on y and a suitable neighbour of v on z and pack the rest of G and the rest of G_{n-1}^+ by Theorem 3.4. In the following probes Algy cannot find a vertex of degree $n - 1$ different from v because we can only answer edge $e' + 1$ times to probes involving another vertex. But that vertex already has degree at most $\frac{1}{2}(n - 3)$ and $\frac{1}{2}(n - 3) + e' + 1 < n - 1$. We can thus force Algy to make at least $2n - 4$ probes.

Note that this strategy need not produce a copy of $\vec{G}$ but only a graph containing $\vec{G}$ so we need the assumption that $\mathcal{P}$ is preserved under addition of edges.

Case 5. $\vec{G}$ has a sink at v and $e' = e - (n - 1) > 0$. We pick one vertex v to be the sink. If vw is probed we say $\overrightarrow{wv}$ is an edge. We say non-edge to the first $n - 3$ other probes. Corollary 3.3(iii) shows that we can then complete the packing of the rest of G. We thus force Algy to make at least $2n - 4$ probes.

We conclude that if $\vec{G}$ contains a sink then $e = n - 1$. By considering the minimal graph with property $\mathcal{P}$ which does not contain a sink we see that either it is a source or every graph with property $\mathcal{P}$ contains a sink. Similarly we conclude that if a graph does not contain a sink or a source then it does not have property $\mathcal{P}$. Since $\bar{\mathcal{P}}$ is monotone, we conclude that either $\mathcal{P}$ is the property of containing a sink (or a source) or $\mathcal{P}$ is the property of containing either a sink or a source. In the former case the theorem is true. In the latter case we can show that $c(\mathcal{P}) \geqslant 2n - \lfloor \log_2 n \rfloor - 2$ using almost the same strategy as we used for the property of containing a sink. The only difference is that if wz is probed and w is a source of degree $n - 2$ then we answer non-edge. ∎

To conclude this chapter we make use of the main packing theorem, Theorem 3.1, to obtain a lower bound on the computational complexity of *graph* properties.

THEOREM 5.3. *If $\mathcal{P}$ is a non-trivial property of graphs of order n then $c(\mathcal{P}) \geqslant 2n - 4$.*

Proof. Let us assume that this is false for $\mathcal{P}$. Exactly as in Case 1 and Case 2 of Theorem 5.2 we can conclude that if $G \in \mathcal{P}$ then $\Delta(G) = n - 1$. As in Case 5 of Theorem 5.2 we see that if $G \in \mathcal{P}$ has minimal number of edges then G

must be a star. Since the result is true if $\mathscr{P} = \{G: \Delta(G) = n - 1\}$ we may assume that there is a graph $H \notin \mathscr{P}$ with $\Delta(H) = n - 1$. Choose H with minimal number, e, of edges. We may assume that $e \leqslant 2n - 4$ for if not we choose a fixed copy of H with a vertex v such that $d_H(v) = n - 1$. Playing the role of the Constructor, we reply edge to the probe ab if one of a and b is v or if ab is one of the first $n - 3$ other pairs of H to be probed. It is easy to see that the pregraph $\tilde{G}_{2n-5}$ always has extensions in $\mathscr{P}$ and extensions in $\bar{\mathscr{P}}$ so $c(\mathscr{P}) \geqslant 2n - 4$. Consequently $n - 1 < e(H) \leqslant 2n - 4$. We choose a vertex v and always answer edge to a probe vw. We answer non-edge to the first $n - 3$ other probes. We can then use Corollary 3.3(iii) to choose a fixed copy of H with v playing the role of the vertex of H with degree $n - 1$. We carry on answering edge to probes vw and answer edge to the first $e(H) - n$ probes of other pairs which are edges of this copy of H. Consequently $e(H) = n - 1$. This contradiction shows that $c(\mathscr{P}) \geqslant 2n - 4$.   ∎

6. EXERCISES, PROBLEMS AND CONJECTURES

1. Give a strategy of the constructor for each of the following properties, proving that the properties are elusive.

  (i) The graph has exactly k edges.

  (ii) The graph is vertex symmetric.

  (iii) The graph is edge symmetric.

(Best, van Emde Boas and Lenstra [BBL1].)

2. Prove that the property $\mathscr{P}_3$ of graphs of order 7, defined in Fig. 1.2, is non-elusive. (Milner and Welsh [MW2].)

3. (Ex. 2 ctd.) Define a property $\mathscr{P}_3$ of graphs of order n for each $n \geqslant 7$ as follows. Replace the edge bc in each of the three graphs in Fig. 1.2 by a path of length $n - 6$ and let $\mathscr{P}_3 \subset G^n$ consist of all graphs isomorphic to one of these three graphs. Prove that $\mathscr{P}_3$ is not elusive.

4. Let $\mathscr{P}_3'$ be the following property of graphs G^n, $n \geqslant 7$. A graph G^n has $\mathscr{P}_3'$ iff there are vertices a, b, c such that $\Gamma(b) \cap \Gamma(c) = \varnothing$ and $\Gamma(b) \cup \Gamma(c) = V(G^n) - \{a\}$. (Note that $\mathscr{P}_3'$ is another extension of $\mathscr{P}_3$ defined on graphs of order 7, cf. Ex. 2 and Fig. 1.2.) Prove that $\mathscr{P}_3'$ is not elusive. (Milner and Welsh [MW2].)

5. Let $\mathscr{P}$ be the following property of graphs of order n: there is a vertex of degree $n - 4$ and the vertices adjacent to this vertex have degree 1. Prove that for $n \geqslant 9$ the property $\mathscr{P}$ is not elusive. (Carter [BBL1].)

6. Let $\mathscr{P}$ be the following property of graphs of order $n = 2k$: there exist two adjacent vertices, say x and y, such that if $X = \Gamma(x) - \{y\}$ and $Y = \Gamma(y) - \{x\}$ then $X \cap Y = \varnothing$, $|X| = |Y| = k - 1$ and the graph does not contain an $X - Y$ edge. Give a strategy proving that

$$c(P) \leqslant 3n^2/8 + n/4 - 1.$$

(D. J. Kleitman [BBL1].)

7. Apply Corollary 2.2 to prove that the following graph properties are elusive.

  (i) The graph is the union of a star and isolated vertices.
  (ii) The graph is connected.
  (iii) The graph has at most $k(<n)$ isolated vertices.

8. Let n be a prime. Deduce from Corollary 2.2 that the property "G contains a Hamiltonian cycle" of graphs of order n is elusive. (Best, van Emde Boas, Lenstra [BBL1].)

9. Deduce from Corollary 2.2 that the property "G contains two adjacent edges" is an elusive property of graphs of order at least 3. (Best, van Emde Boas, Lenstra [BBL1].)

10^+. Let F be a non-elusive property of the subsets of T such that $\Gamma(F)$ is transitive. Use the Möbius inversion formula to prove that the enumerating polynomial $P_T(z)$ of F is divisible by $(1 + z)^2$.

11. Deduce from the result in Ex. 10 that if $\mathscr{P}$ is a non-elusive property of graphs of order n then $|\mathscr{P}|$ is a multiple of 4. (Best, van Emde Boas, Lenstra [BBL1].)

12. CONJECTURE OF RIVEST AND VUILLEMIN. If $\mathscr{F} \subset \{0, 1\}^T$, $\varnothing \in \mathscr{F}$, $T \notin \mathscr{F}$ and $\Gamma(\mathscr{F})$, the group of permutations of T leaving $\mathscr{F}$ invariant, is transitive on T, then $\mathscr{F}$ is elusive. (Rivest and Vuillemin [RV1].)

13^+. Improve the bound $n^2/16$ given in Theorem 2.6 to $n^2/9$, provided n is sufficiently large. (Kleitman and Kwiatkowski [KK2].)

14. Deduce from Theorem 2.7 that if G_0 is a fixed graph or order r and $\mathscr{P}$ is the property of containing G_0 then

$$c(\mathscr{P}) \geqslant \tfrac{1}{2}(n - r + 1)(n - r + 3).$$

(Bollobás [B28].)

15. Give a detailed proof of each of Lemmas 2.11, 2.12 and 2.13. (Bollobás [B28].)

16. Prove the last step of the proof of Theorem 3.1, that is show that if G_1 is the disjoint union of a tree T and a star S with $|S| \geqslant |T| = p$ and $e(G_1) + e(G_2) \leqslant 2n - 3e(G_1) \leqslant e(G_2)$ then there is a packing of G_1 and G_2. (Bollobás and Eldridge [BE2].)

17^-. Let T_1 be a tree of order n and let T_2 be a tree of order $n - 1$. Deduce from Theorem 3.1 that there is a packing of T_1 and T_2 into K^n.

18. CONJECTURE (see Conjecture 4.2). If G_1 and G_2 are graphs of order n and $(\Delta(G_1) + 1)(\Delta(G_2) + 1) \leqslant n + 1$ then there is a packing of G_1 and G_2 into K^n. (Bollobás and Eldridge [BE2], [BE3a].)

19^-. Deduce from Corollary II.4.2 that Conjecture 18 is true when $\Delta(G_1) = 1$.

20. CONJECTURE. Let $0 < c < 1/\sqrt{8}$ and $0 < \alpha < \frac{1}{2}$. If n is sufficiently large and $e(G_1) \leqslant \alpha n$, $e(G_2) \leqslant (c/\sqrt{\alpha})n$ then there is a packing of G_1 and G_2. (Bollobás and Eldridge [BE2].)

21. Let $r \geqslant 4$ and let G_1 and G_2 be graphs of order n satisfying

$$\Delta(G_1) \leqslant \Delta(G_2) = n - r \text{ and } e(G_1) + e(G_2) \leqslant 2(n - r) + (r - 1)r^{\frac{1}{2}}.$$

Eldridge [E3a] proved that if $n \geqslant 9r^{\frac{1}{2}}$ and there is no packing of G_1 and G_2 then r is a perfect square, say $r = (k + 1)^2$, and G_1 and G_2 are copies of $(K^1 + E^m) \cup kK^{k+2}$, where $m = n - (k + 1)^2$.

Check that $(K^1 + E^m) \cup kK^{k+2}$ cannot be packed into its complement.

22. CONJECTURE. Given k, if n is sufficiently large, $|G_i| = n$ and $e(G_i) \leqslant n - k$ for each $i = 1, 2, \ldots, k$ then there is a packing of $G_1, G_2, \ldots, G_k$. (Bollobás and Eldridge [BE2].)

23. CONJECTURE. Let T_k be a tree of order k, $k = 1, 2, \ldots, n$. The there is a packing of $T_1, T_2, \ldots, T_n$ into K^n. (Gyárfás and Lehel [GL1].)

24. PROBLEM. Given n and p, determine the maximal integer m such that if G_1 and G_2 are graphs of order n and $e(G_1) + e(G_2) = m$, then K^n contains copies of G_1 and G_2 with at most p edges in common. (B. Bollobás and P. Erdös.)

25. Let T be a tree that is not a star. Show that it can be packed into its complement. (Eldridge [E3a].)

26^+. Let T be a tree with maximal degree Δ and order $n = 3m$. Show that if $m \geqslant \Delta - 3$ then T has an equipartite 3-colouring (a 3-colouring with equal classes), that is there is a packing of T and $3K^m$.

[Hint. T is bipartite with vertex classes A and B. If $2|A| \leqslant |B|$, consider a maximal subtree S of T such that $|S_A| + n_2 - |\Gamma(S_A)| \geqslant m$, where $S = A \cap V(S)$.] (Bollobás and Guy [GB1].)

27. CONJECTURE. Let Δ and $0 < c < \frac{1}{2}$ be given. Then there exists a constant n_0 with the following properties. If $n \geqslant n_0$, T is a tree of order n and maximal degree Δ and G is a graph of order n and maximal degree $\lfloor cn \rfloor$ then there is a packing of T and G.

28. CONJECTURE. If T is a tree of order k and $e(G^n) \geqslant \lfloor \frac{1}{2}(k - 2)n \rfloor + 1$ then $T \subset G^n$. (Erdös and Sós, see [E17].)

29. CONJECTURE. If F is a forest of size $e(F) = f$ and

$$e(G^n) \geqslant \max\left\{ \binom{2f-1}{2}, \binom{f-1}{2} + (f-1)(n-f+1) \right\} + 1$$

then $F \subset G^n$. Note that, by Corollary II.1.10, the right hand side is exactly the minimal number of edges ensuring that G^n contains f independent edges. (Erdös and Sós, see [E17].)

30. PROBLEM. Given $m \geqslant 1$, determine the minimum size of a graph that contains every graph of size m and minimum degree at least 1.

References

[A1]   Abbott, H. L.   [5.7]
An application of Ramsey's theorem to a problem of Erdös and
Hajnal, *Canad. Math. Bull.* **8** (1965) 515–517.

[A2]   Abbott, H. L.   [5.6, 5.7]
A note on Ramsey's theorem, *Canad. Math. Bull.* **15** (1972)
9–10.

[ABT1]   Albertson, M., Bollobás, B. and Tucker, S.   [5.7]
The independence ratio and maximum degree of a graph, *Proc.
Seventh S-E Conf. Combinatorics, Graph Theory and Computing,
Utilitas Math., Winnipeg*, 1976, 43–50.

[AS1]   Allaire, F. and Swart, E. R.   [5.3]
A systematic approach to the determination of reducible con-
figurations in the four-colour conjecture, *J. Combinatorial
Theory Ser.* B (to appear).

[A3]   Anderson, I.   [2.1, 2.6]
Perfect matchings in a graph, *J. Combinatorial Theory* Ser. B **10**
(1971) 183–186.

[AES1]   Andrásfai, B., Erdös, P. and Sós, V. T.   [6.1]
On the connection between chromatic number, maximal
clique and minimal degree of a graph, *Discrete Math.* **8** (1974)
205–208.

[AH1]   Appel, K. and Haken, W.   [5.3]
Every planar map is four-colorable, Part I: discharging,
University of Illinois, Urbana, 1976.

[AH2]   Appel, K. and Haken, W.   [5.3]
An unavoidable set of configurations in planar triangulations,
J. Combinatorial Theory Ser. B (to appear).

[AH3]   Appel, K. and Haken, W.   [5.3]
The existence of unavoidable sets of geographically good configurations, *Illinois J. Math.,* **20** (1976) 218–297.

[AHK1]   Appel, K., Haken, W. and Koch, J.   [5.3]
Every planar map is four-colorable, Part II: reducibility,
University of Illinois, Urbana, 1976.

[B1]   Baer, R.   [4.5]
       Polarities in finite projective planes, *Bull. Amer. Math. Soc.* **52**
       (1946) 77–93.

[BS1] Baker, B. and Shostak, R.   [4.5]
       Gossips and telephones, *Discrete Math.* **2** (1972) 191–193.

[BI1] Bannai, E. and Ito, T.   [3.1]
       On finite Moore graphs, *J. Fac. Sci. Univ. Tokyo, Sect. I A
       Math.* **20** (1973) 191–208.

[B1a] Bárány, I.   [5.4]
       A short proof of Kneser's conjecture (to appear).

[BC1] Behzad, M. and Chartrand, G.   [1.6]
       "Introduction to the Theory of Graphs", Allyn and Bacon Inc.,
       Boston, 1971.

[B2]   Beineke, L. W.   [2.6]
       Decompositions of complete graphs into forests, *Publ. Math.
       Inst. Hungar. Acad. Sci.* **9** (1964) 589–594.

[BP1] Beineke, L. W. and Plummer, M. D.   [2.2]
       On the 1-factors of a non-separable graph, *J. Combinatorial
       Theory* **2** (1967) 285–289.

[BW1] Beineke, L. W. and Wilson, R. J.   [5.7]
       On the edge-chromatic number of a graph, *Discrete Math.* **5**
       (1973) 15–20.

[B3]   Belck, H. B.   [2.3]
       Reguläre Faktoren von Graphen, *J. Reine Angew. Math.* **188**
       (1950) 228.

[B4]   Benson, C. T.   [3.1]
       Minimal regular graphs of girths eight and twelve, *Canad. J.
       Math.* **26** (1966) 1091–1094.

[B5]   Berge, B.   [2.6]
       Two theorems in graph theory, *Proc. Nat. Acad. Sci. U.S.A.* **43**
       (1957) 842.

[B6]   Berge, C.   [2.1]
       Sur le couplage maximum d'un graphe, *C.R. Acad. Sci. Paris*
       247 (1958) 258–259.

[B7]   Berge, C.   [5.0, 5.5, 5.7]
       Les problèmes de coloration en théorie des graphes, *Publ. Inst.
       Stat. Univ. Paris.* **9** (1960) 123–160.

[B8]   Berge, C.   [5.5]
       Färbung von Graphen deren sämtliche bzw. ungerade Kreise
       starr sind (Zusammenfassung), *Wiss. Z. Martin-Luther Univ.
       Halle-Wittenberg, Math.-Natur. Reihe* (1961) 114.

[B9]   Berge, C.   [5.5]
       "Sur un Conjecture Relative au Problème des Codes Opti-
       maux", Commun. 13-ème Assemblèe Gen. URSI, Tokyo
       (1962).

[B10]   Berge, C.   [4.2, 5.5]
"Graphs and Hypergraphs", North Holland, Amsterdam and London, 1973.

[BL1]   Berge, C. and Las Vergnas, M.   [5.5]
Sur un théorème du type König pour hypergraphes, *Ann. New York Acad. Sci.* **175** (1970) 32–40.

[B11]   Bermond, J. C.   [3.4]
Thesis, University of Paris XI, Orsay 1975.

[B12]   Bermond, J. C.   [3.4, 3.6]
On Hamiltonian walks, *in*: "Proc. Fifth British Combinatorial Conf." (Nash-Williams, C. St. J, A. and Sheehan, J., eds), Utilitas Math., Winnipeg, 1976, pp 41–51.

[B13]   Bermond, J. C.   [4.5]
The telephone disease (to appear).

[B14]   Bernhart, A.   [5.3]
Six rings in minimal five color maps, *Amer. J. Math.* **69** (1947) 391–412.

[B15]   Bernhart, A.   [5.3]
Another reducible edge configuration, *Amer. J. Math.* **70** (1948) 144–146.

[B16]   Bernhart, F. R.   [5.3]
On the characterization of reductions of small order (to appear).

[BBL1]   Best, M. R., van Emde Boas, P and Lenstra, Jr., H. W.   [8.1, 8.2, 8.6]
"A sharpened version of the Aanderaa–Rosenberg conjecture", Math. Centrum, Amsterdam, 1974.

[BLW1]   Biggs, N. L., Lloyd, E. K. and Wilson, R. J.   [5.3]
"Graph Theory 1736–1937", Clarendon Press, Oxford, 1976.

[B17]   Birkhoff, G. D.   [5.3]
The reducibility of maps, *Amer. J. Math.* **35** (1913) 114–128.

[B18]   Bollobás, B.   [3.2]
Graphs without two independent cycles (in Hungarian), *Mat. Lapok* **14** (1963) 311–321.

[B19]   Bollobás, B.   [6.2]
On generalized graphs, *Acta Math. Acad. Sci. Hungar.* **16** (1965) 447–452.

[B20]   Bollobás, B.   [1.5]
On graphs with at most three independent paths connecting any two vertices, *Studia Sci. Math. Hungar.* **1** (1966) 137–140.

[B21]   Bollobás, B.   [6.2]
On a conjecture of Erdös, Hajnal and Moon, *Amer. Math. Monthly* **74** (1967) 178–179.

[B22]   Bollobás, B.   [6.2, 6.6]
Determination of extremal graphs by using weights, *Wiss. Z. Hochsch. Ilmenau* **13** (1967) 419–421.

[B23]   Bollobás, B.   [6.6]
Weakly *k*-saturated graphs *in*: "Beiträge zur Graphentheorie"

(Sachs, H., Voss, H.-J. and Walther, H., eds), Leipzig, 1968, pp. 25–31.

[B24] Bollobás, B.   [4.2, 4.5]
A problem in the theory of communication networks, *Acta. Math. Acad. Sci. Hungar.* **19** (1968) 75–80.

[B25] Bollobás, B.   [4.2, 4.3, 4.5]
Graphs of given diameter, *in*: "Theory of Graphs" (Erdös, P. and Katona, G., eds), Academic Press, New York, 1968, pp. 29–36.

[B26] Bollobás, B.   [4.1, 4.5]
Graphs with a given diameter and maximal valency and with a minimal number of edges, *in*: "Combinatorial Mathematics and its Applications" (Welsh, D. J. A., ed.), Academic Press, London and New York, 1971, pp. 25–37.

[B27] Bollobás, B.   [6.1]
On complete subgraphs of different orders, *Math. Proc. Cambridge Philos. Soc.* **79** (1976) 19–24.

[B28] Bollobás, B.   [8.6, 8.2]
Complete subgraphs are elusive, *J. Combinatorial Theory* Ser. B **21** (1976) 1–7.

[B29] Bollobás, B.   [6.1]
Relations between sets of complete subgraphs, *in*: "Proc. Fifth British Combinatorial Conf." (Nash-Williams, C. St. J. A. and Sheehan, J., eds), Utilitas Math., Winnipeg, 1976, pp. 79–84.

[B30] Bollobás, B.   [4.4, 6.1]
Extremal problems in graph theory, *J. Graph Theory* **1** (1977) 117–123.

[B31] Bollobás, B.   [4.0, 4.2, 4.3]
Graphs with given diameter and minimal degree, *Ars Combinatoria* **2** (1976) 3–9

[B32] Bollobás, B.   [5.1]
Uniquely colorable graphs. *J. Combinatorial Theory* Ser. B **24** (1978).

[B33] Bollobás, B.   [4.0, 4.2]
Strongly two-connected graphs, *in*: "Proc. Seventh S-E Conf. Combinatorics, Graph Theory and Computing", Utilitas Math., Winnipeg, 1976, pp. 161–170.

[B34] Bollobás, B.   [7.3]
Cycles modulo k, *Bull. London Math. Soc.* **9** (1977) 97–98.

[B35] Bollobás, B.   [5.7]
Colouring lattices, *Algebra Universalis* **7** (1977) 313–314.

[B36] Bollobás, B.   [2.2]
The number of 1-factors in $2k$-connected graphs, *J. Combinatorial Theory* Ser. B (to appear).

[B37] Bollobás, B.   [7.3]
Semi-topological subgraphs, *Discrete Math.* **20** (1977) 83–85.

[B38]   Bollobás, B.   [1.5, 2.5]
         Cycles and semi-topological configurations, *in*: "Theory and
         Applications of Graphs" (Y. Alavi and D. R. Lick, eds) Lecture
         Notes in Maths 642, Springer, 1978, pp. 66–74.

[B39]   Bollobás, B.   [7.3, 7.4]
         Nested cycles in graphs, *in*: "Proc. Colloque Intern. CNRS"
         (J.-C. Bermond, J.-C. Fournier, M. das Vergnas and D.
         Sotteau, eds.), 1978.

[BE1]   Bollobás, B. and Eldridge, S. E.   [2.4]
         Maximal matchings in graphs with given minimal and maximal
         degrees, *Math. Proc. Cambridge Philos. Soc.* **79** (1976) 221–234.

[BE2]   Bollobás, B. and Eldridge, S. E.   [81, 8.3, 8.4, 8.6]
         Packings of graphs and applications to computational com-
         plexity, *J. Combinatorial Theory* Ser B **24** (1978).

[BE3]   Bollobás, B. and Eldridge, S. E.   [8.4]
         Problem, *in*: "Proc. Fifth British Combinatorial Conf."
         (Nash-Williams, C. St. J. A. and Sheehan, J., eds.), Utilitas
         Math., Winnipeg, 1976, pp. 689–691.

[BE3a]   Bollobás, B. and Eldridge, S. E.   [8.4, 8.6]
         Problem, *in*: "Proc. Colloque Intern. CNRS" (J.-C. Bermond,
         J.-C. Fournier, M. das Vergnas and D. Sotteau, eds.), 1978.

[BE4]   Bollobás, B. and Eldridge, S. E.   [4.2, 4.3,
         On graphs with diameter 2, *J. Combinatorial Theory* Ser B   4.5]
         21 (1976) 201–205.

[BE5]   Bollobás, B. and Erdös, P.   [7.3]
         On some extremal problems in graph theory (in Hungarian),
         Mat. Lapok **13** (1962) 143–153.

[BE6]   Bollobás, B. and Erdös, P.   [6.0, 6.3]
         On the structure of edge graphs, *Bull. London Math. Soc.* **5**
         (1973) 317–321.

[BE7]   Bollobás, B. and Erdös, P.   [6.6]
         On a Ramsey–Turán type problem, *J. Combinatorial Theory*
         Ser B **21** (1976) 166–168.

[BE8]   Bollobás, B. and Erdös, P.   [5.7]
         Alternating Hamiltonian cycles, *Israel J. Math.* **23** (1976)
         126–131.

[BE9]   Bollobás, B. and Erdös, P.   [4.2, 4.3, 4.5]
         An extremal problem of graphs with diameter 2, *Math. Mag.*
         48 (1975) 281–283.

[BE10]   Bollobás, B. and Erdös, P.   [5.7]
         Cliques in random graphs, *Math. Proc. Cambridge Philos. Soc.*
         80 (1976) 419–427.

[BES1]   Bollobás, B., Erdös, P. and Simonovits, M.   [6.0, 6.3]
         On the structure of edge graphs II, *J. London Math. Soc.* **12** (2)
         (1976) 219–224.

[BESS1] Bollobás, B., Erdös, P., Simonovits, M. and Szemerédi, E. [6.4, 6.6]

Extremal graphs without large forbidden subgraphs, *in*: "Advances in Graph Theory" (Bollobás, B., ed.), North-Holland, 1978, pp. 29–41.

[BES2] Bollobás, B., Erdös, P. and Straus, E. G.   [6.2, 6.6]
Complete subgraphs of chromatic graphs and hypergraphs, *Utilitas Math.* **6** (1974) 343–347.

[BES3] Bollobás, B., Erdös, P. and Szemerédi, E.   [6.2, 6.6]
On complete subgraphs of r-chromatic graphs, *Discrete Math.* **13** (1974) 97–107.

[BGW1] Bollobás, B., Goldsmith, D. L. and Woodall, D. R.   [2.3]
Indestructive deletions of edges from graphs (to appear).

[BG1] Bollobás, B. and Guy, R. K.   [8.6]
Equipartite and proportional colorings of trees (to appear).

[BH1] Bollobás, B and Harary, F.   [5.7]
Point arboricity critical graphs exist, *J. London Math. Soc.* **12** (2) (1975) 97–102.

[BH2] Bollobás, B. and Harary, F.   [4.2, 4.3]
Extremal graphs with given diameter and connectivity, *Ars Combinatoria* **1** (1976) 281–296.

[BH3] Bollobás, B. and Hobbs, A. M.   [3.4]
Hamiltonian cycles in regular graphs, *in*: "Advances in Graph Theory" (Bollobás, B., ed.), North-Holland, 1978, pp. 43–48.

[BM1] Bollobás, B. and Manvel, B.   [5.7]
Optimal partitions of graphs, to appear.

[BS2] Bollobás, B. and Sauer, N.   [5.4]
Uniquely colourable graphs with large girth, *Canad. J. Math.* **28** (1976) 1340–1344.

[BT1] Bollobás, B. and Thomason, A. G.   [5.4, 5.7]
Uniquely partitionable graphs, *J. London Math. Soc.* (2) **16** (1977) 403–410.

[BT2] Bollobás, B. and Thomason, A. G.   [5.7]
Set colourings of graphs, *Discr. Math.* (to appear).

[BV1] Bollobás, B. and Varopoulos, N. Th.   [2.6]
Representation of systems of measurable sets, *Math. Proc. Cambridge Philos. Soc.* **78** (1974) 323–325.

[B40] Bondy, J. A.   [5.1]
Bounds for the chromatic number of a graph, *J. Combinatorial Theory* **7** (1969) 96–98.

[B41] Bondy, J. A.   [3.4, 1.1]
Properties of graphs with constraints on the degrees, *Studia Sci. Math. Hungar.* **4** (1969) 473–475.

[B42] Bondy, J. A.   [3.4]
Large cycles in graphs, "Proc. Louisiana Conf. on Combinatorics, Graph Theory and Computing" (Mullin, R. C., Reid, K. B. and Roselle, D. P., eds), 1970, pp. 47–60.

[B43]   Bondy, J. A.   [3.4, 3.5]
          Large cycles in graphs, *Discrete Math.* 1 (1971) 121–132.

[B44]   Bondy, J. A.   [3.5]
          Pancyclic graphs I, *J. Combinatorial Theory* Ser B 11 (1971)
          80–84.

[B45]   Bondy, J. A.   [1.6]
          Induced subsets, *J. Combinatorial Theory* Ser B 12 (1972)
          201–202.

[B46]   Bondy, J. A.   [3.6]
          Variation on the Hamiltonian theme, *Canad. Math. Bull.* 15
          (1972) 57–62.

[B47]   Bondy, B. A.   [5.1]
          Disconnected orientations and a conjecture of Las Vergnas,
          J. London Math. Soc. 14 (1976) 277–282.

[BC2]   Bondy, J. A. and Chvatal, V.   [3.4, 3.6]
          A method in graph theory, *Discrete Math.* 15 (1976) 111–135.

[BE11]   Bondy, J. A. and Erdös, P.   [5.7]
          Ramsey numbers for cycles in graphs, *J. Combinatorial Theory*
          Ser B 14 (1973) 46–54.

[BM1]   Bondy, J. A. and Murty, U. S. R.   [4.2]
          Extremal graphs of diameter two with prescribed minimum
          degree, *Studia Sci. Math. Hungar.* 7 (1972) 239–241.

[BM2]   Bondy, J. A. and Murty, U. S. R.   [5.6]
          "Graph Theory with Applications", MacMillan Press,
          London, 1976.

[BS3]   Bondy, J. A. and Simonovits, M.   [3.5]
          Cycles of even length in graphs, *J. Combinatorial Theory* Ser B
          16 (1974) 97–105.

[BK1]   Borodin, O. V. and Kostochka, A. V.   [5.7]
          On an upper bound of the graph's chromatic number depend-
          ing on the graph's degree and density, *J. Combinatorial Theory*
          Ser B 23 (1977) 247–250.

[B48]   Bosák, J.   [4.4, 4.5]
          Disjoint factors of diameter two in complete graphs, *J. Combi-
          natorial Theory* Ser B 16 (1974) 57–63.

[BER1]   Bosák, J., Erdös, P. and Rosa, A.   [4.4, 4.5]
          Decomposition of complete graphs into factors with diameter
          two, *Mat. Casopis Sloven. Akad.* 21 (1971) 14–28.

[BRZ1]   Bosák, J., Rosa, A. and Znám, S.   [4.4, 4.5]
          On the decomposition of complete graphs into factors with
          given diameters, *in*: "Theory of Graphs", (Erdös, P. and
          Katona, G., eds), Academic Press, New York, 1968, pp. 37–56.

[B49]   Brooks, R. L.   [5.0, 5.1]
          On colouring the nodes of a network, *Proc. Cambridge Philos.
          Soc.* 37 (1941) 194–197.

[B50]   Brown, W. G.   [3.1, 3.6]

On Hamiltonian regular graphs of girth six, *J. London Math. Soc.* **42** (1967) 514–520.

[B51] Brown, W. G. [6.2]
On graphs that do not contain a Thomsen graph, *Canad. Math. Bull.* **9** (1966) 281–285.

[B52] Brown, W. G. [3.2]
A new proof of a theorem of Dirac, *Canad. Math. Bull.* **8** (1965) 459–463.

[B53] Brown, W. G. [3.1, 3.6]
On the non-existence of a type of regular graphs of girth 5, *Canad. J. Math.* **19** (1967) 644–648.

[BM3] Brown, W. G. and Moon, J. W. [5.2]
Sur les ensembles de sommets indépendants dans les graphes chromatiques minimaux, *Canad. J. Math.* **21** (1969) 274–278.

[BE12] de Bruijn, N. G. and Erdös, P. [5.4]
A colour problem for infinite graphs and a problem in the theory of relations, *Indag. Math.* **13** (1951) 371–373.

[B54] Burr, S. A. [5.6]
Generalized Ramsey theory for graphs—a survey, *in:* "Graphs and Combinatorics" (Bari, R. A. and Harary, F.,eds), Lecture Notes in Mathematics 406, Springer, 1974, pp. 52–75.

[B55] Busolini, D. T. [5.1]
Monochromic paths and circuits in edge-colored graphs, *J. Combinatorial Theory* Ser. B **10** (1971) 299–300.

[B56] Busolini, D. T. [6.6]
"Some Extremal Problems in Graph Theory", Thesis, Reading, 1976.

[B57] Butler, J. [3.4]
Hamiltonian circuits on simple 3-polytopes, *J. Combinatorial Theory* Ser. B **21** (1976) 104–115.

[C1] Caccetta, L. [4.2]
Extremal biconnected graphs of diameter 4, *J. Combinatorial Theory* Ser. B **21** (1976) 104–115.

[C2] Cambridge Mathematical Tripos 1974 [7.3]
Part II, Paper I, No. 10.

[C3] Cambridge Mathematical Tripos 1974 [7.4]
Part II, Paper II, No 11.

[CH1] Cartwright, D. and Harary, F. [5.1]
On the coloring of signed graphs, *Elem. Math.* **23** (1968) 85–89.

[C4] Catlin, P. A. [8.4]
Subgraphs of graphs, I. *Discrete Math.* **10** (1974) 225–233.

[C5] Catlin, P. A. [5.7]
Embedding subgraphs and coloning graphs under extremal degree conditions, Ph.D. Dissertation, Ohio State University, 1976.

[C6]   Catlin, P. A.   [5.7]
          A bound on the chromatic number of a graph, *Discrete Math.*
          22 (1978) 81–83.

[C7]   Chartrand, G.   [1.1]
          A graph-theoretic approach to a communications problem,
          SIAM J. Appl. Math. **14** (1966) 778–781.

[CG1]   Chartrand, G. and Geller, D.   [5.1]
          Uniquely colorable planar graphs, *J. Combinatorial Theory* **6**
          (1969) 271–278.

[CGH1]   Chartrand, G., Geller, D. and Hedetniemi, S.   [7.4]
          Graphs with forbidden subgraphs, *J. Combinatorial Theory*
          Ser. B. **10** (1971) 12–41.

[CH2]   Chartrand, G. and Harary, F.   [7.4]
          Planar permutation graphs, *Ann. Inst. H. Poincaré* Sect. B 3
          (1967) 433–438

[CH3]   Chartrand, G and Harary, F.   [1.1]
          Graphs with prescribed connectivities, *in*: "Theory of Graphs"
          (Erdös, P. and Katona, G., eds), Academic Press, New York,
          1968, pp. 61–63.

[CKL1]   Chartrand, G., Kapoor, S. F. and Lick, D. R.   [3.6]
          n-Hamiltonian graphs, *J. Combinatorial Theory* **9** (1970) 308–
          312.

[CKL2]   Chartrand, G., Kaugars, A. and Lick, D. R.   [1.6]
          Critically *n*-connected graphs, *Proc. Amer. Math. Soc.* **32** (1972)
          63–68.

[CK1]   Chartrand, G. and Kronk, H. V.   [5.7]
          The point arboricity of planar graphs, *J. London Math. Soc.* **44**
          (1969) 612–616.

[CD1]   Chen, C. C. and Daykin, D. E.   [5.7]
          Graphs with Hamiltonian cycles having adjacent lines different
          colors, *J. Combinatorial Theory* Ser. B **21** (1976) 135–139.

[C8]   Chvátal, V.   [3.4]
          On Hamilton's ideals, *J. Combinatorial Theory* Ser. B **12** (1972)
          163–168.

[C9]   Chvátal, V.   [5.1]
          Monochromatic paths in edge-colored graphs, *J. Combina-
          torial Theory* Ser. B **13** (1972) 69–70.

[CE1]   Chvátal, V. and Erdös, P.   [3.6]
          A note on Hamiltonian circuits, *Discrete Math.* **2** (1972)
          111–113.

[CH4]   Chvátal, V. and Hanson, D.   [2.4]
          Degrees and matching, *J. Combinatorial Theory* Ser. B **12**
          (1976) 128–138.

[CH5]   Chvátal, V. and Harary, F.   [5.6]
          Generalized Ramsey theory for graphs, *Bull. Amer. Math.
          Soc.* **78** (1972) 423–426.

[CH6]   Chvátal, V. and Harary, F.   [5.6]
        Generalized Ramsey theory for graphs, I. Diagonal numbers, *Period. Math. Hungar.* **3** (1973) 115–124.

[CK2]   Chvátal, V. and Komlós, J.   [5.1]
        Some combinatorial theorems on monotonicity, *Canad. Math. Bull.* **14** (1971) 151–157.

[C10]   Clapham, C. R. J.   [6.6]
        Triangles in self-complementary graphs, *J. Combinatorial Theory* Ser. B **15** (1973) 74–76.

[C10a]   Cockayne, E. J.   [5.6]
        An application of Ramsey's theorem, *Canad. Math. Bull.* **13** (1970) 145–146

[C11]   Corrádi, K.   [6.5]
        On a problem concerning finite graphs, *Annals Univ. Sci. Budapest, Eötvös Sect. Math.* **9** (1966) 157–165.

[CH7]   Corrádi, K. and Hajnal, A.   [6.0, 6.5, 8.4]
        On the maximal number of independent circuits in a graph, *Acta Math. Acad. Sci. Hungar.* **14** (1963) 423–439.

[C12]   Čulik, K.   [6.2]
        Poznámka k problému K. Zarankiewicze, *Práce Brnenské Základny CSAV* **26** (1955) 341–348.

[C13]   Čulik, K.   [6.6]
        Teilweise Lösung eines verallgemeinerten Problems von K. Zarankiewicze, *Ann. Polon. Math.* **3** (1956) 165–168.

[C14]   Czipszer, J.   [7.3]
        Solution to Problem 127 (in Hungarian), *Mat. Lapok* **14** (1963) 373–374.

[D1]   Damerell, R. M.   [3.1]
        On Moore graphs, *Proc. Cambridge Philos. Soc.* **74** (1973) 227–236.

[DH1]   Dantzig, G. B. and Hoffman, A. J.   [2.1]
        "Dilworth's theorem on partially ordered sets, linear inequalities and related systems", Annals of Math. Studies No. 38, Princeton Univ. Press, Princeton, New York, 1956, pp. 207–214.

[D2]   Deuber, W.   [5.6]
        A generalization of Ramsey's theorem, *in*: "Infinite and Finite Sets", vol. I (Hajnal, A., Rado, R. and Sós, V. T., eds), Coll. Math. Soc. Bolyai 10, North Holland, 1975, pp. 323–332.

[D3]   Descartes, B.   [5.0, 5.4]
        A three colour problem, *Eureka*, April 1947, Solution March 1948.

[D4]   Descartes, B.   [5.4]
        Solution to Advanced Problem No. 4526, proposed by Ungar, P., *Amer. Math. Monthly,* **61** (1954) 352.

[D5]   Dilworth, R. P.   [2.1]
        A decomposition theorem for partially ordered sets, *Ann. of Math.* **51** (1950) 161–166.

[D6] Dirac, G. A. [5.0, 5.2, 5.3]
A property of 4-chromatic graphs and remarks on critical
graphs, *J. London Math. Soc.* **27** (1952) 85–92.

[D7] Dirac, G. A. [3.4, 5.2, 6.5]
Some theorems on abstract graphs, *Proc. London Math. Soc.*
2 (3) (1952) 69–81.

[D8] Dirac, G. A. [5.2]
Circuits in critical graphs, *Monatsh. Math.* **59** (1955) 178–187.

[D9] Dirac, G. A. [5.0, 5.2]
A theorem of R. L. Brooks and a conjecture of H. Hadwiger,
Proc. London Math. Soc. **7** (1957) 161–195.

[D10] Dirac, G. A. [5.3]
Map colour theorems related to the Heawood colour formula
II, *J. London Math. Soc.* **32** (1957) 436–455.

[D11] Dirac, G. A. [5.3]
Short proof of a map-colour theorem, *Canad. J. Math.* **9** (1957)
225–226.

[D12] Dirac, G. A. [1.2]
Généralisations du théorème de Menger, *C.R. Acad. Sci. Paris,*
250 (1960) 4252–4253.

[D13] Dirac, G. A. [1.2, 7.1, 7.2]
In abstrakten Graphen vorhandene vollständige 4-Graphen
und ihre Unterteilungen, *Math. Nach.* **22** (1960) 61–85.

[D14] Dirac, G. A. [3.6]
Note on the structure of graphs, *Canad. Math. Bull.* **5** (1962)
221–227.

[D15] Dirac, G. A. [1.5, 2.6]
Extensions of Menger's theorem, *J. London Math. Soc.* **38** (3)
(1963) 148–161.

[D16] Dirac, G. A. [6.1, 6.6]
Extensions of Turán's theorem on graphs, *Acta Math. Sci.
Hungar.* **14** (1963) 418–422.

[D17] Dirac, G. A. [7.1, 7.4]
On the four-colour conjecture, *Proc. London Math. Soc.* **13** (3)
(1963) 193–218.

[D18] Dirac, G. A. [3.2, 7.1, 7.2, 7.4]
Some results concerning the structure of graphs, *Canad. Math.
Bull.* **6** (1963) 183–210.

[D19] Dirac, G. A. [6.6]
Extensions of theorems of Turán and Zarankiewicz, *in*:
"Theory of Graphs and its Applications" (Fiedler, M., ed.),
Academic Press, New York, 1965, pp. 127–132.

[D20] Dirac, G. A. [7.1, 5.7]
On the structure of 5- and 6-chromatic graphs, *J. Reine Angew.
Math.* **214/215** (1964) 43–52.

[D21] Dirac, G. A. [7.2]

Chromatic number and topological complete subgraphs, *Canad. Math. Bull.* **8** (1965) 711–715.

[D22]   Dirac, G. A.   [1.2, 1.6]
Short proof of Menger's graph theorem, *Mathematika,* **13** (1966) 42–44.

[D23]   Dirac, G. A.   [1.3, 1.6]
Minimally 2-connected graphs, *J. Reine Angew Math.* **228** (1967) 204–216.

[D24]   Dirac, G. A.   [3.4]
On Hamiltonian circuits and Hamiltonian paths, *Math. Ann.* **197** (1972) 57–70.

[D25]   Dirac, G. A.   [5.2]
The number of edges in critical graphs, *J. Reine Angew. Math.* **268/269** (1974) 150–164.

[D26]   Dirac, G. A.   [3.2]
Structural properties and circuits in graphs, *in:* "Proc. Fifth British Combinatorial Conf." (Nash-Williams, C. St. J. A. and Sheehan, J., eds), Utilitas Math., Winnipeg, 1976, pp. 135–140.

[DE1]   Dirac, G. A. and Erdös, P.   [6.6]
On the maximal number of independent circuits in a graph, *Acta Math. Acad. Sci. Hungar.* **14** (1963) 79–94.

[DS1]   Dirac, G. A. and Schuster, S.   [5.3]
A theorem of Kuratowski, *Indag. Math.* **16** (1954) 343–348.

[D27]   Dugundji, J.   [5.4]
"Topology", Allyn and Bacon, Boston, 1966.

[DHM1] Dürre, K., Heesch, H. and Miehe, F.   [5.3]
Eine Figurenliste zur chromatischen Reduktion, Institut für Mathematik, Technische Universität Hannover, Research Paper No. 73, 1977.

[E1]   Edmonds, J.   [2.1]
Paths, trees and flowers, *Canad. J. Math.* **17** (1965) 449–467.

[E2]   Edmonds, J.   [2.5]
Matroid partition, *in:* "Mathematics of the Decision Sciences, Part I" (Dantzig, G. and Veinott, A., eds), Amer. Math. Soc., Providence, R.I., 1968, pp. 335–345.

[E3]   Eggleton, R. B.   [2.3]
Graphic sequences and graphic polynomials; a report, *in:* "Infinite and Finite Sets", vol. I (Hajnal, A., Rado, R. and Sós, V. T., eds), Colloq. Math. Soc. J. Bolyai 10, North-Holland, 1975, pp. 385–292.

[E3a]   Eldridge, S. E.   [8.6]
Packings of Graphs, Ph.D. Thesis, Cambridge, 1976.

[E4]   Erdös, P.   [6.2, 6.6]
On sequences of integers no one of which divides the product of two others and one some related problems, *Mitteilungen des Forschungsinstitutes für Math. und Mechanik,* Tomsk 2 (1938) 74–82.

450   REFERENCES

[E5]   Erdös, P.   [6.6]
Some remarks on the theory of graphs, *Bull. Amer. Math. Soc.*
(1947) 292–294.

[E6]   Erdös, P.   [6.1]
Some theorems on graphs (in Hebrew, English summary),
Riveon Lematematika **9** (1955) 13–17.

[E7]   Erdös, P.   [5.4]
Graph theory and probability, *Canad. J. Math.* **11** (1959) 34–38.

[E8]   Erdös, P.   [3.5, 5.4, 5.6]
Graph theory and probability II, *Canad. J. Math.* (1961)
346–352.

[E9]   Erdös, P.   [6.1]
On a theorem of Rademacher–Turán, *Illinois J. Math.* **6** (1962)
122–127.

[E9a]   Erdös, P.   [6.4]
Über ein Extremalproblem in der Graphentheorie, *Archiv
Math.* **13** (1962) 222–227.

[E10]   Erdös, P.   [5.4]
On circuits and subgraphs of chromatic graphs, *Mathematika* **9**
170–175.

[E11]   Erdös, P.   [6.1, 6.6]
On the number of complete subgraphs contained in certain
graphs, *Publ. Math. Inst. Hungar. Acad. Sci.* **7** (1962) 459–464.

[E12]   Erdös, P.   [3.4]
Remarks on a paper of Pósa, *Publ. Math. Inst. Hungar. Acad.
Sci.* **7** (1962) 227–229.

[E13]   Erdös, P.   [3.5]
On the structure of linear graphs, *Israel J. Math.* **1** (1963)
156–160.

[E14]   Erdös, P.   [8.6]
Extremal problems in graph theory, *in*: "Theory of Graphs and
its Applications" (Fiedler, M., ed.), Academic Press, New York,
1965, pp. 29–36.

[E15]   Erdös, P.   [6.6]
On the number of triangles contained in certain graphs, *Canad.
Math. Bull.* **7** (1964) 53–56.

[E16]   Erdös, P.   [0]
On some extremal problems in graph theory, *Israel J. Math.* **3**
(1965) 113–116.

[E17]   Erdös, P.   [3.5, 8.6]
Extremal problems in graph theory, *in*: "Theory of Graphs and
its Applications" (Fiedler, M., ed.), Academic Press, New
York, 1965, pp. 29–36.

[E18]   Erdös, P.   [6.3]
On extremal problems of graphs and generalized graphs,
Israel J. Math. **2** (1965) 183–190.

[E19] Erdös, P.   [6.5]
       Extremal problems in graph theory, *in*: "A Seminar in Graph
       Theory" (Harary, F., ed.), Holt, Rinehart and Winston, 1967,
       pp. 54–64.

[E20] Erdös, P.   [6.6]
       Some remarks on chromatic graphs, *Colloq. Math.* **16** (1967)
       253–256.

[E21] Erdös, P.   [6.4]
       Some recent results on extremal problems in graph theory, *in*:
       "Theory of Graphs" (Rosenstiehl, P., ed.), Gordon and Breach,
       New York, 1967, pp. 117–130.

[E22] Erdös, P.   [6.0, 6.4]
       On some inequalities concerning extremal properties of
       graphs, *in*: "Theory of Graphs" (Erdös, P. and Katona, G.,
       eds), Academic Press, New York, 1968, pp. 77–81.

[E23] Erdös, P.   [6.6]
       On some applications of graph theory to number theoretic
       problems, *Publ. Ramanujan Inst.* **1** (1969) 131–136.

[E24] Erdös, P.   [6.1]
       On the number of complete subgraphs and circuits contained
       in graphs, *Casopis Pest. Mat.* **94** (1969) 290–296.

[E25] Erdös, P. Some applications of graph theory to number theory,
       in: "The Many Facets of Graph Theory", Lecture Notes in
       Maths 110, Springer, 1969, pp. 77–82.

[E26] Erdös, P.   [6.1]
       On the graph theorem of Turán (in Hungarian), *Mat. Lapok*
       21 (1970) 249–251.

[E27] Erdös, P.   [5.4, 5.6]
       Some new applications of probability methods to combina-
       torial analysis and graph theory, *in*: "Proc. Fifth S-E Conf.
       Combinatorics, Graph Theory and Computing", Utilitas
       Math., Winnipeg, 1974, pp. 39–51.

[E28] Erdös, P.   [2.6, 3.6, 6.6, 7.4]
       Some recent progress on extremal problems in graph theory, *in*:
       "Proc. Sixth S-E Conf. Combinatorics, Graph Theory and
       Computing", Utilitas Math., Winnipeg, 1975, pp. 3–14.

[E29] Erdös, P.   [5.7, 6.6, 7.3, 7.4]
       Problems and results in graph theory and combinatorial
       analysis, *in*: "Proc. Fifth British Combinatorial Conf." (Nash-
       Williams, C. St. J. A. and Sheehan, J. eds), Utilitas Math.,
       Winnipeg, 1976, pp. 169–192.

[E30] Erdös, P.   [7.3]
       Some recent problems and results in graph theory, combina-
       torics and number theory, *in*: "Proc. Seventh S-E Conf.
       Combinatorics, Graph Theory and Computing", Utilitas
       Math., Winnipeg, 1976, pp. 3–14.

[EG1]   Erdös, P. and Gallai, T.   [3.4]
On maximal paths and circuits of graphs, *Acta Math. Acad. Sci. Hungar.* **10** (1959) 337–356.

[EG2]   Erdös, P. and Gallai, T.   [2.3, 2.6]
Graphs with prescribed degrees of vertices (in Hungarian), *Mat. Lapok* **11** (1960) 264–274.

[EG3]   Erdös, P. and Gallai, T.   [2.1]
On the minimal number of vertices representing the edges of a graph, *Publ. Math. Inst. Hungar. Acad. Sci.* **6** (1961) 181–203.

[EGP1]   Erdös, P., Goodman, A. and Pósa, L.   [6.1]
The representation of graphs by set intersections, *Canad. J. Math.* **18** (1966) 106–112.

[EH1]   Erdös, P. and Hajnal, A.   [7.4]
On complete topological subgraphs of certain graphs, *Ann. Univ. Sci. Budapest, Eötvös Sect. Math.* **7** (1964) 143–149.

[EH2]   Erdös, P. and Hajnal, A.   [5.6]
Some remarks on set theory IX, Combinatorial problems in measure theory and set theory, *Michigan Math. J.* **11** (1964) 107–127.

[EH3]   Erdös, P. and Hajnal, A.   [5.1]
On chromatic numbers of graphs and set systems, *Acta Math. Sci. Hungar.* **17** (1966) 61–99.

[EH4]   Erdös, P. and Hajnal, A.   [5.6]
Research Problem 2.5, *J. Combinatorial Theory* **2** (1967) 104.

[EH5]   Erdös, P. and Hajnal, A.   [5.4]
On chromatic graphs (in Hungarian), *Mat. Lapok* **18** (1967) 1–4.

[EHM1]   Erdös, P., Hajnal, A. and Moon, J. W.   [6.1]
A problem in graph theory, *Amer. Math. Monthly* **71** (1964) 1107–1110.

[EHP1]   Erdös, P., Hajnal, A. and Pósa, L.   [5.6]
"Strong Embeddings of Graphs Into Colour Graphs, Infinite and Finite Sets, Vol. I (Hajnal, A., Rado, R. and Sós, V.T., eds), Colloq. Math. Soc. J. Bolyai 10, North Holland, 1975, pp. 585–595.

[EH6]   Erdös, P. and Hobbs, A. M.   [3.4]
Hamiltonian cycles in regular graphs of moderate degree, *J. Combinatorial Theory* Ser B **23** (1977) 139–142.

[EK1]   Erdös, P. and Kelly, P.   [2.6]
The minimal regular graph containing a given graph, *Amer. Math. Monthly* **70** (1963) 1074–1075.

[EK2]   Erdös, P. and Kelly, P.   [2.6]
The minimal regular graph containing a given graph, *in*: "A Seminar in Graph Theory" (Harary, F., ed.), Holt, Rinehart and Winston, 1967, pp. 65–69.

[EP1]   Erdös, P. and Pósa, L.   [2.4, 3.2, 3.3]
On the maximal number of disjoint circuits of a graph, *Publ. Math. Debrecen* **9** (1962) 3–12.

[EP2]   Erdös, P. and Pósa, L.   [3.3, 3.6]
         On independent circuits contained in a graph, *Canad. J. Math.*
         (1965) 347–352.

[ER1]   Erdös, P. and Rado, R.   [5.6]
         A partition calculus in set theory, *Bull. Amer. Math. Soc.* **62**
         (1956) 427–489.

[ER2]   Erdös, P. and Rado, R.   [2.4]
         Intersection theorems for systems of sets, *J. London Math. Soc.*
         35 (1960) 85–90.

[ER3]   Erdös, P. and Rado, R.   [5.4]
         A construction of graphs without triangles having pre-
         assigned order and chromatic number, *J. London Math. Soc.* **35**
         (1960) 445–448.

[ER4]   Erdös, P. and Rényi, A.   [4.0, 4.1, 6.2]
         On a problem in the theory of graphs (in Hungarian), *Publ.*
         Math. Inst. Hungar. Acad. Sci. **7** (1962).

[ERS1]   Erdös, P., Rényi, A. and Sós, V. T.   [4.1, 4.5, 6.2]
         On a problem of graph theory, *Studia Sci. Math. Hungar.* **1**
         (1966) 215–235.

[ES1]   Erdös, P. and Sachs, H.   [3.1, 3.6]
         Reguläre Graphen gegebener Taillenweite mit minimaler
         Knotenzahl, *Wiss. Z. Univ. Halle Martin-Luther Univ.*
         Halle-Wittenberg Math.-Natur. Reihe **12** (1963) 251–257.

[ESSS1] Erdös, P., Sauer, N., Schaer, J. and Spencer, J.   [4.4]
         Factorizing the complete graph into factors with large star
         number, *J. Combinatorial Theory* Ser. B **18** (1975) 180–183.

[ES2]   Erdös, P. and Simonovits, M.   [6.3, 6.4]
         A limit theorem in graph theory, *Studia Sci. Math. Hungar.* **1**
         (1966) 51–57.

[ES3]   Erdös, P. and Simonovits, M.   [6.1, 6.6]
         On a valence problem in extremal graph theory, *Discrete Math.*
         5 (1973) 323–334.

[ES4]   Erdös, P. and Sós, V. T.   [6.3, 6.4]
         Some remarks on Ramsey's and Turán's theorem, *in*:
         "Combinatorial Theory and its Applications", vol. II (Erdös,
         P., Rényi, A. and Sós, V. T., eds), Colloq. Math. Soc. J. Bolyai
         4, North-Holland, 1970, pp. 395–404.

[ES5]   Erdös, P. and Stone, A. H.   [3.5, 6.0, 6.3]
         On the structure of linear graphs, *Bull. Amer. Math. Soc.* **52**
         (1946) 1087–1091.

[ES6]   Erdös, P. and Szekeres, G.   [5.6, 5.7]
         A combinatorial problem in geometry, *Compositio Math.* **2**
         (1935) 463–470.

[ES7]   Erdös, P. and Szemerédi, A.   [6.6]
         On a Ramsey type theorem, *Period. Math. Hungar.* **2** (1972)
         295–299.

[EW1]	Erdös, P. and Wilson, R. J. *J. Combinational Theory* Ser. B **23** (1977) 255–257.	[5.7]
[E31]	Errera, P. Une contribution au problème des quatre couleurs, *Bull. de la Soc. Math. de France* **53** (1925) 42–55.	[5.3]
[EW1]	Everett, C. J. and Whaples, G. Representations of sequences of sets, *Amer. J. Math.* **71** (1949) 287–293.	[2.1, 2.6]
[F1]	Fáry, I. On straight line representation of planar graphs, *Acta Univ. Szeged Sect. Sci. Math.* **11** (1948) 229–233.	[5.3]
[FY1]	Faulkner, G. B. and Younger, D. H. Non-Hamiltonian cubic planar maps, *Discrete Math.* **7** (1974) 67–74.	[3.4]
[FH1]	Feit, W. and Higman, G. The non-existence of certain generalized polygons, *J. Algebra* **1** (1964) 114–131.	[3.1]
[F1a]	Fiorini, S. Some remarks on a paper of Vizing on critical graphs, *Math. Proc. Cambridge Phil. Soc.* **77** (1975), 475–483.	[5.7]
[F2]	Folkman, J. Graphs with monochromatic complete subgraphs in every edge colouring, *SIAM J. Appl. Math.* **18** (1970) 19–29.	[5.6]
[FF1]	Ford, L. R., Jnr. and Fulkerson, D. R. "Flows in Networks", Princeton Univ. Press, Princeton, New York, 1962.	[2.1]
[F3]	Friedman, H. D. On the impossibility of certain Moore graphs, *J. Combinatorial Theory* Ser. B **10** (1971) 245–252.	[3.1]
[F4]	Fulkerson, D. R. Note on Dilworth's decomposition theorem for partially ordered sets, *Proc. Amer. Math. Soc.* **7** (1956) 701–702.	[2.1]
[F5]	Fulkerson, D. R. Anti-blocking polyhedra, *J. Combinatorial Theory* Ser B **12** (1972) 50–71.	[5.0, 5.5]
[G1a]	Gale, D. Neighboring vertices on a convex polyhedron, *in*: "Linear Inequalities and Related Systems" (Kuhn, H. W. and Tucker, A. W., eds), Annals of Math. Studies 38, Princeton, 1956, pp. 255–263.	[5.4, 5.7]
[G1]	Gallai, T. On factorisation of graphs, *Acta Math. Acad. Sci. Hungar.* **1** (1950) 133–153.	[2.1, 2.3]
[G2]	Gallai, T. Über extreme Punkt- und Kantenmengen, *Ann. Univ. Sci. Budapest, Eötvös Sect. Math.* **2** (1959) 133–138.	[2.5]

[G3]   Gallai, T.   [5.5]
        Graphen mit triangulierbaren ungeraden Vielecken, *Publ.*
        Math. Inst. Hungar. Acad. Sci. **7** (1962) 3–36.

[G4]   Gallai, T.   [5.2]
        Kritische Graphen I, *Publ. Math. Inst. Hungar. Acad. Sci.* **8**
        (1963) 165–192.

[G5]   Gallai, T.   [5.2]
        Kritische Graphen II, *Publ. Math. Inst. Hungar. Acad. Sci.* **8**
        (1963) 373–395.

[G6]   Gallai, T.   [2.1]
        Neuer Beweis eines Tutte'schen Satzes, *Publ. Math. Inst.*
        Hungar. Acad. Sci. **8** (1963) 135–139.

[G7]   Gallai, T.   [2.1]
        Maximale Systeme unabhängiger Kanten, *Publ. Math. Inst.*
        Hungar. Acad. Sci. **9** (1964) 401–413.

[G8]   Gallai, T.   [1.6]
        Elementare Relationen bezüglich der Glieder und trennenden
        Punkte von Graphen, *Publ. Math. Inst. Hungar. Acad. Sci.* **9**
        (1964) 235–236.

[G9]   Gallai, T.   [1.6]
        Problems 4 and 5, *in*: "Theory of Graphs" (Erdös, P. and
        Katona, G., eds), Academic Press, New York, 1968, pp. 362.

[G10]   Gallai, T.   [5.0, 5.1]
        On directed paths and circuits, *in*: "Theory of Graphs" (Erdös,
        P. and Katona, G., eds), Academic Press, New York, 1968,
        pp. 115–118.

[GM1]   Gallai, T. and Milgram, A. N.   [2.1]
        Verallgemeinerung eines graphentheoretischen Satzes von
        Rédei, *Acta Sci. Math. Szeged* **21** (1960) 181–186.

[G11]   Gerencsér, L.   [5.1]
        On colouring problems (in Hungarian), *Mat. Lapok* **16** (1965)
        274–277.

[GG1]   Gerencsér, L. and Gyárfas, A.   [5.6]
        On Ramsey-type problems, *Ann. Univ. Sci. Budapest Eötvös*
        Sect. Math. **10** (1967) 167–170.

[G12]   Ghouila-Houri, A.   [4.2]
        Diamètre maximal d'un graphe fortement connexe, *C.R. Acad.*
        Sci. Paris. **250** (1960) 4254–4256.

[G13]   Goldberg, M. K.   [4.2]
        Some applications of contractions to strongly connected
        graphs (in Russian), *Uspekhi AN USSR* **20** (1965) 203–205.

[G14]   Goldberg, M. K.   [4.2, 4.5]
        The diameter of a strongly connected graph (in Russian),
        Doklady AN SSR **170** (1966) 767–769.

[G15]   Goodman, A. W.   [6.1, 6.6]
        On sets of acquaintances and strangers at any party, *Amer.*
        Math. Monthly **66** (1959) 778–783.

[G15a]   Goodey, P. R.   [3.4]
          A class of Hamiltonian polytopes, *J. Graph Theory* **1** (1977)
          181–185.

[G16]   Graham, R. L.   [5.7]
          On edgewise 2-coloured graphs with monochromatic triangles
          and containing no complete hexagon, *J. Combinatorial Theory* **4**
          (1968) 300.

[GY1]   Graver, J. E. and Yackel, J.   [5.6]
          Some graph theoretic results associated with Ramsey's theorem,
          J. Combinatorial Theory **4** (1961) 125–175.

[GM2]   Greene, C. and Magnanti, T. L.   [2.5]
          Some abstract pivot algorithms, *SIAM J. Applied. Math.* **29**
          (1975) 530–539.

[GG2]   Greenwood, R. E. and Gleason, A. M.   [5.6, 5.7]
          Combinatorial relations and chromatic graphs, *Canad. J.
          Math.* **7** (1955) 1–7.

[G17]   Grimmett, G. R. and McDiarmid, C. J. H.   [5.7]
          On colouring random graphs, *Math. Proc. Cambridge Philos.
          Soc.* **77** (1975) 313–324.

[G18]   Grinberg, E.   [3.4]
          Plane homogeneous graphs of degree three without Hamil-
          tonian circuits, *in*: "Latvian Math. Yearbook 4", Izdat
          "Zinatne", Riga, 1968, pp. 51–58 (Russian).

[G19]   Grünbaum, B.   [6.5]
          A result in graph-colouring, *Michigan Math. J.* **13** (1968)
          381–383.

[G20]   Grünbaum, B.   [3.4]
          Polytopes, graphs and complexes, *Bull. Amer. Math. Soc.* **76**
          (1970) 1143–1145.

[GM3]   Grünbaum, B. and Malkevitch, J.   [3.4]
          Pairs of edge-disjoint Hamiltonian circuits (to appear).

[GZ1]   Grünbaum, B. and Zaks, J.   [3.4]
          The existence of certain planar maps, *Discrete Math.* **10** (1974)
          93–115.

[G21]   Guy, R. K.   [6.2]
          A problem of Zarankiewicz, *in*: "Theory of Graphs" (Rosen-
          stiehl, P. ed.), Gordon and Breach, New York, 1967, pp.
          139–142.

[G22]   Guy, R. K.   [6.2]
          A problem of Zarankiewicz, *in*: "Theory of Graphs" (Erdös, P.
          and Katona, G., eds), Academic Press, New York, 1968, pp.
          119–150.

[G23]   Guy, R. K.   [6.2]
          The many faceted problem of Zarankiewicz, *in*: "The Many
          Facets of Graph Theory", Lecture Notes in Maths 110,
          Springer, 1969, pp. 129–148.

[GZ2]   Guy, R. K. and Znám, S.   [6.2]
        A problem of Zarankiewicz, in: "Recent Progress in Combina-
        torics" (Tutte, W. T., ed.), Academic Press, New York, 1969,
        pp. 237–243.

[GL1]   Gyárfás, A. and Lehel, J.   [8.4, 8.6]
        Packing trees of different order in a complete graph (to appear).

[H1]   Hadwiger, H.   [5.3]
        Über eine Klassifikation der Streckenkomplexe, Vierteljschr.
        Naturforsch. Ges. Zürich 88 (1943) 133–142.

[H2]   Hadwiger, H.   [5.3]
        Ungelöste Probleme Nr. 26, Elem. Math. 13 (1958) 128–128.

[HMS1] Hajnal, A., Milner, E. C. and Szemerédi, E.   [4.5]
        A cure for the telephone disease, Canad. Math. Bull. 15 (1972),
        447–450.

[HS1]   Hajnal, A. and Sós, V. T.   [2.6]
        Problem, in: "Combinatorial Theory and its Applications",
        vol. II (Erdös, P., Rényi, A. and Sós, V. T., eds), Colloq. Math.
        Soc. J. Bolyai 4, North-Holland, 1970, pp. 1163–1164.

[HS2]   Hajnal, A. and Surányi, J.   [5.5]
        Über die Auflösung von Graphen in vollständige Teilgraphen,
        Ann. Univ. Budapest Eötvös Sect. Math. 1 (1958) 113–121.

[HS1]   Hajnal, A. and Szemerédi, E.   [3.2, 6.5, 8.4]
        Proof of a conjecture of Erdös, in: "Combinatorial Theory and
        its Applications", vol. II (Erdös, P., Renyi, A. and Sós, V. T.,
        eds), Colloq. Math. Soc. J. Bolyai 4, North-Holland, Amster-
        dam, 1970, pp. 601–623.

[H3]   Hajós, G.   [5.3, 7.3]
        Über ein Konstruktion nicht n-färbbarer Graphen, Wiss. Z.
        Martin-Luther Univ. Halle-Wittenberg Math.-Natur. Reihe 10
        (1961) 116–117.

[H4]   Hakimi, S.   [2.3, 2.6]
        On the realizability of a set of integers as degrees of the vertices
        of a graph, J. Soc. Indust. Appl. Math. 10 (1962) 496–506.

[H5]   Halin, R.   [7.4]
        Übereinen graphentheoretischenBasisbegriff und seine Anwen-
        dung auf Färbungsprobleme, Doctoral Thesis, Univ. zu Köln,
        1962.

[H6]   Halin, R.   [5.3]
        Über einen Satz von K. Wagner zum Vierfarbenproblem,
        Math. Ann. 153 (1964) 47–62.

[H7]   Halin, R.   [5.3]
        Zur Klassifikation der endlichen Graphen nach H. Hadwiger
        und K. Wagner, Math. Ann. 172 (1967) 46–78.

[H8]   Halin, R.   [5.1, 7.2]
        Unterteilungen vollständiger Graphen in Graphen mit
        unendlicher chromatischer Zahl, Abh. Math. Sem. Hamburg.
        Univ. 31 (1967) 156–165.

[H9]   Halin, R.   [1.4]
On the classification of finite graphs according to H. Hadwiger and K. Wagner, *in*: "Theory of Graphs" (Erdös, P. and Katona, G., eds), Academic Press, New York, 1968, pp. 161–167.

[H10]   Halin, R.   [1.4]
A theorem on n-connected graphs, *J. Combinatorial Theory* **7** (1969) 150–154.

[H11]   Halin, R.   [1.4]
On the structure of n-connected graphs, *in*: "Recent Progress in Combinatorics" (Tutte, W. T., ed.) Academic Press, New York, 1969, pp. 91–102.

[H12]   Halin, R.   [1.4]
Untersuchungen über minimale n-fach zusammenhängende Graphen, *Math. Ann.* **182** (1969) 175–188.

[H13]   Halin, R.   [1.2, 1.3, 1.4]
Zur Theorie der n-fach zusammenhängenden Graphen. *Abh. Math. Sem. Hamburg Univ.* **33** (1969) 133–164.

[H14]   Halin, R.   [1.4]
Ecken n-ten Grades in minimalen n-fach zusammenhängenden Graphen, *Abh. Math. Sem. Hamburg Univ.* **35** (1970) 39–53.

[H15]   Halin, R.   [1.4]
Studies on minimally n-connected graphs, *in*: "Combinatorial Mathematics and its Applications" (Welsh, D. J. A., ed.). Academic Press, London and New York, 1971, pp. 129–136.

[HJ1]   Halin, R. and Jung, H. A.   [1.2, 7.1]
Über Minimalstrukturen von Graphen, insbesondere von n-fach zusammenhängenden Graphen, *Math. Ann.* **152** (1963) 75–94.

[H16]   Hall, M.   [2.1, 2.6]
Dinstinct representatives of subsets, *Bull. Amer. Math. Soc.* **54** (1948) 922.

[H17]   Hall, M. Jr.   [2.6]
A combinatorial problem on abelian groups, *Proc. Amer. Math. Soc.* **3** (1952) 584–587.

[H18]   Hall, P.   [1.2, 2.6]
On representatives of subsets, *J. London Math. Soc.* **10** (1935) 26–30.

[HV1]   Halmos, P. R. and Vaughan, H. E.   [2.1, 2.6]
The marriage problem, *Amer. J. Math.* **72** (1950) 214–215.

[H19]   Harary, F.   [1.2, 5.3]
"Graph Theory", Addison Wesley, Reading, Mass., Menlo Park, Calif. and London, 1969.

[H20]   Harary F.   [1.1]
The maximum connectivity of a graph, *Proc. Nat. Acad. Sci. U.S.A.* **48** (1962) 1142–1146.

[HHR1]   Harary, F., Hedetniemi, S. T. and Robinson, R. W.   [5.4]

Uniquely colourable graphs, *J. Combinatorial Theory* **6** (1969) 264–270.

[HK1] Harary, F. and Kodama, Y. [1.1, 1.6]
On the genus of an *n*-connected graph, *Fund. Math.* **54** (1964) 7–13.

[HP1] Harary, F. and Prins, G. [1.6, 1.4]
The block-cutpoint-tree of a graph, *Publ. Math. Debrecen* **13** (1966) 103–107.

[HT1] Harary, F. and Tutte, W. T. [5.3]
A dual form of Kuratowski's theorem. *Canad. Math. Bull.* **8** (1965) 17–20.

[HW1] Hardy, G. H. and Wright, E. M. [0]
"An introduction to the Theory of Numbers", Fourth Ed., Oxford, 1960.

[HMR1] Hartman, S., Mycielski, J. and Ryll-Nardzewski, C. [6.2]
Systèmes spéciaux de points à coordonnées entières, *Colloq. Math.* **3** (1954) 84–85.

[H21] Havel, V. [2.3, 2.6]
A remark on the existence of finite graphs, *Casopis Pest. Mat.* **80** (1955) 477–480.

[H22] Heawood, P. J. [5.3]
Map colour theorems, *Quart. J. Math.* **24** (1890) 332–338.

[H23] Hedetniemi, S. T. [2.5]
Hereditary properties of graphs, *J. Combinatorial Theory* Ser B **14** (1973) 94–99.

[H24] Heesch, H. [5.3]
"Untersuchungen zum Vierfarbenproblem", B-I-Hochschulskripten 810/810a/810b, Bibliographisches Institut, Mannheim, Vienna, Zürich, 1969.

[H25] Heesch, H. [5.3]
Chromatic reduction of the triangulations, T_e, $e = e_5 + e_7$, *J. Combinatorial Theory* Ser B **13** (1972) 46–55.

[H26] Heffter, L. [5.3]
Über das Problem der Nachbargebiete, *Ann. Math.* **38** (1891) 477–508.

[HDV1] Herz, J.-C., Duby, J.-J. and Vigue, F. [3.6]
Recherche systématique des graphes hypo-Hamiltoniens, *in*: "Theory of Graphs" (Rosenstiehl, P., ed.), Gordon and Breach, New York, 1967, pp. 153–159.

[HS4] Hoffman, A. J. and Singleton, R. R. [3.1]
On Moore graphs with diameters 2 and 3, *IBM J. Res. Develop.* **4** (1960) 497–504.

[H27] Hoheisel, G. [0]
Primzahlprobleme in der Analysis, *Sitz. Preuss. Akad. Wiss.* **2** (1930) 1–13.

[HR1]   Holt, R. C. and Reingold, E. M.   [8.1]
         On the time required to detect cycles and connectivity in directed
         graphs, Math. Systems Theory 6 (1972) 103–107.

[HO1]   Homenko, N. P. and Ostroverhii, N. A.   [4.1, 4.5]
         Diameter-critical graphs (in Russian), Ukrainian Math. J. 22
         (1970) 637–646.

[HS5]   Homenko, N. P. and Strok, V. V.   [4.1]
         Some combinatorial identities for sums of composition coeffi-
         cients (in Russian), Ukrainian Math. J. 23 (1971) 830–837.

[HT2]   Hopcroft, J. and Tarjan, R.   [8.1]
         "Efficient Planarity Testing", TR 73-165, Dept. of Computer
         Sci., Cornell Univ. April, 1973.

[H28]   Huxley, M. N.   [0]
         On the difference between consecutive primes, Invent. Math. 15
         (1972) 164–170.

[H29]   Hylten-Cavallius, C.   [0]
         On a combinatorial problem, Colloq. Math. 6 (1958) 59–65.

[I1]   Ingham, A. E.   [6.2]
         On the difference between consecutive primes, Quart. J. Math.
         Oxford (1937) 255–266.

[J1]   Jakobsen, I. T.   [5.7]
         On critical graphs with chromatic index 4, Discrete Math. 9
         (1974) 265–276.

[J2]   Jung, H. A.   [7.2]
         Anwendung einer Methode von K. Wagner bei Färbungsprob-
         lemen für Graphen, Math. Ann. 161 (1965) 325–326.

[J3]   Jung, H. A.   [7.2]
         Zusammenzüge und Unterteilungen von Graphen, Math.
         Nachr. 35 (1967) 241–267.

[J4]   Jung, H. A.   [3.4]
         On maximal circuits in finite graphs, in: "Advances in Graph
         Theory" (Bollobás, B., ed.), North-Holland, 1978.

[K1]   Kárteszi, F.   [3.1, 3.6]
         On a combinatorial minimum problem (in Hungarian), Mat.
         Lapok 11 (1960) 323–329.

[K2]   Kárteszi, F.   [3.1, 3.6]
         Piani finiti ciclici come risoluzione di un certo problema di
         minimo, Boll. Un. Mat. Ital. 15 (1906) 522–528.

[K3]   Katona, G.   [6.1]
         A theorem of finite sets, in: "Theory of Graphs" (Erdös, P. and
         Katona, G., eds), Academic Press, New York, 1968, pp.
         187–207.

[K4]   Kaugars, A.   [1.6]
         A theorem on the removal of vertices from blocks, Senior
         Thesis, Kalamazoo College, 1968.

[KK1]   Kelly, J. B. and Kelly, L. M.   [5.4]
         Paths and circuits in critical graphs, *Amer. J. Math.* **76** (1954)
         786–792.

[K5]   Kempe, A. B.   [5.3]
         On the geographical problems of the four colours, *Amer. J.
         Math.* **2** (1879) 193–200.

[K6]   Kirkpatrick, D.   [8.1, 8.2, 8.3]
         Determining graph properties from matrix representations, *in*:
         "Proc. Sixth ACM Symp. in the Theory of Computing",
         Seattle, 1974, pp. 84–90.

[KK2]   Kleitman, D. and Kwiatkowski, D. J.   [8.2, 8.6]
         Further results on the Aanderaa–Rosenberg conjecture (to
         appear).

[KW1]   Kleitman, D. J. and Wang, D. L.   [2.6]
         Algorithms for constructing graphs and digraphs with given
         valencies and factors, *Discrete Math.* **6** (1973) 79–88.

[K7]   Kneser, M.   [5.7]
         Aufgabe 360, *Jahresbericht. Deutschen Math. Ver.* **58** (2) (1955–
         56) 27.

[K8]   König, D.   [2.6]
         Über Graphen und ihre Anwendungen, *Math. Ann.* **77** (1915)
         453.

[K9]   König, D.   [2.1, 2.5]
         Graphen und Matrizen, *Math. Lapok* **38** (1931) 116–119.

[K10]   König, D.   [2.6]
         "Theorie der endlichen und unendlichen Graphen", Leipzig,
         1936; reprinted Chelsea and New York, 1950.

[K11]   Kotzig, A.   [2.1]
         On the theory of finite graphs with a linear factor I, II, III
         (in Slovak, with German summary), *Mat. Fiz. Casopis* **9** (1959)
         73–91, 136–159 and **10** (1960) 205–2115.

[K12]   Kotzig, A.   [2.6]
         Problem No. 6, *in*: "Combinatorial Structures and their
         Applications" (Guy, R., Hanani, H., Sauer, N. and Schönheim,
         J., eds), Gordon and Breach, New York, 1970, p. 494.

[KST1]   Kövári, P., Sós, V. T. and Turán, P.   [6.2]
         On a problem of K. Zarankiewicz, *Colloq. Math.* **3** (1954)
         50–57.

[KM1]   Kronk, H. V. and Mitchem, J.   [5.2]
         On Dirac's generalisation of Brooks' theorem, *Canad. J. Math.*
         24 (1972) 805–807.

[KM2]   Kronk, H. V. and Mitchem, J.   [5.2, 5.7]
         Critical point-arboritic graphs, *J. London Math. Soc.* **9** (2)
         (1975) 459–466.

[K13]   Kruskal, J. B.   [6.1]
         The number of simplices in a complex, *in*: "Math. Optimiza-
         tion Techniques", Univ. Calif. Press, Berkeley and Los
         Angeles, 1963, pp. 251–278.

[K14]   Kundu, S.   [2.6]
The k-factor conjecture is true, *Discrete Math.* **6** (1973) 367–376.

[K15]   Kuratowski, K.   [5.3]
Sur le problème des courbes gauches en topologie, *Fund. Math.* **15** (1930) 271–283.

[LM1]   Larman, D. G. and Mani, P.   [7.4]
On the existence of certain configurations within graphs and the 1-skeletons of polytopes, *Proc. London Math. Soc.* **20** (3) (1970) 144–160.

[L1]   Las Vergnas, M.   [3.4, 3.6]
Sur une propriété des arbres maximaux dans un graphe, *C.R. Acad. Sci. Paris* Ser A-B **272** (1971) 1297–1300.

[L2]   Las Vergnas, M.   [5.1]
Problem, *in*: "Proc. Fifth British Combinatorial Conference" (Nash-Williams, C. St. J. A. and Sheehan, J., eds), Utilitas Math., Winnipeg, 1976, p. 689.

[L3]   Leonard, J. L.   [1.5, 1.4]
On a conjecture of Bollobás and Erdös, *Period. Math. Hungar.* **3** (1973) 281–284.

[L4]   Leonard, J. L.   [1.5]
On graphs with at most four line-disjoint paths connecting any two vertices, *J. Combinatorial Theory* Ser. B **13** (1972) 242–250.

[L5]   Leonard, J. L.   [1.5]
Graphs with 6-ways, *Canad. J. Math.* **25** (1973) 687–692.

[L6]   Lesniak, L.   [1.6]
Results on the edge-connectivity of graphs, *Discrete Math.* **8** (1974) 351–354.

[L7]   Lewin, M.   [2.6]
A note on line coverings of graphs, *Discrete Math.* **5** (1973) 283–285.

[L8]   Lick, D. R.   [1.2]
Characterizations of n-connected and n-line connected graphs, *J. Combinatorial Theory* Ser. B **14** (1973) 122–124.

[LW1]   Lick, D. R. and White, A. T.   [5.7]
k-degenerate subgraphs, *Canad. J. Math.* **22** (1970) 1082–1096.

[L9]   Lindgren, W. F.   [3.6]
An infinite class of hypo-Hamiltonian graphs, *Amer. Math. Monthly* **74** (1967) 1087–1088.

[LS1]   Lipton, R. J. and Snyder, L.   [8.2]
On the Aanderaa–Rosenberg conjecture, *SIGACT News* **6** (Jan. 1974) 30–31.

[L10]   Longyear, J. Q.   [3.1]
Regular d-valent graphs of girth 6 and $2(d^2 - d + 1)$ vertices, *J. Combinatorial Theory* **9** (1970) 420–422.

[L11]   Lorden, G.   [6.6]
Blue-empty chromatic graphs, *Amer. Math. Monthly* **69** (1962)
114–120.

[L12]   Lovász, L.   [3.2]
On graphs not containing independent circuits (in Hungarian),
Mat. Lapok **16** (1965) 289–299.

[L13]   Lovász, L.   [5.7]
On decomposition of graphs, *Studia Sci. Math. Hungar.* **1**
(1966) 237–238.

[L14]   Lovász, L.   [5.4]
Graphs and set systems, *in*: "Beiträge zur Graphentheorie
(Sachs, H., Voss, H. J. and Walther, H., eds), Teubner
Verlagsgesellschaft, Leipzig, 1968, pp. 99–106.

[L15]   Lovász, L.   [2.5]
On covering of graphs, *in*: "Theory of Graphs" (Erdös, P.
and Katona, G., eds), Academic Press, New York, 1968, pp.
231–236.

[L16]   Lovász, L.   [5.4]
On chromatic number of finite set-systems, *Acta Math. Acad.
Sci. Hungar.* **19** (1968) 59–67.

[L17]   Lovász, L.   [2.1]
Subgraphs with prescribed valencies, *J. Combinatorial Theory*
8 (1970) 391–416.

[L18]   Lovász, L.   [5.0, 5.5]
Normal hypergraphs and the perfect graph conjecture, *Discrete
Math.* **2** (1972) 253–267.

[L19]   Lovász, L.   [5.5]
A characterization of perfect graphs, *J. Combinatorial Theory*
Ser. B **13** (1972) 95–98.

[L20]   Lovász, L.   [6.1]
On the sieve formula (in Hungarian), *Mat. Lapok* **23** (1972)
53–69.

[L21]   Lovász, L.   [2.1, 2.2]
On the structure of factorizable graphs, *Acta Math. Acad. Sci.
Hungar.* **23** (1972) 179–195.

[L22]   Lovász, L.   [5.2]
Independent sets in critical chromatic graphs, *Studia Sci.
Math. Hungar.* **8** (1973) 165–168.

[L23]   Lovász, L.   [5.1]
Three short proofs in graph theory, *J. Combinatorial Theory*
Ser. B **19** (1975) 269–271.

[L24]   Lovász, L.   [1.6]
Problem II, *in*: "Proc. Fifth British Combinatorial Conf."
(Nash-Williams, C. St. J. A. and Sheehan, J., eds), Utilitas
Math., Winnipeg, 1976, pp. 684–685.

[LNP1]   Lovász, L., Neumann-Lara, V. and Plummer, M.   [1.2, 1.6]

Mengerian theorems for paths of bounded length, *J. Combinatorial Theory* Ser. B (to appear).

[L25]   Lovász, L.   [1.4]
On some connectivity properties of eulerian graphs, *Acta Math. Acad. Sci. Hungar.* **28** (1976) 129–138.

[L26]   Lovász, L.   [5.4, 5.7]
Kneser's conjecture, chromatic number and homotopy, *J. Combinatorial Theory* (to appear).

[LS2]   Lovász, L. and Simonovits, M.   [6.1]
On the number of complete subgraphs of a graph, *in*: "Proc. Fifth British Combinatorial Conf." (Nash-Williams, C. St. J. A. and Sheehan, J., eds), Utilitas Math., Winnipeg, 1976, pp. 431–441.

[M1]   MacLane, S.   [5.3]
A structural characterization of planar combinatorial graphs, *Duke Math. J* **3** (1937) 340–372.

[M2]   Mader, W.   [7.1, 7.2, 7.4]
Homomorphieeigenschaften und mittlere Kantendichte von Graphen, *Math. Ann.* **174** (1967) 265–268.

[M3]   Mader, W.   [7.1, 7.2, 7.4]
Homomorphiesätze für Graphen, *Math. Ann.* **178** (1968) 154–168.

[M4]   Mader, W.   [1.5]
Existenz gewisser Konfigurationen in *n*-gesättigten Graphen. und in Graphen genügend grosser Kantendichte, *Math. Ann.* **194** (1971) 295–312.

[M5]   Mader, W.   [1.4]
Minimale *n*-fach kantenzusammenhängende Graphen, *Math. Ann.* **191** (1971) 21–28.

[M6]   Mader, W.   [1.4]
Minimale *n*-fach zusammenhängende Graphen mit maximaler Kantenzahl, *J. Reine Angew. Math.* **249** (1971) 201–207.

[M7]   Mader, W.   [1.4, 1.6]
Ecken vom Grad *n* in minimalen *n*-fach zusammenhängenden Graphen, *Archiv. Math.* **23** (1972) 219–224.

[M8]   Mader, W.   [1.5]
Existenz *n*-fach zusammenhängender Teilgraphen in Graphen genügend grossen Kantendichte, *Abh. Math. Sem. Hamburg Univ.* **37** (1972) 86–97.

[M9]   Mader, W.   [7.0, 7.2]
Hinreichende Bedingungen für die Existenz von Teilgraphen, die zu einem vollständigen Graphen homöomorph sind, *Math. Nachr.* **53** (1972) 145–150.

[M10]   Mader, W.   [1.4]
Über minimal *n*-fach zusammenhängende, unendliche Graphen und ein Extremalproblem, *Arch. Math.* (*Basel*) **23** (1972) 553–560.

[M11] Mader, W.   [1.5, 1.6]
       Ein Extremalproblem des Zusammenhangs von Graphen,
       Math. Z. **131** (1973) 223–231.

[M12] Mader, W.   [1.5, 2.1, 2.2]
       Grad und lokaler Zusammenhang in endlichen Graphen, *Math.
       Ann.* **205** (1973) 9–11.

[M13] Mader, W.   [2.6]
       Über die Anzahl der von den 1-Faktoren eines Graphen unter-
       deckten Ecken, *Math. Nachr.* **56** (1973) 195–200.

[M14] Mader, W.   [1.5]
       Kreuzungsfreie *a, b*-Wege in endlichen Graphen, *Abh. Math.
       Sem. Hamburg Univ.* **42** (1974) 187–204.

[M15] Mader, W.   [1.5]
       Extremal connectivity problems, *in*: "Infinite and Finite Sets",
       vol. II (Hajnal, A., Rado, R. and Sós, V. T., eds), Colloq. Math.
       Soc. J. Bolyai 10, North-Holland, Amsterdam, 1975, pp. 1089–
       1094.

[M15a] Mader, W.   [2.4]
       Edge-connectivity preserving reductions, *in*: "Advances in
       Graph Theory" (Bollobás, B., ed.), North-Holland, 1978.

[MR1] Mann, H. B. and Ryser, H. J.   [2.1]
       Systems of distinct representatives, *Amer. Math. Monthly*
       60 (1953) 397–401.

[M16] Mantel, W.   [6.1]
       Problem 28, soln. by H. Gouwentak, W. Mantel, J. Teixeira de
       Mattes, F. Schuh and W. A. Wythoff, *Wiskundige Opgaven* **10**
       (1907) 60–61

[M16a] Marczewski, E.   [6.1]
       Sur deux propriétés des classes d'ensembles, *Fund. Math.* **33**
       (1945) 303–307.

[M17] Martin, P.   [3.4]
       Cycles Hamiltoniens dans les graphes 4-reguliers 4-connexes,
       Aequationes Math. (to appear).

[M18] Matula, D. W.   [5.1]
       A min-max theorem for graphs with application to graph
       colouring, *SIAM Rev.* **10** (1968) 481–482.

[M19] Matula, D. W.   [5.7]
       On the complete subgraphs of a random graph, *in*: "Combi-
       natorial Mathematics and its Applications" (Bose, R. C. *et al.*,
       eds.), Chapel Hill, N.C., 1970, pp. 356–369.

[M20] Matula, D. W.   [5.7]
       The employee party problem (to appear).

[M21] Maunsell, F. G.   [2.1]
       A note on Tutte's paper "The Factorization of linear graphs",
       J. London Math. Soc. **27** (1952) 127–128.

[M22]   Mayer, J.   [5.3]
          Le problème des regions voisines sur les surfaces closes orient-
          ables, *J. Combinatorial Theory* **6** (1969) 177–195.

[M23]   Mayer, J.   [5.3]
          Inégalités nouvelles dans le problème des quatre couleurs, *J.
          Combinatorial Theory* Ser. B **19** (1975) 119–149.

[M24]   Mayer, J.   [5.3]
          Problème des quatre couleurs, un contre-exemple doit avoir
          au moins 96 sommets, *J. Combinatorial Theory* Ser. B (to
          appear).

[M25]   McGee, W. F.   [3.6]
          A minimal cubic graph of girth seven,
          Canad. Math. Bull. **3** (1960) 149–152.

[MV1]   Melnikov, L. S. and Vizing, V. G.   [5.1]
          New proof of Brooks' theorem, *J. Combinatorial Theory* **7**
          (1969) 289–290.

[M26]   Menger, K.   [1.2]
          Zur allgemeinen Kurventheorie, *Fund. Math.* **10** (1927) 96–115.

[M27]   Meredith, G. H. J.   [3.4]
          Regular n-valent n-connected non-Hamiltonian non-n-edge
          colorable graphs, *J. Combinatorial Theory* Ser. B **14** (1973) 55–60.

[M28]   Milgram, M.   [5.3]
          Irreducible graphs, *J. Combinatorial Theory* Ser. B **12** (1972)
          6–31.

[M29]   Milgram, M.   [5.3]
          Irreducible graphs—Part 2, *J. Combinatorial Theory* Ser. B **14**
          (1973) 7–45.

[MW1]   Milner, E. C. and Welsh, D. J. A.   [8.1, 8.3]
          "On the Computational Complexity of Graph Theoretical
          Properties", Univ. of Calgary, Res. Paper No. 232, June, 1974.

[MW2]   Milner, E. C. and Welsh, D. J. A.   [8.1, 8.6]
          On the computational complexity of graph theoretical proper-
          ties", *in*: "Proc. Fifth British Combinatorial Conf." (Nash-
          Williams, C. St. J. A. and Sheehan, J., eds), Utilitas Math.,
          Winnipeg, 1976, pp. 471–487.

[M30]   Mirsky, L.   [2.1]
          "Transversal Theory", Academic Press, New York, 1971.

[MP1]   Mirsky, L. and Perfect, H.   [2.1]
          Systems of representatives, *J. Math. Anal. Applications* **15**
          15 (1966) 520–568.

[M31]   Mitchem, J.   [5.7]
          An extension of Brooks' theorem to n-degenerate graphs (to
          appear).

[M31a]   Mitchem, J.   [5.7]
          A short proof of Catlin's extension of Brooks' theorem, *Discrete
          Math.* **21** (1978) 213–214.

[M32]   Montgomery, H. L.   [0]
          "Topics in Multiplicative Number Theory, Lecture Notes in
          Math. 227, Springer, Berlin, 1971.

[M33]	Moon, J. W. On edge-disjoint cycles in a graph, *Canad. Math. Bull.* **7** (1964) 519–523.	[3.3]
[M34]	Moon, J. W. On the number of complete subgraphs of a graph, *Canad. Math. Bull.* **8** (1965) 831–834.	[6.6]
[M35]	Moon, J. W. On independent complete subgraphs in a graph, *in*: "Intern. Congress of Math.", Moscow, 1966, Abstracts, Section 13.	[6.4]
[MM1]	Moon, J. W. and Moser, L. On a problem of Turán, *Publ. Math. Inst. Hungar. Acad. Sci.* **7** (1962) 283–286.	[6.1]
[MM2]	Moon, J. W. and Moser, L. On Hamiltonian bipartite graphs, *Israel J. Math.* **1** (1963) (1963) 163–165.	[3.4]
[MM3]	Moon, J. W. and Moser, L. On cliques in graphs, *Israel J. Math.* **3** (1965) 23–28.	[6.6]
[MS1]	Motzkin, T. S. and Straus, E. G. Maxima for graphs and a new proof of a theorem of Turán, *Canad. J. Math.* **17** (1965) 533–540.	[6.6]
[M36]	Murty, U. S. R. On critical graphs of diameter 2, *Math. Mag.* **41** (1968) 138–140.	[4.2, 4.3]
[M37]	Murty, U. S. R. On some extremal graphs, *Acta Math. Acad. Sci. Hungar.* **19** (1968) 69–74.	[4.2, 4.5]
[M38]	Murty, U. S. R. Extremal non-separable graphs of diameter 2, *in*: "Proof Techniques in Graph Theory" (Harary, F., ed.), Academic Press, New York, 1969, pp. 111–117.	[4.2, 4.5]
[MV2]	Murty, U. S. R. and Vijayan, K. On accessibility in graphs, *Sankhya* Ser. A **26** (1964) 299–302.	[4.0, 4.2, 4.3]
[M39]	Mycielski, J. Sur le coloriage des graphes, *Colloq. Math.* **3** (1955) 161–162.	[5.0, 5.4]
[N1]	Nash-Williams, C. St. J. A. Edge-disjoint spanning trees in finite graphs, *J. London Math. Soc.* **36** (1961) 445–450.	[2.5]
[N2]	Nash-Williams, C. St. J. A. Decomposition of finite graphs into forests, *J. London Math. Soc.* **39** (1964) 12.	[2.5]
[N3]	Nash-Williams, C. St. J. A. On Hamiltonian circuits in finite graphs, *Proc. Amer. Math. Soc.* **17** (1966) 466–467.	[3.4]
[N4]	Nash-Williams, C. St. J. A. Hamiltonian circuits in graphs and digraphs, *in*: "The many facets of graph theory", Lecture Notes in Math. 110, Springer, Berlin, 1969, pp. 237–243.	[3.4]

[N5]   Nash-Williams, C. St. J. A.   [3.4]
          in: "Combinatorial Mathematics and its Applications" (Welsh,
          D. J. A., ed.), Academic Press, London and New York, 1971,
          pp. 191–200.

[N6]   Nash-Williams, C. St. J. A.   [3.6]
          Edge-disjoint Hamiltonian circuits in graphs with vertices of
          large valency, *in*: "Studies in Pure Mathematics" (volume
          presented to Richard Rado; Mirsky, L., ed.), Academic Press,
          Press, London, 1971, pp. 157–183.

[N7]   Nash-Williams, C. St. J. A.   [2.6]
          Unexplored and semi-explored territories in graph theory, *in*:
          "New Directions in the Theory of Graphs", (Harary, F., ed.),
          Academic Press, New York, 1973, pp. 149–186.

[N8]   Nešetřil, J.   [5.4]
          k-chromatic graphs without cycles of length at most 7 (in
          Czech), *Comment. Math. Univ. Carolinae* **7** (1966) 373–376.

[NR1]   Nešetřil, J. and Rödl, V.   [5.6]
          A Ramsey graph without triangles exists for any graph without
          triangles, *in*: "Infinite and Finite Sets", vol. II (Hajnal, A.,
          Rado, R. and Sós, V. T., eds), Colloq. Math. Soc. J. Bolyai 10,
          North Holland, 1975, pp. 1127–1132.

[NR2]   Nešetřil, J. and Rödl, V.   [5.6]
          The Ramsey property for graphs with forbidden complete
          subgraphs, *J. Combinatorial Theory* Ser. B **20** (1976) 243–249.

[NR3]   Nešetřil, J. and Rödl, V.   [5.6]
          A structural generalization of the Ramsey theorem, *Bull. Amer.
          Math. Soc.* **83** (1977) 127–128.

[NR4]   Nešetřil, J. and Rödl, V.   [5.6]
          Partitions of relational and set systems, *J. Combinatorial Theory*
          Ser. A **22** (1977) 289–312.

[NR5]   Nešetřil, J. and Rödl, V.   [5.4]
          Construction of sparse graphs of high chromatic number, to
          appear

[NR6]   Nešetřil, J. and Rödl, V.   [5.7]
          Ordered induced subgraphs, to appear.

[NR7]   Nešetřil, J. and Rödl, V.   [5.6]
          Selective graphs and hypergraphs, *in*: "Advances in Graph
          Theory" (Bollobás, B., ed.), North-Holland, 1978.

[NG1]   Nordhaus, E. A. and Gaddum, J. W.   [5.1]
          On complementary graphs, *Amer. Math. Monthly* **63** (1956)
          175–177.

[NS1]   Nordhaus, E. A. and Stewart, B. M.   [6.1]
          Triangles in an ordinary graph, *Canad. J. Math.* **15** (1963)
          33–41.

[NR3]   Norman, R. Z. and Rabin, M.   [2.6]
          Algorithm for a minimal cover of a graph, *Proc. Amer. Math.*
          10 (1959) 315–319.

[O1]   Ore, O.   [2.1]
        Graphs and matching theorems, *Duke Math. J.* **22** (1955)
        625–639.

[O2]   Ore, O.   [3.4]
        Note on Hamiltonian circuits, *Amer. Math. Monthly* **67** (1960)
        55.

[O3]   Ore, O.   [3.6]
        Arc coverings of graphs, *Ann. Mat. Pura Appl.* **55** (1961) 315–
        321.

[O4]   Ore, O.   [5.1, 5.3]
        The four-colour problem, Academic Press, New York, 1967.

[O5]   Ore, O.   [3.6]
        On a graph theorem of Dirac, *J. Combinatorial Theory* **2** (1967)
        35–42.

[O6]   Ore, O.   [4.1, 4.5]
        Diameters in graphs, *J. Combinatorial Theory* **5** (1968) 75–81.

[OS1]   Ore, O. and Stemple, J.   [5.3]
        Numerical calculations on the four-colour problem, *J. Com-
        binatorial Theory* **8** (1970) 65–78.

[O7]   Osgood, T.   [5.3]
        "An Existence Theorem for Planar Triangulations with
        Vertices of Degree Five, Six and Eight", Ph.D. Thesis, Uni-
        versity of Illinois, 1974.

[OSH1] Ostroverhii, N. A., Strok, V. V. and Homenko, N. P.   [4.1]
        Diameter-critical graphs and digraphs (in Russian), *in*:
        "Topological Aspects of Graph Theory", Math. Inst. AN
        USSR, K., 1971, pp. 214–271.

[P1]   Palumbiny, D.   [4.4]
        On a certain type of decomposition of complete graphs into
        factors with equal diameters, *Mat. Casopis Sloven. Akad.* **22**
        (1972) 235–242.

[P2]   Palumbiny, D.   [4.4]
        On decompositions of complete graphs into factors with equal
        diameters, *Boll. Univ. Math. Ital.* **7** (1973) 420–428.

[P3]   Pelikán, J.   [1.5, 7.2]
        Valency conditions for the existence of certain subgraphs, *in*:
        "Theory of Graphs" (Erdös, P. and Katona, G., eds), Academic
        Press, New York, 1968, pp. 251–259.

[P4]   Perles, M. A.   [2.1]
        A proof of Dilworth's decomposition theorem for partially
        ordered sets, *Israel J. Math.* **1** (1963) 105–107.

[P5]   Petersen, J.   [2.3, 2.4, 2.6]
        Die Theorie der regularen Graphen, *Acta Math.* **15** (1891)
        193–220.

[P6]   Plummer, M. D.   [1.3]
        On minimal blocks, *Trans. Amer. Math. Soc.* **134** (1968)
        85–94.

470   REFERENCES

[P7]   Ponstein, J.   [5.1]
A new proof of Brooks' chromatic number theorem for graphs, *J. Combinatorial Theory* **7** (1969) 255–257.

[P8]   Pósa, L.   [7.3]
Problem No. 127 (in Hungarian), *Mat. Lapok* **12** (1961) 254.

[P9]   Pósa, L.   [3.4]
A theorem concerning Hamiltonian lines, *Publ. Math. Inst. Hungar. Acad. Sci.* **7** (1962) 225–226.

[P10]   Pósa, L.   [3.4]
On the circuits of finite graphs, *Publ. Math. Inst. Hungar. Acad. Sci.* **8** (1963) 355–361.

[P11]   Pym, J. S.   [1.2]
A proof of Menger's theorem, *Monatshefte für Mathematik* **73** (1969) 81–88.

[R1]   Rado, R.   [2.6]
A theorem on general measure functions, *Proc. London Math. Soc.* **44** (1938) 61–91.

[R2]   Rado, R.   [2.3]
Factorization of even graphs, *Quart. J. Math. Oxford* **20** (1949) 95–104.

[R3]   Rado, R.   [2.1]
Note on the transfinite case of Hall's theorem on representatives, *J. London Math. Soc.* **42** (1967) 321–324.

[R4]   Rado, R.   [2.6]
On the number of systems of distinct representatives, *J. London Math. Soc.* **42** (1967) 107–109.

[R5]   Rado, R.   [5.1]
Theorems on the colouring of the edges of a graph, *in*: "Combinatorial Mathematics and its Applications" (Bose, R. C. *et al.*, eds.), Chapel Hill, N. C., 1970, pp. 385–390.

[LR1]   Ramachandra Rao, A. and Rao, S. B.   [2.6]
On factorable degree sequences, *J. Combinatorial Theory* Ser. B **13** (1972) 185–191.

[R6]   Ramsey, F. P.   [5.0, 5.6]
On a problem of formal logic, *Proc. London Math. Soc.* **30** (2) (1929) 264–286.

[R7]   Reiman,.I.   [6.2]
Über ein Problem von K. Zarankiewicz, *Acta. Math. Acad. Sci. Hungar.* **9** (1958) 269–279.

[R8]   Ringel, G.   [5.3]
"Färbungsprobleme auf Flächen und Graphen", Deutscher Verlag der Wissenschaften, Berlin, 1962.

[R8a]   Ringel, G.   [4.5]
Selbstkomplementäre Graphen, *Archive. Math.* **14** (1963) 354–358.

[R9]	Ringel, G. "The Map Color Theorem", Grundlehren der math. Wiss 209, Springer Verlag, Berlin, 1969.	[5.3]
[RY1]	Ringel, G. and Youngs, J. W. T. Solution of the Heawood map-coloring problem, *Proc. Nat. Acad. Sci. U.S.A.* **60** (1968) 438–445.	[5.3]
[RV1]	Rivest, R. L. and Vuillemin, J. A generalization and proof of the Aanderaa–Rosenberg conjecture, *in*: "Proc. SIGACT Conf.", Albuquerque, May, 1975.	[8.2, 8.6]
[RV2]	Rivest, R. L. and Vuillemin, J. On recognising graph properties from adjacency matrices, *Theor. Comput. Sci.* **3** (1976/77) 371–384.	[8.2]
[R10]	Robertson, N. The smallest graph of girth 5 and valency 4, *Bull. Amer. Math. Soc.* **70** (1964) 824–825.	[3.1, 3.6]
[R11]	Rödl, V. The dimension of graph and a generalised Ramsey theorem, Thesis, Charles University, Praha, 1973.	[5.6]
[R12]	Rödl, V. On the chromatic number of subgraphs of a given graph, *Proc. Amer. Math. Soc.* **64** (1977) 370–371.	[5.7]
[R13]	Rosenberg, A. L. On the time required to recognise properties of graphs: a problem, *SIGACT News* **5** (1973) 15–16.	[8.2]
[R14]	Roy, B. Nombre chromatique et plus longs chemins d'un graphe, *Revue AFIRO* **1** (1967) 127–132.	[5.1]
[R15]	Ryser, H. J. "Combinatorial Mathematics", Carus Math. Monographs 14, Amer. Math. Assoc. and John Wiley, New York, 1963.	[2.1]
[S1]	Sachs, H. Über selbstkomplementäre Graphen, *Publ. Math. Debrecen* **9** (1962) 270–288.	[4.5]
[S2]	Sachs, H. Regular graphs with given girth and restricted circuits, *J. London Math. Soc.* **38** (1963) 423–429).	[3.6]
[S3]	Sachs, H. On regular graphs with given girth, *in*: "Theory of Graphs and its Applications", (Fiedler, M., ed.), Academic Press, New York, 1965, pp. 91–97.	[3.6]
[S4]	Sachs, H. Construction of non-Hamiltonian planar regular graphs of degrees 3, 4 and 5 with highest possible connectivity, *in*: "Theory of Graphs" (Rosenstiehl, P., ed.), Gordon and Breach, New York, 1967, pp. 373–382.	[3.4]

[S5]   Sachs, H.   [3.4]
Ein von Kozyrev und Grinberg angegebener nicht-hamiltonischer planarer Graph, *in*: "Beiträge zur Graphentheorie" (Sachs, H., Voss, H.-J. and Walther, H., eds), Teubner Verlagsgesellschaft, Leipzig, 1968, pp. 127–130.

[S6]   Sachs, H.   [5.5]
On the Berge conjecture concerning perfect graphs, *in*: "Combinatorial Structures and their Applications" (Guy, R., Hanani, H., Sauer, N. and Schönheim, J., eds), Gordon and Breach, New York, 1969, pp. 377–384.

[S7]   Sauer, N.   [3.1]
Extremaleigenschaften regulärer Graphen gegebener Taillenweite, I and II, *Sitzungsberichte Österreich. Akad. Wiss. Math. Natur. Kl.*, S-B II, 176 (1967) 9–25, ibid 176 (1967) 27–43.

[S8]   Sauer, N.   [4.4]
On the factorization of the complete graph into factors of diameter 2, *J. Combinatorial Theory* 9 (1970) 423–426.

[S9]   Sauer, N.   [2.4]
The largest number of edges of a graph such that not more than g intersect in a point or more than n are independent, *in*: "Combinatorial Mathematics and its Applications" (Welsh, D. J. A., ed.), Academic Press, London and New York, 1971, pp. 253–257.

[S10]   Sauer, N.   [6.1]
A generalization of a theorem of Turán, *J. Combinatorial Theory* Ser. B (1971) 109–112.

[SS1]   Sauer, N. and Schaer, J.   [4.4]
On the factorization of the complete graph, *J. Combinatorial Theory* Ser. B 14 (1973) 1–6.

[SS2]   Sauer, N. and Spencer, J.   [8.3, 8.4]
Edge disjoint placement of graphs, *J. Combinatorial Theory* Ser. B (to appear).

[S11]   Sauve, L.   [6.1, 6.6]
On chromatic graphs, *Amer. Math. Monthly* 68 (1961) 107–111.

[S12]   Schweitzer Competition 1959   [1.5]
Problem 10, Solution by Bártfai, P. (in Hungarian), *Mat. Lapok* 11 (1960) 175–176.

[S13]   Sheehan, J.   [6.6]
Non-bipartite graphs of girth 4, *Discrete Math.* 8 (1974) 383–402.

[S14]   Sierpinski, W.   [6.2]
Sur un problème concernant un reseau à 36 points, *Ann. Polon. Math.* 24 (1951) 173–174.

[S15]   Sierpinski, W.   [6.2]
"Problems in the Theory of Numbers", Pergamon, Oxford, 1964, p. 16.

[S16] Simoes-Pereira, J. M. S.   [5.7]
       A note on graphs with prescribed clique and point-partition
       numbers, *J. Combinatorial Theory* Ser. B **14** (1973) 256–258.

[S17] Simoes-Pereira, J. M. S.   [5.7]
       On graphs uniquely partitionable into n-degenerate subgraphs,
       in: "Infinite and Finite Sets", vol. III (Hajnal, A., Rado, R.
       and Sós, V. T., eds), Colloq. Math. Soc. J. Bolyai 10, North
       Holland, 1973, pp. 1351–1364.

[S18] Simonovits, M.   [3.2, 3.3]
       A new proof and generalization of a theorem of Erdös and
       Pósa on graphs without $k + 1$ independent circuits, *Acta. Math.
       Acad. Sci. Hungar.* **18** (1967) 191–206.

[S19] Simonovits, M.   [6.0, 6.3, 6.4]
       A method for solving extremal problems in graph theory,
       stability problems, *in*: "Theory of Graphs" (Erdös, P. and
       Katona, G., eds), Academic Press, New York, 1968, pp. 279–
       319.

[S20] Simonovits, M.   [5.2]
       On colour-critical graphs, *Studia Sci. Math. Hungar.* **7** (1971)
       67–81.

[S21] Singleton, R. R.   [3.1, 3.6]
       On minimal graphs of maximum even girth, *J. Combinatorial
       Theory* **1** (1966) 306–332.

[S22] Slater, P. J.   [1.3, 1.4]
       A classification of 4-connected graphs, *J. Combinatorial Theory*
       Ser. B **17** (1974) 281–298.

[ST1] Sørensen, B. A. and Thomassen, C.   [1.5, 1.6]
       On k-rails in graphs, *J. Combinatorial Theory* **17** (1974) 143–159.

[ST2] Stone, N.   [1.6, 8.6]
       Whatever Austria achieved against the Turks, she failed to save
       western civilization from the Hungarians, *The Eastern Front*,
       1914–17, Hodder and Stoughton, 1975, p. 125.

[S23] Stromquist, W.   [5.3]
       "Some Aspects of the Four Color Problem", Ph.D. Thesis,
       Harvard University 1975.

[S24] Stromquist, W.   [5.3]
       The four-colour problem for small maps, *J. Combinatorial
       Theory* Ser. B **19** (1975) 256–268.

[S25] Sumner, D. P.   [6.5]
       On a problem of Erdös, *in*: "Recent Progress in Combinatorics"
       (Tutte, W. T., ed.), Academic Press, New York and London,
       1969, pp. 319–322.

[S26] Surányi, L.   [5.5]
       The covering of graphs by cliques, *Studia Sci. Math. Hungar.* **3**
       (1968) 345–349.

[SW1] Szekeres, G. and Wilf, H. S.   [5.1]
       An inequality for the chromatic number of a graph, *J. Com-
       binatorial Theory* **4** (1968) 1–3.

[S27] Szemerédi, E.   [6.6]
       On graphs without complete quadrilaterals (in Hungarian),
       Mat. Lapok **23** (1972) 113–116.

[S28] Szemerédi, E.   [6.3]
       On a set containing no k elements in arithmetic progression,
       Acta Arithmetica **27** (1975) 199–245.

[S29] Szemerédi, E.   [6.3]
       Regular partitions of graphs, in "Proc. Colloque Inter. CNRS"
       (J.-C. Bermond, J.-C. Fournier, M. das Vergnas, D. Sotteau,
       eds.), 1978.

[T1]   Tait, P. G.   [5.4]
       Listing's topologie, *Phil. Mag.* **17** (5) (1884) 30–46.

[T1a] Thomason, A. G.   [5.7]
       Critically partitionable graphs I, *J. Combinatorial Theory* Ser
       B (to appear)

[T1b] Thomason, A. G.   [3.4, 5.1, 5.7]
       Hamiltonian cycles and uniquely edge colourable graphs, *in*:
       "Advances in Graph Theory" (Bollobás, B. ed.), North-
       Holland, 1978, pp. 259–268.

[T1c] Thomason, A. G.   [5.7]
       Critically partitionable graphs II (to appear)

[T2]   Thomassen, C.   [7.2, 7.4]
       Some homomorphism properties of graphs, *Math. Nachr.* **64**
       (1974) 119–133.

[T2a] Thomassen, C.   [7.2, 7.3]
       A minimal condition implying a special K_4-subdivision in a
       graph, *Archiv Math.* **25** (1974) 210–215.

[T3]   Toft, B.   [5.2]
       On the maximal number of edges of critical k-chromatic
       graphs, *Studia Sci. Math. Hungar.* **5** (1970) 461–470.

[T4]   Toft, B.   [5.2]
       Two theorems on critical 4-chromatic graphs, *Studia Sci. Math.
       Hungar.* **7** (1972) 83–99.

[T4a] Toft, B.   [5.2]
       Colour-critical graphs, *in*: "Advances in Graph Theory"
       (Bollobás, B., ed.), North-Holland, 1978.

[T5]   Turán, P.   [6.0, 6.1]
       On an extremal problem in graph theory (in Hungarian), *Mat.
       Fiz. Lapok* **48** (1941) 436–452.

[T6]   Turán, P.   [6.1]
       On the theory of graphs, *Colloq. Math.* **3** (1954) 19–30.

[T7]   Tutte, W. T.   [3.4]
       On Hamiltonian circuits, *J. London Math. Soc.* **21** (1946)
       98–101.

[T8]   Tutte, W. T.   [3.1]
       A family of cubical graphs, *Proc. Cambridge Philos. Soc.* **43**
       (1947) 459–474.

[T9]   Tutte, W. T.   [2.1]
         The factorization of linear graphs, *J. London Math. Soc.* **22**
         (1947) 107–111.

[T10]   Tutte, W. T.   [2.3]
         The factors of graphs, *Canad. J. Math.* **4** (1952) 314–328.

[T11]   Tutte, W. T.   [2.3]
         A short proof of the factor theorem for finite graphs, *Canad.
         J. Math.* **6** (1954) 347–352.

[T12]   Tutte, W. T.   [3.4]
         A non-Hamiltonian graph, *Canad. Math. Bull.* **3** (1960) 1–5.

[T13]   Tutte, W. T.   [3.4]
         A non-Hamiltonian planar graph, *Acta Math. Acad. Sci.
         Hungar.* **11** (1960) 371–375.

[T14]   Tutte, W. T.   [2.5]
         On the problem of decomposing a graph into n connected
         factors, *J. London Math. Soc.* **36** (1961) 221–230.

[T15]   Tutte, W. T.   [1.3]
         A theory of 3-connected graphs, *Indag. Math.* **23** (1961)
         441–455.

[T16]   Tutte, W. T.   [5.3]
         How to draw a graph, *Proc. London Math. Soc.* **13** (1963)
         743–767.

[T17]   Tutte, W. T.   [3.6]
         "Connectivity in Graphs", Math. Exposition 15, Univ. of
         Toronto Press, Toronto, Ont. Oxford Univ. Press, London,
         1966.

[T18]   Tutte, W. T.   [3.4]
         Hamiltonian circuits, *in*: "Graph Theory and Computing"
         (Read, R. C. ed.), Academic Press, 1972, pp. 295–301.

[T19]   Tutte, W. T.   [2.3, 2.6]
         Spanning subgraphs with specified valencies, *Discrete Math.*
         9 (1974) 97–108.

[T20]   Tutte, W. T.   [5.1, 5.7]
         Hamiltonian circuits, "Teorie Combinatorie", Atti dei Con-
         veni dincei 17, Roma, 1976, pp. 193–199.

[T21]   Tutte, W. T.   [2.3]
         The subgraph problem, *in*: "Advances in Graph Theory"
         (Bollobás, B., ed.), North-Holland, 1978.

[TW1]   Tutte, W. T. and Whitney, H.   [5.3]
         Kempe chains and the four color problem, *Utilitas Mathe-
         matica* **2** (1972) 241–281.

[T22]   Tverberg, H.   [2.1]
         On Dilworth's decomposition theorem for partially ordered
         sets, *J. Combinatorial Theory* **3** (1967) 305–306.

[V1]　　Van der Waerden, B. L.　　　　　　　　　　　　　　[2.6]
　　　　Ein Satz über Klasseneinteilung von endlichen Mengen,
　　　　Abh. Math. Sem. Hamburg Univ. **5** (1927) 185–188.

[V2]　　Van der Waerden, B. L.　　　　　　　　　　　　　　[5.6]
　　　　Beweis einer Bundetschen Vermutung, *Nieuw Arch. Wisk.* **16**
　　　　(1927) 212–216.

[V3]　　van Lint, J. H.　　　　　　　　　　　　　　　　[4.3, 4.5]
　　　　Problem 350, solns. by D. Bolton, P. Erdös, J. H. van Lint,
　　　　P. Seymour, J. Spencer and R. Tijdeman, *Nieuw Arch. Wisk.* (3)
　　　　22 (1974) 94–109.

[V4]　　van Lint, J. H.　　　　　　　　　　　　　　　　[4.3, 4.6]
　　　　A problem on bipartite graphs, *Amer. Math. Monthly* **82**
　　　　(1975) 55–56.

[V5]　　Vizing, V. G.　　　　　　　　　　　　　　　　　[5.0, 5.1]
　　　　On an estimate of the chromatic class of a p-graph (in Russian),
　　　　Diskret. Analiz **3** (1964) 23–30.

[V6]　　Vizing, V. G.　　　　　　　　　　　　　　　　　　[5.7]
　　　　Critical graphs with a given chromatic class (in Russian),
　　　　Diskret. Analiz **5** (1965) 9–17.

[V7]　　Vizing, V. G.　　　　　　　　　　　　　　　　　　[4.2]
　　　　The number of edges in a graph of given radius (in Russian),
　　　　Doklady AN USSR **173** (1967) 1245–1246.

[V8]　　Vizing, V. G.　　　　　　　　　　　　　　　　　　[5.7]
　　　　Some unsolved problems in graph theory, *Russian Math.
　　　　Surveys* **23** (1968) 125–141.

[V9]　　Vollmerhaus, H.　　　　　　　　　　　　　　　　　[5.3]
　　　　Über die Einbettung von Graphen in zweidimensionale orien-
　　　　tierbare Manningfaltigkeiten kleinsten Geschlechts, *in*:
　　　　"Beiträge zur Graphentheorie" (Sachs, H., Voss, H. and
　　　　Walther, H. eds), Teubner Verlagsgesellschaft, Leipzig, 1968.

[V10]　Voss, H.-J.　　　　　　　　　　　　　　　　　　　[3.3]
　　　　Über die Taillenweite in Graphen, die maximal k unabhängige
　　　　Kreise enthalten, und über die Anzahl der Knotenpunkte, die
　　　　alle Kreise repräsentieren, X. Internat. Wiss. Koll.TH Ilmenau
　　　　1965, Vortragsreihe, Math. Probl. Ok. u. Rechentechnik, 23–27.

[V11]　Voss, H.-J.　　　　　　　　　　　　　　　　　　　[3.3]
　　　　Some properties of graphs containing k independent circuits,
　　　　in: "Theory of Graphs" (Erdös, P. and Katona, G., eds),
　　　　Academic Press, New York, 1968, pp. 321–335.

[W1]　　Wagner, K.　　　　　　　　　　　　　　　　　　[5.3, 7.4]
　　　　Über eine Eigenschaft der Ebenen Komplexe, *Math. Ann.* **114**
　　　　(1937) 570–590.

[W2]　　Wagner, K.　　　　　　　　　　　　　　　　　　　[5.3]
　　　　Über eine Erweiterung eines Satzes von Kuratowski, *Deut.
　　　　Math.* **2** (1937) 280–285.

[W3]   Wagner, K.   [7.1]
Bemerkungen zu Hadwigers Vermutung, *Math. Ann.* **141** (1960) 433–451.

[W4]   Wagner, K.   [7.1, 7.4]
Beweis einer Abschwächung der Hadwiger-Vermutung, *Math. Ann.* **153** (1964) 139–141.

[W5]   Walther, H.   [3.1]
Eigenschaften von regulären Graphen gegebener Taillenweite und minimaler Knotenzahl, *Wiss. Z. Ilmenau* **11** (1965) 167–168.

[W6]   Walther, H.   [3.4]
Ein kubischer, planarer, zyklisch fünffach zusammenhängender Graph, der keinen Hamilton-kreis besitzt, *Wiss. Z. Techn. Hochsch. Ilmenau* **11** (1965) 163–166.

[W7]   Walther, H.   [3.1]
Über reguläre Graphen gegebener Taillenweite und minimaler Knotenzahl, *Wiss. Z. Techn. Hochsch. Ilmenau* **11** (1965) 93–96.

[W8]   Walther, H.   [3.4]
"Über das Problem der Existenz von Hamiltonkreisen in planaren, regularen Graphen der Grade 3, 4 und 5", Dissertation, T. H. Ilmenau, 1966.

[W9]   Walther, H.   [3.4]
On the problem of the existence of Hamilton-lines in planar regular graphs, *in*: "Theory of Graphs" (Erdös, P. and Katona, G., eds), Academic Press, New York, 1968, pp. 341–343.

[W10]   Walther, H.   [3.4]
Über das Problem der Existenz von Hamiltonkreisen in planaren regulären Graphen, *Math. Nachr.* **39** (1969) 277–296.

[WV1] Walther, H. and Vos, H.-J.   [3.3]
Über Kreise in Graphen, *VEB Deutscher Verlag*, 1974.

[W11]   Watkins, M. E.   [4.1, 4.5]
A lower bound for the number of vertices of a graph, *Amer. Math. Monthly* **74** (1976) 297.

[W12]   Watkins, M.   [1.6]
On the existence of certain disjoint arcs in graphs, *Duke Math. J.* **35** (1968) 231–246.

[W13]   Wegner, G.   [3.1]
A smallest graph of girth 5 and valency 5, *J. Combinatorial Theory* Ser. **B 14** (1973) 203–208.

[WP1] Welsh, D. J. A. and Powell, M. B.   [5.1]
An upper bound for the chromatic number of a graph and its applications to timetabling problems, *Computer J.* **10** (1967) 85–87.

[W14]   Weinstein, J. H.   [3.2]
On the number of disjoint edges in a graph, *Canad. J. Math.* **15** (1963) 106–111.

[W14a]   Weinstein, J.   [5.2, 5.7]
          Excess in critical graphs, *J. Combinatorial Theory* Ser. B **18**
          (1975) 24–31.

[W15]   Wessel, W.   [6.2]
          Über eine Klasse paarer Graphen, I: Beweis einer Vermutung
          von Erdös, Hajnal and Moon, *Wiss. Z. Techn. Hochsch.*
          Ilmenau **12** (1966) 253–256.

[W16]   Wessel, W.   [6.2, 6.6]
          Über eine Klasse paarer Graphen, II: Bestimmung der Minimal-
          graphen, *Wiss Z. Techn. Hochsch. Ilmenau* **13** (1976) 423–426.

[W17]   Whitney, H.   [3.4]
          A theorem on graphs, *Ann. Math.* **32** (1931) 378–390.

[W18]   Whitney, H.   [1.1]
          Congruent graphs and the connectivity of graphs, *Amer. J.*
          Math. **54** (1932) 150–168.

[W19]   Whitney, H.   [1.1, 5.3]
          Non-separable and planar graphs, *Trans. Amer. Math. Soc.* **34**
          (1932) 339–362.

[W20]   Whitney, H.   [5.3]
          Planar graphs, *Fund. Math.* **21** (1933) 73–84.

[W21]   Wilson, R. J.   [5.7]
          Problem 2, *in*: "Proc. Fifth British Combinatorial Conf."
          (Nash-Williams, C. St. J. A. and Sheehan, J. eds), Utilitas
          Math., Winnipeg, 1967, p. 696.

[WB1]   Wilson, R. J. and Beineke, L. W.   [5.7]
          Three conjectures on critical graphs, *Amer. Math. Monthly* **83**
          (1976) 128–129.

[W22]   Woodall, D. R.   [7.4]
          Thrackles and deadlock, *in*: "Combinatorial Mathematics and
          and its Applications" (Welsh, D. J. A., ed.), Academic Press,
          London and New York, 1971, pp. 335–347.

[W23]   Woodall, D. R.   [3.4, 3.5]
          Sufficient conditions for circuits in graphs, *Proc. London Math.*
          Soc. **24** (3) (1972) 739–755.

[W24]   Woodall, D. R.   [2.6, 6.6]
          The binding number of a graph and its Anderson number,
          J. Combinatorial Theory **15** (1973) 225–255.

[Y1]   Yackel, J.   [5.7]
          Inequalities and asymptotic bounds for Ramsey numbers, *J.*
          Combinatorial Theory Ser B **13** (1972) 56–58.

[Y2]   Yackel, J.   [5.7]
          Inequalities and asymptotic bounds for Ramsey numbers II, *in*:
          "Infinite and Finite Sets", vol. III (Erdös, P., Hajnal, A. and
          Rado, R., eds), Colloq. Math. Soc. J. Bolyai 10, Academic
          Press, New York, 1973, pp. 1537–1545.

[Y3]   Young, H. P.   [5.3]
        A quick proof of Wagner's equivalence theorem, *J. London
        Math. Soc.* **3** (2) (1971) 661–664.

[Y4]   Youngs, J. W. T.   [5.3]
        The Heawood map colouring conjecture, *in*: "Graph Theory
        and Theoretical Physics" (Harary, F., ed.), Academic Press,
        London, 1967, pp. 313–354.

[Z1]   Zaks, J.   [2.2, 2.6]
        On the 1-factors of *n*-connected graphs, *in*: "Combinatorial
        Structures and their Applications" (Guy, R. K., Hanani, H.,
        Sauer, N. and Schönheim, J., eds), Gordon and Breach, New
        York, 1970, pp. 481–488.

[Z2]   Zaks, J.   [3.4]
        Pairs of Hamiltonian circuits in 5-connected planar graphs,
        J. Combinatorial Theory Ser. B **21** (1976) 116–131.

[Z3]   Zarankiewicz, K.   [6.2]
        Problem P 101, *Colloq. Math.* **2** (1951) 301.

[Z3a]   Zarankiewicz, K.   [6.1]
        On a problem of Turán concerning graphs, *Fund. Math.* **41**
        (1954) 137–145.

[Z4]   Zeidl, B.   [5.2]
        Über 4- und 5-chrome Graphen, *Monatsh. Math.* **62** (1958)
        212–218.

[Z5]   Zelinka, B.   [6.5]
        On the number of independent complete subgraphs, *Publ.
        Math. Debrecen* **13** (1966) 95–97.

[Z6]   Znám, Š.   [6.2]
        On a combinatorial problem of K. Zarankiewicz, *Colloq.
        Math.* **11** (1963) 81–84.

[Z7]   Znám, Š.   [6.2]
        Two improvements of a result concerning a problem of K.
        Zarankiewicz, *Colloq. Math.* **13** (1965) 255–258.

[Z8]   Zykov, A. A.   [5.0, 5.4, 6.1]
        On some properties of linear complexes (in Russian), *Mat.
        Sbornik N.S.* **24** (1949) 163–188; *Amer. Math. Soc. Transl.*
        79 (1952).

Index of Symbols

Most of the symbols occurring in the text are listed here, separated into three groups: symbols for graphs and operations on graphs, Roman letters and Greek letters.

481

Index of Definitions

kernel, 387
Kneser graph, 259

length, xv
H line, 252
linear colouring, 274
local connectivity, 7
loop, xviii

map, 249
matching, 50
maximum degree, xiv
minimum degree, xiv
monotone property, 91, 405, 411
Moore graph, 106
multigraph, xviii
multiple edge, xviii
multiplying a graph, 264

neighbourhood complex, 260
neighbouring countries, 249
normal set, 69
null graph, xix

odd component, 55, 67
odd cycle, xv
order of a property, 408
orientation of a graph, xviii
oriented graph, xviii, 429
oriented pregraph, 429

packing of graphs, 401
pancyclic graph, 103
r-partite graph, xvi
partition theory, 275
(*k, P*)-partition, 261
path, xv
    directed, xviii, 225
    Hamiltonian, 102
    x–y, X–Y, xv
perfect graph, 219, 263
Petersen graph, 106
planar graph, 244
plane graph, 244
plane map, 249

pregraph, 429
    oriented, 429
preset, 404
probe, 404
property, 402
    complement of, 405
    dual, 405
    elusive, 403
    k-elusive, 403
    monotone, 91, 405, 411
    of graphs, 402
    of subsets, 405
    order of, 408
    k-stable, 131
    trivial, 403
pseudograph, xviii

radius, xvi
Ramsey number, 272
    generalized, 275
realization of a graph, 243
regular of degree *k*, xiv
k-regular, xiv
representation, 50

saturated graph, 103, 308, 325
Seeker, 403
segment, xv
semi-topological graph, 387
r-set, xiii
r-set colouring, 289
set of distinct representatives, 53
set system, xvii
simple strategy, 406
sink, 429
size of a graph, xiv
f-soluble graph, 74
source, 430
spanned subgraph, xiv
k-stable property, 131
star, xix
strategy, 405
strongly saturated, 308, 325
subcontraction, xix, 251
subdivision, xviii
subgraph, xiv
substituting a graph, 264

A CATALOG OF SELECTED
DOVER BOOKS
IN SCIENCE AND MATHEMATICS

Astronomy

BURNHAM'S CELESTIAL HANDBOOK, Robert Burnham, Jr. Thorough guide to the stars beyond our solar system. Exhaustive treatment. Alphabetical by constellation: Andromeda to Cetus in Vol. 1; Chamaeleon to Orion in Vol. 2; and Pavo to Vulpecula in Vol. 3. Hundreds of illustrations. Index in Vol. 3. 2,000pp. 6⅛ x 9¼.
Vol. I: 23567-X
Vol. II: 23568-8
Vol. III: 23673-0

EXPLORING THE MOON THROUGH BINOCULARS AND SMALL TELESCOPES, Ernest H. Cherrington, Jr. Informative, profusely illustrated guide to locating and identifying craters, rills, seas, mountains, other lunar features. Newly revised and updated with special section of new photos. Over 100 photos and diagrams. 240pp. 8¼ x 11. 24491-1

THE EXTRATERRESTRIAL LIFE DEBATE, 1750–1900, Michael J. Crowe. First detailed, scholarly study in English of the many ideas that developed from 1750 to 1900 regarding the existence of intelligent extraterrestrial life. Examines ideas of Kant, Herschel, Voltaire, Percival Lowell, many other scientists and thinkers. 16 illustrations. 704pp. 5⅜ x 8½. 40675-X

THEORIES OF THE WORLD FROM ANTIQUITY TO THE COPERNICAN REVOLUTION, Michael J. Crowe. Newly revised edition of an accessible, enlightening book recreates the change from an earth-centered to a sun-centered conception of the solar system. 242pp. 5⅜ x 8½. 41444-2

A HISTORY OF ASTRONOMY, A. Pannekoek. Well-balanced, carefully reasoned study covers such topics as Ptolemaic theory, work of Copernicus, Kepler, Newton, Eddington's work on stars, much more. Illustrated. References. 521pp. 5⅜ x 8½. 65994-1

A COMPLETE MANUAL OF AMATEUR ASTRONOMY: Tools and Techniques for Astronomical Observations, P. Clay Sherrod with Thomas L. Koed. Concise, highly readable book discusses: selecting, setting up and maintaining a telescope; amateur studies of the sun; lunar topography and occultations; observations of Mars, Jupiter, Saturn, the minor planets and the stars; an introduction to photoelectric photometry; more. 1981 ed. 124 figures. 26 halftones. 37 tables. 335pp. 6½ x 9¼. 42820-6

AMATEUR ASTRONOMER'S HANDBOOK, J. B. Sidgwick. Timeless, comprehensive coverage of telescopes, mirrors, lenses, mountings, telescope drives, micrometers, spectroscopes, more. 189 illustrations. 576pp. 5⅝ x 8¼. (Available in U.S. only.) 24034-7

STARS AND RELATIVITY, Ya. B. Zel'dovich and I. D. Novikov. Vol. 1 of *Relativistic Astrophysics* by famed Russian scientists. General relativity, properties of matter under astrophysical conditions, stars, and stellar systems. Deep physical insights, clear presentation. 1971 edition. References. 544pp. 5⅝ x 8¼. 69424-0

Chemistry

THE SCEPTICAL CHYMIST: The Classic 1661 Text, Robert Boyle. Boyle defines the term "element," asserting that all natural phenomena can be explained by the motion and organization of primary particles. 1911 ed. viii+232pp. 5⅜ x 8½.
42825-7

RADIOACTIVE SUBSTANCES, Marie Curie. Here is the celebrated scientist's doctoral thesis, the prelude to her receipt of the 1903 Nobel Prize. Curie discusses establishing atomic character of radioactivity found in compounds of uranium and thorium; extraction from pitchblende of polonium and radium; isolation of pure radium chloride; determination of atomic weight of radium; plus electric, photographic, luminous, heat, color effects of radioactivity. ii+94pp. 5⅜ x 8½.   42550-9

CHEMICAL MAGIC, Leonard A. Ford. Second Edition, Revised by E. Winston Grundmeier. Over 100 unusual stunts demonstrating cold fire, dust explosions, much more. Text explains scientific principles and stresses safety precautions. 128pp. 5⅜ x 8½.   67628-5

THE DEVELOPMENT OF MODERN CHEMISTRY, Aaron J. Ihde. Authoritative history of chemistry from ancient Greek theory to 20th-century innovation. Covers major chemists and their discoveries. 209 illustrations. 14 tables. Bibliographies. Indices. Appendices. 851pp. 5⅜ x 8½.   64235-6

CATALYSIS IN CHEMISTRY AND ENZYMOLOGY, William P. Jencks. Exceptionally clear coverage of mechanisms for catalysis, forces in aqueous solution, carbonyl- and acyl-group reactions, practical kinetics, more. 864pp. 5⅜ x 8½.
65460-5

ELEMENTS OF CHEMISTRY, Antoine Lavoisier. Monumental classic by founder of modern chemistry in remarkable reprint of rare 1790 Kerr translation. A must for every student of chemistry or the history of science. 539pp. 5⅜ x 8½.   64624-6

THE HISTORICAL BACKGROUND OF CHEMISTRY, Henry M. Leicester. Evolution of ideas, not individual biography. Concentrates on formulation of a coherent set of chemical laws. 260pp. 5⅜ x 8½.   61053-5

A SHORT HISTORY OF CHEMISTRY, J. R. Partington. Classic exposition explores origins of chemistry, alchemy, early medical chemistry, nature of atmosphere, theory of valency, laws and structure of atomic theory, much more. 428pp. 5⅜ x 8½. (Available in U.S. only.)   65977-1

GENERAL CHEMISTRY, Linus Pauling. Revised 3rd edition of classic first-year text by Nobel laureate. Atomic and molecular structure, quantum mechanics, statistical mechanics, thermodynamics correlated with descriptive chemistry. Problems. 992pp. 5⅜ x 8½.   65622-5

FROM ALCHEMY TO CHEMISTRY, John Read. Broad, humanistic treatment focuses on great figures of chemistry and ideas that revolutionized the science. 50 illustrations. 240pp. 5⅜ x 8½.   28690-8

Engineering

DE RE METALLICA, Georgius Agricola. The famous Hoover translation of greatest treatise on technological chemistry, engineering, geology, mining of early modern times (1556). All 289 original woodcuts. 638pp. 6¾ x 11. 60006-8

FUNDAMENTALS OF ASTRODYNAMICS, Roger Bate et al. Modern approach developed by U.S. Air Force Academy. Designed as a first course. Problems, exercises. Numerous illustrations. 455pp. 5⅜ x 8½. 60061-0

DYNAMICS OF FLUIDS IN POROUS MEDIA, Jacob Bear. For advanced students of ground water hydrology, soil mechanics and physics, drainage and irrigation engineering, and more. 335 illustrations. Exercises, with answers. 784pp. 6⅛ x 9¼. 65675-6

THEORY OF VISCOELASTICITY (Second Edition), Richard M. Christensen. Complete, consistent description of the linear theory of the viscoelastic behavior of materials. Problem-solving techniques discussed. 1982 edition. 29 figures. xiv+364pp. 6⅛ x 9¼. 42880-X

MECHANICS, J. P. Den Hartog. A classic introductory text or refresher. Hundreds of applications and design problems illuminate fundamentals of trusses, loaded beams and cables, etc. 334 answered problems. 462pp. 5⅜ x 8½. 60754-2

MECHANICAL VIBRATIONS, J. P. Den Hartog. Classic textbook offers lucid explanations and illustrative models, applying theories of vibrations to a variety of practical industrial engineering problems. Numerous figures. 233 problems, solutions. Appendix. Index. Preface. 436pp. 5⅜ x 8½. 64785-4

STRENGTH OF MATERIALS, J. P. Den Hartog. Full, clear treatment of basic material (tension, torsion, bending, etc.) plus advanced material on engineering methods, applications. 350 answered problems. 323pp. 5⅜ x 8½. 60755-0

A HISTORY OF MECHANICS, René Dugas. Monumental study of mechanical principles from antiquity to quantum mechanics. Contributions of ancient Greeks, Galileo, Leonardo, Kepler, Lagrange, many others. 671pp. 5⅜ x 8½. 65632-2

STABILITY THEORY AND ITS APPLICATIONS TO STRUCTURAL MECHANICS, Clive L. Dym. Self-contained text focuses on Koiter postbuckling analyses, with mathematical notions of stability of motion. Basing minimum energy principles for static stability upon dynamic concepts of stability of motion, it develops asymptotic buckling and postbuckling analyses from potential energy considerations, with applications to columns, plates, and arches. 1974 ed. 208pp. 5⅜ x 8½. 42541-X

METAL FATIGUE, N. E. Frost, K. J. Marsh, and L. P. Pook. Definitive, clearly written, and well-illustrated volume addresses all aspects of the subject, from the historical development of understanding metal fatigue to vital concepts of the cyclic stress that causes a crack to grow. Includes 7 appendixes. 544pp. 5⅜ x 8½. 40927-9

ROCKETS, Robert Goddard. Two of the most significant publications in the history of rocketry and jet propulsion: "A Method of Reaching Extreme Altitudes" (1919) and "Liquid Propellant Rocket Development" (1936). 128pp. 5⅜ x 8½.   42537-1

STATISTICAL MECHANICS: Principles and Applications, Terrell L. Hill. Standard text covers fundamentals of statistical mechanics, applications to fluctuation theory, imperfect gases, distribution functions, more. 448pp. 5⅜ x 8½.   65390-0

ENGINEERING AND TECHNOLOGY 1650–1750: Illustrations and Texts from Original Sources, Martin Jensen. Highly readable text with more than 200 contemporary drawings and detailed engravings of engineering projects dealing with surveying, leveling, materials, hand tools, lifting equipment, transport and erection, piling, bailing, water supply, hydraulic engineering, and more. Among the specific projects outlined–transporting a 50-ton stone to the Louvre, erecting an obelisk, building timber locks, and dredging canals. 207pp. 8⅜ x 11¼.   42232-1

THE VARIATIONAL PRINCIPLES OF MECHANICS, Cornelius Lanczos. Graduate level coverage of calculus of variations, equations of motion, relativistic mechanics, more. First inexpensive paperbound edition of classic treatise. Index. Bibliography. 418pp. 5⅜ x 8½.   65067-7

PROTECTION OF ELECTRONIC CIRCUITS FROM OVERVOLTAGES, Ronald B. Standler. Five-part treatment presents practical rules and strategies for circuits designed to protect electronic systems from damage by transient overvoltages. 1989 ed. xxiv+434pp. 6⅛ x 9¼.   42552-5

ROTARY WING AERODYNAMICS, W. Z. Stepniewski. Clear, concise text covers aerodynamic phenomena of the rotor and offers guidelines for helicopter performance evaluation. Originally prepared for NASA. 537 figures. 640pp. 6⅛ x 9¼.
        64647-5

INTRODUCTION TO SPACE DYNAMICS, William Tyrrell Thomson. Comprehensive, classic introduction to space-flight engineering for advanced undergraduate and graduate students. Includes vector algebra, kinematics, transformation of coordinates. Bibliography. Index. 352pp. 5⅜ x 8½.   65113-4

HISTORY OF STRENGTH OF MATERIALS, Stephen P. Timoshenko. Excellent historical survey of the strength of materials with many references to the theories of elasticity and structure. 245 figures. 452pp. 5⅜ x 8½.   61187-6

ANALYTICAL FRACTURE MECHANICS, David J. Unger. Self-contained text supplements standard fracture mechanics texts by focusing on analytical methods for determining crack-tip stress and strain fields. 336pp. 6⅛ x 9¼.   41737-9

STATISTICAL MECHANICS OF ELASTICITY, J. H. Weiner. Advanced, self-contained treatment illustrates general principles and elastic behavior of solids. Part 1, based on classical mechanics, studies thermoelastic behavior of crystalline and polymeric solids. Part 2, based on quantum mechanics, focuses on interatomic force laws, behavior of solids, and thermally activated processes. For students of physics and chemistry and for polymer physicists. 1983 ed. 96 figures. 496pp. 5⅜ x 8½.   42260-7

Mathematics

FUNCTIONAL ANALYSIS (Second Corrected Edition), George Bachman and Lawrence Narici. Excellent treatment of subject geared toward students with background in linear algebra, advanced calculus, physics, and engineering. Text covers introduction to inner-product spaces, normed, metric spaces, and topological spaces; complete orthonormal sets, the Hahn-Banach Theorem and its consequences, and many other related subjects. 1966 ed. 544pp. 6⅛ x 9¼. 40251-7

ASYMPTOTIC EXPANSIONS OF INTEGRALS, Norman Bleistein & Richard A. Handelsman. Best introduction to important field with applications in a variety of scientific disciplines. New preface. Problems. Diagrams. Tables. Bibliography. Index. 448pp. 5⅜ x 8½. 65082-0

VECTOR AND TENSOR ANALYSIS WITH APPLICATIONS, A. I. Borisenko and I. E. Tarapov. Concise introduction. Worked-out problems, solutions, exercises. 257pp. 5⅝ x 8¼. 63833-2

THE ABSOLUTE DIFFERENTIAL CALCULUS (CALCULUS OF TENSORS), Tullio Levi-Civita. Great 20th-century mathematician's classic work on material necessary for mathematical grasp of theory of relativity. 452pp. 5⅝ x 8¼. 63401-9

AN INTRODUCTION TO ORDINARY DIFFERENTIAL EQUATIONS, Earl A. Coddington. A thorough and systematic first course in elementary differential equations for undergraduates in mathematics and science, with many exercises and problems (with answers). Index. 304pp. 5⅜ x 8½. 65942-9

FOURIER SERIES AND ORTHOGONAL FUNCTIONS, Harry F. Davis. An incisive text combining theory and practical example to introduce Fourier series, orthogonal functions and applications of the Fourier method to boundary-value problems. 570 exercises. Answers and notes. 416pp. 5⅜ x 8½. 65973-9

COMPUTABILITY AND UNSOLVABILITY, Martin Davis. Classic graduate-level introduction to theory of computability, usually referred to as theory of recurrent functions. New preface and appendix. 288pp. 5⅜ x 8½. 61471-9

ASYMPTOTIC METHODS IN ANALYSIS, N. G. de Bruijn. An inexpensive, comprehensive guide to asymptotic methods–the pioneering work that teaches by explaining worked examples in detail. Index. 224pp. 5⅜ x 8½ 64221-6

APPLIED COMPLEX VARIABLES, John W. Dettman. Step-by-step coverage of fundamentals of analytic function theory–plus lucid exposition of five important applications: Potential Theory; Ordinary Differential Equations; Fourier Transforms; Laplace Transforms; Asymptotic Expansions. 66 figures. Exercises at chapter ends. 512pp. 5⅜ x 8½. 64670-X

INTRODUCTION TO LINEAR ALGEBRA AND DIFFERENTIAL EQUATIONS, John W. Dettman. Excellent text covers complex numbers, determinants, orthonormal bases, Laplace transforms, much more. Exercises with solutions. Undergraduate level. 416pp. 5⅜ x 8½. 65191-6

CALCULUS OF VARIATIONS WITH APPLICATIONS, George M. Ewing. Applications-oriented introduction to variational theory develops insight and promotes understanding of specialized books, research papers. Suitable for advanced undergraduate/graduate students as primary, supplementary text. 352pp. 5⅜ x 8½.
64856-7

COMPLEX VARIABLES, Francis J. Flanigan. Unusual approach, delaying complex algebra till harmonic functions have been analyzed from real variable viewpoint. Includes problems with answers. 364pp. 5⅜ x 8½.
61388-7

AN INTRODUCTION TO THE CALCULUS OF VARIATIONS, Charles Fox. Graduate-level text covers variations of an integral, isoperimetrical problems, least action, special relativity, approximations, more. References. 279pp. 5⅜ x 8½.
65499-0

COUNTEREXAMPLES IN ANALYSIS, Bernard R. Gelbaum and John M. H. Olmsted. These counterexamples deal mostly with the part of analysis known as "real variables." The first half covers the real number system, and the second half encompasses higher dimensions. 1962 edition. xxiv+198pp. 5⅜ x 8½.
42875-3

CATASTROPHE THEORY FOR SCIENTISTS AND ENGINEERS, Robert Gilmore. Advanced-level treatment describes mathematics of theory grounded in the work of Poincaré, R. Thom, other mathematicians. Also important applications to problems in mathematics, physics, chemistry, and engineering. 1981 edition. References. 28 tables. 397 black-and-white illustrations. xvii+666pp. 6⅛ x 9¼.
67539-4

INTRODUCTION TO DIFFERENCE EQUATIONS, Samuel Goldberg. Exceptionally clear exposition of important discipline with applications to sociology, psychology, economics. Many illustrative examples; over 250 problems. 260pp. 5⅜ x 8½.
65084-7

NUMERICAL METHODS FOR SCIENTISTS AND ENGINEERS, Richard Hamming. Classic text stresses frequency approach in coverage of algorithms, polynomial approximation, Fourier approximation, exponential approximation, other topics. Revised and enlarged 2nd edition. 721pp. 5⅜ x 8½.
65241-6

INTRODUCTION TO NUMERICAL ANALYSIS (2nd Edition), F. B. Hildebrand. Classic, fundamental treatment covers computation, approximation, interpolation, numerical differentiation and integration, other topics. 150 new problems. 669pp. 5⅜ x 8½.
65363-3

THREE PEARLS OF NUMBER THEORY, A. Y. Khinchin. Three compelling puzzles require proof of a basic law governing the world of numbers. Challenges concern van der Waerden's theorem, the Landau-Schnirelmann hypothesis and Mann's theorem, and a solution to Waring's problem. Solutions included. 64pp. 5⅜ x 8½.
40026-3

THE PHILOSOPHY OF MATHEMATICS: An Introductory Essay, Stephan Körner. Surveys the views of Plato, Aristotle, Leibniz & Kant concerning propositions and theories of applied and pure mathematics. Introduction. Two appendices. Index. 198pp. 5⅜ x 8½.
25048-2

INTRODUCTORY REAL ANALYSIS, A.N. Kolmogorov, S. V. Fomin. Translated by Richard A. Silverman. Self-contained, evenly paced introduction to real and functional analysis. Some 350 problems. 403pp. 5⅜ x 8½.   61226-0

APPLIED ANALYSIS, Cornelius Lanczos. Classic work on analysis and design of finite processes for approximating solution of analytical problems. Algebraic equations, matrices, harmonic analysis, quadrature methods, more. 559pp. 5⅜ x 8½.   65656-X

AN INTRODUCTION TO ALGEBRAIC STRUCTURES, Joseph Landin. Superb self-contained text covers "abstract algebra": sets and numbers, theory of groups, theory of rings, much more. Numerous well-chosen examples, exercises. 247pp. 5⅜ x 8½.   65940-2

QUALITATIVE THEORY OF DIFFERENTIAL EQUATIONS, V. V. Nemytskii and V.V. Stepanov. Classic graduate-level text by two prominent Soviet mathematicians covers classical differential equations as well as topological dynamics and ergodic theory. Bibliographies. 523pp. 5⅜ x 8½.   65954-2

THEORY OF MATRICES, Sam Perlis. Outstanding text covering rank, nonsingularity and inverses in connection with the development of canonical matrices under the relation of equivalence, and without the intervention of determinants. Includes exercises. 237pp. 5⅜ x 8½.   66810-X

INTRODUCTION TO ANALYSIS, Maxwell Rosenlicht. Unusually clear, accessible coverage of set theory, real number system, metric spaces, continuous functions, Riemann integration, multiple integrals, more. Wide range of problems. Undergraduate level. Bibliography. 254pp. 5⅜ x 8½.   65038-3

MODERN NONLINEAR EQUATIONS, Thomas L. Saaty. Emphasizes practical solution of problems; covers seven types of equations. ". . . a welcome contribution to the existing literature. . . ."–*Math Reviews.* 490pp. 5⅜ x 8½.   64232-1

MATRICES AND LINEAR ALGEBRA, Hans Schneider and George Phillip Barker. Basic textbook covers theory of matrices and its applications to systems of linear equations and related topics such as determinants, eigenvalues, and differential equations. Numerous exercises. 432pp. 5⅜ x 8½.   66014-1

MATHEMATICS APPLIED TO CONTINUUM MECHANICS, Lee A. Segel. Analyzes models of fluid flow and solid deformation. For upper-level math, science, and engineering students. 608pp. 5⅜ x 8½.   65369-2

ELEMENTS OF REAL ANALYSIS, David A. Sprecher. Classic text covers fundamental concepts, real number system, point sets, functions of a real variable, Fourier series, much more. Over 500 exercises. 352pp. 5⅜ x 8½.   65385-4

SET THEORY AND LOGIC, Robert R. Stoll. Lucid introduction to unified theory of mathematical concepts. Set theory and logic seen as tools for conceptual understanding of real number system. 496pp. 5⅜ x 8¼.   63829-4

TENSOR CALCULUS, J.L. Synge and A. Schild. Widely used introductory text covers spaces and tensors, basic operations in Riemannian space, non-Riemannian spaces, etc. 324pp. 5⅜ x 8¼. 63612-7

ORDINARY DIFFERENTIAL EQUATIONS, Morris Tenenbaum and Harry Pollard. Exhaustive survey of ordinary differential equations for undergraduates in mathematics, engineering, science. Thorough analysis of theorems. Diagrams. Bibliography. Index. 818pp. 5⅜ x 8½. 64940-7

INTEGRAL EQUATIONS, F. G. Tricomi. Authoritative, well-written treatment of extremely useful mathematical tool with wide applications. Volterra Equations, Fredholm Equations, much more. Advanced undergraduate to graduate level. Exercises. Bibliography. 238pp. 5⅜ x 8½. 64828-1

FOURIER SERIES, Georgi P. Tolstov. Translated by Richard A. Silverman. A valuable addition to the literature on the subject, moving clearly from subject to subject and theorem to theorem. 107 problems, answers. 336pp. 5⅜ x 8½. 63317-9

INTRODUCTION TO MATHEMATICAL THINKING, Friedrich Waismann. Examinations of arithmetic, geometry, and theory of integers; rational and natural numbers; complete induction; limit and point of accumulation; remarkable curves; complex and hypercomplex numbers, more. 1959 ed. 27 figures. xii+260pp. 5⅜ x 8½. 42804-4

POPULAR LECTURES ON MATHEMATICAL LOGIC, Hao Wang. Noted logician's lucid treatment of historical developments, set theory, model theory, recursion theory and constructivism, proof theory, more. 3 appendixes. Bibliography. 1981 ed. ix+283pp. 5⅜ x 8½. 67632-3

CALCULUS OF VARIATIONS, Robert Weinstock. Basic introduction covering isoperimetric problems, theory of elasticity, quantum mechanics, electrostatics, etc. Exercises throughout. 326pp. 5⅜ x 8½. 63069-2

THE CONTINUUM: A Critical Examination of the Foundation of Analysis, Hermann Weyl. Classic of 20th-century foundational research deals with the conceptual problem posed by the continuum. 156pp. 5⅜ x 8½. 67982-9

CHALLENGING MATHEMATICAL PROBLEMS WITH ELEMENTARY SOLUTIONS, A. M. Yaglom and I. M. Yaglom. Over 170 challenging problems on probability theory, combinatorial analysis, points and lines, topology, convex polygons, many other topics. Solutions. Total of 445pp. 5⅜ x 8½. Two-vol. set.
Vol. I: 65536-9 Vol. II: 65537-7

INTRODUCTION TO PARTIAL DIFFERENTIAL EQUATIONS WITH APPLICATIONS, E. C. Zachmanoglou and Dale W. Thoe. Essentials of partial differential equations applied to common problems in engineering and the physical sciences. Problems and answers. 416pp. 5⅜ x 8½. 65251-3

THE THEORY OF GROUPS, Hans J. Zassenhaus. Well-written graduate-level text acquaints reader with group-theoretic methods and demonstrates their usefulness in mathematics. Axioms, the calculus of complexes, homomorphic mapping, p-group theory, more. 276pp. 5⅜ x 8½. 40922-8

Math–Decision Theory, Statistics, Probability

ELEMENTARY DECISION THEORY, Herman Chernoff and Lincoln E. Moses. Clear introduction to statistics and statistical theory covers data processing, probability and random variables, testing hypotheses, much more. Exercises. 364pp. 5⅜ x 8½. 65218-1

STATISTICS MANUAL, Edwin L. Crow et al. Comprehensive, practical collection of classical and modern methods prepared by U.S. Naval Ordnance Test Station. Stress on use. Basics of statistics assumed. 288pp. 5⅜ x 8½. 60599-X

SOME THEORY OF SAMPLING, William Edwards Deming. Analysis of the problems, theory, and design of sampling techniques for social scientists, industrial managers, and others who find statistics important at work. 61 tables. 90 figures. xvii +602pp. 5⅜ x 8½. 64684-X

LINEAR PROGRAMMING AND ECONOMIC ANALYSIS, Robert Dorfman, Paul A. Samuelson and Robert M. Solow. First comprehensive treatment of linear programming in standard economic analysis. Game theory, modern welfare economics, Leontief input-output, more. 525pp. 5⅜ x 8½. 65491-5

PROBABILITY: An Introduction, Samuel Goldberg. Excellent basic text covers set theory, probability theory for finite sample spaces, binomial theorem, much more. 360 problems. Bibliographies. 322pp. 5⅜ x 8½. 65252-1

GAMES AND DECISIONS: Introduction and Critical Survey, R. Duncan Luce and Howard Raiffa. Superb nontechnical introduction to game theory, primarily applied to social sciences. Utility theory, zero-sum games, n-person games, decision-making, much more. Bibliography. 509pp. 5⅜ x 8½. 65943-7

INTRODUCTION TO THE THEORY OF GAMES, J. C. C. McKinsey. This comprehensive overview of the mathematical theory of games illustrates applications to situations involving conflicts of interest, including economic, social, political, and military contexts. Appropriate for advanced undergraduate and graduate courses; advanced calculus a prerequisite. 1952 ed. x+372pp. 5⅜ x 8½. 42811-7

FIFTY CHALLENGING PROBLEMS IN PROBABILITY WITH SOLUTIONS, Frederick Mosteller. Remarkable puzzlers, graded in difficulty, illustrate elementary and advanced aspects of probability. Detailed solutions. 88pp. 5⅜ x 8½. 65355-2

PROBABILITY THEORY: A Concise Course, Y. A. Rozanov. Highly readable, self-contained introduction covers combination of events, dependent events, Bernoulli trials, etc. 148pp. 5⅜ x 8¼. 63544-9

STATISTICAL METHOD FROM THE VIEWPOINT OF QUALITY CONTROL, Walter A. Shewhart. Important text explains regulation of variables, uses of statistical control to achieve quality control in industry, agriculture, other areas. 192pp. 5⅜ x 8½. 65232-7

Math–Geometry and Topology

ELEMENTARY CONCEPTS OF TOPOLOGY, Paul Alexandroff. Elegant, intuitive approach to topology from set-theoretic topology to Betti groups; how concepts of topology are useful in math and physics. 25 figures. 57pp. 5⅜ x 8½. 60747-X

COMBINATORIAL TOPOLOGY, P. S. Alexandrov. Clearly written, well-organized, three-part text begins by dealing with certain classic problems without using the formal techniques of homology theory and advances to the central concept, the Betti groups. Numerous detailed examples. 654pp. 5⅜ x 8½. 40179-0

EXPERIMENTS IN TOPOLOGY, Stephen Barr. Classic, lively explanation of one of the byways of mathematics. Klein bottles, Moebius strips, projective planes, map coloring, problem of the Koenigsberg bridges, much more, described with clarity and wit. 43 figures. 210pp. 5⅜ x 8½. 25933-1

CONFORMAL MAPPING ON RIEMANN SURFACES, Harvey Cohn. Lucid, insightful book presents ideal coverage of subject. 334 exercises make book perfect for self-study. 55 figures. 352pp. 5⅝ x 8¼. 64025-6

THE GEOMETRY OF RENÉ DESCARTES, René Descartes. The great work founded analytical geometry. Original French text, Descartes's own diagrams, together with definitive Smith-Latham translation. 244pp. 5⅜ x 8½. 60068-8

PRACTICAL CONIC SECTIONS: The Geometric Properties of Ellipses, Parabolas and Hyperbolas, J. W. Downs. This text shows how to create ellipses, parabolas, and hyperbolas. It also presents historical background on their ancient origins and describes the reflective properties and roles of curves in design applications. 1993 ed. 98 figures. xii+100pp. 6½ x 9¼. 42876-1

THE THIRTEEN BOOKS OF EUCLID'S ELEMENTS, translated with introduction and commentary by Thomas L. Heath. Definitive edition. Textual and linguistic notes, mathematical analysis. 2,500 years of critical commentary. Unabridged. 1,414pp. 5⅜ x 8½. Three-vol. set. Vol. I: 60088-2 Vol. II: 60089-0 Vol. III: 60090-4

GEOMETRY OF COMPLEX NUMBERS, Hans Schwerdtfeger. Illuminating, widely praised book on analytic geometry of circles, the Moebius transformation, and two-dimensional non-Euclidean geometries. 200pp. 5⅝ x 8¼. 63830-8

DIFFERENTIAL GEOMETRY, Heinrich W. Guggenheimer. Local differential geometry as an application of advanced calculus and linear algebra. Curvature, transformation groups, surfaces, more. Exercises. 62 figures. 378pp. 5⅜ x 8½. 63433-7

CURVATURE AND HOMOLOGY: Enlarged Edition, Samuel I. Goldberg. Revised edition examines topology of differentiable manifolds; curvature, homology of Riemannian manifolds; compact Lie groups; complex manifolds; curvature, homology of Kaehler manifolds. New Preface. Four new appendixes. 416pp. 5⅜ x 8½. 40207-X

History of Math

THE WORKS OF ARCHIMEDES, Archimedes (T. L. Heath, ed.). Topics include the famous problems of the ratio of the areas of a cylinder and an inscribed sphere; the measurement of a circle; the properties of conoids, spheroids, and spirals; and the quadrature of the parabola. Informative introduction. clxxxvi+326pp; supplement, 52pp. 5⅜ x 8½. 42084-1

A SHORT ACCOUNT OF THE HISTORY OF MATHEMATICS, W. W. Rouse Ball. One of clearest, most authoritative surveys from the Egyptians and Phoenicians through 19th-century figures such as Grassman, Galois, Riemann. Fourth edition. 522pp. 5⅜ x 8½. 20630-0

THE HISTORY OF THE CALCULUS AND ITS CONCEPTUAL DEVELOP-MENT, Carl B. Boyer. Origins in antiquity, medieval contributions, work of Newton, Leibniz, rigorous formulation. Treatment is verbal. 346pp. 5⅜ x 8½. 60509-4

THE HISTORICAL ROOTS OF ELEMENTARY MATHEMATICS, Lucas N. H. Bunt, Phillip S. Jones, and Jack D. Bedient. Fundamental underpinnings of modern arithmetic, algebra, geometry, and number systems derived from ancient civilizations. 320pp. 5⅜ x 8½. 25563-8

A HISTORY OF MATHEMATICAL NOTATIONS, Florian Cajori. This classic study notes the first appearance of a mathematical symbol and its origin, the competition it encountered, its spread among writers in different countries, its rise to popularity, its eventual decline or ultimate survival. Original 1929 two-volume edition presented here in one volume. xxviii+820pp. 5⅜ x 8½. 67766-4

GAMES, GODS & GAMBLING: A History of Probability and Statistical Ideas, F. N. David. Episodes from the lives of Galileo, Fermat, Pascal, and others illustrate this fascinating account of the roots of mathematics. Features thought-provoking references to classics, archaeology, biography, poetry. 1962 edition. 304pp. 5⅜ x 8½. (Available in U.S. only.) 40023-9

OF MEN AND NUMBERS: The Story of the Great Mathematicians, Jane Muir. Fascinating accounts of the lives and accomplishments of history's greatest mathematical minds—Pythagoras, Descartes, Euler, Pascal, Cantor, many more. Anecdotal, illuminating. 30 diagrams. Bibliography. 256pp. 5⅜ x 8½. 28973-7

HISTORY OF MATHEMATICS, David E. Smith. Nontechnical survey from ancient Greece and Orient to late 19th century; evolution of arithmetic, geometry, trigonometry, calculating devices, algebra, the calculus. 362 illustrations. 1,355pp. 5⅜ x 8½. Two-vol. set. Vol. I: 20429-4 Vol. II: 20430-8

A CONCISE HISTORY OF MATHEMATICS, Dirk J. Struik. The best brief history of mathematics. Stresses origins and covers every major figure from ancient Near East to 19th century. 41 illustrations. 195pp. 5⅜ x 8½. 60255-9

CATALOG OF DOVER BOOKS

Physics

OPTICAL RESONANCE AND TWO-LEVEL ATOMS, L. Allen and J. H. Eberly. Clear, comprehensive introduction to basic principles behind all quantum optical resonance phenomena. 53 illustrations. Preface. Index. 256pp. 5⅜ x 8½.   65533-4

QUANTUM THEORY, David Bohm. This advanced undergraduate-level text presents the quantum theory in terms of qualitative and imaginative concepts, followed by specific applications worked out in mathematical detail. Preface. Index. 655pp. 5⅜ x 8½.   65969-0

ATOMIC PHYSICS: 8th edition, Max Born. Nobel laureate's lucid treatment of kinetic theory of gases, elementary particles, nuclear atom, wave-corpuscles, atomic structure and spectral lines, much more. Over 40 appendices, bibliography. 495pp. 5⅜ x 8½.   65984-4

A SOPHISTICATE'S PRIMER OF RELATIVITY, P. W. Bridgman. Geared toward readers already acquainted with special relativity, this book transcends the view of theory as a working tool to answer natural questions: What is a frame of reference? What is a "law of nature"? What is the role of the "observer"? Extensive treatment, written in terms accessible to those without a scientific background. 1983 ed. xlviii+172pp. 5⅜ x 8½.   42549-5

AN INTRODUCTION TO HAMILTONIAN OPTICS, H. A. Buchdahl. Detailed account of the Hamiltonian treatment of aberration theory in geometrical optics. Many classes of optical systems defined in terms of the symmetries they possess. Problems with detailed solutions. 1970 edition. xv+360pp. 5⅜ x 8½.   67597-1

PRIMER OF QUANTUM MECHANICS, Marvin Chester. Introductory text examines the classical quantum bead on a track: its state and representations; operator eigenvalues; harmonic oscillator and bound bead in a symmetric force field; and bead in a spherical shell. Other topics include spin, matrices, and the structure of quantum mechanics; the simplest atom; indistinguishable particles; and stationary-state perturbation theory. 1992 ed. xiv+314pp. 6⅛ x 9¼.   42878-8

LECTURES ON QUANTUM MECHANICS, Paul A. M. Dirac. Four concise, brilliant lectures on mathematical methods in quantum mechanics from Nobel Prize–winning quantum pioneer build on idea of visualizing quantum theory through the use of classical mechanics. 96pp. 5⅜ x 8½.   41713-1

THIRTY YEARS THAT SHOOK PHYSICS: The Story of Quantum Theory, George Gamow. Lucid, accessible introduction to influential theory of energy and matter. Careful explanations of Dirac's anti-particles, Bohr's model of the atom, much more. 12 plates. Numerous drawings. 240pp. 5⅜ x 8½.   24895-X

ELECTRONIC STRUCTURE AND THE PROPERTIES OF SOLIDS: The Physics of the Chemical Bond, Walter A. Harrison. Innovative text offers basic understanding of the electronic structure of covalent and ionic solids, simple metals, transition metals and their compounds. Problems. 1980 edition. 582pp. 6⅛ x 9¼.   66021-4

HYDRODYNAMIC AND HYDROMAGNETIC STABILITY, S. Chandrasekhar. Lucid examination of the Rayleigh-Benard problem; clear coverage of the theory of instabilities causing convection. 704pp. 5⅜ x 8¼. 64071-X

INVESTIGATIONS ON THE THEORY OF THE BROWNIAN MOVEMENT, Albert Einstein. Five papers (1905–8) investigating dynamics of Brownian motion and evolving elementary theory. Notes by R. Fürth. 122pp. 5⅜ x 8½. 60304-0

THE PHYSICS OF WAVES, William C. Elmore and Mark A. Heald. Unique overview of classical wave theory. Acoustics, optics, electromagnetic radiation, more. Ideal as classroom text or for self-study. Problems. 477pp. 5⅜ x 8½. 64926-1

PHYSICAL PRINCIPLES OF THE QUANTUM THEORY, Werner Heisenberg. Nobel Laureate discusses quantum theory, uncertainty, wave mechanics, work of Dirac, Schroedinger, Compton, Wilson, Einstein, etc. 184pp. 5⅜ x 8½. 60113-7

ATOMIC SPECTRA AND ATOMIC STRUCTURE, Gerhard Herzberg. One of best introductions; especially for specialist in other fields. Treatment is physical rather than mathematical. 80 illustrations. 257pp. 5⅜ x 8½. 60115-3

AN INTRODUCTION TO STATISTICAL THERMODYNAMICS, Terrell L. Hill. Excellent basic text offers wide-ranging coverage of quantum statistical mechanics, systems of interacting molecules, quantum statistics, more. 523pp. 5⅜ x 8½. 65242-4

THEORETICAL PHYSICS, Georg Joos, with Ira M. Freeman. Classic overview covers essential math, mechanics, electromagnetic theory, thermodynamics, quantum mechanics, nuclear physics, other topics. xxiii+885pp. 5⅜ x 8½. 65227-0

PROBLEMS AND SOLUTIONS IN QUANTUM CHEMISTRY AND PHYSICS, Charles S. Johnson, Jr. and Lee G. Pedersen. Unusually varied problems, detailed solutions in coverage of quantum mechanics, wave mechanics, angular momentum, molecular spectroscopy, more. 280 problems, 139 supplementary exercises. 430pp. 6½ x 9¼. 65236-X

THEORETICAL SOLID STATE PHYSICS, Vol. I: Perfect Lattices in Equilibrium; Vol. II: Non-Equilibrium and Disorder, William Jones and Norman H. March. Monumental reference work covers fundamental theory of equilibrium properties of perfect crystalline solids, non-equilibrium properties, defects and disordered systems. Total of 1,301pp. 5⅜ x 8½. Vol. I: 65015-4 Vol. II: 65016-2

WHAT IS RELATIVITY? L. D. Landau and G. B. Rumer. Written by a Nobel Prize physicist and his distinguished colleague, this compelling book explains the special theory of relativity to readers with no scientific background, using such familiar objects as trains, rulers, and clocks. 1960 ed. vi+72pp. 23 b/w illustrations. 5⅜ x 8½. 42806-0 $6.95

A TREATISE ON ELECTRICITY AND MAGNETISM, James Clerk Maxwell. Important foundation work of modern physics. Brings to final form Maxwell's theory of electromagnetism and rigorously derives his general equations of field theory. 1,084pp. 5⅜ x 8½. Two-vol. set. Vol. I: 60636-8 Vol. II: 60637-6

CATALOG OF DOVER BOOKS

QUANTUM MECHANICS: Principles and Formalism, Roy McWeeny. Graduate student–oriented volume develops subject as fundamental discipline, opening with review of origins of Schrödinger's equations and vector spaces. Focusing on main principles of quantum mechanics and their immediate consequences, it concludes with final generalizations covering alternative "languages" or representations. 1972 ed. 15 figures. xi+155pp. 5⅜ x 8½. 42829-X

INTRODUCTION TO QUANTUM MECHANICS WITH APPLICATIONS TO CHEMISTRY, Linus Pauling & E. Bright Wilson, Jr. Classic undergraduate text by Nobel Prize winner applies quantum mechanics to chemical and physical problems. Numerous tables and figures enhance the text. Chapter bibliographies. Appendices. Index. 468pp. 5⅜ x 8½. 64871-0

METHODS OF THERMODYNAMICS, Howard Reiss. Outstanding text focuses on physical technique of thermodynamics, typical problem areas of understanding, and significance and use of thermodynamic potential. 1965 edition. 238pp. 5⅜ x 8½. 69445-3

TENSOR ANALYSIS FOR PHYSICISTS, J. A. Schouten. Concise exposition of the mathematical basis of tensor analysis, integrated with well-chosen physical examples of the theory. Exercises. Index. Bibliography. 289pp. 5⅜ x 8½. 65582-2

THE ELECTROMAGNETIC FIELD, Albert Shadowitz. Comprehensive undergraduate text covers basics of electric and magnetic fields, builds up to electromagnetic theory. Also related topics, including relativity. Over 900 problems. 768pp. 5⅝ x 8¼. 65660-8

GREAT EXPERIMENTS IN PHYSICS: Firsthand Accounts from Galileo to Einstein, Morris H. Shamos (ed.). 25 crucial discoveries: Newton's laws of motion, Chadwick's study of the neutron, Hertz on electromagnetic waves, more. Original accounts clearly annotated. 370pp. 5⅜ x 8½. 25346-5

RELATIVITY, THERMODYNAMICS AND COSMOLOGY, Richard C. Tolman. Landmark study extends thermodynamics to special, general relativity; also applications of relativistic mechanics, thermodynamics to cosmological models. 501pp. 5⅜ x 8½. 65383-8

STATISTICAL PHYSICS, Gregory H. Wannier. Classic text combines thermodynamics, statistical mechanics, and kinetic theory in one unified presentation of thermal physics. Problems with solutions. Bibliography. 532pp. 5⅜ x 8½. 65401-X

Paperbound unless otherwise indicated. Available at your book dealer, online at **www.doverpublications.com**, or by writing to Dept. GI, Dover Publications, Inc., 31 East 2nd Street, Mineola, NY 11501. For current price information or for free catalogs (please indicate field of interest), write to Dover Publications or log on to **www.doverpublications.com** and see every Dover book in print. Dover publishes more than 500 books each year on science, elementary and advanced mathematics, biology, music, art, literary history, social sciences, and other areas.